Grand Challenges in Biology and Biotechnology

Emerging Issues and Trends

Series Editor

Pabulo Henrique Rampelotto, Hospital de Clínicas de Porto Alegre (HCPA), Porto Alegre, Brazil

This subseries of "Grand Challenges in Biology and Biotechnology" publishes volumes that take an in-depth look at selected emerging topics and trends in various areas of human, animal, plant, environmental and micro- biology and biotechnology. Each book addresses a current topic in a specialized manner, covering new technologies, new areas of research, or a new concept or idea that may be in its infancy today but could develop into an important mainstream topic in the future.

Andriy A. Sibirny
Editor

Biotechnology of Yeasts and Filamentous Fungi

Second Edition

Springer

Editor
Andriy A. Sibirny
Institute of Biotechnology
University of Rzeszow
Rzeszow, Poland

ISSN 2367-1017 ISSN 2367-1025 (electronic)
Grand Challenges in Biology and Biotechnology
ISBN 978-3-031-74725-0 ISBN 978-3-031-74726-7 (eBook)
ISSN 2948-2461 ISSN 2948-247X (electronic)
Emerging Issues and Trends
https://doi.org/10.1007/978-3-031-74726-7

This Springer imprint is published by the registered company Springer Nature Switzerland AG
The registered company address is: Gewerbestrasse 11, 6330 Cham, Switzerland

Preface

The reader is referred to the 2nd edition of Springer book "Biotechnology of Yeasts and Filamentous Fungi". The book contains both revised and renewed chapters from the first edition, as well as completely new reviews. These updates resemble the continuous developments of biotechnology of yeasts and filamentous fungi over the last years, marking its impact of an important field of industrial biotechnology. Suffice it to tell that yeast is responsible for production of more than 120 billion liters of fuel ethanol, including also fuel ethanol from lignocellulosic feedstock, e.g. from sugarcane bagasse. Fungal antibiotics have the largest market among all anti-infectives. Huge market relies on the traditional use of *S. cerevisiae* for baking. Yeasts and fungi are responsible for production of glycerol, lipids, organic acids, vitamins, antibiotics, carotenoids and other low-molecular-weight compounds as well as enzymes and recombinant proteins. Indeed, there are several popular yeast platforms for expression of heterologous proteins of pharmaceutical, food, agricultural and industrial importance, which have large and steadily growing market. Finally, yeast and fungal enzymes and intact cells are widely used for analytical purposes as components of enzymatic kits and biosensors. The presented book contains 14 chapters which cover the latest achievements in traditional and modern use of yeasts and filamentous fungi. Authors, the leaders in their research fields, represent 9 countries from Europe and Asia: Germany, Italy, Japan, Latvia, Poland, Spain, Sweden, Thailand and Ukraine. The reviews describe history and current state of the particular fields of yeast and fungal biotechnology; moreover, perspectives of further development of research and implementation into industry are also discussed. It became clear that new achievements in yeast and fungal biotechnology are possible and they largely depend on search for new, still unknown species of yeasts and filamentous fungi with some novel features (see, e.g., chapter by P. Buzzini on *Mrakia* yeasts), as well as on development of the new tools of molecular research.

First three chapters deal with yeast alcoholic fermentation and describe construction of efficient producers of 1st and 2nd generation ethanol, other biofuels as well as high-temperature alcoholic fermentation. Yeast metabolic engineering was started a hundred years ago by Carl Neuberg and was directed for glucose conversion to glycerol instead of ethanol production. Consequently, one of the chapters of the current book describes modern approaches of metabolic engineering of *S. cerevisiae* for anaerobic glycerol production as well as glycerol bioconversion. The next chapter describes yeasts producing lactate and other organic acids constructed due to yeast metabolic engineering. Four chapters devoted to the wild or mutant strains of yeasts and filamentous fungi producing low-molecular-weight metabolites: amino acids, lipids, carotenoids or riboflavin. One chapter describes production of recombinant proteins with methylotrophic yeasts, another one is devoted to mentioned earlier extremophilic *Mrakia* yeasts. The next chapter describes modern approaches to traditional use of yeast for baking, namely, yeast dehydration-rehydration. Finally, two chapters describe using yeast and fungal cells for development of biosensors used as the analytical tools for assaying various chemical substances.

I hope that the presented book will be interested and find reader among biotechnologists, microbiologists, geneticists and biochemists.

Rzeszów, Poland Andriy A. Sibirny

Contents

1 **Construction of the Improved Producers of First- and Second-Generation Ethanol in Conventional and Non-conventional Yeasts** 1
Roksolana Vasylyshyn, Kostyantyn Dmytruk, Justyna Ruchala, and Andriy A. Sibirny

2 **Thermotolerant Yeasts and High-Temperature Fermentation** 49
Hisashi Hoshida and Rinji Akada

3 **Development of Yeast *Saccharomyces* Platform for the Production of Biofuels and Bio-based Products** 77
Pattanan Songdech and Nitnipa Soontorngun

4 **Glycerol Production and Conversion in Yeasts and Biotechnological Significance of These Processes** 91
Kostyantyn Dmytruk, Marta Semkiv, and Andriy A. Sibirny

5 **Production of Industrially Relevant Organic Acids by Yeasts and Filamentous Fungi** 123
Riccardo Milanesi, Danilo Porro, Paola Branduardi, and Immacolata Serra

6 **Functional Amino Acid Engineering: A New Breeding Technology for Brewer's Yeasts** 141
Hiroshi Takagi

7 **Lipids of Yeasts and Filamentous Fungi and Their Importance to Biotechnology** 163
Volkmar Passoth and Bettina Müller

8 **Biotechnological Production and Applications of Fungal Carotenoids** 229
Javier Avalos, Javier Pardo-Medina, Julia Marente, and M. Carmen Limón

9 **Riboflavin Overproduction in Yeasts and Filamentous Fungi** 263
Justyna Ruchala, Dariya Fedorovych, Dominik Wojdyla,
Kostyantyn V. Dmytruk, and Andriy A. Sibirny

10 **Production of Recombinant Proteins in the Methylotrophic Yeasts** 291
Olena Stasyk and Oleh Stasyk

11 ***Mrakia* spp. Yeasts: Extremophilic Organisms with Underexplored Biotechnological Potential** 321
B. Turchetti, G. Mugnai, D. Andreani, C. Sannino, and P. Buzzini

12 **Dehydration of Yeasts and Anhydrobiosis** 343
Alexander Rapoport

13 **Biosensors Based on Yeast/Fungal Cells** 379
Alexandre Chamas, Martin Jähne, Uwe Wegner, Ha Thi Minh Pham,
Kirsten Simon, Steffen Uhlig, and Gotthard Kunze

14 **Yeast-Based Biosensors for Clinical Diagnostics, Food Control, and Environmental Safety** 405
G. Gayda, N. Stasyuk, O. Smutok, M. Gonchar,
and Andriy A. Sibirny

Chapter 1
Construction of the Improved Producers of First- and Second-Generation Ethanol in Conventional and Non-conventional Yeasts

Roksolana Vasylyshyn, Kostyantyn Dmytruk, Justyna Ruchala, and Andriy A. Sibirny

Contents

1.1 Introduction 2
1.2 *Saccharomyces cerevisiae* 3
1.3 *Scheffersomyces (Pichia) stipitis* 16
1.4 *Ogataea (Hansenula) polymorpha* 23
1.5 Concluding Remarks 36
References 37

Abstract The production of fuel ethanol represents one of the largest sectors in industrial biotechnology, driven by the demand for sustainable and renewable energy sources. Currently, the majority of fuel ethanol is derived from conventional feedstock (first-generation ethanol); however, lignocellulose-derived (second-generation) ethanol holds significant promise due to its potential to utilize non-food biomass, reducing competition with food supplies and lowering greenhouse gas emissions. This review outlines the advancements in engineering *Saccharomyces*

R. Vasylyshyn (✉)
Department of Molecular Genetics and Biotechnology, Institute of Cell Biology, NAS of Ukraine, Lviv, Ukraine

Department of Biotechnology and Microbiology, University of Rzeszow, Rzeszow, Poland

K. Dmytruk
Department of Molecular Genetics and Biotechnology, Institute of Cell Biology, NAS of Ukraine, Lviv, Ukraine

J. Ruchala
Department of Biotechnology and Microbiology, University of Rzeszow, Rzeszow, Poland

A. A. Sibirny
Institute of Biotechnology, University of Rzeszow, Rzeszow, Poland

A. A. Sibirny (ed.), *Biotechnology of Yeasts and Filamentous Fungi*, Grand Challenges in Biology and Biotechnology,
https://doi.org/10.1007/978-3-031-74726-7_1

cerevisiae strains to enhance ethanol yield from glucose and other quantitatively significant sugars in lignocellulosic hydrolysates, focusing on genetic modifications and optimizations of metabolic pathways. The review also includes comprehensive data on two non-conventional yeast species—*Scheffersomyces stipitis*, recognized as one of the most efficient natural xylose-fermenting yeasts, and the thermotolerant yeast *Ogataea polymorpha*, noted for its robustness in high-temperature fermentation processes. Both species are promising candidates for participation in second-generation ethanol production.

1.1 Introduction

Global energy and environmental problems have stimulated an increase in efforts for biofuels production from renewable resources as a drop in replacements for the currently used fossil fuels such as gasoline produced from oil (Broda et al. 2022). Due to the current political instability in oil-producing nations, and the concerns about climate change (attributed to harmful greenhouse gas emissions), there has been a call for greater reliance on domestic energy sources for the development of economically viable and environmentally friendly biofuels (Cavelius et al. 2023).

Fuel ethanol is a promising renewable liquid fuel for transportation that currently occupies a significant portion of the renewable biofuel market in the transport sector (Cavelius et al. 2023). Globally, ethanol production has generally increased, though there was a decline in 2020 due to the COVID-19 pandemic. While production has since grown, it has not yet returned to pre-pandemic levels. The United States remains the world's largest producer of ethanol, with production exceeding approximately 56.78 billion liters (https://afdc.energy.gov/data/10331) in both 2021 and 2022. Combined, the United States and Brazil account for 80% of global ethanol production. Ethanol can be blended with gasoline or used as a neat alcohol in dedicated engines, leveraging its higher octane number and greater heat of vaporization. Additionally, it is considered an excellent fuel for modern flex-fuel hybrid vehicles (Watanabe et al. 2022)

The current review summarizes the recent progress in the construction of more robust efficient yeast producers of ethanol from conventional and non-conventional feedstocks or what is referred to as first- and second-generation ethanol (1G, 2G). *Saccharomyces cerevisiae* is considered as the major industrial producer of 1G ethanol. Construction of the producers of 2G ethanol will be described for three organisms: *S. cerevisiae*, one of the best natural xylose fermenters, *Scheffersomyces (Pichia) stipitis*, and the most thermotolerant yeast known that also ferments xylose, *Ogataea (Hansenula) polymorpha*. Each of these organisms has some advantages and drawbacks. *S. cerevisiae* is the primary industrial 1G ethanol producer and is the most ethanol-tolerant natural yeast known. However, this yeast cannot metabolize xylose at all. *S. stipitis* can effectively ferment both glucose and xylose; however, it has much lower ethanol tolerance than *S. cerevisiae* and requires oxygen for growth. *O. polymorpha* can grow and ferment glucose and xylose at temperatures of up to 50 °C and for this reason can be promising for use in the process of simultaneous saccharification and fermentation (SSF). This yeast also shows relatively high ethanol tolerance, but the wild-type strains of *O. polymorpha* produce very low amounts of ethanol from xylose. The drawbacks of all three yeasts that are discussed can be successfully overcome by combining metabolic and evolutionary engineering in some cases with classical mutation and selection approaches.

1.2 *Saccharomyces cerevisiae*

Today, industrial-scale production of fuel ethanol uses primarily conventional food-based feedstocks such as glucose (derived from starch) and sucrose and is known as the first-generation (1G) ethanol (Cavelius et al. 2023).

The yeast *Saccharomyces cerevisiae* is the organism of choice for the industrial production of ethanol. The production of ethanol by fermentation represents the largest industrial biotechnological application of yeast. *S. cerevisiae* is characterized by several desirable industrial properties which include fast growth, efficient glucose anaerobic metabolism, high ethanol productivity, high yield, and high tolerance to several environmental stress factors, such as high ethanol concentration, low pH, and low oxygen level. In addition to this, yeast is insensitive to bacteriophage infection, which is particularly relevant in large industrial processes that use bacteria as the production microorganism. During alcoholic fermentation, *S. cerevisiae* produces ethanol with a yield close to the theoretical maximum (0.51 g *ethanol* per g of consumed glucose) (Gombert and van Maris 2015). At industrial scale, ethanol is produced with a yield that is higher than 90% of the theoretical maximum yield possible from glucose (Della-Bianca et al. 2013). Taking into account the large worldwide ethanol production, an increase of ethanol yield of even 1% can provide an additional estimated profit that approaches hundreds of millions of dollars annually.

The yeast *S. cerevisiae* catabolizes glucose via the Embden-Meyerhof-Parnas (EMP) pathway yielding anaerobically of 2 moles ATP per mole of consumed glucose. The efficiency of this pathway for anabolic processes is low with a maximal biomass yield of around 7% and an ethanol yield in the range of 90–93% from the glucose consumed (Ingledew 1999). In contrast to *S. cerevisiae*, the bacterium, *Zymomonas mobilis,* ferments glucose through the Entner-Doudoroff (ED) pathway. This pathway provides only 1 mole of ATP per mole of glucose, and consequently directs only 3% of glucose used to biomass production while the remaining 97% is converted to ethanol at almost the possible theoretical value (Sprenger 1996). When ATP is used for growth, cell biomass is formed at the expense of glucose that is not converted to ethanol. In other words, cell biomass is a by-product that is produced during alcoholic fermentation. Therefore, the lowering of ATP yield during alcoholic fermentation increases ethanol yield with reduced substrate conversion to cell mass. To achieve this goal several strategies can be applied such as: (1) substitution of EMP pathway in yeast by ED pathway; (2) activation of energy-consuming plasma membrane sugars symporters; (3) generation of futile cycles; and (4) elevation activity of ATP-degrading enzymes (De Kok et al. 2012). An alternative possibility to increasing ethanol yield during alcoholic fermentation is based on decreasing production of glycerol, which is another major by-product of ethanol production in yeast (Gombert and van Maris 2015).

Notwithstanding the earlier patent application by Lancashire et al. (1998), which describes the functional integration of the ED pathway to bypass glycolysis in *S. cerevisiae* with a positive impact on ethanol yield, more recent work demonstrates the inability to express one of the two unique enzymes of the ED pathway, namely 6-phosphogluconate dehydratase (PGDH), in this yeast (Benisch and Boles 2014). Several attempts to improve the availability of iron–sulfur clusters in the yeast cells and to attract the iron–sulfur cluster assembly machinery to the PGDH protein did not result in improved enzyme activities (Benisch and Boles 2014).

The impact of the disaccharide-proton symport on biomass and ethanol yields was studied in *S. cerevisiae* by comparing anaerobic growth on maltose, which is transported by a maltose-proton symporter and intracellularly hydrolyzed by maltase, to that of glucose (Weusthuis et al. 1993). It was demonstrated that biomass and ethanol yields in anaerobic maltose-limited cultures were 25% lower and 8% higher, respectively, than that of glucose-limited cultures. Relocation of sucrose hydrolysis from the extracellular space to the cytosol with additional evolutionary engineering resulted in a strain that had elevated sucrose uptake kinetics with a 30% decrease in the biomass yield and an 11% increase in the ethanol yield relative to the reference strain. Such evolved strain was characterized by increased transcript level of *AGT1* gene encoded sucrose-proton symporter (Basso et al. 2011). Therefore, the improved kinetics of sucrose transport with concomitant ATP reduction can be achieved by targeted overexpression of *AGT1* (Kaliterna et al. 1995).

In theory, the ethanol yield in anaerobic *S. cerevisiae* cultures can be further improved by lowering the ATP yield of sucrose dissimilation in this yeast even more. Introduction of glucose-proton symporter (*MAL11* or *KmHGT1*), fructose-proton symporter (*KlFRT1* or *SeFSY1*), and extracellular invertase (*SUC2*) under the context of deletion of genes encoding hexose transporters, disaccharide transporters, and disaccharide hydrolases resulting in a 16.6% increased anaerobic ethanol yield (from 1.51 to 1.76 mol mol hexose equivalent−1) and 46.6% decreased biomass yield on sucrose (de Valk et al. 2020). These studies provide a proof-of-concept for the replacement of the endogenous hexose transporters of *S. cerevisiae* by hexose-proton symport, and the concomitant decrease in ATP yield, to greatly improve the anaerobic yield of ethanol on sugar.

A futile cycle is a set of biochemical reactions of at least two metabolic pathways that run simultaneously in opposite directions which result in energy dissipation.

Phosphofructokinase is a glycolytic enzyme that phosphorylates fructose 6-phosphate to fructose-1,6-bisphosphate in an ATP-dependent manner. Fructose-1,6-biphosphatase (FBPase) is one of the gluconeogenesis enzymes. This enzyme hydrolyzes fructose-1,6-bisphosphate to fructose-6-phosphate in an ATP-dependent reaction. The simultaneous action of both enzymes—phosphofructokinase and fructose-1,6-biphosphatase, leads to a futile cycle generation between fructose-1,6-biphosphate and fructose-6-phosphate with ATP dissipation.

Phosphofructokinase is primarily regulated at the level of activity and the differences in the amount of the enzyme in glycolytic and gluconeogenic conditions do not exceed twofold (Bafuelos and Gancedo 1978; Clifton and Fraenkel 1981). In contrast, the FBPase activity is tightly regulated by catabolic repression, inactivation through ubiquitination, inhibition by AMP and fructose-2,6-biphosphate (Navas et al. 1993; Navas and Gancedo 1996). Therefore, intracellular FBPase activity level is maintained at basal level in cells grown in media containing fermentable carbon sources. To overcome the tight regulation of the yeast FBPase, the bacterial FBPase from *E. coli* that is insensitive to fructose-2,6-biphosphate inhibition was constitutively expressed in yeast (Navas and Gancedo 1996; Semkiv et al. 2016). The resulting recombinant strain was characterized by a threefold increase of FBPase activity, a 30% reduced intracellular ATP level and up to 9% increase of ethanol production relative to that of parental strain (Semkiv et al. 2016).

An alternate futile cycle was generated via the simultaneous activation of two enzymes, pyruvate carboxylase and the *gluconeogenic enzyme*, phosphoenolpyruvate carboxykinase (Semkiv et al. 2016). Pyruvate carboxylase catalyzes the conversion of pyruvate into oxaloacetate in an ATP-dependent reaction, and phosphoenolpyruvate carboxykinase uses ATP energy to convert oxaloacetate into phosphoenolpyruvate. ATP is also synthesized when phosphoenolpyruvate is converted into pyruvate by pyruvate kinase. Therefore, the resulting total loss in ATP is one molecule for each one turn of the cycle. The specific activity of pyruvate carboxylase was increased 3–5-fold via substitution of target gene promoter with a strong constitutive one.

The activity of phosphoenolpyruvate carboxykinase is carefully regulated at the post-translational level (Wilson and Bhattacharjee 1986). In order to avoid such regulation, a heterologous gene coding for the corresponding enzyme from *E. coli* was overexpressed. As a result, the specific activity of phosphoenolpyruvate carboxykinase was elevated 6–7-fold. Ethanol production by the constructed recombinant strains revealed a twofold increase over the parental strain by the end of the first day of fermentation (Semkiv et al. 2016).

Potential futile cycle based on the interconversion between glucose and trehalose, which theoretically could lead to ATP dissipation in cells of *S. cerevisiae*, was generated by the activation of trehalose-6-phosphate synthase and the neutral trehalase responsible for the synthesis and degradation of trehalose (Semkiv et al. 2016). Despite the increase in the activities of both enzymes, biomass accumulation remained unchanged.

Lowering intracellular ATP content in yeast cells, along with an increase in ethanol yield, can be achieved by activating certain cytosolic ATPases. A decrease in cellular ATP pool and activation of alcoholic fermentation was achieved by the overexpression of the soluble part (F_1) of H^+-ATPase or a portion of F_1 exhibiting ATPase from different origin in *S. cerevisiae* (Jensen et al. 2006). Similar results were obtained after the overexpression of *PHO5* coding for acid phosphatase which is a non-specific enzyme that also hydrolyzes ATP (Rogers and Szostak 1993). Overexpression of the vacuolar alkaline phosphatase Pho8 increases ethanol yield and production from glucose in both laboratory and industrial strains of *S.*

cerevisiae (Fig. 1.1), whereas the expression of the truncated, cytosol-localized form is detrimental to cell growth (Semkiv et al. 2014).

The galactose-induced expression of ATP hydrolysis region of the ribosome-associated molecular chaperon encoded by gene *SSB1* of *S. cerevisiae* and the ATP-diphosphohydrolases also known as apyrases from *E. coli* resulted in an increase in ethanol yield of 39 and 29%, respectively, during fermentation of the corresponding recombinant *S. cerevisiae* strains in galactose containing media (Semkiv et al. 2016).

S. cerevisiae generally consumes glucose to produce ethanol accompanied by the main by-products of glycerol, acetic acid, and lactic acid.

Glycerol is the second primary by-product after cell biomass during the ethanol production. Glycerol is formed at the expense of sugar that is not converted to ethanol (Gombert and van Maris 2015). In yeast, the reduction of glycerol formation can most probably result in an increase in ethanol yields. Therefore, significant research efforts were directed to reduce glycerol formation during fermentation. This can be accomplished by deletion of one or both genes *GPD1* and *GPD2*, coding for glycerol-3-phosphate dehydrogenase (Ansell et al. 1997). Deletion of both genes affected anaerobic growth. Deletion of *GPD2* resulted in an increase in ethanol

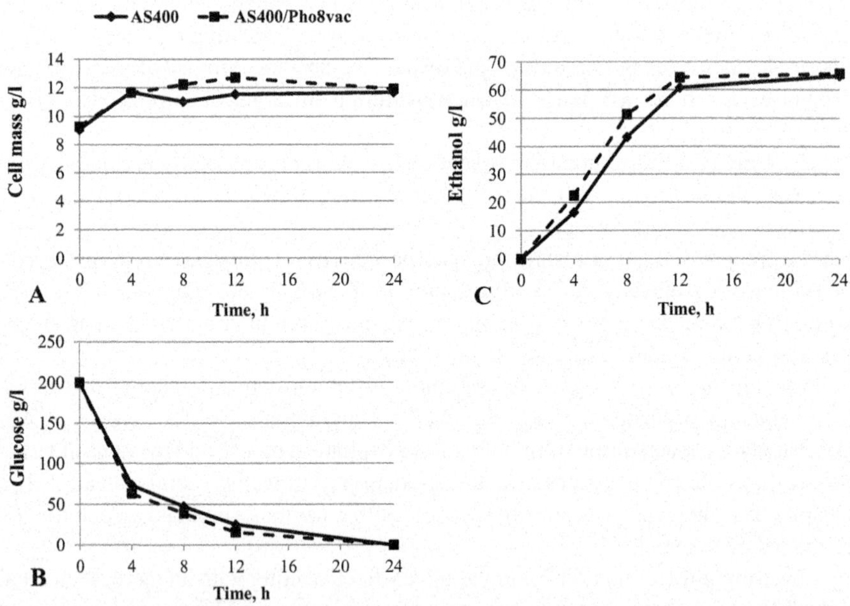

Fig. 1.1 Growth, glucose consumption, and ethanol production during alcoholic fermentation of recombinant strains overexpressing vacuolar form of alkaline phosphatase. Strains were cultured under semi-anaerobic conditions in YNB medium with 200 g/l glucose. Cultures were incubated at 30 °C, with limited aeration using a gyratory shaker at a setting of 120 rpm. AS400—WT, recipient strain. AS400/Pho8vac—strain containing vector pUC57-delta1_2-ADHpr-PHO8-CYCt-kanMX. (**a**) Biomass accumulation (g of biomass/l of medium); (**b**) glucose consumption (g of glucose/l of medium); (**c**) ethanol production (g of ethanol/l of medium)

yield with concomitant decrease of glycerol production; however, such a mutation also reduced growth and ethanol productivity (Valadi et al. 1998). Glycerol formation is caused by the oxidation of the excess NADH, which is produced from glycolysis under anaerobic conditions. In order to decrease cytosolic NADH formation, the gene *GDH1* encoding NADPH-dependent glutamate dehydrogenase was deleted, while *GLN1* and *GLT1* coding for glutamine synthetase and NAD-dependent glutamate synthase were overexpressed. During ammonium assimilation that is linked with NADH and ATP consumption, a recombinant strain showed a decrease in the glycerol yield to 38%, while ethanol yield was increased by 10% (Nissen et al. 2000). Another approach for reducing the intracellular pool of NADH and ATP production consisted of replacing the glyceraldehyde-3-phosphate dehydrogenase gene by that of the non-phosphorylating heterologous NADP-dependent analog from *Bacillus cereus*, or *Streptococcus mutans* (Guo et al. 2011; Zhang et al. 2011). The combination of this approach with the overexpression of NAD-dependent fumarate reductase or acetaldehyde dehydrogenase in both cases increased ethanol yield to 95% of the theoretical maximum (Zhang et al. 2011).

Glycerol is mainly exported across the plasma membrane in *S. cerevisiae* through the protein channel Fps1 regulated by extracellular osmolarity (Remize et al. 2001; Tamas et al. 1999). *S. cerevisiae fps1Δ* mutant exhibits the intracellular accumulation of glycerol. The glycerol accumulation triggers other regulatory systems for the reduction of glycerol biosynthesis, which then results in an increase of ethanol yield (Zhang et al. 2007). Moreover, disrupting the alcohol dehydrogenase II (*ADH2*) gene via complete deletion of the gene and introducing a frameshift mutation in the *ADH2* locus via CRISPR/Cas9 technology showed that the ethanol yield improved by up to 74.7% compared with the yield obtained using the native strain (Xue et al. 2018). *ADH2* deletion is involved in the increased demand for NAD^+ regeneration in the carbon metabolism. Both cofactor removal and consumption reduction could result in the metabolic cofactor imbalance (Hou et al. 2009). To improve ethanol yield and reduce the formation of by-products, a Cas9 (CRISPR-Cas9) approach was employed to knock out the *S. cerevisiae GPD2*, *FPS1*, *ADH2*, and *DLD3* genes. D-lactate dehydrogenase 3 (Dld3) in *S. cerevisiae* is involved in the conversion of D-lactate to pyruvate (Lee et al. 2016). The resulting recombinant strains exhibited an 18.58% increase in ethanol yield, along with reduced glycerol content and organic acids by 22.32%, 8.87% (acetic acid), and 16.82% (lactic acid), respectively (Yang et al. 2022).

Other work has also demonstrated that the expression of the NAD-dependent acetaldehyde dehydrogenase from an *E. coli* in *gpd1Δ gpd2Δ* strain supports the growth in the medium with glucose supplemented with acetate under anaerobic conditions (Guadalupe Medina et al. 2010). The constructed strains can be used for alcoholic fermentation of acetate containing substrates, as they are able to convert acetate into ethanol. The main drawback of these *S. cerevisiae* strains is that the reduced glycerol production resulted also in reduced osmotolerance and overall viability (Hohmann 2002). This finding imposes certain restrictions on the application of these strains in industrial fermentation processes that normally run at high concentration of fermented substrate.

Significant other research efforts have focused on improving the stress resistance of yeast strains with reduced glycerol production. Osmotolerance, thermotolerance, and tolerance to high concentrations of ethanol of several target strains were elevated by applying methods of metabolic engineering and genomic shuffling (Wang et al. 2012; Hou 2010). The combination of the deletion of the *GPD1* gene with the overexpression of the *nadF* gene from *B. cereus*, that encodes for an NADP-dependent glyceraldehyde-3-phosphate dehydrogenase with the derepression of homologous genes of trehalose synthesis *TPS1* and *TPS2* (encoding trehalose-6-phosphate synthase and trehalose-6-phosphate phosphatase), resulted in increased ethanol production, reduced glycerol formation, but did not exhibit a negative influence on strains viability during alcoholic fermentation (Guo et al. 2011). Trehalose protects cells from stress and its intracellular concentration correlates with resistance to high temperatures and high concentrations of ethanol in the medium (Tao et al. 2012). Overexpression of the *TPS1* gene in *S. cerevisiae* resulted in increased thermotolerance thereby allowing for the possible reduction in energy costs for cooling of fermentation vessels as well as for savings in energy used for heating due to reduced temperature difference between fermentation and distillation processes (An et al. 2011). Moreover, high-temperature fermentation has advantages during the process of simultaneous saccharification and fermentation, as high temperature (around 50 °C) is optimal for hydrolytic enzymes involved in substrates saccharification (Shahsavarani et al. 2012).

The development of improved ethanol-producing strains can be achieved by applying traditional selection and adaptive evolution as a useful alternative to metabolic engineering approaches. The mutants that are resistant to toxic concentrations of oxythiamine, trehalose, 3-bromopyruvate, glyoxylic acid, and glucosamine have been isolated. Some of these are characterized by 5–8% increase in ethanol yield when compared to the parental industrial ethanol-producing strain (Table 1.1) (Dmytruk et al. 2016). By applying adaptive evolution, useful yeast strains with enhanced maltose utilization and osmotolerance (Higgins et al. 2001), increased ethanol tolerance (Stanley et al. 2010), and yeast with improved ethanol production rate and decreased formation of acetate were selected (Cadiere et al. 2011).

In a recent study, evolutionary engineering methods were effectively applied to create thermo-acidic and acid-resistant strains of *S. cerevisiae*. The obtained strains demonstrated the ability to thrive in minimal media containing 12 g/l acetic acid at pH 4 and 30 °C, while also showing high levels of ethanol production up to 29 mmol/gDCW/h. Whole-genome sequencing analysis identified mutations and altered expression profiles in key genes involved in the RAS-cAMP-PKA signaling pathway (such as *RAS2*) and the heat shock transcription factor (HSF1). Reverse engineering experiments confirmed that the RAS2 mutation confers acid resistance and the HSF1 mutation confers thermotolerance (Salas-Navarrete et al. 2022)

In contrast to 1G, the 2G ethanol is an ethanol produced from non-food feedstocks such as plant biomass. The utilization of lignocellulosic biomass for 2G ethanol production would be preferable over sugar and starch-based 1G ethanol production because of the absence of competition with food production and minimal changes to land use (Sánchez and Cardona 2008; Nigam and Singh 2011). 2G

Table 1.1 Growth rate, ethanol production, productivity, specific productivity, and yield of *S. cerevisiae* mutants resistant to oxythiamine, trehalose, bromopyruvate, glyoxylic acid, glucosamine and initial industrial strain AS400 during alcoholic fermentation on YNB medium supplemented with 20% glucose and corn steep liquor (CSL) medium with hydrolyzed meal indicated in brackets

Strain	Selective agent	Specific growth rate g/l/h	Ethanol g/l	Ethanol productivity g/l/h	Specific ethanol productivity g/g biomass/h	Ethanol yield g/g of consumed glucose
AS400	–	0.133 ± 0.003	80.3 ± 1.5 (84.6 ± 1.0)	4.46 ± 0.08 (3.53 ± 0.03)	1.86 ± 0.03	0.402 ± 0.007 (0.423 ± 0.005)
AS400-567	Oxythiamine	0.133 ± 0.004	85.2 ± 1.6 (90.5 ± 0.9)	4.73 ± 0.09 (3.77 ± 0.04)	1.97 ± 0.04	0.426 ± 0.008 (0.453 ± 0.004)
AS400-543	Trehalose	0.083 ± 0.002	84.3 ± 1.6 (88.8 ± 0.9)	4.68 ± 0.09 (3.70 ± 0.03)	2.75 ± 0.06	0.422 ± 0.008 (0.444 ± 0.005)
AS400-617	Bromopyruvate	0.133 ± 0.003	84.1 ± 1.5 (89.7 ± 1.0)	4.67 ± 0.09 (3.74 ± 0.03)	1.95 ± 0.03	0.421 ± 0.007 (0.449 ± 0.005)
AS400-510	Glyoxylic acid	0.133 ± 0.004	85.1 ± 1.6 (91.4 ± 0.9)	4.73 ± 0.09 (3.81 ± 0.03)	1.97 ± 0.04	0.426 ± 0.007 (0.457 ± 0.005)
AS400-128	Glucosamine	0.117 ± 0.003	84.6 ± 1.7 (89.6 ± 0.9)	4.70 ± 0.09 (3.73 ± 0.04)	2.24 ± 0.04	0.423 ± 0.008 (0.448 ± 0.005)
AS400-510-42	Glyoxylic acid, glucosamine	0.133 ± 0.004	86.7 ± 1.7 (93.0 ± 0.9)	4.82 ± 0.09 (3.88 ± 0.03)	2.01 ± 0.04	0.434 ± 0.008 (0.465 ± 0.004)
AS400-510-42-214	Glyoxylic acid, glucosamine, bromopyruvate	0.117 ± 0.003	88.4 ± 1.5 (94.8 ± 1.0)	4.91 ± 0.08 (3.95 ± 0.04)	2.34 ± 0.04	0.442 ± 0.008 (0.474 ± 0.005)

(±), absolute error

For alcoholic fermentation, cells of AS400 strain and its derivatives were grown overnight in 100 ml of YPD medium in 300 ml Erlenmeyer flasks and then used to inoculate 20 ml aliqouts of mineral YNB medium supplemented with 200 g/l glucose or CSL medium supplemented with hydrolyzed meal in 50 ml Erlenmeyer flasks. An initial biomass concentration of 8 g (dry weight)/l was used for fermentation in YNB medium. For CSL medium, an initial biomass concentration of 10 g (dry weight)/l was used. Fermentation was carried out at a temperature of 34 °C with limited aeration using a gyratory shaker at a setting of 120 rpm. Samples were taken every 3 h for YNB medium or 12 h for CSL medium ethanol can utilize a range of different types of lignocellulosic material substrates (Saini et al. 2015; Kucharska et al. 2018). For a long time, a limited amount of 2G ethanol was produced in demonstration plants around the world. The first commercial ethanol plant in Romania started production in 2022, planning to convert 250,000 tons of locally sourced agricultural residues into 50,000 tons of ethanol per year. (https://biooekonomie.de/en/news/clariant-end-commercial-biorefinery-romania). However, due to higher cost the large-scale ethanol production from lignocellulosics and the annual production volume of ≈99 billion liters of ethanol in 2021 (https://ethanolrfa.org/markets-and-statistics/annual-ethanol-production), even small improvements of the ethanol yield would have large economic benefits.

The technologies to produce 2G ethanol do exist, however there are many improvements that are needed. The production of 2G ethanol from lignocellulosics will require the development of robust microbial strains that can grow and produce ethanol from at least glucose and xylose, which are the major fermentable sugars produced from the hydrolysis of lignocellulosic biomass. Other sugars that are produced during hydrolysis include cellobiose, the hexose sugars—fructose, mannose, galactose, and the pentose sugar arabinose.

Wild-type strains of *S. cerevisiae* are incapable of fermenting intermediate sugars, specifically cellodextrins like cellobiose and cellotriose, during the process of cellulose hydrolysis (Turner et al. 2016). Cellobiose-fermenting *S. cerevisiae* strains have been developed by the introduction of a cellobiose metabolic pathway based on either intracellular β-glucosidase (GH1-1) or cellobiose phosphorylase (CBP), along with either an energy-consuming active cellodextrin transporter (CDT-1) or a non-energy-consuming passive cellodextrin facilitator (CDT-2) (Choi et al. 2022). Cellobiose fermentation performance of the phosphorolytic cellobiose-fermenting *S. cerevisiae* strains can be significantly improved by employing mutant CDTs, such as CDT-1 (F213L) and CDT-2 (N306I), which exhibit threefold and sixfold enhanced cellobiose transporting activities compared to those of the wild-type CDT-1 and CDT-2, respectively (Ha et al. 2013; Kim et al. 2018). Another valuable strategy to make lignocellulosic biomass more accessible to the target microorganisms is the heterologous expression of cellulases. Expression of the cassette carrying a cellulase gene from *Ampullaria gigas Spix* into the *S. cerevisiae* genome resulted in a 23.03-fold increase in endo-1,4-β-glucanase (EG) activity, a 17-fold increase in exo-1,4-β-glucanase (CBH) activity, along with a 37.7-fold higher ethanol yield compared to the wild-type strain (Yang et al. 2018).

In another investigation, the *CWP2* gene, responsible for the major cell wall mannoprotein in the GPI-protein family essential for cell wall stability, and the *YGP1* gene, which codes for a secretory glycoprotein involved in yeast cell wall biogenesis, were deliberately disrupted in the *S. cerevisiae* INVSc1 strain. Subsequently, this strain underwent transformation with a plasmid containing the *BGL* gene (encoding β-glucosidase, a crucial enzyme in ethanol production from cellulose) from *Periconia* sp. BCC 2871 fused with an anchoring protein gene, thereby facilitating the integration of the *BGL* gene product into the yeast cell wall. The outcomes demonstrated that disrupting the *YGP1* and *CWP2* genes increased β-glucosidase activity by 63% and 24%, respectively. Moreover, the strain with a disrupted *YGP1* gene yielded 59% more ethanol from cellobiose compared to the original strain (Arnthong et al. 2022).

The yeast *S. cerevisiae* efficiently ferments hexose, glucose, fructose, and mannose but to a lesser extent galactose. In this yeast, galactose transporters are subjected to catabolic repression by glucose thereby limiting co-fermentation of these sugars. In contrast, mannose enters yeast cells using the glucose transport system allowing glucose/mannose co-fermentation to ethanol (Madhavan et al. 2012). The combined deletion of the genes *GAL6*, *GAL80,* and *MIG1* involved in negative regulation of galactose catabolism in a laboratory *S. cerevisiae* strain resulted in a partial co-consumption of glucose and galactose in aerobic batch cultures (Ostergaard et al. 2001). Normally, galactose metabolism in *S. cerevisiae* requires respiration. However, a *cox9Δ gal80Δ* double mutant has been isolated which effectively fermented galactose anaerobically (Quarterman et al. 2016). A natural strain of *S. cerevisiae* (NRRL Y-1528), which catabolizes galactose more effectively than glucose or mannose and able to simultaneous fermentation of all hexose presented in the hydrolyzed biomass was described. This strain may have a defect in carbon catabolic repression which alleviates glucose repression of galactose utilization (Keating et al. 2004).

Wild-type strains of *S. cerevisiae* are unable to catabolize and ferment pentoses (xylose or arabinose) which are major constituents of hydrolysates of plant biomass. This physiological feature of *S. cerevisiae* is due to the lack of enzymes that catalyze the initial stages of the pentose catabolism. Since the xylose content in the plant biomass hydrolysates is significantly higher than that of arabinose, most of the work has been directed to the construction of *S. cerevisiae* strains which are capable of xylose fermentation. These efforts have focused on the functional expression of heterologous genes for xylose catabolism. Some efforts have focused on the expression of the gene encoding xylose isomerase (XI) from different microbial origins. This enzyme does not require cofactors and catalyzes the isomerization of xylose in xylulose (Fig. 1.2). Several successful attempts to express XI from the bacteria *Thermus thermophilus* (Walfridsson et al. 1996), *Clostridium phytofermentans* (Brat et al. 2009), *Bacteroides stercoris* (Ha et al. 2011), or from the anaerobic fungus *Piromyces* sp. E2 (Karhumaa et al. 2007) or from *Orpinomyces* sp. (Madhavan et al. 2009) were performed. Expression of the codons optimized XI of *C. phytofermentans* in *S. cerevisiae* resulted in a 46% increase in specific growth rate on xylose as compared to the strain expressing a non-optimized version of the gene (Brat et al. 2009).

Other research efforts have focused on the expression of genes encoding xylose reductase (XR) and xylitol dehydrogenase (XDH), which catalyze the conversion of xylose into xylitol and subsequently to xylulose (Fig. 1.2). XR is encoded by gene *XYL1* and requires the cofactors NADH or NADPH with higher affinity to NADPH. XDH is an NAD-dependent enzyme that is encoded by *XYL2*. The co-expression of both genes results in an imbalance of cofactors with a deficiency in NAD and an excess of NADP. It is assumed that this imbalance causes the low efficiency of xylose alcoholic fermentation, primarily due to the xylitol accumulation (Jin et al. 2005; Krahulec et al. 2012). A genome-wide scale model of *S.*

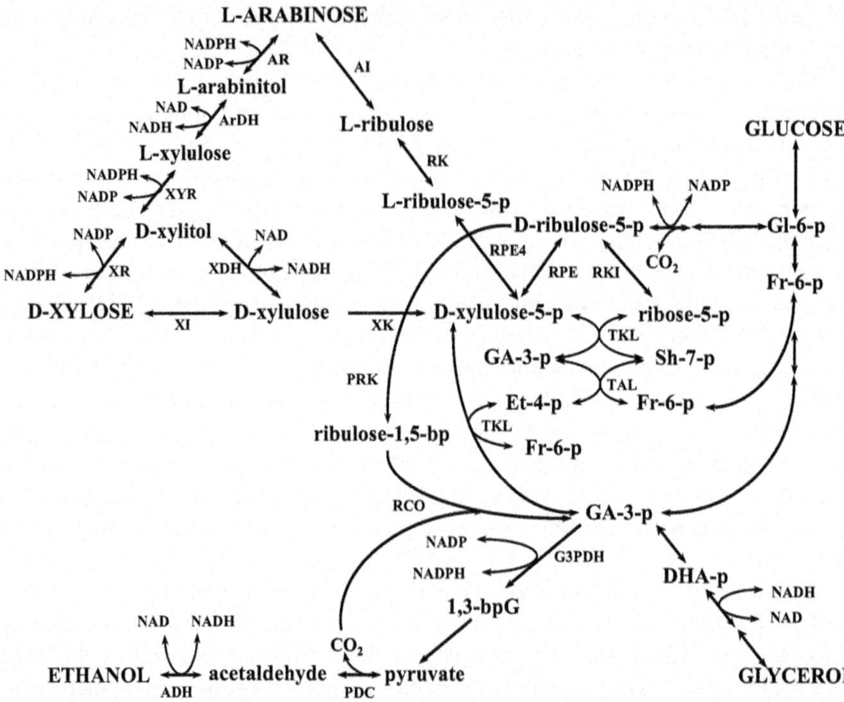

Fig. 1.2 Pathways of xylose, L-arabinose, and glucose fermentation to ethanol in yeasts. *Gl-6-p* glucose 6-phosphate, *Fr-6-p* fructose 6-phosphate, *GA-3-p* glyceraldehyde 3-phosphate, *SH-7-p* sedoheptulose 7-phosphate, *ET-4-p* erythrose 4-phosphate, *1,3-bpG* 1,3-bisphospho-glycerate, *DHA-p* dihydroxyacetone phosphate, *XR* xylose reductase, *XDH* xylitol dehydrogenase, *XI* xylose isomerase, *XK* xylulokinase, *AI* arabinose isomerases, *RK* ribulokinase, *RPE4* ribulose-5-phosphate-4-epimerase, *AR* aldose reductase, *ArDH* arabinitol dehydrogenase, *XYR* xylulose reductase, *RKI* ribose-5-phosphate ketol-isomerase, *RPE* ribulose-5-phosphate-3-epimerase, *TKL* transketolase, *TAL* transaldolase, *PRK* phosphoribulokinase, *RCO* ribulose-1,5-bisphosphate carboxylase/oxygenase, *G3PDH* glyceraldehyde 3-phosphate dehydrogenase, *PDC* pyruvate decarboxylase, *ADH* alcohol dehydrogenase

cerevisiae was used to predict the maximal achievable growth rate for cofactor balanced xylose utilization pathway. By applying the use of dynamic modeling with experimental results, it was shown that a balanced cofactor system XR/XDH stimulated an increase in ethanol batch production of 24.7% while simultaneously reducing the predicted substrate utilization time by up to 70% (Ghosh et al. 2011). The combination of computational design (Khoury et al. 2009) and site-specific mutagenesis of domains responsible for binding these cofactors (Watanabe et al. 2005) resulted in a modification of the cofactor specificity of XR and XDH and for balanced action. The co-expression of NADH-dependent mutant forms of XR from *S. stipitis* (K270M, K270R, K270R/N272D, N272D/P275Q, R276H) in part with a native NAD-dependent XDH increased productivity and ethanol yield with concomitant reduction of xylitol yield in recombinant strains of *S. cerevisiae* (Bengtsson et al. 2009; Lee et al. 2012). A decrease in xylitol yield and an increase in ethanol yield were also reported as a result of co-expression of the native XR with the modified NADP-dependent XDH (D207A/I208R/F209S/N211R). Xylose consumption was also 32% faster when compared to the strain expressing wild-type alleles of XR and XDH (Matsushika and Sawayama 2008). The ratio between XR and XDH activities is essential to improve xylose alcoholic fermentation. The low level of *XYL2* gene expression is the main reason for the accumulation of xylitol in strains expressing *XYL1*, *XYL2*, and *XYL3* (Kim et al. 2013a).

Xylulokinase (XK) catalyzes the ATP-dependent phosphorylation of xylulose and is an important enzyme in xylose catabolism (Fig. 1.2). XK catalyzes the limiting step in xylose alcoholic fermentation as *S. cerevisiae* strains expressing XR/XDH or XI have a significant reduction in xylitol accumulation when XK was overexpressed (Parachin et al. 2011). The activation of XK was accomplished by the derepression of the homologous gene *XKS1 S. cerevisiae* or heterologous *XYL3 S. stipitis* (Jin et al. 2003). However, an extreme increase in XK activity has a negative impact on the growth of recombinant strains on xylose, which is probably due to a rapid depletion of ATP that is used in xylulose phosphorylation (Du et al. 2010).

Two different L-arabinose catabolizing pathways were identified in bacteria (Wisselink et al. 2007) and fungi (Richard et al. 2001; Richard et al. 2002). In the bacterial pathway, arabinose is isomerased to ribulose by arabinose isomerase (*araA*), then ribulose is phosphorylated to ribulose-5-phosphate with ribulokinase (*araB*) that in turn is converted to xylulose-5-phosphate by ribulose-5-phosphate-4-epimerase (*araD*) (Wisselink et al. 2007). Arabinose fermentation was observed when *araA, araB,* and *araD* from *Lactobacillus plantarum* were co-expressed in *S. cerevisiae* (Wisselink et al. 2007). In fungi, L-arabinose is reduced to L-arabitol by aldose reductase, then converted to L-xylulose with arabinitol dehydrogenase. L-xylulose is reduced to xylitol by xylulose reductase, which in turn is oxidized to xylulose by xylitol dehydrogenase. The expression of the fungal arabinose utilization pathway (aldose reductase Gre3 from *S. cerevisiae*, arabinitol dehydrogenase and xylulose reductase from *Trichoderma reesei*, XDH from *S. stipitis*, and XK from *S. cerevisiae*) in *S. cerevisiae* resulted in low ethanol production with substantial arabinitol accumulation. This is apparently due to the imbalance caused by the cofactor mismatch among the utilized reductases/dehydrogenases (Richard et al. 2001, 2002; Bettiga et al. 2009). Based on published results, the expression of genes from the bacterial redox-independent pathway for L-arabinose utilization in *S. cerevisiae* is a more promising approach (Zhang et al. 2015a).

To increase the ethanol production, xylose-fermenting strains of *S. cerevisiae* that expressed the genes for xylose catabolism were subjected to other metabolic modifications. Mutations in several native genes (e.g., *ISU1, SSK2, HOG1, IRA2*), as well as the introduction of xylose transporters from natural xylose-utilizing yeasts (e.g., *S. stipitis* and *Candida intermedia*) also contributed to xylose utilization (Kwak and Jin 2017). In particular, the heterologous expression of sugar transporters Sut1 and Sut2 from *S. stipitis* was carried out (Du et al. 2010; Hector et al. 2008).

A system for the selection of glucose transporters with high affinity for xylose was also developed. As a result, a modified version of Gal2 *S. cerevisiae* transporter that is able to transport only xylose was selected (Farwick et al. 2014). It was also shown that endogenous *S. cerevisiae* hexose transporters (Hxt) can be engineered into specific xylose transporters (Farwick et al. 2014; Nijland et al. 2014). However, the expression of these genes remains subjected to glucose-regulated protein degradation. As a result, some Hxt proteins with high xylose transport capacity are rapidly degraded in the absence of glucose or when glucose is exhausted from the medium (Nijland et al. 2016). The N-terminal lysine residues of the Hxt proteins were predicted to be the target of ubiquitination. The substitution of N-terminal lysine residues in the hexose transporters Hxt1 and Hxt36 that are subjected to catabolite degradation resulted in an improved retention of these transporters at the cytoplasmic membrane in the absence of glucose and causes improved xylose fermentation upon the depletion of glucose and when cells are grown in xylose alone (Nijland et al. 2016).

In order to achieve efficient glucose/xylose co-fermentation, a modified xylose-specific, glucose-insensitive transporter Mgt05196 (N360F) from *Meyerozyma guilliermondii* was expressed on the background of a xylose-fermenting *S. cerevisiae* strain overproducing XI, xylulokinase Xks1 and the non-oxidative pentose phosphate pathway (PPP), and the inactivation of the aldose reductase Gre3p and the alkaline phosphatase Pho13. These rationally designed genetic modifications, combined with alternating adaptive evolution in xylose and lignocellulosic hydrolysates, resulted in a final strain, with excellent xylose fermentation that had an enhanced resistance to inhibitors (Li et al. 2016).

The co-expression of genes of non-oxidative part of PPP (*RKI1*, *RPE1*, *TKL1*, and *TAL1*, encoding ribose-5-phosphate isomerase, ribulose-5-phosphate epimerase, transketolase, and transaldolase, respectively) (Fig. 1.2) on the background of a *S. cerevisiae* strain expressing XI and XK showed improved growth on xylose (Karhumaa et al. 2005). However, expression of genes of the non-oxidative part of PPP in a *S. cerevisiae* strain expressing XR, XDH, and XK did not improve its growth on a medium supplemented with xylose (Bera et al. 2011).

It is known that the regeneration of NADPH mainly occurs in the oxidative part of PPP. A series of studies were directed to reduce the activity of the enzymes participating in the oxidative branch of PPP. This approach will facilitate the reaction catalyzed by XR using NADH instead of NADPH, with a reduced production of CO_2. The deletion of genes *ZWF1* (encoding glucose-6-phosphate dehydrogenase) and/or *GND1* (decarboxylating 6-phosphogluconate dehydrogenase) with the simultaneous expression of the genes *XYL1* and *XYL2* increased ethanol yield during xylose fermentation (Jeppsson et al. 2002; Verho et al. 2003). Another successful approach describes the combination of fungal NADP-dependent glyceraldehyde-3-phosphate dehydrogenase overexpression in part with the deletion of the *ZWF1* gene resulted in an increase in ethanol yield and productivity during xylose fermentation (Verho et al. 2003). To reduce xylitol production and increase ethanol yield during fermentation of glucose/xylose mixture, a modification of the redox balance of recombinant strains of *S. cerevisiae* was carried out by the deletion of gene *GDH1* (coding for NADPH-dependent glutamate dehydrogenase) and the overexpression of *GDH2* gene (NADH-dependent glutamate dehydrogenase) responsible for ammonium assimilation (Roca et al. 2003). The reduction of xylitol accumulation was achieved by the overexpression of the *noxE* gene from *L. lactis*, coding for the water forming NADH oxidase (Zhang et al. 2012). The modification of the acetate biosynthetic pathway that involved the deletion of the *ALD6* gene encoding NADP-dependent aldehyde dehydrogenase increased the efficiency of xylose alcohol fermentation (Lee et al. 2012).

To increase the production of ethanol during xylose fermentation, recombinant strains of *S. cerevisiae* were subjected to adaptive evolution and genome shuffling (Kim et al. 2013b; Jingping et al. 2012; Ren et al. 2016). Sequencing of the genomes of these strains after long-term culturing in xylose containing media revealed new potential target genes for metabolic engineering approach, e.g., *XKS1*, *SOL3* (6-phosphogluconolactonase), *GND1* (6-phosphogluconate dehydrogenase), *TAL1*, *TKL1*, *YCR020C*, *YBR083W*, and *YPR199C* (Wahlbom et al. 2003). The genes for *TAL1* and *PHO13* (non-specific alkaline phosphatase) were identified by transposon mutagenesis as a new targets. The truncation of these genes increased production of ethanol during xylose alcoholic fermentation (Ni et al. 2007).

Carbon dioxide is a major by-product of carbohydrate alcoholic fermentation. A potential target for manipulation is the conservation of carbon by re-assimilating carbon dioxide released during the decarboxylation of pyruvate to acetaldehyde in the final stages of the ethanol production pathway. This approach was described by the introduction a synthetic reductive PPP into a xylose-fermenting *S. cerevisiae* strain to achieve simultaneous lignocellulosic ethanol production and carbon dioxide recycling. The heterologous enzymes phosphoribulokinase from *Spinacia oleracea* and ribulose-1,5-bisphosphate carboxylase/oxygenase from *Rhodospirillum rubrum* were introduced into the *S. cerevisiae* strain harboring XR/XDH pathway, up-regulated PPP, and *knocked out ALD6* and *PHO13* genes (Kim et al. 2013b). The phosphoribulokinase enzyme catalyzes ribulose-5-phosphate conversion to ribulose-1,5-bisphosphate, and the ribulose-1,5-bisphosphate carboxylase/oxygenase converts one ribulose-1,5-bisphosphate and CO_2 to two molecules of glycerate-3-phosphate (Fig. 1.2). The constructed strain with the synthetic reductive PPP revealed a higher yield of ethanol with lower yields of xylitol and glycerol than that of the parental strain. Moreover, a reduced release of CO_2 by the engineered strain was observed during xylose fermentation, suggesting that the carbon dioxide generated by pyruvate decarboxylase was partially reassimilated through the synthetic reductive PPP (Xia et al. 2016). The strategy of carbon dioxide recycling from the ethanol fermentation pathway in yeast has a great potential in alleviating greenhouse gas emissions during the production of 2G ethanol.

Fermentation of biomass-derived hydrolysates is accompanied with the inhibition of ethanol production due to the presence of toxic compounds produced mainly during lignocelluloses pretreatment, e.g., furfural, hydroxymethylfurfural (HMF), weak acids, and phenols (Palmqvist and Hahn-Hägerdal 2001). To overcome this limitation, the strains of *S. cerevisiae* that are able to produce ethanol in the presence of these inhibitors have been selected through directed evolution and adaptation (Hawkins et al. 2013; Parreiras et al. 2014). Another alternative molecular approaches were also applied. Furfural and HMF were converted to less toxic furfuryl alcohol and furan dimethanol, respectively, by the overexpression of the endogenous *S. cerevisiae* oxidoreductases such as alcohol dehydrogenases (*ADH1*, *ADH6*, and *ADH7*) (Almeida et al. 2008; Liu et al. 2008; Petersson et al. 2006) and aldo-keto reductases (*GRE2*) (Moon and Liu 2012). It was shown (Gorsich et al. 2006) that tolerance to furfural-induced stress is associated with genes of the pentose phosphate pathway *ZWF1*, *GND1*, *RPE1*, and *TKL1*. The overexpression of the gene *PAD1* encoding phenylacrylic acid decarboxylase in *S. cerevisiae* resulted in improved growth rate and ethanol productivity in dilute acid hydrolysates (Larsson et al. 2001). An original approach to improve tolerance to fermentation inhibitors and ethanol was proposed by modulation of the polyamine (spermidine) content in *S. cerevisiae* (Kim et al. 2015). Intracellular spermidine contents was increased by a double gene deletion *OAZ1* (ornithine decarboxylase antizyme) and *TPO1* (polyamine transport protein) genes and the overexpression of *SPE3* (spermidine synthase) (Kim et al. 2015). In a later study, Liu and Ma also investigated the transcriptomic responses of evolved strains tolerant to furfural and HMF to the influence of these inhibitors. The results of comparative transcriptomic analysis identified the cell wall response, endogenous and exogenous detoxification pathways, and specific transcription factors such as Yap1, Met4, Msn2/4, and Pdr1/3 as key components necessary for inhibitor tolerance (Liu and Ma 2020).

In summary, 1G ethanol production is a profitable technology that can be further improved by increasing ethanol yield and productivity by applying molecular and classical techniques to industrial ethanol-producing strains of *S. cerevisiae.* Several novel strains, including SA1, CAT1, BG1, and PE2, originally isolated from first-generation bioethanol distilleries in Brazil, are now used as commercial strains in the production of fuel ethanol and are capable of producing higher ethanol titers compared to traditional bioethanol-producing strains (Secches et al. 2022). Despite significant efforts, existing technologies of 2G ethanol production are still non-profitable. Although recombinant *S. cerevisiae* strains capable of hexoses/pentoses co-fermentation were developed, none of the engineered strains are able to ferment glucose/xylose mixture as fast as the rate of glucose fermentation by the parental strain (Zhang et al. 2015a). Another important limitation is the sensitivity of ethanol-producing strains to inhibitors in lignocellulosic hydrolysates. Further studies are needed to develop recombinant *S. cerevisiae* strains that are capable of rapid fermentation of mixed sugars with improved resistance to fermentation inhibitors.

1.3 *Scheffersomyces (Pichia) stipitis*

Scheffersomyces (Pichia) stipitis belongs to the group of naturally xylose-fermenting yeasts that is characterized by rather low amounts of accumulated by-products from this fermentation like xylitol (Kurtzman and Suzuki 2010). Other known representatives of the group of native xylose-fermenting yeasts are *Candida shehatae,*

Pachysolen tannophilus, Spathaspora passalidarum, and *Ogataea polymorpha* (Schneider et al. 1981; du Preez and van der Walt 1983; Ryabova et al. 2003; Nguyen et al. 2006). As a rule, most natural xylose-fermenting yeasts inhabit the guts of passalid beetles that degrade white-rotted hardwood (Nigam 2001; Nardi et al. 2006 with the exception of *O. polymorpha* strains that have been isolated from other natural habitats (Gellisen 2002). *S. stipitis* is closely related to other yeast endosymbionts of passalid beetles (Suh et al. 2003). This yeast has one of the highest native capacities for xylose fermentation among yeast species described so far (du Preez et al. 1989). In addition to utilizing xylose, *S. stipitis* has the capability to use all of the other major sugars found in wood. It also transforms low-molecular-weight lignin moieties, reduces acyclic enones to the corresponding alcohols, and forms various esters and aroma components. It has been engineered to produce lactic acid or xylitol in high yield (Lee et al. 1986; Targonski 1992; Kim et al. 2001; Conceicao et al. 2003; Ilmen et al. 2007). The published fermented sugar yields from lignocellulose hydrolysates by this yeast approach 80% of theoretical yield (Nigam 2001). The availability of genetic tools and capability for fermentation of hydrolysates has made *S. stipitis* an attractive microorganism for bioconversion of lignocellulose to fuels and chemicals. The major drawbacks of *S. stipitis* are the low fermentation rates, low ethanol tolerance, and the inability to grow anaerobically (du Preez et al. 1989; Grootjen et al. 1990; Shi and Jeffries 1998).

S. stipitis is a predominantly haploid, homothallic, hemiascomycetous yeast that forms buds along with pseudomycelia during vegetative growth, and two hat-shaped ascospores from each ascus (Kurtzman 1990). Genetic manipulation of *S. stipitis* is much more complicated relative to those of *S. cerevisiae*. This is because *S. stipitis* is resistant to most common antibiotics and the number of strains with convenient auxotrophic markers is limited. This yeast uses an alternative codon system that decodes CUG for serine instead of leucine as common in the classical genetic code. Thus, the expression of foreign proteins, including those used as drug resistance markers, requires codon modification. Unfortunately, random (non-homologous) integration prevails in *S. stipitis* which makes targeted deletions much more difficult to obtain (Jeffries and Van Vleet 2009). Nevertheless, many efficient genetic methods have been developed for *S. stipitis* that include methods of sexual mating and sporulation (Melake et al. 1996). Auxotrophic strains have been created and methods for high efficiency transformation have been developed for the auxotrophic mutants *ura3*, *leu2*, *trp5*, and *his3* of *S. stipitis* (Yang et al. 1994; Lu et al. 1998; Piontek et al. 1998). Genetic tools based on a *loxP/Cre* recombination system and the dominant marker for zeocin resistance have been developed for *S. stipitis* (Laplaza et al. 2006). Deletion of the *KU80* gene that is responsible for non-homologous end joining significantly increases the fraction of homologous recombinant transformants, albeit at the expense of transformation frequency (Maassen et al. 2008).

The 15.4-Mbp genome of *S. stipitis* was sequenced (Jeffries et al. 2007). *S. stipitis* CBS 6054 is known to have eight chromosomes, of which two pairs are very similar in size, accounting for the earlier results that suggested the presence only of six chromosomes (Passoth et al. 1992). *S. stipitis* genome annotation predicted that

5841 genes are present in this of which a majority of 72% have a single exon. The average gene density is 56% and the average gene, transcript, and protein lengths are 1.6 kb, 1.5 kb and 493 amino acids, respectively. Expressed sequence tags (ESTs) confirmed the identity of 40% of the predicted genes with 84% showing strong similarity to proteins in other fungi (Jeffries et al. 2007). So far, protein function can be tentatively assigned to about 70% of the genes. Synteny analysis of *P. stipitis* with its nearest completely sequenced yeast genome neighbor, *Debaryomyces hansenii*, shows extensive recombination and shuffling of the chromosomes, which appear to be a common feature. *S. stipitis* and *D. hansenii* share 151 gene families that are not found in the other genomes. The *S. stipitis* gene set is missing 81 gene families (442 proteins) relative to the other yeast genomes in the analysis.

The most frequent domains characterized so far include protein kinases, helicases, transporters (sugar and MFS), and domains involved in transcriptional regulation (fungal specific transcription factors, RNA recognition motifs, and WD40 domains). A majority of these is shared with other hemiascomycota. These range from 1534 domains in common with *Schizosaccharomyces pombe* and 1639 domains with *D. hansenii*. One of the few *S. stipitis*-specific domains belongs to one of glycosyl hydrolase families, a subgroup of cellulases and xylanases. All of the genes for xylose assimilation, including the oxidative PPP, glycolytic cycle, the tricarboxylic acid cycle (TCA), and ethanol production were present in isoforms similar to those found in other yeasts (Jeffries et al. 2007). Genes of first three enzymes of xylose metabolism, *XYL1*, *XYL2*, *XYL3* and that of PPP (*ZWF1*, *GND1*, *TKL1*, *TAL1*, *RPI1*) are also induced by xylose and the expression of *XYL2* yields one of the most abundant transcripts in xylose-grown cells (Jeffries and Van Vleet 2009). Another interesting pattern of the regulation of gene expression was found by oxygen limitation. Such limitation led to strong derepression of some glycolytic genes that include two genes of glyceraldehyde-3-phosphate dehydrogenase *TDH1* and *TDH2*, pyruvate decarboxylase *PDC1* and *PDC2*, and of the secondary alcohol dehydrogenase *SAD2* (the function of the last gene/enzyme in metabolism is not known) (Jeffries and Van Vleet 2009). Genome displayed genes for sensing and regulatory proteins (< 200 putative Zn-finger regulatory proteins) are similar to that in *S. cerevisiae*. The genome *S. stipitis* also revealed many gene clusters representing either pairs/clusters of non-homologous genes in which each cluster has a single function such as galactose metabolism or tandem repeats of paralogous genes. Gene clusters seem to be particularly abundant in *S. stipitis* as there are at least 35 clusters of functionally related genes (Jeffries et al. 2007). The studies of structure, function, and regulation in *S. stipitis* genome are likely to have important impact on understanding its physiology and could be used for metabolic engineering of this organism.

The study of *S. stipitis* has attracted scientists and technologists primarily due to its natural ability to produce large amounts of ethanol during xylose fermentation with small or no production of xylitol. It was found that fed-batch cultures of *S. stipitis* produce around 47 g/l of ethanol with yields of 0.36 g/g xylose at 30 °C (van Dijken et al. 1986). However, *S. stipitis* fermentation rate on xylose is low relative to that of *S. cerevisiae* on glucose, so increasing the rate of fermentation of xylose by *S. stipitis* could greatly improve its usefulness in commercial applications (Jeffries and Van Vleet 2009). Another drawback to using *S. stipitis* is that it is much more susceptible to ethanol inhibition relative to *S. cerevisiae* (Shi and Jeffries 1998).

There are many detailed publications on the physiology of *S. stipitis*. Oxygen plays an important role in cell growth, redox balance, functioning of the mitochondria and generation of energy for xylose transport in *S. stipitis* (Skoog and Hahn-Hagerdahl 1990). Fermentation in *S. stipitis* is activated by oxygen limitation (Passoth et al. 1996, 2003; Klinner et al. 2005). It is interesting to note that *S. stipitis* can metabolize xylose anaerobically, even though it does not grow under anaerobic conditions (Wijsman et al. 1985). The optimal temperature for *S. stipitis* fermentation is between 25 and 33 °C and the optimal pH is in the range of 4.5–5.5 (du Preez et al. 1986). The nutrients in the fermentation media play an important part in the growth and ethanol production in *S. stipitis*. Ethanol production increased with the addition of amino acids and nitrogen was required for non-growth associated ethanol production (Slininger et al. 2006). Ammonium salts increased the ethanol productivity and the ethanol to biomass yield in *S. stipitis* (Guebel et al. 1992; Agbogbo and Wenger 2006). Magnesium has also been shown to play an important role in redox balance and therefore has an effect on xylitol production (Mahler and Nudel 2000). Low levels of Mg^{2+} resulted in xylitol accumulation and a high intracellular NADH content. Corn steep liquor is a viable nutrient source for *S. stipitis* fermentation when used as a sole nitrogen source compared to amino acids, vitamins, and other nutrients (Amartey and Jeffries 1994). The initial xylose concentration has an effect on the fermentation parameters of *S. stipitis* with maximum ethanol productivities occurring at a xylose concentration of 50 g/l (du Preez et al. 1985).

The conversion of xylose to ethanol in *P. stipitis* consists of three stages: (1) xylose transport and the initial reaction to enter the PPP; (2) non-oxidative reactions of PPP; (3) glycolysis (Fig. 1.2). Little is known about xylose transport in *S. stipitis*. The low-affinity transport system is shared between glucose and xylose for sugar transport. Glucose inhibits xylose transport by noncompetitive inhibition in the low-affinity proton symport system (Kilian and van Uden 1988). The low-affinity transport is used when sugar concentrations are high and the high-affinity systems are used when sugar concentrations are low. Repression of xylose uptake occurs in fermentation media containing glucose and xylose. Therefore, glucose is the preferred sugar by *S. stipitis* in ethanol production. The rate of glucose consumption is higher than xylose under similar growth conditions (Agbogbo et al. 2006). The transport of sugars into the cells is the rate-limiting step in the utilization of sugars for ethanol production in *S. stipitis* (Legthelm et al. 1988). A high-affinity xylose-transporting system has been described that is specific for xylose in this yeast (Hahn-Hägerdal et al. 2001). Three genes, *SUT1*, *SUT2,* and *SUT3*, have been characterized that encode glucose transporters in *S. stipitis* (Weierstall et al. 1999). Sut2 and Sut3 are highly similar to the *S. cerevisiae* glucose transporter family and the Sut2 and Sut3 transporters have a higher affinity for glucose than for xylose. Transcription of *SUT1* is induced in *S. stipitis* independently of oxygen supply, whereas *SUT2* and *SUT3* are expressed only under aerobic conditions, but independently of the carbon source. Disruption of *SUT1* eliminates the low-affinity xylose transport system in *S. stipitis* (Weierstall et al. 1999).

Initial metabolism of xylose in *S. stipitis* is similar to other natural xylose-fermenting yeasts (Fig. 1.2). Xylose first is reduced by xylose reductase (aldose reductase, gene *XYL1*) to xylitol (Veduyn et al. 1985). This enzyme has affinity to both NADH and NADPH but shows much higher affinity toward NADPH. By comparison, *S. stipitis* xylitol dehydrogenase (gene *XYL2*), which converts xylitol to xylulose, has affinity only for NAD (Metzger and Hollenberg 1995). The third reaction of xylose metabolism is catalyzed by xylulokinase (gene *XYL3*), which converts xylulose to the PPP intermediate xylulose-5-phosphate (Jin et al. 2002). The cofactor unbalance resulting from the first two reactions involving xylose reductase and xylitol dehydrogenase leads to xylitol accumulation in most natural xylose-fermenting yeasts, but not in *S. stipitis* (Skoog and Hahn-Hagerdahl 1990). In *P. stipitis* there are efficient systems for NADH reoxidation to NAD and NADPH regeneration (Balagurunathan et al. 2012). One particular pathway that has been observed for xylose metabolism that is induced under oxygen-limited conditions is effective in tackling the cofactor imbalance caused by the first two steps in xylose utilization (Jeffries et al. 2007; Jeffries and Van Vleet 2009). This pathway involves four enzymes: NAD-dependent glutamate dehydrogenase (*GDH2*), which converts 2-oxoglutarate to L-glutamate consuming NADH; glutamate decarboxylase (*GAD2*), which decarboxylates L-glutamate to 4-aminobutyrate; 4-aminobutyrate aminotransferase (*UGA1*), which transaminates 4-aminobutyrate to succinate semialdehyde; and succinate semialdehyde dehydrogenase (*UGA2*), which oxidizes succinate semialdehyde to succinate using NADP. The net result of the four reactions is the conversion of NADH to NADPH.

The PPP plays an important role in xylose fermentation (Fig. 1.2) and the corresponding genes (*ZWF1*, *GND1*, *TKL1*, *TAL1*, *RPI1*) are known to be induced by xylose (Jeffries and Van Vleet 2009). Alcohol dehydrogenase is also an important enzyme in ethanol production with the deletion of either *ADH1* or *ADH2* gene significantly reducing ethanol formation, and the deletion of both entirely abolishes ethanol production (Cho and Jeffries 1998; Passoth et al. 1998).

Relatively little is known about the rate-limiting steps in ethanol production from xylose. It has been shown that XR and XDH are repressed by glucose and are induced during growth on xylose. Xylose is generally not consumed in the presence of glucose; hence, under glucose repression, these activities, along with xylose transport, are rate limiting (Bicho et al. 1988). The XK does not, however, appear to be rate limiting in *S. stipitis* once its activity is induced by xylose (Jin et al. 2002).

In addition to the ultimate goal of using *S. stipitis* for ethanol production from lignocellulosic feedstock, this yeast can also be successfully used for xylitol production from xylose using a mutant defective in *XYL2* that codes for xylitol dehydrogenase (Kim et al. 2001; Rodrigues et al. 2008). Efficient producers of lactic acid have also been constructed on *S. stipitis* after expression of the lactate dehydrogenase *LDH* gene from *Lactobacillus helveticus* under control of the yeast *ADH1* promoter. It is interesting to note that xylose was more efficient substrate for lactate synthesis than glucose (Ilmen et al. 2007). A strain of *S. stipitis* hyperaccumulating

S-adenosylmethionine has been isolated (Križanovic et al. 2015). Other *S. stipitis* recombinant strains have been constructed that produce fumaric acid from xylose after acquiring heterologous reductive pathway from *Rhizopus oryzae* (Wei et al. 2008).

In spite of the fact that *S. stipitis* is one of the best natural xylose-fermenting yeasts with no xylitol accumulation, it has several drawbacks which include a low rate of fermentation, low tolerance to ethanol, and requirement of oxygen for growth. When grown on lignocellulosic hydrolysates, other limitations to the use of this yeast are observed. Among these is the need for simultaneous fermentation of glucose and xylose and the poor tolerance to inhibitors that are present in hydrolysates. Additionally as oxygen is required, *S. stipitis* has a tendency to utilize the ethanol produced with a considerable amount of xylose remaining in the medium (Harner et al. 2015). One of the most popular methods used to increase S. stipitis tolerance toward inhibitors is adaptation through repeated sub-culturing or recycling of yeast cells, gradually increasing the concentration of inhibitor(s) by stepwise addition of more concentrated lignocellulosic hydrolysate solutions (Amartey and Jeffries 1996; Huang et al. 2009; Watanabe et al. 2011; Yang et al. 2011). Strains adapted to inhibitors present in a specific hydrolysate may exhibit cross-tolerance to other hydrolysates. Adaptation has been done to individual inhibitor(s) as well as to mixtures of inhibitors, with the latter approach being most common. Ability to tolerate inhibitory compounds in lignocellulose hydrolysates could reduce the need for detoxification procedures, and this can decrease the overall production cost of ethanol. Adaptation is frequently substituted by random mutagenesis (Bajwa et al. 2009) or by using genome shuffling via protoplast fusion of strains with different genotypes (Bajwa et al. 2010, 2011). To improve ethanol tolerance and production, similar approaches of random mutagenesis and genome shuffling were combined and used (Harner et al. 2015). Selection of higher ethanol tolerance strains has been accomplished using UV mutagenesis, which has also led to improved ethanol production (Watanabe et al. 2011).

Protoplast fusion of *S. stipitis* with *S. cerevisiae* has allowed the isolation of a hybrid strain with higher ethanol productivity from xylose relative to the *S. stipitis* parental strain; however, such hybrid was unstable (Yoon et al. 1996). A popular method for genetic improvement that consist of genome shuffling uses transformation of a yeast strain with total DNA isolated from another strain of the same species or even from different yeast species. In one series of experiments, total DNA of *S. stipitis* was introduced in *S. cerevisiae* by electroporation and the best xylose-fermenting strain isolated was used as the source of DNA for the next round of transformation. This strain showed improvement in ethanol production from xylose and higher ethanol tolerance (Zhang and Geng 2012). To isolate *S. stipitis* strains with improved growth and fermentation characteristics on the xylose/glucose mixture, several fast growing mutants were isolated on a xylose medium with respiration inhibitors antimycin A and salicyl hydroxamate. Several other mutants have also been isolated which produced more ethanol on xylose/glucose mixture (Sreenath and Jeffries 1997). In another approach, mutants of *S. stipitis* which grow anaerobically on xylose plates were isolated. In contrast to the wild-type strain, these isolated mutants grew and fermented xylose and glucose anaerobically though very slowly (Hughes et al. 2012).

Glucose prevents xylose utilization as it competes with xylose for transport and its use is subject to glucose catabolite repression. To obtain strains of *S. stipitis* that can ferment simultaneously glucose and xylose, 2-deoxyglucose resistant mutants were isolated (Sreenath and Jeffries 1999). This was also accomplished by deletion of the gene *HXK1* coding for hexokinase I (Dashtban et al. 2015). No strains with glucose-insensitive xylose transport have been reported for *S. stipitis*. The last experiment could be considered as the only one in which a metabolic engineering approach was applied (Dashtban et al. 2015). As a rule, experiments in metabolic engineering of *S. stipitis* are hampered by our limited knowledge of the limiting steps of the fermentation process. In another approach, the mutant of *S. stipitis* with disruption of cytochrome *c* was isolated. Due to defects in respiration, this mutant appeared to be superior when compared to the wild-type strain in xylose alcoholic fermentation as it accumulated elevated amounts of ethanol (Shi et al. 1999).

In order to determine the gene(s) affecting *S. stipitis* fermentative capabilities, insertional mutants with altered ethanol production from glucose and xylose have been isolated (Berezka et al. 2021). Mutants obtained by random insertional mutagenesis were screened for their growth abilities on solid media with different sugars and for resistance to the glycolysis inhibitor, 3-bromopyruvate. Some of the isolated mutants were shown to have a substantial increase or decrease in the rate of alcoholic fermentation of glucose, xylose, or both these sugars. Insertion was found to occur within the ORF of a homolog to the *S. cerevisiae* gene *HEM25* (*YDL119C*), encoding a mitochondrial glycine transporter required for heme synthesis. Data suggest that the phenotype of obtained strain may be caused not only by HEM25 disruption but also by side mutation that additionally impairs respiration activity and in this way reinforces the effect of the *HEM25* gene deletion on the efficiency of glucose and xylose alcoholic fermentation in *S. stipitis* (Berezka et al. 2021).

Additional experimental studies were conducted to evaluate whether inhibiting the respiratory chain enhances ethanol production in *S. stipitis* (Acevedo et al. 2017). Inhibitors targeting the respiratory chain were introduced into batch cultures of *S. stipitis* utilizing xylose under conditions of limited oxygen supply. Specifically, salicylhydroxamic acid (SHAM) was used to inhibit alternative oxidase (AOX), while potassium cyanide and sodium azide were employed to inhibit cytochrome oxidase (COX). The inhibitors resulted in a decrease in ethanol yield, with SHAM causing a reduction of over 25%, and potassium cyanide and sodium azide leading to a reduction of more than 50%. Given that all previously identified in-silico deletions were related to the respiratory chain, the inhibition of the NADH dehydrogenase complex I was also investigated. The addition of rotenone, which inhibits Complex I, increased the ethanol yield by up to 13% (Acevedo et al. 2017). The combination of SHAM and antimycin A (AA) abolished respiration, depleting the glycolytic flux using both carbon sources tested, leading to increased ethanol production of 21.05 g/l for glucose, and 8.3 g/l ethanol using xylose. In contrast, inhibition of only the AA-sensitive respiration caused increased ethanol production to 30 g/l using glucose and 11.3 g/l from xylose (Granados-Arvizu et al. 2021).

In summary, *S. stipitis* remains as one of the most efficient organisms for xylose and in general, lignocellulose fermentation. Access to genome sequence and the development of methods of molecular genetics have been used to alleviate some of the shortcomings of this yeast (low tolerance to ethanol, glucose inhibition of xylose metabolism, inability to grow anaerobically). This can be accomplished by using the combination of rational design methods of metabolic engineering and random selection. Once the aforementioned shortcomings have been addressed, this yeast organism has the potential for use in industrial fermentation

1.4 *Ogataea (Hansenula) polymorpha*

Ogataea (Hansenula) polymorpha is one of the most thermotolerant species of yeast known. This yeast has the ability to grow up to a temperature of 50 °C. Wild-type strains of this yeast are able to ferment glucose, xylose, mannose, maltose, and cellobiose into ethanol, but are not able to ferment galactose and L-arabinose (Ryabova et al. 2003). Sugar fermentation in this yeast is most efficient under conditions of limited aeration and even at 45–48 °C it is still sufficiently vigorous. At industrial scale high fermentation temperature can reduce cooling costs and the risk of contamination with more energy-efficient removal of ethanol. Several successful attempts have been made to improve this yeast thermotolerance even further (Ishchuk et al. 2009). It is known that, similar to other fungi, *O. polymorpha* accumulates trehalose and expresses heat shock proteins (Hsps) under heat shock conditions (Guerra et al. 2005). The increase in the intracellular level of trehalose in *O. polymorpha* following the knockout of acid trehalase gene *ATH1* resulted in a sixfold higher ethanol production of xylose fermentation at 50 °C. The overexpression of the heat shock proteins Hsp16 and Hsp104 has also led to three to six times improved ethanol production at 50 °C (Ishchuk et al. 2009).

High ethanol tolerance is another important feature of ethanol producers for industrial applications. *O. polymorpha* appear to be more resistant to the toxic concentrations of ethanol when compared to *S. stipites,* but it was more sensitive than *S. cerevisiae* (Dmytruk and Sibirny 2013). The overexpression of the endogeneous *ETT1* gene (a homolog of *S. cerevisiae MPE1* gene) significantly increased the resistance of *O. polymorpha* to ethanol, resulting in ten- and threefold improvements in the growth on agar and in liquid media with ethanol, respectively. The resistance of *O. polymorpha* to ethyl alcohol was also enhanced by the overexpression of the heterologous *S. cerevisiae MPR1* gene that codes for acetyltransferase (Ishchuk et al. 2010).

Direct microbial conversion of carbon polymers into ethanol is a promising technology for the production of alcohols from lignocellulosic raw material. The optimal temperature for the activity of hydrolytic enzymes used during microbial conversion of polymers into ethanol is about 50 °C. Recombinant *O. polymorpha* strains that fermented starch were constructed by the expression of heterologous secretory α-amylase and glucoamylase that code for the *SWA2* and *GAM1* genes of *Schwanniomyces occidentalis* (Wang et al. 1989). The heterologous expression of *XYN2* gene of *Trichoderma reesei* and *xlnD* of *Aspergillus niger* that code for secretory endoxylanase and secretory β-xylosidase were successfully performed in *O. polymorpha* resulting in strains with the ability of direct xylan fermentation at high temperature (Voronovsky et al. 2009).

O. polymorpha has a good potential to be used for the efficient process of simultaneous saccharification and fermentation (SSF) due to its temperature tolerance and ability to ferment xylose to ethanol (Radecka et al. 2015). However, the ethanol yield and productivity from xylose of wild-type strains of *O. polymorpha* are very low. Identification of rate-limiting enzymes for xylose conversion to ethanol is necessary for rational strain modification to improve the fermenting efficiency.

The molecular tools for this yeast species are well-developed (Gellisen 2000) and a complete genome sequence of the strain NCYC495 of *O. polymorpha* is publically available (http://genome.jgi-psf.org/Hanpo2/Hanpo2.home.html). As a consequence, a combination of metabolic engineering and classical selection approaches was successfully used for improvement of the pathway for xylose alcoholic fermentation in *O. polymorpha* (Kurylenko et al. 2014).

The difference in cofactor specificity in the first steps of xylose metabolism results in cofactor imbalance causing a substantially reduced ethanol production with accumulation of xylitol as a by-product in yeasts. The problem arises during consecutive action of NADPH-dependent XR and NAD-dependent XDH that catalyze the reduction of xylose to xylitol and the oxidation of xylitol to xylulose (Fig. 1.2). To avoid the cofactor imbalance, XR and XDH were replaced with bacterial XI, which immediately converts xylose into xylulose with no cofactors required. The bacterial gene *xylA* from *E. coli* or *Streptomyces coelicolor* coding for XI was successfully expressed in *O. polymorpha* strain with a knockout of *XYL1* gene coding for XR and two paralogs *XYL2A* and *XYL2B* of XDH. The recombinant strains were able to grow on xylose as carbon source; however, the amount of accumulated ethanol remained at the level of the wild-type strain CBS4732 (0.15 g/l) (Voronovsky et al. 2005). The overexpression of *E. coli xylA* together with *O. polymorpha XYL3* that codes for XK has led to fourfold increase of ethanol production, but still the maximal ethanol accumulation did not exceed 0.6 g/l at 48 °C (Dmytruk et al. 2008b).

The XR of *O. polymorpha* can use both NADPH and NADH as cofactors. However, the affinity of XR to NADH is significantly lower than that to NADPH. Therefore, another approach to eliminate the imbalance of these two cofactors consists of engineering of the actions of XR and XDH using an *O. polymorpha* XR with reduced affinity toward NADPH. Using site-specific mutagenesis, a modified XR was constructed by the substitution of lysine and asparagine for arginine and aspartic acid at amino acid positions 341 and 343. As a result of the modification of the primary structure of the protein, the affinity of XR to NADPH was 17-fold decreased as compared to the native enzyme, while the affinity of modified XR to NADH remained unchanged. A recombinant strain of *O. polymorpha* with enhanced expression of modified XR and a native XDH and XK was characterized by a fivefold decrease of xylitol accumulation as compared to the wild-type strain and twofold higher ethanol production reaching 1.3 g of ethanol/l (Dmytruk et al. 2008b).

The wild-type *O. polymorpha* strain NCYC495 has been shown to be a more efficient xylose fermenter compared to the strain CBS4732, accumulating up to 0.5 g of ethanol/l during high-temperature xylose alcoholic fermentation. Pyruvate decarboxylase (PDC) is a key enzyme in the final steps of alcoholic fermentation, catalyzing the conversion of pyruvate into acetaldehyde and CO2. Subsequently, acetaldehyde is reduced to ethanol by alcohol dehydrogenase (ADH), encoded by the *ADH1* gene (Fig. 1.2). Under conditions of limited aeration, sufficient PDC activity is important for redirecting pyruvate toward ethanol formation instead of respiration. Overexpression of the endogenous *PDC1* gene under the control of a strong constitutive promoter using a plasmid with multicopy integration led to a twofold increase in ethanol production from xylose in the wild-type strain NCYC495 (Ishchuk et al. 2008). Overexpression of both *PDC1* and *ADH1* genes in *O. polymorpha* resulted in an additional twofold activation of xylose alcoholic fermentation compared to the strain expressing solely *PDC1* (Kurylenko et al. 2016b). Further improvement of these strains was achieved by overexpressing genes involved in either the oxidative pathway of glycerol catabolism (through dihydroxyacetone, specifically the *GCY1* and *DAK1* genes) or the phosphorylative pathway (through glycerol-3-phosphate, specifically the *GUT1* and *GUT2* genes). Additionally, the heterologous gene coding for glycerol transporter *FPS1* from *Komagataella phaffii* (formerly *Pichia pastoris*) was introduced. Obtained recombinant strains with overexpression of two genes (*GUT1* and *GPD1* or *GCY1* and *DAK1*) produced up to 10.7 g/l of ethanol from pure glycerol. From crude glycerol, the strains produced up to 3.55 g/l of ethanol. This represents approximately 15 times more ethanol production compared to the wild-type *O. polymorpha* strain and 2.2 times more than the previously constructed strain (Semkiv et al. 2019).

The *O. polymorpha* mutant 2EthOH^{-} unable to utilize ethanol as a sole carbon source was isolated from strain NCYC495 and characterized by a threefold increase in ethanol production from xylose. Subsequently, the overexpression of the gene *PDC1* in this mutant^{-} resulted in further improvement of ethanol accumulation, reaching 2.5 g of ethanol/l at 48 °C (Ishchuk et al. 2008). To achieve higher amount of ethanol from xylose, several successful metabolic engineering approaches were combined to modify the genome of the 2EthOH^{-} mutant. The overexpression of the genes *XYL1*, *XYL2*, and *XYL3*, which encode a modified XR, native XDH, and XK, respectively, in the 2EthOH- mutant strain harboring an unidentified mutation, resulted in a significant increase in ethanol accumulation during xylose fermentation, reaching 7.4 g/L at 45°C compared to 0.6 g/L in the wild-type strain NCYC495 (Kurylenko et al. 2014). The additional activation of PDC did not lead to any further improvement of xylose conversion to ethanol, though the overexpression of *PDC1* in the background of *XYL1* and *XYL2* overexpressed strain increased ethanol production. The impact of XK on ethanol production during xylose alcoholic fermentation is more pronounced relative to PDC suggesting that PDC does not limit xylose conversion in strain with higher activities of XR, XDH, and XK.

An additional increase in ethanol production from xylose (to 10 g/l at 45 °C) was obtained in mutants selected on a medium supplemented with toxic concentrations of 3-bromopyruvate (BrPA), as described for ethanol accumulation from glucose in *S. cerevisiae* (Dmytruk et al. 2015).

While mutation(s) causing resistance to BrPA in the ethanol overproducing strain remain to be identified, a corresponding mutation was mapped in the strain with the wild-type background. Insertional mutagenesis was used for NCYC495 strain of *O. polymorpha* with subsequent selection of transformants using a mineral medium supplemented with 25 mM of BrPA. Sequencing of the flanking regions revealed that the insertional cassette disrupted the ORF of a gene homologous to the *S. cerevisiae* autophagy-related gene *ATG13* (Dmytruk et al. 2018). This gene encodes a regulatory subunit of the Atg1 signaling complex, stimulating Atg1 kinase activity, which is required for vesicle formation during autophagy and the cytoplasm-to-vacuole targeting pathway. Identified mutation led to a 40% increase in ethanol production from xylose as compared to the parental strain. However, the mechanism of such regulation remains unknown (Kurylenko et al. 2016b).

The potential to increase ethanol yield from xylose in *O. polymorpha* is limited by insufficient knowledge of xylose metabolism and fermentation regulation. Identifying and modulating key genes is essential. Xylose can be fermented to ethanol and converted into glucose and other hexoses, mainly via the pentose phosphate pathway, with some input from gluconeogenesis. The *CAT8* gene that codes for a zinc-finger cluster protein regulates at least 30 genes that are involved in gluconeogenesis, ethanol utilization, glyoxylate cycle, and diauxic shift from fermentation to respiration (Haurie et al. 2001). The roles of *CAT8* gene in the regulation of cell metabolism are well understood in *S. cerevisiae*. It was shown that the deletion of this gene in *S. cerevisiae* and *P. guilliermondii* activated glucose alcoholic fermentation though maximally achieved level of ethanol in the latter species was still very low (Watanabe et al. 2013; Qi et al. 2014). The role of *CAT8* in regulation of xylose metabolism was poorly understood. Transcriptome analysis of the natural xylose-metabolizing yeast *O. polymorpha* did not find changes in *CAT8* expression between xylose- and glucose-containing media (Kim et al. 2013a).

To define the role of Cat8 transcriptional factor in xylose fermentation, *cat8Δ* knock-out mutants in *O. polymorpha* were constructed and analyzed (Ruchala et al. 2017). In *O. polymorpha*, deletion of *CAT8* did not lead to any significant changes in ethanol production from glucose, whereas a considerable increase in xylose alcoholic fermentation was observed (Fig. 1.3). The cell respiration of *cat8Δ* mutants on xylose was impaired to a much higher extent relative to that on glucose as a substrate. Moreover, the impaired ethanol and glycerol utilization in *cat8Δ* mutants were observed, suggesting the involvement of *CAT8* in the regulation of gluconeogenesis in *O. polymorpha*, similar to that described for *S. cerevisiae*. Remarkably, growth on xylose of the *cat8Δ* mutant strain was also partially impaired which suggests that xylose can be considered, at least partially, as a gluconeogenic substrate. The slight decrease in the specific activity of fructose-1,6-bisphosphatase in *cat8Δ* mutants was observed suggesting differences in Cat8 action between *S. cerevisiae* and *O. polymorpha*. The drop in other enzymes of gluconeogenesis, e.g.,

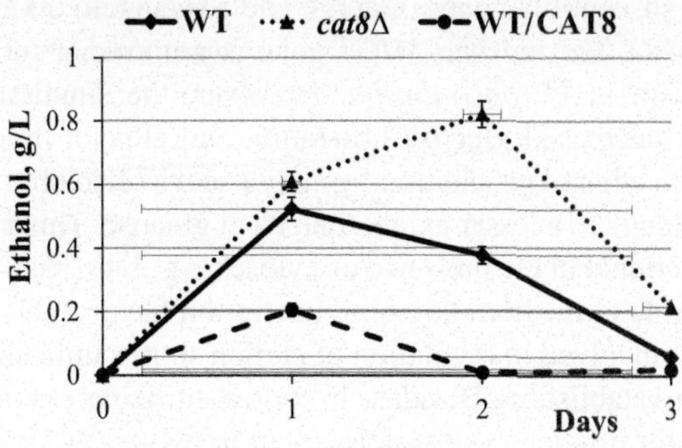

Fig. 1.3 The ethanol production during xylose fermentation of *O. polymorpha* WT, *cat8*Δ and WT/CAT8 strains. Alcoholic fermentation of yeast strains was fulfilled by cultivation in liquid mineral medium at oxygen-limited conditions at 45 °C. The conditions were provided by agitation at 140 rpm. 9% xylose was added into the medium used for the fermentation. The cells were pre-grown in 100 ml of liquid YPX medium (1% yeast extract, 2% peptone, and 4% xylose) in 300 ml Erlenmeyer flasks at 220 rpm till the mid-exponential growth phase. Then the cells were precipitated by centrifugation, washed by water, and inoculated into 40 ml of the fermentation medium in 100 ml Erlenmeyer flasks covered with cotton plugs. The initial biomass concentration for fermentation experiments was 2 g (dry weight)/l

phosphoenolpyruvate carboxykinase, quite possibly determines growth impairment of *O. polymorpha cat8Δ* mutants on gluconeogenic substrates. Overexpression of *CAT8* had opposite effect on xylose alcoholic fermentation as compared to that in *cat8Δ* mutants, as transformants overexpressing *CAT8* gene were characterized by a decrease in ethanol production from xylose (Fig. 1.3). Thus, the *CAT8* gene is one of the first identified genes that are specifically involved in the regulation of xylose alcoholic fermentation in natural xylose-fermenting yeasts. Mutant *O. polymorpha cat8Δ* isolated from an advanced ethanol producer accumulated 30% more ethanol relative to the parental strain, reaching 12.5 g ethanol/l at 45 °C, which is the highest ethanol titer for high-temperature xylose fermentation (Ruchala et al. 2017). These data show important role of transcription factors in xylose alcoholic fermentation in *O. polymorpha*.

Transcription activator Hap4 is the key regulator of respiratory activity in response to carbon sources in *S. cerevisiae* (Olesen and Guarente 1990). A total of two of its functional homologs have been found in *O. polymorpha*, named Hap4-A and Hap4-B (Sybirna et al. 2010). The expression level of Hap4-A is significant for metabolism and fermentation of xylose rather than glucose. The deletion of *HAP4-A* gene increased and overexpression of *HAP4-A* gene reduced the metabolic flux into ethanol production from xylose (Fig. 1.4) (Kurylenko et al. 2021). Transcriptional repressor Tup1 acts as the global repressor and is involved in regulation of large number of genes including those participating in carbon metabolism (Courey and Jia 2001). The deletion of *TUP1* gene activated xylose fermentation but reduced ethanol production from glucose. However, the high expression level of *TUP1* negatively affected both glucose and xylose fermentation (Fig. 1.4) (Kurylenko et al. 2021). Transcriptional repressors Mig1 and Mig2 are involved in glucose catabolite repression of the susceptible genes (George and Santangelo 2006). The deletion or overexpression of *O. polymorpha MIG1* gene is not crucial for the regulation of glucose metabolism in *O. polymorpha*. However, the simultaneous deletion of *MIG1* and *MIG2* decreased glucose utilization and ethanol production from glucose. A decrease in ethanol production from xylose was also observed in the mig1Δ mig2Δ mutant, though to a lesser extent than from glucose. This indicates that Mig1 and Mig2 are functional in the presence of xylose (Fig. 1.4) (Kurylenko et al. 2021).

The potential role of the homolog of transcription factor Azf1 which is known in *S. cerevisiae* to be involved in regulation of carbon source utilization and in glucose sensing has been established. Besides, in xylose-utilizing strain of *S. cerevisiae*, Azf1 is required for efficient xylose fermentation (Myers et al. 2019). In *O. polymorpha*, it was found that *AZF1* deletion retards both glucose and xylose fermentation. In the advanced ethanol producer from xylose, *AZF1* overexpression elevates ethanol production from glucose and especially from xylose (Table 1.2) (Semkiv et al. 2022).

It was also found on the important role of sugar transporters in xylose metabolism and fermentation.

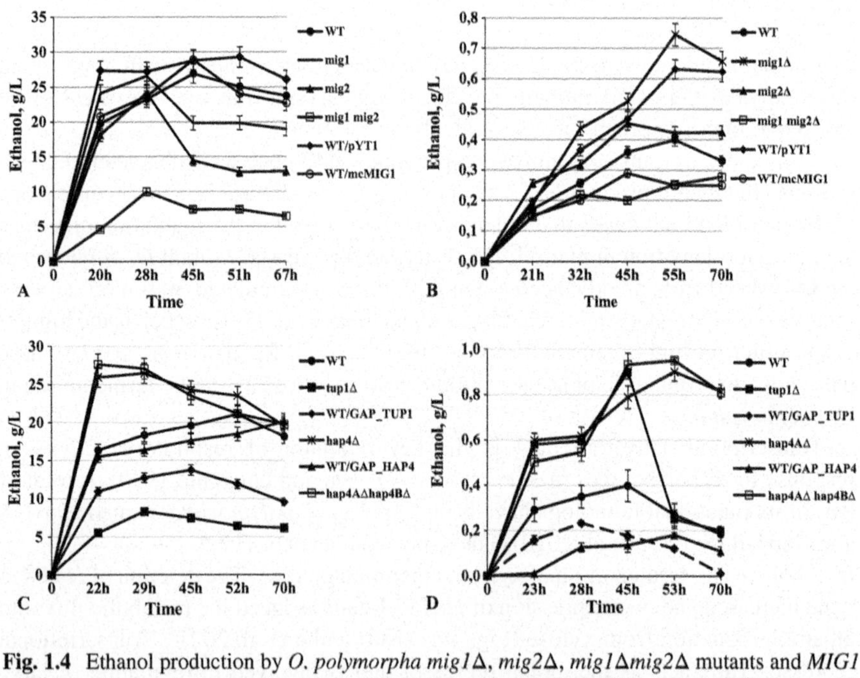

Fig. 1.4 Ethanol production by *O. polymorpha mig1Δ*, *mig2Δ*, *mig1Δmig2Δ* mutants and *MIG1* overexpressing strain as compared to the wild-type strains (WT and WT carrying pYT1 plasmid) during alcoholic fermentation of 10% glucose (**a**) or 10% xylose (**b**) at 45 °C. The impact on glucose and xylose fermentation was studied under semi-anaerobic conditions at 45 °C. Ethanol production by *O. polymorpha* the wild-type strain (WT), *tup1Δ, hap4-AΔ, hap4-AΔhap4-BΔ* mutants and obtained WT/GAP_HAP4-A, WT/GAP_TUP1 strains during alcoholic fermentation at 45 °C in the medium with 10% glucose (**c**) or 10% xylose (**d**) at 45 °C. The mean values and standard deviations (error bars) from two independent experiments are presented

Table 1.2 Main parameters of xylose fermentation at 37 °C by the tested *O. polymorpha* strains with *HXS1* and *AZF1* genes deletion and overexpression

Strain	Ethanol (g/l)	Ethanol yield (mg/g of consumed xylose)	Ethanol specific production rate (mg/g biomass/h)	Ethanol productivity (mg/l/h)
WT[b]	0.49 ± 0.03	15 ± 2	4 ± 1	7 ± 1
WT/LEU[b]	0.56 ± 0.07	19 ± 2	5 ± 1	8 ± 1
Ins[b]	0.02 ± 0.00	1 ± 0	0 ± 0	0 ± 0
WT/Δhxs1[b]	0.03 ± 0.00	2 ± 0	0 ± 0	0 ± 0
WT/Δhxs1/LEU[b]	0.01 ± 0.00	0 ± 0	0 ± 0	0 ± 0
WT/Δhxs/LEU/HXS1[b]	0.93 ± 0.06	20 ± 2	7 ± 1	13 ± 2
WT/LEU/HXS1[b]	0.83 ± 0.07	25 ± 4	8 ± 1	12 ± 3
WT/AZF1[a]	1.20 ± 0.05	41 ± 8	10 ± 3	25 ± 2
WT/Δazf1[b]	0.33 ± 0.03	20 ± 4	3 ± 1	4 ± 1
BEP/Δcat8[b]	9.10 ± 0.10	154 ± 20	49 ± 11	126 ± 22
BEP/Δcat8/AZF1[c]	11.02 ± 0.11	354 ± 31	40 ± 19	114 ± 19
BEP/Δcat8/HXS1[b]	13.04 ± 0.12	352 ± 29	63 ± 17	181 ± 27

Ethanol production during 10% xylose alcoholic fermentation of *O. polymorpha* wild-type strains (WT); wild-type strain with a heterologous selection marker ScLEU2, to restore leucine prototrophy (WT/LEU); insertion mutant (Ins); advanced ethanol producer from xylose (BEP/Δcat8) and recombinant strains with *AZF1* and *HXS1* genes overexpression and deletion at 37 °C. Data are shown as mean of three independent experiments

[a]Data of ethanol yield (mg/g of consumed xylose), ethanol-specific production rate (mg/g biomass/h), ethanol productivity (mg/l/h), and ethanol concentration (g/l) are represented on YNB medium supplemented with 10% of xylose on 48 h of fermentation

[b]72 h of fermentation

[c]96 h of fermentation

Earlier, was described hexose transporter-like sensor gene *HXS1* in *O. polymorpha* (Stasyk et al. 2008). This gene was not involved in glucose repression or catabolite inactivation though its deletion led to significantly impaired transient transcriptional repression in response to fructose. In later studies, it was found that this sensor is important not only for glucose utilization but also for xylose metabolism and alcoholic fermentation of both sugars. Deletion of the *HXS1* gene strongly suppressed utilization and fermentation of both sugars whereas its overexpression activated these processes. It is important to note that overexpression of *HXS1* gene also enhanced ethanol production from xylose on background of the advanced ethanol producer from this pentose (Table 1.2) (Semkiv et al. 2022).

Own hexose transporter Hxt1 of *O. polymorpha* was engineered to eliminate glucose inhibition and endocytosis. Resulted strain utilized simultaneously glucose and xylose from their mixture, in contrast to parental strain, and accumulated elevated amounts of ethanol in the mixture of glucose (70%) and xylose (30%) (Fig. 1.5). Modified versions of the heterologous transporters Hxt7 and Gal2 from

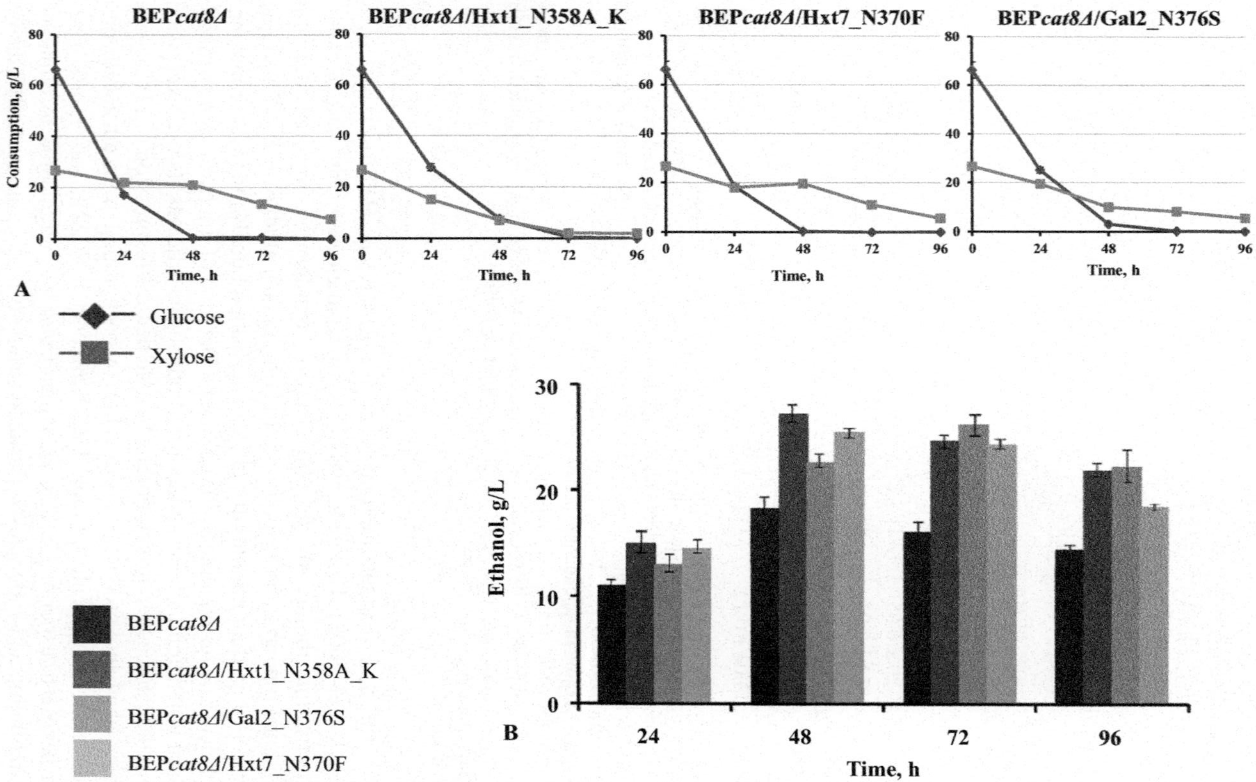

Fig. 1.5 Glucose and xylose consumption (**a**) and ethanol production (**b**) by the parental BEP/cat8 strain and obtained strains Hxt1_N358A_K, Gal2_N376S, Hxt7_N370F during alcoholic fermentation at 45 °C in the medium with 7% glucose and 3% xylose

S. cerevisiae were constructed and overexpressed in the BEP/cat8Δ strain. The overexpression of heterologous Gal2 had a less pronounced effect on the sugar uptake rate, while the overexpression of Hxt7 showed the least impact on sugar consumption (Vasylyshyn et al. 2020).

Further increase in ethanol production form xylose was achieved by selection for mutants forming colonies on plates with L-arabinose as sole carbon source (Vasylyshyn et al., unpublished results). L-Arabinose-growing strains were characterized by further elevated ethanol production from xylose (to 20 g/l and more) and from L-arabinose (still very low, near 0.2 g/l) at 45 °C (Fig. 1.6). To investigate the genetic and metabolic changes that led to increased ethanol production from xylose in mutants capable of growing well on L-arabinose, whole-genome sequencing was employed. Two mutations were identified. The probable product of the first gene mutation is arabinose-5-phosphate isomerase, an enzyme catalyzing the conversion of D-arabinose-5-phosphate to D-ribulose-5-phosphate, potentially facilitating improved growth on L-arabinose. The second mutation is an insertion causing a frameshift and premature stop codons. The likely product of this gene is an activating Ras-GTPase protein domain homologous to RasGAP proteins Ira1 and Ira2 in *S. cerevisiae*, which are negative regulators of the Ras-cAMP signaling pathway. Strains with a deletion of the *IRA1* gene in *O. polymorpha* demonstrated significant advantages under conditions of aerobic fermentation of both pentose sugars and resistance to high concentrations of xylose in the environment. Overexpression of the *IRA1* gene results in impaired growth or death of recombinant strains of *O. polymorpha* (Fig. 1.6) (Vasylyshyn et al., unpublished results). Achieved ethanol titer from xylose is highest at high temperature, however should be further elevated to reach the highest level obtained for mesophilic yeasts.

O. polymorpha belongs to the methylotrophic yeasts that are capable for growth on methanol as sole source of carbon and energy. To enable methanol utilization these organisms have evolved highly specialized metabolic pathways that are partly compartmentalized in peroxisomes. The first enzyme of methanol catabolism, alcohol oxidase, catalyzes oxidation of methanol into the two reactive compounds, formaldehyde and hydrogen peroxide. Alcohol oxidase is localized in peroxisomes together with catalase, which decomposes hydrogen peroxide into water and oxygen. A third peroxisomal enzyme of methanol metabolism is dihydroxyacetone synthase (DHAS). DHAS is a component of the xylulose-5-phosphate cycle and catalyzes the transfer of two-carbon moieties from xylulose-5-phosphate to formaldehyde with the production of glyceraldehyde-3-phosphate (an intermediate of glycolysis) and dihydroxyacetone, which after phosphorylation is converted to a glycolytic intermediate (van der Klei et al. 2006). DHAS can also display classical transketolase activity using aldose phosphates (such as ribose 5-phosphate) as the acceptors for the glycolyl group from the donor substrate xylulose 5-phosphate, therefore playing role in xylose utilization through PPP (Lindley et al. 1981).

Screening for other enzymes that are putatively involved in xylose utilization in *O. polymorpha* has revealed a peroxisomal transaldolase coded for the gene *TAL2* (Kurylenko et al. 2018). The functional roles of peroxisome localized transaldolase and the specific peroxisomal transketolase in xylose utilization and fermentation in

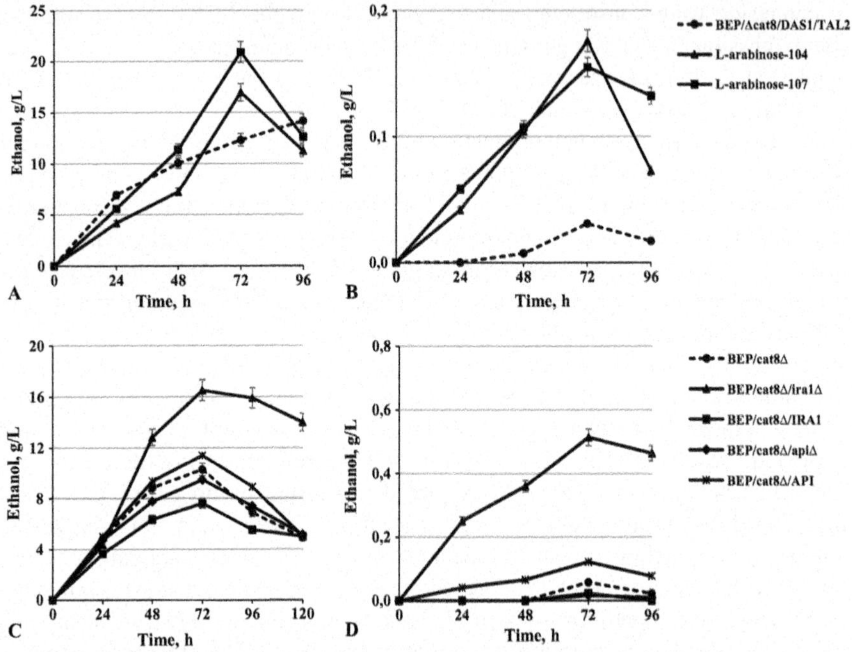

Fig. 1.6 Ethanol production from 10% xylose (**a**) and 10% L-arabinose (**b**) by yeast strains *O. polymorpha* BEP/cat8Δ/DAS1/TAL2 and UV-induced mutants during alcoholic fermentation at 45 °C. Production of ethanol by the best ethanol-producing strain from xylose (BEP/cat8Δ) and recombinant strains with overexpression or deletion of genes *API* or *IRA1*, previously identified through whole-genome sequencing of an L-arabinose mutant (107), during high-temperature alcoholic fermentation of 10% xylose (**c**) and 10% L-arabinose (**d**) at 45 °C

O. polymorpha remain unclear. To investigate the role of these enzymes in ethanol production during xylose fermentation, the corresponding genes DAS1 and TAL2 were overexpressed in *O. polymorpha* NCYC495 strain. The recombinant strains revealed 4.6- and 1.5-fold increase in the specific activity of the corresponding enzymes. The overexpression of *TAL2* gene resulted in a 1.5-fold increase in ethanol production at the fourth day of xylose fermentation as compared to the wild-type strain (Fig. 1.7). The effect of the overexpression of *DAS1* gene was more pronounced. A strain overexpressing *DAS1* gene synthesized 2.3-fold higher amount of ethanol than that of the parental strain after 4 days of xylose fermentation (Fig. 1.7) (Kurylenko et al. 2015). Both *das1Δ* and *tal2Δ* mutants did not show growth retardation on xylose as carbon source but were impaired in xylose alcoholic fermentation as compared to the wild-type strain (Fig. 1.7). Co-overexpression of *DAS1* and *TAL2* in the best previously isolated *O. polymorpha* xylose to ethanol producer led to increase in ethanol accumulation up to 16.5 g/l at 45 °C; or 30–40

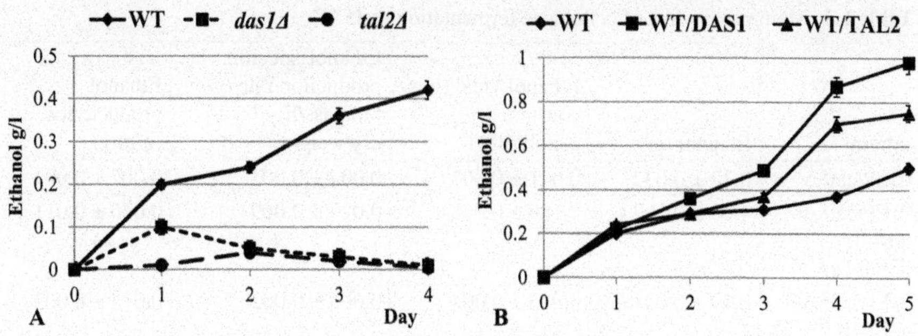

Fig. 1.7 The ethanol production during fermentation of *O. polymorpha* WT, *das1Δ*, *tal2Δ*, WT/DAS1 and WT/TAL2 strains on xylose containing medium

times more ethanol than is produced by the wild-type strain. Both peroxisomal transketolase and transaldolase are not necessary for growth on xylose as a sole carbon and energy source, in contrast to their cytosolic counterparts, Tkl1 and Tal1, but are required for xylose-to-ethanol fermentation. Conversely, the knockout of genes coding for cytosolic transketolase, *TKL1*, and transaldolase, *TAL1*, completely abolished growth on xylose while only partially suppressing the fermentation of this sugar. Overexpression of *TKL1* and *TAL1* also elevated ethanol production from xylose. As a consequence, it was shown for the first time that peroxisomal enzymes Das1 and Tal2 are involved in the xylose alcoholic fermentation in *O. polymorpha*; however, the functions of peroxisomes during alcoholic fermentation of xylose require further investigation (Kurylenko et al. 2018).

Wild-type *O. polymorpha* NCYC495 cannot assimilate cellobiose. Cellobiose-fermenting strains, derivatives of an improved ethanol producer from xylose, *O. polymorpha* BEP/cat8Δ, were constructed by the introduction of heterologous genes *gh1-1* of β-glucosidase, *CDT-1m* and *CDT-2m* of cellodextrin transporters from *Neurospora crassa*, and the *CBP* gene coding for cellobiose phosphorylase from *Saccharophagus degradans*. Through metabolic engineering, laboratory evolution, and mutagenesis of the obtained recombinant strains, mutants were developed, showing improved parameters for high-temperature alcoholic fermentation of cellobiose (Table 1.3) and simultaneous consumption of all quantitatively significant sugars in lignocellulose hydrolysates (Fig. 1.8) (Vasylyshyn et al. 2024).

Based on published data, the ethanol yield of the constructed *O. polymorpha* recombinant strain (0.39 g/g xylose) (Kurylenko et al. 2018) is close to that described for *S. stipitis* (0.35–0.44 g/g xylose) (Jeffries et al. 2007) and *Spathaspora passalidarum* (0.42 g/g xylose) (Long et al. 2012). However, this yield was achieved for *O. polymorpha* at 45 °C, whereas the compared organisms are mesophilic and therefore unable to grow and ferment at such a high temperature. Among the thermotolerant ethanol-producing strains, the most promising one is an engineered

Table 1.3 Main parameters of cellobiose fermentation at 45 °C

Strain	Ethanol (g/l)	Ethanol yield (g/g consumed cellobiose)	Ethanol-specific production rate (g/g biomass/h) dry weight	Ethanol productivity (g/l/h)
BEP/cat8Δ[b]	0.47 ± 0.032	0.041 ± 0.002	0.004 ± 0.001	0.007 ± 0.001
BEP/cat8Δ/gh1-1/CDT-1m[c]	4.60 ± 0.102	0.180 ± 0.028	0.027 ± 0.009	0.076 ± 0.011
BEP/cat8Δ/CBP-1/CDT-1m[c]	0.79 ± 0.041	0.045 ± 0.004	0.006 ± 0.002	0.013 ± 0.003
BEP/cat8Δ/gh1-1/CDT-2mA[c]	0.68 ± 0.062	0.047 ± 0.004	0.005 ± 0.001	0.011 ± 0.001
BEP/cat8Δ/CBP-1/CDT-2mAM[a]	4.20 ± 0.108	0.183 ± 0.031	0.025 ± 0.010	0.070 ± 0.009

Main parameters of cellobiose fermentation at 45 °C by the tested *O. polymorpha* strains with *gh1-1/CDT-1m, CBP-1/CDT-1m, gh1-1/CDT-2mA, CBP-1/CDT-2mAM* simultaneous overexpression genes. "A"—mutants obtained by laboratory evolution, "AM"—mutants obtained in two stages as a result of laboratory evolution and UV—mutagenesis with 2-DG as a glycolysis inhibitor

[a]Data of ethanol yield and ethanol (g/l) are represented on YNB medium supplemented with 10% of cellobiose on 48 h of fermentation

[b]72 h of fermentation

[c]96 h of fermentation

Kluyveromyces marxianus strain with an ethanol yield of 0.38 g/g xylose at 42 °C, but a lower yield at 45 °C (0.27 g/g xylose). An additional advantage of the *O. polymorpha* recombinant strain, in contrast to the recombinant *K. marxianus*, is that no xylitol accumulation was observed. Today, a number of microorganisms are also known to have a natural or acquired ability to ferment cellobiose into ethanol. For example, *S. cerevisiae*, which accumulates 38 g/l of ethanol at 30 °C (Choi et al. 2022), *Myceliophthora thermophila*—11.3 g/l of ethanol at 45–50 °C, (Li et al. 2020), *Zymobacter palmae*—10 g/l of ethanol at 30 °C, (Yanase et al. 2005). The *O. polymorpha* currently produces a maximum of only 5 g/l of ethanol at 45 °C (Vasylyshyn et al. 2024). Thus, the ethanol yield from the best *O. polymorpha* strain, developed through metabolic engineering and classical selection approaches, is still insufficient for economic feasibility. Further improvement of the parameters of alcoholic fermentation of xylose in *O. polymorpha* could be obtained by optimization of the transport of this pentose into cells, as well as that of the genes determining the resistance to toxic and inhibitory compounds derived after pretreatment of lignocellulosic biomass. At present, the resistance of *O. polymorpha* to toxic products (aldehydes, phenols, acetic and formic acids) accumulated in lignocellulose hydrolysates under the conditions of acidic hydrolysis has not been precisely studied.

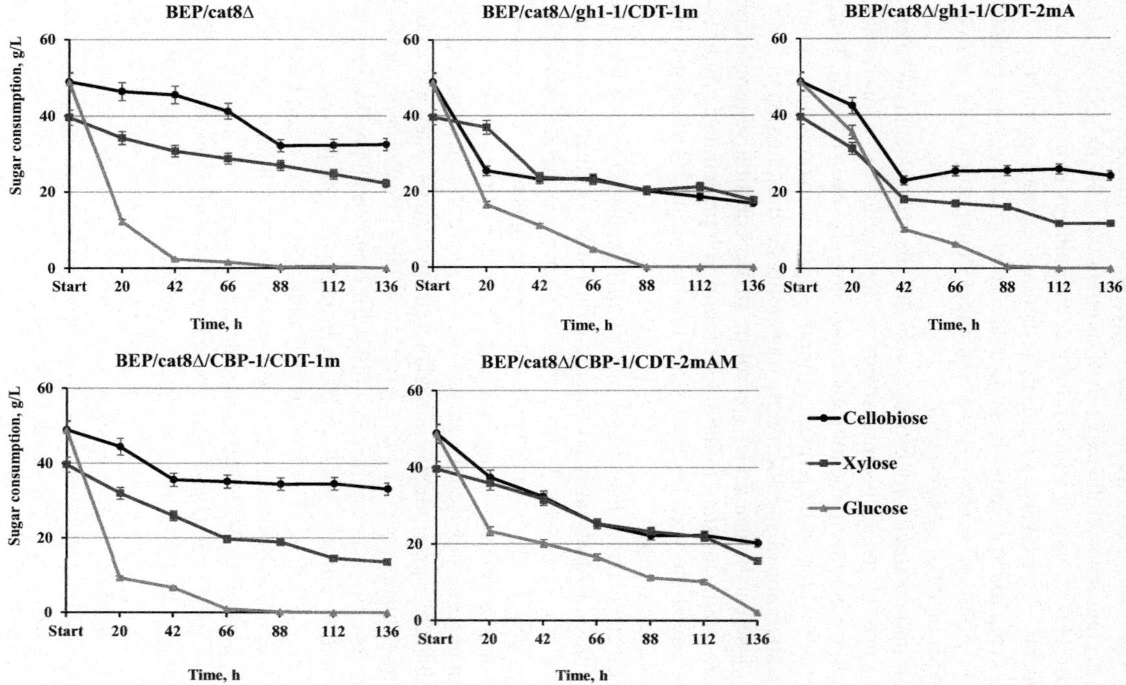

Fig. 1.8 Cellobiose, glucose, and xylose consumption by the parental strain BEP/cat8Δ, evolutionary mutants BEP/cat8Δ/gh1-1/CDT-2mA, BEP/cat8Δ/CBP-1/CDT-2mAM and obtained strains BEP/cat8Δ/gh1-1/CDT-1m, BEP/cat8Δ/CBP-1/CDT-1m during alcoholic fermentation at 45 °C in the media with 5% cellobiose / 4% xylose / 5% glucose. "A"—mutants obtained by laboratory evolution, "AM"—mutants obtained in two stages as a result of laboratory evolution and UV—mutagenesis with 2-DG as a glycolysis inhibitor

1.5 Concluding Remarks

Commercial strains of *S. cerevisiae*, known for their enhanced ethanol production from glucose and sucrose (first-generation ethanol), have been widely used in industrial applications. However, these strains often come with limitations. Strains that prioritize ethanol accumulation at the expense of biomass typically exhibit reduced robustness and struggle to compete with wild-type contaminants in non-sterile production environments. Conversely, strains that increase ethanol production by reducing glycerol synthesis may suffer from decreased performance under osmotic stress compared to wild-type strains. These challenges could potentially be overcome through further metabolic engineering.

S. cerevisiae strains constructed for production of 2G ethanol efficiently ferment, in addition to glucose, abundant pentose sugars of lignocellulosic hydrolysates, xylose and L-arabinose. Strains fermenting galactose have also been isolated. There are also known strains which could ferment different sugars of hydrolysates simultaneously due to elaboration of specific xylose transporters and expression of genes responsible for xylose catabolism under control of strong constitutive promoters. Moreover, there are strains resistant to inhibitors found in lignocellulosic hydrolysates, which have paved the way for pilot-scale production of 2G ethanol using engineered *S. cerevisiae* strains.

Similarly, improvements have been made in *S. stipitis* strains for xylose fermentation on lignocellulosic hydrolysates, addressing issues such as glucose catabolite repression and inhibitor resistance. Despite these advantages, challenges persist. *S. stipitis* strains typically have lower ethanol tolerance compared to *S. cerevisiae*, which can limit their efficiency in industrial-scale fermentations. Furthermore, *S. stipitis* is known to require oxygen for growth, which adds complexity to fermentation processes and may necessitate specialized bioreactor designs. Plans are underway to establish pilot plants for 2G ethanol production based on *S. stipitis* strains, aiming to leverage its unique metabolic capabilities for commercial applications.

On the other hand, *O. polymorpha* is another yeast species with promising attributes for 2G ethanol production. *O. polymorpha* exhibits robust growth and fermentation capabilities under a wide range of environmental conditions, including temperatures as high as 45 °C. This thermotolerance is advantageous for simultaneous saccharification and fermentation (SSF) processes, where lignocellulosic biomass is enzymatically hydrolyzed at elevated temperatures. However, ethanol production from xylose and other pentose sugars in *O. polymorpha* remains a challenge. Efforts are focused on optimizing metabolic pathways and genetic engineering to develop tailored strains that maximize ethanol production from lignocellulosic biomass while overcoming inherent metabolic limitations.

Acknowledgment This work was supported by the National Science Centre, project number UMO-2020/37/B/NZ1/02232; the Subcarpathian Centre for Innovation, project number N3_53; MSCA4Ukraine consortium funded under the EU's Marie Skłodowska-Curie Actions, project number 1232295, and partially supported by the "Presidential Discretionary-Ukraine Support Grants" from the Simons Foundation, award no 1030281.

References

Acevedo A, Conejeros R, Aroca G (2017) Ethanol production improvement driven by genome-scale metabolic modeling and sensitivity analysis in *Scheffersomyces stipitis*. PLoS One 12(6):e0180074

Agbogbo FK, Wenger KS (2006) Effect of pretreatment chemicals on xylose fermentation by *P. stipitis*. Biotechnol Lett 28(24):2065–2069

Agbogbo FK, Coward-Kelly G, Torry-Smith M, Wenger KS (2006) Fermentation of glucose/xylose mixtures using *P. stipitis*. Process Biochem 41(11):2333–2336

Almeida JR, Roder A, Modig T, Laadan B, Liden G, Gorwa-Grauslund MF (2008) NADH- vs NADPH-coupled reduction of 5-hydroxymethyl furfural (HMF) and its implications on product distribution in *Saccharomyces cerevisiae*. Appl Microbiol Biotechnol 78(6):939–945

Amartey SA, Jeffries TW (1994) Comparison of corn steep liquor with other nutrients in the fermentation of D-xylose by *Pichia stipitis* CBS6054. Biotechnol Lett 16(2):211–214

Amartey S, Jeffries T (1996) An improvement in *Pichia stipitis* fermentation of acid-hydrolysed hemicellulose achieved by overliming (calcium hydroxide treatment) and strain adaptation. World J Microbiol Biotechnol 12(3):281–283

An MZ, Tang YQ, Mitsumasu K, Liu ZS, Shigeru M, Kenji K (2011) Enhanced thermotolerance for ethanol fermentation of *Saccharomyces cerevisiae* strain by overexpression of the gene coding for trehalose-6-phosphate synthase. Biotechnol Lett 33(7):1367–1374

Ansell R, Granath K, Hohmann S, Thevelein JM, Adler L (1997) The two isoenzymes for yeast NAD^+-dependent glycerol 3–phosphate dehydrogenase encoded by *GPD1* and *GPD2* have distinct roles in osmoadaptation and redox regulation. EMBO J 16(9):2179–2187

Arnthong J, Ponjarat J, Bussadee P, Deenarn P, Prommana P, Phienluphon A, Charoensri S et al (2022) Enhanced surface display efficiency of β-glucosidase in *Saccharomyces cerevisiae* by disruption of cell wall protein-encoding genes *YGP1* and *CWP2*. Biochem Eng J 179:108305

Bafuelos M, Gancedo C (1978) In situ study of the glycolytic pathway in *Saccharomyces cerevisiae*. Arch Microbiol 117(2):197–201

Bajwa PK, Shireen T, D'Aoust F, Pinel D, Martin VJ, Trevors JT, Lee H (2009) Mutants of the pentose-fermenting yeast *Pichia stipitis* with improved tolerance to inhibitors in hardwood spent sulfite liquor. Biotechnol Bioeng 104(5):892–900

Bajwa PK, Pinel D, Martin VJ, Trevors JT, Lee H (2010) Strain improvement of the pentose-fermenting yeast *Pichia stipitis* by genome shuffling. J Microbiol Methods 81:179–186

Bajwa PK, Phaenark C, Grant N, Zhang X, Paice M, Martin VJ, Trevors JT, Lee H (2011) Ethanol production from selected lignocellulosic hydrolysates by genome shuffled strains of *Scheffersomyces stipitis*. Bioresour Technol 102(21):9965–9969

Balagurunathan B, Jonnalagadda S, Tan L, Srinivasan R (2012) Reconstruction and analysis of a genome-scale metabolic model for *Scheffersomyces stipitis*. Microb Cell Factories 11:27

Basso TO, de Kok S, Dario M, do Espirito-Santo JC, Müller G, Schlölg PS, Silva CP, Tonso A, Daran JM, Gombert AK, van Maris AJ, Pronk JT, Stambuk BU (2011) Engineering topology and kinetics of sucrose metabolism in *Saccharomyces cerevisiae* for improved ethanol yield. Metab Eng 13(6):694–703

Bengtsson O, Hahn-Hägerdal B, Gorwa-Grauslund MF, Bengtsson O (2009) Xylose reductase from *Pichia stipitis* with altered coenzyme preference improves ethanolic xylose fermentation by recombinant *Saccharomyces cerevisiae*. Biotechnol Biofuels 2:9

Benisch F, Boles E (2014) The bacterial Entner-Doudoroff pathway does not replace glycolysis in *Saccharomyces cerevisiae* due to the lack of activity of iron-sulfur cluster enzyme 6-phosphogluconate dehydratase. J Biotechnol 171:45–55

Bera AK, Ho NW, Khan A, Sedlak M (2011) A genetic overhaul of *Saccharomyces cerevisiae* 424A(LNH–ST) to improve xylose fermentation. J Ind Microbiol Biotechnol 38(5):617–626

Berezka K, Semkiv M, Borbuliak M, Blomqvist J, Linder T, Ruchała J, Dmytruk K, Passoth V, Sibirny A (2021) Insertional tagging of the *Scheffersomyces stipitis* gene HEM25 involved in regulation of glucose and xylose alcoholic fermentation. Cell Biol Int 45(3):507–517

Bettiga M, Bengtsson O, Hahn-Hagerdal B, Gorwa-Grauslund MF (2009) Arabinose and xylose fermentation by recombinant *Saccharomyces cerevisiae* expressing a fungal pentose utilization pathway. Microb Cell Fact 8:40

Bicho PA, Runnals PL, Cunningham JD, Lee H (1988) Induction of xylose reductase and xylitol dehydrogenase activities in *Pachysolen tannophilus* and *Pichia stipitis* on mixed sugars. Appl Environ Microbiol 54(1):50–54

Brat D, Boles E, Wiedemann B (2009) Functional expression of a bacterial xylose isomerase in *Saccharomyces cerevisiae*. Appl Environ Microbiol 75(8):2304–2311

Broda M, Yelle DJ, Serwańska K (2022) Bioethanol production from lignocellulosic biomass-challenges and solutions. Molecules 27(24):8717

Cadiere A, Ortiz-Julien A, Camarasa C, Dequin S (2011) Evolutionary engineered *Saccharomyces cerevisiae* wine yeast strains with increased in vivo flux through the pentose phosphate pathway. Metab Eng 13(3):263–271

Cavelius P, Engelhart-Straub S, Mehlmer N, Lercher J, Awad D, Brück T (2023) The potential of biofuels from first to fourth generation. PLoS Biol 21(3):e3002063

Cho JY, Jeffries TW (1998) *Pichia stipitis* genes for alcohol dehydrogenase with fermentative and respiratory functions. Appl Environ Microbiol 64:1350–1358

Choi HJ, Jin YS, Lee WH (2022) Effects of engineered S*accharomyces cerevisiae* fermenting cellobiose through low-energy-consuming phosphorolytic pathway in simultaneous saccharification and fermentation. J Microbiol Biotechnol 32(1):117–125

Clifton D, Fraenkel DG (1981) The gcr (glycolysis regulation) mutation of *Saccharomyces cerevisiae*. J Biol Chem 256(24):13074–13078

Conceicao GJA, Moran PJS, Rodrigues JAR (2003) Highly efficient extractive biocatalysis in the asymmetric reduction of an acyclic enone by the yeast *Pichia stipitis*. Tetrahedron Asymmetr 14:43–45

Courey AJ, Jia S (2001) Transcriptional repression: the long and the short of it. Genes Dev 15:2786–2796

Dashtban M, Wen X, Bajwa PK, Ho CY, Lee H (2015) Deletion of *hxk1* gene results in derepression of xylose utilization in *Scheffersomyces stipitis*. J Ind Microbiol Biotechnol 42(6):889–896

De Kok S, Kozak B, Pronk J, Van Maris A (2012) Energy coupling in *Saccharomyces cerevisiae*: selected opportunities for metabolic engineering. FEMS Yeast Res 12:387–397

Della-Bianca BE, Basso TO, Stambuk BU, Basso LC, Gombert AK (2013) What do we know about the yeast strains from the Brazilian fuel ethanol industry. Appl Microbiol Biotechnol 97:979–991

de Valk SC, Bouwmeester SE, de Hulster E et al (2020) Engineering proton-coupled hexose uptake in *Saccharomyces cerevisiae* for improved ethanol yield. Biotechnol Biofuels 15:47

Dmytruk KV, Sibirny AA (2013) Metabolic engineering of yeast *Hansenula polymorpha* for construction of efficient ethanol producers. Tsitol Genet 47(6):3–21

Dmytruk OV, Voronovsky AY, Abbas CA, Dmytruk KV, Ishchuk OP, Sibirny AA (2008b) Overexpression of bacterial xylose isomerase and yeast host xylulokinase improves xylose alcoholic fermentation in the thermotolerant yeast *Hansenula polymorpha*. FEMS Yeast Res 8(1):165–173

Dmytruk K, Kshanovska B, Sibirny A (2015) New methods for selection of *Saccharomyces cerevisiae* ethanol overproducing mutants resistant to oxythiamine, trehalose, bromopyruvate, glyoxylic acid and glucosamine. In: Living organisms and bioanalytical approaches for detoxification and monitoring of toxic compounds. University of Rzeszow, Rzeszow, pp 237–246

Dmytruk K, Kshanovska B, Abbas C, Sibirny A (2016) New methods for positive selection of yeast ethanol overproducing mutants. Bioethanol 2:24–31

Dmytruk KV, Ruchala J, Grabek-Lejko D, Puchalski C, Bulbotka NV, Sibirny AA (2018) Autophagy-related gene *ATG13* is involved in control of xylose alcoholic fermentation in the thermotolerant methylotrophic yeast *Ogataea polymorpha*. FEMS Yeast Res 18:foy010

du Preez JC, van der Walt JP (1983) Fermentation of D-xylose to ethanol by a strain of *Candida shehatae*. Biotechnol Lett 5:357–362 du Preez JC, Bosch M, Prior BA (1985) Xylose fermentation by *Candida shehatae* and *Pichia stipitis*: effects of pH, temperature and substrate concentration. Enzym Microb Technol 8:360–364

du Preez JC, Bosch M, Prior B (1986) The fermentation of hexose and pentose sugars by *Candida shehatae* and *Pichia stipitis*. Appl Microbiol Biotechnol 23(3):228–233

du Preez JC, van Driessel B, Prior BA (1989) Ethanol tolerance of *Pichia stipitis* and *Candida shehatae* strains in fed-batch cultures at controlled low dissolved-oxygen levels. Appl Microbiol Biotechnol 30:53–58

Du J, Li S, Zhao H (2010) Discovery and characterization of novel D-xylose-specific transporters from *Neurospora crassa* and *Pichia stipitis*. Mol BioSyst 6(11):2150–2156

Farwick A, Bruder S, Schadeweg V, Oreb M, Boles E (2014) Engineering of yeast hexose transporters to transport D-xylose without inhibition by D-glucose. Proc Natl Acad Sci USA 111(14):5159–5164

Gellisen G (2000) Heterologous protein production in methylotrophic yeasts. Appl Microbiol Biotechnol 54:741–750

Gellisen G (2002) *Hansenula polymorpha*—biology and applications. Wiley VCH, Weinheim, p 347

George M, Santangelo (2006) Glucose Signaling in Saccharomyces cerevisiae Microbiology and Molecular Biology Reviews 70(1):253–282. https://doi.org/10.1128/MMBR.70.1.253-282.2006

Ghosh A, Zhao H, Price ND (2011) Genome-scale consequences of cofactor balancing in engineered pentose utilization pathways in *Saccharomyces cerevisiae*. PLoS One 6(11):e27316

Gombert AK, van Maris AJ (2015) Improving conversion yield of fermentable sugars into fuel ethanol in 1st generation yeast-based production processes. Curr Opin Biotechnol 33:81–86

Gorsich SW, Dien BS, Nichols NN, Slininger PJ, Liu ZL, Skory CD (2006) Tolerance to furfural-induced stress is associated with pentose phosphate pathway genes *ZWF1*, *GND1*, *RPE1*, and *TKL1* in *Saccharomyces cerevisiae*. Appl Microbiol Biotechnol 71(3):339–349

Granados-Arvizu JA, Canizal-García M, Madrigal-Pérez LA, González-Hernández JC, Regalado-González C (2021) Inhibition of alternative respiration system of *Scheffersomyces stipites* and effect on glucose or xylose fermentation. FEMS Yeast Res 21(2)

Grootjen DRJ, van der Lans RGJM, Luyben KCA (1990) Effects of the aeration rate on the fermentation of glucose and xylose by *Pichia stipitis* CBS 5773. Enzym Microb Technol 12:20–23

Guadalupe Medina V, Almering MJH, van Maris AJA, Pronk JT (2010) Elimination of glycerol production in anaerobic cultures of a *Saccharomyces cerevisiae* strain engineered to use acetic acid as an electron acceptor. Appl Environ Microbiol 76(1):190–195

Guebel DV, Cordenons A, Cascone O, Giulietti AM, Nudel C (1992) Influence of the nitrogen source on growth and ethanol production by *Pichia stipitis* NRRL Y-7124. Biotechnol Lett 14(12):1193–1198

Guerra E, Chye PP, Berardi E, Piper PW (2005) Hypoxia abolishes transience of the heat-shock response in the methylotrophic yeast *Hansenula polymorpha*. Microbiology 151:805–811

Guo ZP, Zhang L, Ding ZY, Shi GY (2011) Minimization of glycerol synthesis in industrial ethanol yeast without influencing its fermentation performance. Metab Eng 13(1):49–59

Ha SJ, Wei Q, Kim SR, Galazka JM, Cate JH, Jin YS (2011) Cofermentation of cellobiose and galactose by an engineered *Saccharomyces cerevisiae* strain. Appl Environ Microbiol 77(16):5822–5825

Ha SJ, Galazka JM, Oh EJ, Kordić V, Kim H, Jin YS et al (2013) Energetic benefits and rapid cellobiose fermentation by *Saccharomyces cerevisiae* expressing cellobiose phosphorylase and mutant cellodextrin transporters. Metab Eng 15:134–143

Hahn-Hägerdal B, Wahlbom CF, Gárdonyi M, van Zyl WH, Cordero Otero RR, Jönsson LJ (2001) Metabolic engineering of *Saccharomyces cerevisiae* for xylose utilization. Adv Biochem Eng Biotechnol 73:53–84

Harner NK, Wen X, Bajwa PK, Austin GD, Ho CY, Habash MB, Trevors JT, Lee H (2015) Genetic improvement of native xylose-fermenting yeasts for ethanol production. J Ind Microbiol Biotechnol 42(1):1–20

Haurie V, Perrot M, Mini T, Jenö P, Sagliocco F, Boucherie H (2001) The transcriptional activator Cat8p provides a major contribution to the reprogramming of carbon metabolism during the diauxic shift in *Saccharomyces cerevisiae*. J Biol Chem 276:76–85

Hawkins GM, Ghose D, Russel J, Peterson J (2013) Production of ethanol from high dry matter of pretreated loblolly pine by an evolved strain of *Saccharomyces cerevisiae*. J Bioremediat Biodegrad 4:195

Hector RE, Qureshi N, Hughes SR, Cotta MA (2008) Expression of a heterologous xylose transporter in a *Saccharomyces cerevisiae* strain engineered to utilize xylose improves aerobic xylose consumption. Appl Microbiol Biotechnol 80(4):675–684

Higgins VJ, Bell PJL, Dawes IW, Attfield PV (2001) Generation of a novel *Saccharomyces cerevisiae* strain that exhibits strong maltose utilization and hyperosmotic resistance using nonrecombinant techniques. Appl Environ Microbiol 67:4346–4348

Hohmann S (2002) Osmotic stress signaling and osmoadaptation in yeasts. Microbiol Mol Biol Rev 66(2):300–372

Hou L (2010) Improved production of ethanol by novel genome shuffling in *Saccharomyces cerevisiae*. Appl Microbiol Biotechnol 160(4):1084–1093

Hou J, Lages NF, Oldiges M, Vemuri GN (2009) Metabolic impact of redox cofactor perturbations in *Saccharomyces cerevisiae*. Metab Eng 11:253–261

Huang CF, Lin TH, Guo GL, Hwang WS (2009) Enhanced ethanol production by fermentation of rice straw hydrolysate without detoxification using a newly adapted strain of *Pichia stipitis*. Bioresour Technol 100:3914–3920

Hughes SR, Gibbons WR, Bang SS, Pinkelman R, Bischoff KM, Slininger PJ, Qureshi N, Kurtzman CP, Liu S, Saha BC, Jackson JS, Cotta MA, Rich JO, Javers JE (2012) Random UV-C mutagenesis of *Scheffersomyces* (formerly *Pichia*) *stipitis* NRRL Y-7124 to improve anaerobic growth on lignocellulosic sugars. J Ind Microbiol Biotechnol 39:163–173

Ilmen M, Koivuranta K, Ruohonen L, Suominen P, Penttilä M (2007) Efficient production of L-lactic acid from xylose by *Pichia stipitis*. Appl Environ Microbiol 73:117–123

Ingledew WM (1999) Alcohol production by *Saccharomyces cerevisiae*: a yeast primer. In: Lyons TP, Kelsall DR (eds) The alcohol textbook, 3rd edn. Nottingham University Press, Nottingham, UK, pp 49–87

Ishchuk OP, Voronovsky AY, Stasyk OV, Gayda GZ, Gonchar MV, Abbas CA, Sibirny AA (2008) Overexpression of pyruvate decarboxylase in the yeast *Hansenula polymorpha* results in increased ethanol yield in high-temperature fermentation of xylose. FEMS Yeast Res 8(7):1164–1174

Ishchuk OP, Voronovsky AY, Abbas CA, Sibirny AA (2009) Construction of *Hansenula polymorpha* strains with improved thermotolerance. Biotechnol Bioeng 104(5):911–919

Ishchuk OP, Abbas CA, Sibirny AA (2010) Heterologous expression of *Saccharomyces cerevisiae MPR1* gene confers tolerance to ethanol and L-azetidine-2-carboxylic acid in *Hansenula polymorpha*. J Ind Microbiol Biotechnol 37(2):213–218

Jeffries TW, Van Vleet JR (2009) *Pichia stipitis* genomics, transcriptomics, and gene clusters. FEMS Yeast Res 9(6):793–807

Jeffries TW, Grigoriev IV, Grimwood J, Laplaza JM, Aerts A, Salamov A, Schmutz J, Lindquist E, Dehal P, Shapiro H, Jin YS, Passoth V, Richardson PM (2007) Genome sequence of the lignocellulose-bioconverting and xylose-fermenting yeast *Pichia stipitis*. Nat Biotechnol 25(3):319–326

Jensen P, Snoep J, Westerhoff H (2006) Method of improving the production of biomass or a desired product from a cell. US Patent 20060094078: A1

Jeppsson M, Johansson B, Hahn-Hägerdal B, Gorwa-Grauslund MF (2002) Reduced oxidative pentose phosphate pathway flux in recombinant xylose–utilizing *Saccharomyces cerevisiae* strains improves the ethanol yield from xylose. Appl Environ Microbiol 68(4):1604–1609

Jin YS, Jones S, Shi NQ, Jeffries TW (2002) Molecular cloning of *XYL3* (D-xylulokinase) from *Pichia stipitis* and characterization of its physiological function. Appl Environ Microbiol 68:1232–1239

Jin YS, Ni H, Laplaza JM, Jeffries TW (2003) Optimal growth and ethanol production from xylose by recombinant *Saccharomyces cerevisiae* require moderate D-xylulokinase activity. Appl Environ Microbiol 69:495–503

Jin YS, Cruz J, Jeffries TW (2005) Xylitol production by a Pichia stipitis D-xylulokinase mutant. Appl Microbiol Biotechnol 68(1):42–45. https://doi.org/10.1007/s00253-004-1854-5

Jingping G, Hongbing S, Gang S, Hongzhi L, Wenxiang P (2012) A genome shuffling-generated *Saccharomyces cerevisiae* isolate that ferments xylose and glucose to produce high levels of ethanol. J Ind Microbiol Biotechnol 39(5):777–787

Jinyang, Li Yongli, Zhang Jingen, Li Tao, Sun Chaoguang, Tian (2020) Metabolic engineering of the cellulolytic thermophilic fungus Myceliophthora thermophila to produce ethanol from cellobiose Abstract Biotechnology for Biofuels 13(1). https://doi.org/10.1186/s13068-020-1661-y

Kaliterna J, Weusthuis RA, Castrillo JI, van Dijken JP, Pronk JT (1995) Coordination of sucrose uptake and respiration in the yeast *Debaryomyces yamadae*. Microbiology 141:1567–1574

Karhumaa K, Hahn-Hägerdal B, Gorwa-Grauslund MF (2005) Investigation of limiting metabolic steps in the utilization of xylose by recombinant *Saccharomyces cerevisiae* using metabolic engineering. Yeast 22:359–368

Karhumaa K, Garcia Sanchez R, Hahn-Hägerdal B, Gorwa-Grauslund MF (2007) Comparison of the xylose reductase-xylitol dehydrogenase and the xylose isomerase pathways for xylose fermentation by recombinant *Saccharomyces cerevisiae*. Microb Cell Fact 6:5

Keating JD, Robinson J, Cotta MA, Saddler JN, Mansfield SD (2004) An ethanologenic yeast exhibiting unusual metabolism in the fermentation of lignocellulosic hexose sugars. J Ind Microbiol Biotechnol 31(5):235–244

Khoury GA, Fazelinia H, Chin JW, Pantazes RJ, Cirino PC, Maranas CD (2009) Computational design of *Candida boidinii* xylose reductase for altered cofactor. Protein Sci 18(10):2125–2138

Kilian SG, van Uden N (1988) Transport of xylose and glucose in the xylose fermenting yeast *Pichia stipitis*. Appl Microbiol Biotechnol 27:545–548

Kim MS, Chung YS, Seo JH, Jo DH, Park YH, Ryu YW (2001) High-yield production of xylitol from xylose by a xylitol dehydrogenase defective mutant of *Pichia stipitis*. J Microbiol Biotechnol 11:564–569

Kim OC, Suwannarangsee S, Oh DB, Kim S, Seo JW, Kim CH, Kang HA, Kim JY, Kwon O (2013a) Transcriptome analysis of xylose metabolism in the thermotolerant methylotrophic yeast *Hansenula polymorpha*. Bioprocess Biosyst Eng 36:1509–1518

Kim SR, Skerker JM, Kang W, Lesmana A, Wei N, Arkin AP, Jin Y-S (2013b) Rational and evolutionary engineering approaches uncover a small set of genetic changes efficient for rapid xylose fermentation in Saccharomyces cerevisiae. PLoS One 8:e57048

Kim SK, Jin YS, Choi IG, Park YC, Seo JH (2015) Enhanced tolerance of *Saccharomyces cerevisiae* to multiple lignocellulose-derived inhibitors through modulation of spermidine contents. Metab Eng 29:46–55

Kim H, Oh EJ, Lane ST, Lee WH, Cate JH, Jin YS (2018) Enhanced cellobiose fermentation by engineered *Saccharomyces cerevisiae* expressing a mutant cellodextrin facilitator and cellobiose phosphorylase. J Biotechnol 275:53–59

Klinner U, Fluthgraf S, Freese S, Passoth V (2005) Aerobic induction of respiro-fermentative growth by decreasing oxygen tensions in the respiratory yeast *Pichia stipitis*. Appl Microbiol Biotechnol 67:247–253

Krahulec S, Klimacek M, Nidetzky B (2012) Analysis and prediction of the physiological effects of altered coenzyme specificity in xylose reductase and xylitol dehydrogenase during xylose fermentation by *Saccharomyces cerevisiae*. J Biotechnol 158(4):192–202

Križanović S, Butorac A, Mrvčić J, Krpan M, Cindrić M, Bačun-Družina V, Stanzer D (2015) Characterization of a S-adenosyl-l-methionine (SAM)-accumulating strain of *Scheffersomyces stipitis*. Int Microbiol 18(2):117–125

Kucharska K, Rybarczyk P, Hołowacz I, Łukajtis R, Glinka M, Kamiński M (2018) Pretreatment of lignocellulosic materials as substrates for fermentation processes. Molecules 23(11):2937

Kurtzman CP (1990) *Candida shehatae*—genetic diversity and phylogenetic relationships with other xylose-fermenting yeasts. Antonie Van Leeuwenhoek 57:215–222

Kurtzman CP, Suzuki M (2010) Phylogenetic analysis of ascomycete yeasts that form coenzyme Q-9 and the proposal of the new genera *Babjeviella*, *Meyerozyma*, *Millerozyma*, *Priceomyces*, and *Scheffersomyce*. Mycoscience 51:2–14

Kurylenko OO, Ruchala J, Hryniv OB, Abbas CA, Dmytruk KV, Sibirny AA (2014) Metabolic engineering and classical selection of the methylotrophic thermotolerant yeast *Hansenula polymorpha* for improvement of high-temperature xylose alcoholic fermentation. Microb Cell Fact 13:122

Kurylenko O, Ruchala J, Dmytruk K, Sibirny A (2015) New targets for improvement of xylose alcoholic fermentation in the methylotrophic yeast *Hansenula polymorpha*. In: Living organisms and bioanalytical approaches for detoxification and monitoring of toxic compounds. University of Rzeszow, Rzeszow, pp 247–257

Kurylenko O, Vasylyshyn R, Kata I, Ruchala J, Semkiv M, Dmytruk K, Sibirny A (2016b) Improvement of high-temperature xylose and glucose alcoholic fermentation in methylotrophic yeast *Ogataea (Hansenula) polymorpha* by overexpression of *PDC1* and *ADH1* genes. Visnyk Lviv Univ 73:147–153

Kurylenko O, Ruchala J, Vasylyshyn R, Stasyk O, Dmytruk O, Dmytruk K, Sibirny AA (2018) Peroxisomes and peroxisomal transketolase and transaldolase enzymes are essential for xylose alcoholic fermentation by the methylotrophic thermotolerant yeast, *Ogataea (Hansenula) polymorpha*. Biotechnol Biofuels 11:197

Kurylenko O, Ruchala J, Kruk B, Vasylyshyn R, Szczepaniak J, Dmytruk K, Sibirny A (2021) The role of Mig1, Mig2, Tup1 and Hap4 transcription factors in regulation of xylose and glucose fermentation in the thermotolerant yeast *Ogataea polymorpha*. FEMS Yeast Res 21–24

Kwak S, Jin YS (2017) Production of fuels and chemicals from xylose by engineered *Saccharomyces cerevisiae:* a review and perspective. Microb Cell Fact 16:82

Lancashire WE, Dickinson JR, Malloch RA (1998) DNA encoding enzymes of the glycolytic pathway for use in alcohol producing yeast. US Patent 5786186 A

Laplaza JM, Torres BR, Jin YS, Jeffries TW (2006) *Sh ble* and Cre adapted for functional genomics and metabolic engineering of *Pichia stipitis*. Enzyme Microb Tech 38:741–747

Larsson S, Nilvebrant NO, Jonsson LJ (2001) Effect of overexpression of *Saccharomyces cerevisiae* Padlp on the resistance to phenylacrylic acids and lignocellulose hydrolysates under aerobic and oxygen limited conditions. Appl Microbiol Biotechnol 57:167–174

Lee H, James AP, Zahab DM, Mahmourides G, Maleszka R, Schneider H (1986) Mutants of *Pachysolen tannophilus* with improved production of ethanol d-xylose. Appl Environ Microbiol 51:1252–1258

Lee SH, Kodaki T, Park YC, Seo JH (2012) Effects of NADH–preferring xylose reductase expression on ethanol production from xylose in xylose–metabolizing recombinant *Saccharomyces cerevisiae*. J Biotechnol 158(4):184–191

Lee JY, Ishida Y, Takahashi T, Naganuma A, Hwang GW (2016) Transport of pyruvate into mitochondria is involved in methylmercury toxicity. Sci Rep 6:21528

Legthelm ME, Prior JC, du Preez JC, Brandt V (1988) An investigation of D-xylose metabolism in *Pichia stipitis* under aerobic and anaerobic conditions. Appl Microbiol Biotechnol 28:293–296

Li H, Shen Y, Wu M, Hou J, Jiao C, Li Z, Liu X, Bao X (2016) Engineering a wild-type diploid *Saccharomyces cerevisiae* strain for second-generation bioethanol production. Bioresour Bioprocess 3(1):51

Li J, Zhang Y, Li J, Sun T, Tian C (2020) Metabolic engineering of the cellulolytic thermophilic fungus Myceliophthora thermophila to produce ethanol from cellobiose. Biotechnol Biofuels 13:23. https://doi.org/10.1186/s13068-020-1661-y

Lindley ND, Waites MJ, Quayle JR (1981) Simultaneous assay of dihydroxyacetone synthase and transketolase in a methylotrophic yeast grown in continuous culture. A cautionary note. J Gen Microbiol 126(2):253–259

Liu ZL, Ma M (2020) Pathway-based signature transcriptional profiles as tolerance phenotypes for the adapted industrial yeast *Saccharomyces cerevisiae* resistant to furfural and HMF. Appl Microbiol Biotechnol 104:3473–3492

Liu ZL, Moon J, Andersh BJ, Slininger PJ, Weber S (2008) Multiple gene-mediated NAD(P) H-dependent aldehyde reduction is a mechanism of in situ detoxification of furfural and 5-hydroxymethylfurfural by *Saccharomyces cerevisiae*. Appl Microbiol Biotechnol 81(4):743–753

Long TM, Su YK, Headman J, Higbee A, Willis LB, Jeffries TW (2012) Cofermentation of glucose, xylose and cellobiose by the beetle-associated yeast *Spathaspora passalidarum*. Appl Environ Microbiol 78(16):5492–5500

Lu P, Davis BP, Hendrick J, Jeffries TW (1998) Cloning and disruption of the beta-isopropylmalate dehydrogenase gene (*LEU2*) of *Pichia stipitis* with *URA3* and recovery of the double auxotroph. Appl Microbiol Biotechnol 49:141–146

Maassen N, Freese S, Schruff B, Passoth V, Klinner U (2008) Nonhomologous end joining and homologous recombination DNA repair pathways in integration mutagenesis in the xylose-fermenting yeast *Pichia stipitis*. FEMS Yeast Res 8:735–743

Madhavan A, Tamalampudi S, Ushida K, Kanai D, Katahira S, Srivastava A, Fukuda H, Bisaria VS, Kondo A (2009) Xylose isomerase from polycentric fungus *Orpinomyces*: gene sequencing, cloning, and expression in *Saccharomyces cerevisiae* for bioconversion of xylose to ethanol. Appl Microbiol Biotechnol 82(6):1067–1078

Madhavan A, Srivastava A, Kondo A, Bisaria VS (2012) Bioconversion of lignocellulose–derived sugars to ethanol by engineered *Saccharomyces cerevisiae*. Crit Rev Biotechnol 32(1):22–48

Mahler G, Nudel C (2000) Effect of magnesium ions on the fermentative and respirative functions in *Pichia stipitis* under oxygen-restricted growth. Microbiol Res 155:31–35

Matsushika A, Sawayama S (2008) Efficient bioethanol production from xylose by recombinant *Saccharomyces cerevisiae* requires high activity of xylose reductase and moderate xylulokinase activity. J Biosci Bioeng 106:306–309

Melake T, Passoth V, Klinner U (1996) Characterization of the genetic system of the xylose-fermenting yeast Pichia stipitis. Curr Microbiol 33:237–242

Metzger MH, Hollenberg CP (1995) Amino acid substitutions in the yeast *Pichia stipitis* xylitol dehydrogenase coenzyme-binding domain affect the coenzyme specificity. Eur J Biochem 228(1):50–54

Moon J, Liu ZL (2012) Engineered NADH-dependent *GRE2* from *Saccharomyces cerevisiae* by directed enzyme evolution enhances HMF reduction using additional cofactor NADPH. Enzym Microb Technol 50(2):115–120

Myers KS, Riley NM, MacGilvray ME, Sato TK, McGee M, Heilberger J et al (2019) Rewired cellular signaling coordinates sugar and hypoxic responses for anaerobic xylose fermentation in yeast. PLoS Genet 15(3):e1008037

Nardi JB, Bee CM, Miller LA, Nguyen NH, Suh SO, Blackwell M (2006) Communities of microbes that inhabit the changing hindgut landscape of a subsocial beetle. Arthropod Struct Dev 35:57–68

Navas MA, Gancedo JM (1996) The regulatory characteristics of yeast fructose-1,6-bisphosphatase confer only a small selective advantage. J Bacteriol 178:1809–1812

Navas MA, Cerdan S, Gancedo JM (1993) Futile cycles in *Saccharomyces cerevisiae* strains expressing the gluconeogenic enzymes during growth on glucose. Prot Natl Acad Sci USA 90:1290–1294

Nguyen NH, Suh SO, Marshall CJ, Blackwell M (2006) Morphological and ecological similarities: wood-boring beetles associated with novel xylose-fermenting yeasts, *Spathaspora passalidarum gen. sp. nov.* and *Candida jeffriesii sp. nov.* Mycol Res 110:1232–1241

Ni H, Laplaza JM, Jeffries TW (2007) Transposon mutagenesis to improve the growth of recombinant *Saccharomyces cerevisiae* on D-xylose. Appl Environ Microbiol 73:2061–2066

Nigam JN (2001) Development of xylose-fermenting yeast *Pichia stipitis* for ethanol production through adaptation on hardwood hemicellulose acid prehydrolysate. J Appl Microbiol 90:208–215

Nigam PS, Singh A (2011) Production of liquid biofuels from renewable resources. Prog Energy Combust Sci 37:52–68

Nijland JG, Shin HY, de Jong RM, de Waal PP, Klaassen P, Driessen AJ (2014) Engineering of an endogenous hexose transporter into a specific d-xylose transporter facilitates glucose-xylose co-consumption in *Saccharomyces cerevisiae*. Biotechnol Biofuels 7:168

Nijland JG, Vos E, Shin HY, de Waal PP, Klaassen P, Driessen AJ (2016) Improving pentose fermentation by preventing ubiquitination of hexose transporters in *Saccharomyces cerevisiae*. Biotechnol Biofuels 9:158

Nissen TL, Kielland-Brandt MC, Nielsen J, Villadsen J (2000) Optimization of ethanol production in *Saccharomyces cerevisiae* by metabolic engineering of the ammonium assimilation. Metab Eng 2(1):69–77

Olesen JT, Guarente L (1990) The HAP2 subunit of yeast CCAAT transcriptional activator contains adjacent domains for subunit association and DNA recognition: model for the HAP2/3/4 complex. Genes Dev 4:1714–1729

Ostergaard S, Walloe KO, Gomes SG, Olsson L, Nielsen J (2001) The impact of *GAL6*, *GAL80*, and *MIG1* on glucose control of the GAL system in *Saccharomyces cerevisiae*. FEMS Yeast Res 1:47–55

Palmqvist E, Hahn-Hägerdal B (2001) Fermentation of lignocellulosic hydrolysates. I: inhibition and detoxification. Bioresour Technol 74:17–24

Parachin NS, Bergdahl B, van Niel EWJ, Gorwa-Grauslund MF (2011) Kinetic modelling reveals current limitations in the production of ethanol from xylose by recombinant *Saccharomyces cerevisiae*. Metab Eng 13:508–517

Parreiras LS, Breuer RJ, Avanasi Narasimhan R, Higbee AJ, La Reau A, Tremaine M, Qin L, Willis LB, Bice BD et al (2014) Engineering and two-stage evolution of a lignocellulosic hydrolysate-tolerant *Saccharomyces cerevisiae* strain for anaerobic fermentation of xylose from AFEX pretreated corn stover. PLoS One 9(9):e107499

Passoth V, Hansen M, Klinner U, Emeis CC (1992) The electrophoretic banding patterns of the chromosomes of *Pichia stipitis* and *Candida shehatae*. Curr Genet 22:429–431

Passoth V, Zimmermann M, Klinner U (1996) Peculiarities of the regulation of fermentation and respiration in the crabtree-negative, xylose-fermenting yeast *Pichia stipitis*. Appl Biochem Biotechnol 57(58):201–212

Passoth V, Schafer B, Liebel B, Weierstall T, Klinner U (1998) Molecular cloning of alcohol dehydrogenase genes of the yeast *Pichia stipitis* and identification of the fermentative *ADH*. Yeast 14:1311–1325

Passoth V, Cohn M, Schafer B, Hahn-Hägerdal B, Klinner U (2003) Analysis of the hypoxia-induced *ADH2* promoter of the respiratory yeast *Pichia stipitis* reveals a new mechanism for sensing of oxygen limitation in yeast. Yeast 20:39–51

Petersson A, Almeida JR, Modig T, Karhumaa K, Hahn-Hägerdal B, Gorwa-Grauslund MF, Lidén G (2006) A 5-hydroxymethyl furfural reducing enzyme encoded by the *Saccharomyces cerevisiae ADH6* gene conveys HMF tolerance. Yeast 23(6):455–464

Piontek M, Hagedorn J, Hollenberg CP, Gellissen G, Strasser AW (1998) Two novel gene expression systems based on the yeasts *Schwanniomyces occidentalis* and *Pichia stipitis*. Appl Microbiol Biotechnol 50:331–338

Qi K, Zhong JJ, Xia XX (2014) Triggering respirofermentative metabolism in the Crabtree-negative yeast *Pichia guilliermondii* by disrupting the *CAT8* gene. Appl Environ Microbiol 80:3879–3887

Quarterman J, Skerker JM, Feng X, Liu IY, Zhao H, Arkin AP, Jin YS (2016) Rapid and efficient galactose fermentation by engineered *Saccharomyces cerevisiae*. J Biotechnol 229:13–21

Radecka D, Mukherjee V, Mateo RQ, Stojiljkovic M, Foulquié-Moreno MR, Thevelein JM (2015) Looking beyond Saccharomyces: the potential of non-conventional yeast species for desirable traits in bioethanol fermentation. FEMS Yeast Res 15(6):pii: fov053

Remize F, Barnavon L, Dequin S (2001) Glycerol export and glycerol-3-phosphate dehydrogenase, but not glycerol phosphatase, are rate limiting for glycerol production in *Saccharomyces cerevisiae*. Metab Eng 3:301–312

Ren X, Wang J, Yu H, Peng C, Hu J, Ruan Z, Zhao S, Liang Y, Peng N (2016) Anaerobic and sequential aerobic production of high-titer ethanol and single cell protein from NaOH-pretreated corn stover by a genome shuffling-modified *Saccharomyces cerevisiae* strain. Bioresour Technol 218:623–630

Richard P, Londesborough J, Putkonen M, Kalkkinen N, Penttila M (2001) Cloning and expression of a fungal L-arabinitol-4-dehydrogenase gene. J Biol Chem 276(44):40631–40637

Richard P, Putkonen M, Vaananen R, Londesborough J, Penttila M (2002) The missing link in the fungal L-arabinose catabolic pathway, identification of the L-xylulose reductase gene. Biochemistry 41(20):6432–6437

Roca C, Nielsen J, Olsson L (2003) Metabolic engineering of ammonium assimilation in xylose-fermenting *Saccharomyces cerevisiae* improves ethanol production. Appl Environ Microbiol 69(8):4732–4736

Rodrigues R, Lu CF, Lin B, Jeffries TW (2008) Fermentation kinetics for xylitol production by a *Pichia stipitis* D-xylulokinase mutant previously grown in spent sulfite liquor. Appl Biochem Biotechnol 148:199–209

Rogers DT, Szostak JW (1993) Strains of yeast with increased rates of glycolysis. US Patent 5268285:A1

Ruchala J, Kurylenko O, Soontorngun N, Dmytruk K, Sibirny A (2017) Transcriptional activator Cat8 is involved in regulation of xylose alcoholic fermentation in the thermotolerant yeast *Ogataea* (*Hansenula*) *polymorpha*. Microb Cell Fact 16:36

Ryabova OB, Chmil OM, Sibirny AA (2003) Xylose and cellobiose fermentation to ethanol by the thermotolerant methylotrophic yeast *Hansenula polymorpha*. FEMS Yeast Res 4(2):157–164

Saini JK, Saini R, Tewari L (2015) Lignocellulosic agriculture wastes as biomass feedstocks for second-generation bioethanol production: concepts and recent developments. 3. Biotech 4:337–353

Salas-Navarrete PC, de Oca Miranda AI, Martínez A, Caspeta L (2022) Evolutionary and reverse engineering to increase *Saccharomyces cerevisiae* tolerance to acetic acid, acidic pH, and high temperature. Appl Microbiol Biotechnol 106:383–399

Sánchez ÓJ, Cardona CA (2008) Trends in biotechnological production of fuel ethanol from different feedstocks. Bioresour Technol 99(13):5270–5295

Schneider H, Wang PY, Chan YK, Kurtzman CP (1981) Conversion of d-xylose into ethanol by the yeast *Pachysolen tannophilus*. Biotechnol Lett 3:89–92

Secches TO, Santos Viera CF, Pereira TKE, Santos VTO, Ribeirodos Santos J, Pereira GAG, Carazzolle MF (2022) Brazilian industrial yeasts show high fermentative performance in high solids content for corn ethanol process. Bioresour Bioprocess 9(1):97

Semkiv MV, Dmytruk KV, Abbas CA, Sibirny AA (2014) Increased ethanol accumulation from glucose via reduction of ATP level in a recombinant strain of *Saccharomyces cerevisiae* overexpressing alkaline phosphatase. BMC Biotechnol 14:42

Semkiv MV, Dmytruk KV, Abbas CA, Sibirny AA (2016) Activation of futile cycles as an approach to increase ethanol yield during glucose fermentation in *Saccharomyces cerevisiae*. Bioengineered 7(2):106–111

Semkiv M, Kata I, Ternavska O, Sibirny W, Dmytruk K, Sibirny A (2019) Overexpression of the genes of glycerol catabolism and glycerol facilitator improves glycerol conversion to ethanol in the methylotrophic thermotolerant yeast *Ogataea polymorpha*. Yeast 36(5):329–339

Semkiv MV, Ruchala J, Tsaruk AY, Zazulya AZ, Vasylyshyn RV, Dmytruk OV, Zuo M, Kang Y, Dmytruk KV, Sibirny AA (2022) The role of hexose transporter-like sensor hxs1 and transcription activator involved in carbohydrate sensing azf1 in xylose and glucose fermentation in the thermotolerant yeast *Ogataea polymorpha*. Microb Cell Fact 21(1):162

Shahsavarani H, Sugiyama M, Kaneko Y, Chuenchit B, Harashima S (2012) Superior thermotolerance of *Saccharomyces cerevisiae* for efficient bioethanol fermentation can be achieved by overexpression of *RSP5* ubiquitin ligase. Biotechnol Adv 30(6):1289–1300

Shi NQ, Jeffries TW (1998) Anaerobic growth and improved fermentation of *Pichia stipitis* bearing a *URA1* gene from *Saccharomyces cerevisiae*. Appl Microbiol Biotechnol 50(3):339–345

Shi NQ, Davis B, Sherman F, Cruz J, Jeffries TW (1999) Disruption of the cytochrome c gene in xylose-utilizing yeast *Pichia stipitis* leads to higher ethanol production. Yeast 15:1021–1030

Skoog K, Hahn-Hagerdahl B (1990) Effect of oxygenation on xylose fermentation by *Pichia stipitis*. Appl Environ Microbiol 56(11):3389–3394

Slininger PJ, Dien BS, Gorsick SW, Liu ZL (2006) Nitrogen source and mineral optimization enhances D-xylose conversion to ethanol by the yeast *Pichia stipitis* NRRL Y-7124. Appl Microbial Cell Physiol 72(6):1285–1296

Sprenger GA (1996) Carbohydrate metabolism in *Zymomonas mobilis*: a catabolic highway with some scenic routes. FEMS Microbiol Lett 145:301–307

Sreenath HK, Jeffries TW (1997) Diminished respirative growth and enhanced assimilative sugar uptake result in higher specific fermentation rates by the mutant *Pichia stipitis* FPL-061. Appl Biochem Biotechnol 63–65:109–116

Sreenath HK, Jeffries TW (1999) 2-deoxyglucose as a selective agent for derepressed mutants of *Pichia stipitis*. Appl Biochem Biotechnol 77:211–222

Stanley D, Fraser S, Chambers PJ, Rogers P, Stanley GA (2010) Generation and characterisation of stable ethanol-tolerant mutants of *Saccharomyces cerevisiae*. J Ind Microbiol Biotechnol 37:139–149

Stasyk OG, Maidan MM, Stasyk OV, Van Dijck P, Thevelein JM, Sibirny AA (2008) Identification of hexose transporter-like sensor HXS1 and functional hexose transporter HXT1 in the methylotrophic yeast *Hansenula polymorpha*. Eukaryot Cell 7(4):735–746

Suh SO, Marshall CJ, McHugh JV, Blackwell M (2003) Wood ingestion by passalid beetles in the presence of xylose-fermenting gut yeasts. Mol Ecol 12(11):3137–3145

Sybirna K, Petryk N, Zhou YF et al (2010) A novel *Hansenula polymorpha* transcriptional factor HpHAP4-B, able to functionally replace the *S. cerevisiae* HAP4 gene, contains an additional bZip motif. Yeast 27:941–954

Tamas MJ, Luyten K, Sutherland FC, Hernandez A, Albertyn J, Valadi H, Li H, Prior BA, Kilian SG, Ramos J et al (1999) Fps1p controls the accumulation and release of the compatible solute glycerol in yeast osmoregulation. Mol Microbiol 31:1087–1104

Tao X, Zheng D, Liu T, Wang P, Zhao W, Zhu M, Jiang X, Zhao Y, Wu X (2012) A novel strategy to construct yeast *Saccharomyces cerevisiae* for very high gravity fermentation. PLoS One 7(2):e31235

Targonski Z (1992) Biotransformation of lignin-related aromatic-compounds by *Pichia stipitis* Pignal. Zentralbl Mikrobiol 147:244–249

Turner TL, Kim H, Kong II, Liu JJ, Zhang GC, Jin YS (2016) Engineering and evolution of *Saccharomyces cerevisia* to produce biofuels and chemicals. Synth Biol 162

Valadi H, Larsson C, Gustafsson L (1998) Improved ethanol production by glycerol-3-phosphate dehydrogenase mutants of *Saccharomyces cerevisiae*. Appl Microbiol Biotechnol 50(4):434–439

van der Klei IJ, Yurimoto H, Sakai Y, Veenhuis M (2006) The significance of peroxisomes in methanol metabolism in methylotrophic yeast. Biochim Biophys Acta 1763(12):1453–1462

van Dijken JP, van den Bosch E, Hermans JJ, de Miranda LR, Scheffers WA (1986) Alcoholic fermentation by 'nonfermentative' yeasts. Yeast 2:123–127

Vasylyshyn R, Kurylenko O, Ruchala J, Shevchuk N, Kuliesiene N, Khroustalyova G et al (2020) Engineering of sugar transporters for improvement of xylose utilization during high-temperature alcoholic fermentation in *Ogataea polymorpha* yeast. Microb Cell Fact 19:96

Vasylyshyn R, Dmytruk O, Sybirnyy A, Ruchała J (2024) Engineering of *Ogataea polymorpha* strains with ability for high-temperature alcoholic fermentation of cellobiose. FEMS Yeast Res 24

Veduyn C, Van Kleef R, Frank J, Schreuder H, Van Dijken JP, Scheffers WA (1985) Properties of the NAD(P)H-dependent xylose reductase from the xylose-fermenting yeast Pichia stipitis. Biochem J 226(3):669–677

Verho R, Londesborough J, Penttilä M, Richard P (2003) Engineering redox cofactor regeneration for improved pentose fermentation in *Saccharomyces cerevisiae*. Appl Environ Microbiol 69(10):5892–5897

Voronovsky AY, Ryabova OB, Verba OV, Ishchuk OP, Dmytruk KV, Sibirny AA (2005) Expression of *xylA* genes encoding xylose isomerases from *Escherichia coli* and *Streptomyces coelicolor* in the methylotrophic yeast *Hansenula polymorpha*. FEMS Yeast Res 5(11):1055–1062

Voronovsky AY, Rohulya OV, Abbas CA, Sibirny AA (2009) Development of strains of the thermotolerant yeast *Hansenula polymorpha* capable of alcoholic fermentation of starch and xylan. Metab Eng 11:234–242

Wahlbom CF, Cordero Otero RR, van Zyl WH, Hahn-Hägerdal B, Jönsson LJ (2003) Molecular analysis of a *Saccharomyces cerevisiae* mutant with improved ability to utilize xylose shows enhanced expression of proteins involved in transport, initial xylose metabolism, and the pentose phosphate pathway. Appl Environ Microbiol 69(2):740–746

Walfridsson M, Bao X, Anderlund M, Lilius G, Bülow L, Hahn-Hägerdal B (1996) Ehanolic fermentation of xylose with *Saccharomyces cerevisiae* harboring the *Thermus thermophilus xylA* gene, which expresses an active xylose (glucose) isomerase. Appl Environ Microbiol 62(12):4648–4651

Wang TT, Lin LL, Hsu WH (1989) Cloning and expression of a *Schwanniomyces occidentalis* α-amylase gene in *Saccharomyces cerevisiae*. Appl Environ Microbiol 55:3167–3172

Wang PM, Zheng DQ, Liu TZ, Tao XL, Feng MG, Min H, Jiang XH, Wu XC (2012) The combination of glycerol metabolic engineering and drug resistance marker-aided genome shuffling to improve very-high-gravity fermentation performances of industrial *Saccharomyces cerevisiae*. Bioresour Technol 108:203–210

Watanabe S, Kodaki T, Makino K (2005) Complete reversal of coenzyme specificity of xylitol dehydrogenase and increase of thermostability by the introduction of structural zinc. J Biol Chem 280(11):10340–10349

Watanabe T, Watanabe I, Yamamoto M, Ando A, Nakamura T (2011) A UV-induced mutant of *Pichia stipitis* with increased ethanol production from xylose and selection of a spontaneous mutant with increased ethanol tolerance. Bioresour Technol 102:1844–1848

Watanabe D, Hashimoto N, Mizuno M, Zhou Y, Akao T, Shimoi H (2013) Accelerated alcoholic fermentation caused by defective gene expression related to glucose derepression in *Saccharomyces cerevisiae*. Biosci Biotechnol Biochem 77:2255–2262

Watanabe M, Cherubini F, Tisserant A, Cavalett O (2022) Drop-in and hydrogen-based biofuels for maritime transport: Country-based assessment of climate change impacts in Europe up to 2050. Energy Convers Manag 273:116403

Wei P, Li Z, He P, Lin Y, Jiang N (2008) Genome shuffling in the ethanologenic yeast *Candida krusei* to improve acetic acid tolerance. Biotechnol Appl Biochem 49:113–120

Weierstall T, Hollenberg CP, Boles E (1999) Cloning and characterization of three genes (*SUT1-3*) encoding glucose transporters of the yeast *Pichia stipitis*. Mol Microbiol 31:871–883

Weusthuis RA, Adams H, Scheffers WA, van Dijken JP (1993) Energetics and kinetics of maltose transport in *Saccharomyces cerevisiae*: a continuous culture study. Appl Environ Microbiol 59(9):3102–3109

Wijsman MR, Bruinenberg PM, van Diken JP, Scheffers WA (1985) Incapacity for anaerobic growth in xylose-fermenting yeasts. Antonie Van Leeuwenhoek 51:563–564

Wilson AJ, Bhattacharjee JK (1986) Regulation of phosphoenolpyruvate carboxykinase and pyruvate kinase in *Saccharomyces cerevisiae* grown in the presence of glycolytic and gluconeogenic carbon sources and the role of mitochondrial function on gluconeogenesis. Can J Microbiol 32(12):969–972

Wisselink HW, Toirkens MJ, del Rosario Franco Berriel M, Winkler AA, van Dijken JP, Pronk JT, van Maris AJ (2007) Engineering of *Saccharomyces cerevisiae* for efficient anaerobic alcoholic fermentation of L-arabinose. Appl Environ Microbiol 73(15):4881–4891

Xia PF, Zhang GC, Walker B, Seo SO, Kwak S, Liu JJ, Kim H, Ort DR, Wang SG, Jin YS (2016) Recycling carbon dioxide during xylose fermentation by engineered *Saccharomyces cerevisiae*. ACS Synth Biol 6(2)

Xue T, Liu K, Chen D et al (2018) Improved bioethanol production using CRISPR/Cas9 to disrupt the *ADH2* gene in *Saccharomyces cerevisiae*. World J Microbiol Biotechnol 34:154

Yanase H, Yamamoto K, Sato D et al (2005) Ethanol production from cellobiose by *Zymobacter palmae* carrying the *Ruminocuccus albus* beta-glucosidase gene. J Biotechnol 118:35–43

Yang VW, Marks JA, Davis BP, Jeffries TW (1994) High efficiency transformation of *Pichia stipitis* based on its *URA3* gene and a homologous autonomous replication sequence, *ARS2*. Appl Environ Microbiol 60:4245–4254

Yang XS, Zhang SJ, Zuo Z, Men X, Tian S (2011) Ethanol production from the enzymatic hydrolysis of non-detoxified steamexploded corn stalk. Bioresour Technol 102:7840–7844

Yang P, Wu Y, Zheng Z, Cao L, Zhu X, Mu D, Jiang S (2018) CRISPR-Cas9 approach constructing cellulase sestc-engineered *Saccharomyces cerevisiae* for the production of orange peel ethanol. Front Microbiol 9:2436

Yang P, Jiang S, Lu S et al (2022) Ethanol yield improvement in *Saccharomyces cerevisiae GPD2 Delta FPS1 Delta ADH2 Delta DLD3 Delta* mutant and molecular mechanism exploration based on the metabolic flux and transcriptomics approaches. Microb Cell Fact 21:160

Yoon GS, Lee TS, Kim C, Seo JH, Ryu YW (1996) Characterization of alcohol fermentation and segregation of protoplast fusant of *Saccharomyces cerevisiae* and *Pichia stipitis*. J Microbiol Biotechnol 6:286–291

Yong-Su, Jin Jose, Cruz Thomas W., Jeffries (2005) Xylitol production by a Pichia stipitis D-xylulokinase mutant Applied Microbiology and Biotechnology 68(1):42–45 https://doi.org/10.1007/s00253-004-1854-5

Zhang W, Geng AL (2012) Improved ethanol production by a xylose-fermenting recombinant yeast strain constructed through a modified genome shuffling method. Biotechnol Biofuels 5:46–57

Zhang A, Kong Q, Cao L, Chen X (2007) Effect of FPS1 deletion on the fermentation properties of *Saccharomyces cerevisiae*. Lett Appl Microbiol 44:212–217

Zhang L, Tang Y, Guo ZP, Ding ZY, Shi GY (2011) Improving the ethanol yield by reducing glycerol formation using cofactor regulation in *Saccharomyces cerevisiae*. Biotechnol Lett 33(7):1375–1380

Zhang GC, Liu JJ, Ding WT (2012) Decreased xylitol formation during xylose fermentation in *Saccharomyces cerevisiae* due to overexpression of water-forming NADH oxidase. Appl Environ Microbiol 78(4):1081–1086

Zhang GC, Liu JJ, Kong II, Kwak S, Jin YS (2015a) Combining C6 and C5 sugar metabolism for enhancing microbial bioconversion. Curr Opin Chem Biol 29:49–57

Chapter 2
Thermotolerant Yeasts and High-Temperature Fermentation

Hisashi Hoshida and Rinji Akada

Contents

2.1 Introduction ... 50
2.2 Advantages of High-Temperature Fermentation ... 51
2.3 Isolation of Natural Thermotolerant Ethanol-Producing Yeasts ... 52
2.4 Yeast Immobilization for High-Temperature Fermentation ... 58
2.5 High-Temperature SSF ... 58
2.6 Pilot-Scale High-Temperature SSF Using *K. marxianus* ... 62
2.7 Consolidated Bioprocessing by Natural and Genetically Engineered Yeasts ... 64
2.8 Strain Improvement for High-Temperature Fermentation ... 65
2.9 Thermotolerant Mechanism of Yeasts ... 68
2.10 Conclusion ... 70
References ... 70

Abstract Fuel ethanol production has dramatically increased in the world. Ethanol fermentation for alcohol beverages is a traditional and mature technology. However, the fermentation technology for fuel ethanol should be reconsidered because it must be large scale and low cost. High-temperature fermentation is the most promising approach for efficient fuel ethanol production from the viewpoints of cost-reduction and robustness. In addition, high temperature is helpful to establish simultaneous saccharification and fermentation process because sugar hydrolyzing enzyme activ-

H. Hoshida (✉) · R. Akada
Division of Applied Chemistry, Graduate School of Sciences and Technology for Innovation, Yamaguchi University, Ube, Japan

Research Center for Thermotolerant Microbial Resources, Yamaguchi University, Yamaguchi, Japan

Yamaguchi University Biomedical Engineering Center, Ube, Japan
e-mail: hoshida@yamaguchi-u.ac.jp

A. A. Sibirny (ed.), *Biotechnology of Yeasts and Filamentous Fungi*,
Grand Challenges in Biology and Biotechnology,
https://doi.org/10.1007/978-3-031-74726-7_2

ities can be enhanced. For the high-temperature fermentation, thermotolerant ethanol-producing yeasts are indispensable. In this chapter, yeast strains suitable for high-temperature bioethanol fermentation are reviewed. Among these, *Kluyveromyces marxianus* achieved sufficient ethanol production in a pilot-scale high-temperature fermentation from starch contained in waste cassava pulp. Strain improvement has also been conducted to establish further cost-effective high-temperature ethanol fermentation. In addition, the studies of thermotolerant mechanisms will be helpful to get the strains suitable for high-temperature fermentation through genetic engineering.

2.1 Introduction

Fermentation is a biological process by which microorganisms convert carbohydrates to ethanol or other organic compounds. Humans have used yeasts for fermentation since ancient times to produce beer, wine, and bread; however, only since the late nineteenth century has the nature and mechanism of yeast fermentation started to be understood. Louis Pasteur's and Eduard Buchner's findings led to a complete understanding of the enzymatic reactions in the fermentation pathway. Glucose is metabolized to pyruvate by the enzymes in the glycolysis pathway, and the pyruvate is used by the fermentation pathway to produce ethanol under anaerobic conditions or in the presence of high concentrations of sugar. These ethanol fermentation studies have been performed unexceptionally using the yeast *Saccharomyces cerevisiae* because *S. cerevisiae* is commonly used in traditional yeast fermentation.

In addition to traditional fermentation, ethanol has become one of the most important industrial products these days because ethanol is an alternative fuel for automobiles instead of gasoline. Ethanol produced via fermentation from plant biomass is called bioethanol. Bioethanol is a renewable energy because plant biomass is produced by the fixation of carbon dioxide in the atmosphere. The carbon dioxide generated by the combustion of bioethanol is the material originally fixed by plants. Therefore, bioethanol will be the best alternative liquid fuel to establish a carbon-neutral society. Ethanol fermentation has been developed based on brewing and bread fermentation, which require taste and flavor considerations. Instead, the fermentation of bioethanol as fuel only requires efficient production. Therefore, bioethanol fermentation for fuel should be reconsidered from the viewpoint of cost, speed, type of biomass, location, and operation conditions. New technologies are being developed to increase ethanol production in more economically and environmentally favorable manners, in which nonconventional yeasts will play important roles in addition to the conventional yeast *S. cerevisiae*. Especially in the bioethanol production process, fermentation at a high-temperature environment will be the most desirable process. In this chapter, we describe the advantages of high-temperature fermentation, the yeasts capable of high-temperature fermentation, simultaneous saccharification and fermentation (SSF) at high-temperature, genetic engineering for consolidated bioprocess (CBP), development of more thermotolerant strains, and the studies to understand thermotolerant mechanisms for further efficient high-temperature fermentation.

2.2 Advantages of High-Temperature Fermentation

Cooling Cost

During fermentation, heat is released from microorganisms and reactors. As a result, the fermentation temperature of the broth rises above an optimal temperature for a microorganism as fermentation progresses. Optimal temperatures for various kinds of fermentation depend on microorganisms used in the reactors. In the case of *S. cerevisiae*, the optimal temperature for ethanol fermentation is 30–35 °C. For stable and efficient ethanol production, the reactor must be cooled to maintain the temperature below 35 °C. The biomass used for ethanol fermentation is rich in tropical and subtropical areas, where environmental temperature is not cold enough for cooling. Therefore, fermentation plants require an expensive refrigerant system using additional energy. High-temperature fermentation is defined as fermentation conducted at 5–10 °C higher temperatures than the conventional fermentation. If the bioethanol fermentation is conducted at 45 °C, various costs in the fermentation process could be reduced, including investment for refrigerant systems and the cooling energy.

Contamination and Sterilization

Most microorganisms in the environment prefer 25–35 °C for growth, which is also the conventional fermentation temperature for *S. cerevisiae*. Contamination of other microorganisms leads to loss of ethanol production. Complete sterilization of huge amounts of biomass and fermentation facilities for bioethanol production is usually impractical. The high temperature in bioethanol fermentation will reduce contamination risk.

Viscosity of Fermentation Broth

Fermentation broth containing a large amount of polysaccharides, such as starch, to yield high concentration of ethanol has high viscosity. In the case of lignocellulose, it is often mixed with a small amount of water, and the mixture is a kind of mash. In these types of broths, thermal diffusion and dissipation was low, and more energy is required for mixing and cooling. By fermenting at a higher temperature, the viscosity of the broth can be reduced compared to the conventional fermentation.

Robustness Against Accidental Temperature Elevation

During a long-time large-scale fermentation process, unexpected problems, such as an electricity outage and operation mistake, can happen. Generally, the fermentation system is operated at the maximum temperature to achieve optimal production. Therefore, accidental temperature elevation impairs yeast cell activity and decreases or halts ethanol production. In high-temperature fermentation, using a thermotolerant microorganism would have a higher limiting temperature than general *S. cerevisiae* strains, resulting in the robust fermentation process.

Enzyme Activity

Temperature affects enzyme activities. Biomass consisting of polysaccharides, such as starch and cellulose, must be degraded to glucose by enzymes. In the case of starch, it is digested by α-amylases and glucoamylases, and the resulting glucose is used for fermentation. In the case of cellulose, several types of cellulases, such as endo-1,4-β-glucanases and β-glucosidases, are used to generate glucose. Generally, the optimal temperatures for these enzymes are higher than 50 °C, which is much higher than the optimal fermentation temperature of the yeast *S. cerevisiae*. The degradation process of oligosaccharides to glucose is called saccharification. To conduct saccharification and fermentation in optimal conditions, these processes are generally separated. In contrast to this, saccharification and fermentation can be performed simultaneously in the same reactor, which is called simultaneous saccharification and fermentation (SSF). However, the ability of the enzymes to catalyze saccharification is suboptimal because activities of the enzymes are generally about half of their maximal activity at 30–35 °C compared with their optimal temperature. In high-temperature fermentation, the temperature is closer to the optimal temperatures for the sugar-degrading enzymes. In high-temperature SSF, saccharification proceeds faster or needs a lower concentration of enzyme (Matsushita et al. 2015). Increased enzymatic activity at high temperatures shortens the fermentation period and/or reduces the cost of enzymes.

The concept of high-temperature fermentation for ethanol production was proposed by Hacking et al. (1984), but it is not yet used in commercial ethanol fermentation. To establish high-temperature ethanol fermentation, thermotolerant and ethanol-producing yeasts are essential.

2.3 Isolation of Natural Thermotolerant Ethanol-Producing Yeasts

Many natural isolates, including strains from culture collections, have been tested for thermotolerance and high-temperature ethanol fermentation. In these studies, primary screening conditions to select strains were categorized into four groups: thermotolerant yeast, ethanol-producing yeast, ethanol-producing yeast at high temperature, and one of these with a specific condition. Regardless of the primary condition, strains of *K. marxianus* and *S. cerevisiae* were often isolated as ethanol-producing yeasts at high temperatures, likely because *K. marxianus* is a highly thermotolerant yeast and *S. cerevisiae* is an efficient ethanol-producing yeast. The ethanol production of thermotolerant ethanol-producing yeasts was summarized in Table 2.1.

Hacking et al. (1984) compared 55 efficient ethanol-producing strains from seven genera and found that *K. marxianus* was more thermotolerant, but *S. cerevisiae*, *Saccharomyces uvarum*, and *Candida pseudotropicalis* produced much higher amounts of ethanol at 40 °C. Hughes et al. (1984) evaluated the ethanol fermentation ability of a *K. marxianus* strain at 30–48 °C because *K. marxianus* was found

Table 2.1 Ethanol production by yeast species at high temperature

Species	Strain	Initial sugar conc. % (w/v)	Type of sugars	Ethanol production[a] (%) Temperature (°C) 35	36	37	38	39	40	41	42	43	44	45	46	50	Unit of ethanol	Ref
Kluyveromyces marxianus																		
	NCYC587	14.5	Glc						***6.3***								v/v	Hacking et al. (1984)
	974	15	Glc											***6.3***			w/v	Anderson et al. (1986)
	FT23[b]	14	Glc						5.4			**5.6**			3.2		w/v	Szczodrak and Targoński (1998)
	FT5[b]	14	Glc						**5.0**			3.6			2.4		w/v	Szczodrak and Targoński (1998)
	CCY51-1-1[c]	14	Glc						**5.6**			**5.6**			3.5		w/v	Szczodrak and Targoński (1998)
	2713	5	Glc						2.1		**2.2**			2.0			w/v	Ballesteros et al. (1991)
	2671[b]	5	Glc						**2.1**		**2.1**			**2.1**			w/v	Ballesteros et al. (1991)
	DMKU3-1042	18	Suc[f]			**7.4**			7.2					4.9			w/v	Limtong et al. (2007)
	NBRC1777	10	Glc						(4.5)		**(4.6)**			(4.2)		(2.0)	w/v	Yanase et al. (2010)
	IIPE453[d]	20	Glc													***8.2***	w/v	Kumar et al. (2009)
	DBKKUY-103	19.4	SSJ[g]			8.2			**8.3**								w/v	Pilap et al. (2018)
	MSS6.3	15	Glc											**4.7**		3.4	w/v	Avchar et al. (2022)
Saccharomyces cerevisiae																		
	YSa86	13.5	Glc						***8.0***								v/v	Hacking et al. (1984)
	FTJa(a)	14	Glc						**4.3**			3.7			1.9		v/v	Szczodrak and Targoński (1998)
	LB62[e]	15	Glc						**6.4**			(4.8)		(1.2)			w/v	D'Amore et al. (1989)
	VS1	15	Glc	**6.6**					5.5		5.0		4.0				w/v	Kiran Sree et al. (2000)

(continued)

Table 2.1 (continued)

Species	Strain	Initial sugar conc. % (w/v)	Type of sugars	Ethanol production[a] (%) Temperature (°C) 35	36	37	38	39	40	41	42	43	44	45	46	50	Unit of ethanol	Ref
	VS3	15	Glc	**7.5**					6.0		5.8		5.8				w/v	Kiran Sree et al. (2000)
	DBKKU Y-53	22	SSJ[g]			**7.1**			5.8		3.2			0.91			w/v	Nuanpeng et al. (2016)
	C3867	10	Glc			3.0				**3.9**							w/v	Auesukaree et al. (2012)
	DMKU 3-S087	20	Glc			**7.2**			**7.2**								w/v	Pattanakittivorakul et al. (2019)
Saccharomyces uvarum																		
	YSa85	14.5	Glc						***8.2***								v/v	Hacking et al. (1984)
Saccharomyces carlsbergensis																		
	FT14	14	Glc						**5.4**			2.1			1.5		w/v	Szczodrak and Targoński (1998)
Ogataea polymorpha																		
	356	4	Glc			**1.3**			**1.3**			0.79			0.24		w/v	Ryabova et al. (2003)
Candida pseudotropicalis																		
	YCa9	14.2	Glc						***8.6***								v/v	Hacking et al. (1984)
Candida acidothermophilum																		
	ATCC 20381	5	Glc			(2.2)			**(2.5)**					(2.3)		(0.6)	w/v	Kadam and Schmidt (1997)

Pichia kudriavzevii																		
	KVMP10	10	Mixed[h]			3.9			4.0		**5.4**			1.0			w/v	Koutinas et al. (2016)
	DMKU 3-ET15	18	Glc	7.5					**7.9**		7.7			3.8			w/v	Yuangsaard et al. (2013)
	RZ8-1	16	Glc			6.0			**7.0**					3.5			w/v	Chamnipa et al. (2018)
	DSA3.2	10	Glc											***3.9***			w/v	Avchar et al. (2022)
Unknown																		
	RND13	15	Glc						**6.6**			6.0					w/v	Ueno et al. (2002)

[a]Bold: maximum ethanol production was achieved at the temperature; parentheses: value was read from a graph; bold-italic: data was shown only at the temperature in a condition
[b]*K. fragilis*
[c]*Fabospora fragilis*
[d]*Kluyveromyces* sp.
[e]*Saccharomyces cerevisiae* var. *diastaticus*, Labatt Brewing culture number
[f]Sugar cane juice supplemented with sucrose
[g]Sweet sorghum juice
[h]Galactose, fructose, glucose, and sucrose

to be a thermotolerant strain. The strain showed maximum ethanol production (approximately 65 g/L from approximately 150 g/L glucose) at 30–40 °C. *K. marxianus* strains isolated from sugar cane mills produced 5.50–6.26% (w/v) ethanol at 45 °C (Anderson et al. 1986). The *K. marxianus* strain CCY51-1-1 (former *Fabospora fragilis*) was isolated as the best ethanol producer at high temperatures among 58 yeast strains belonging to the genera *Aureobasidium, Candida, Cryptococcus, Pichia, Schizosaccharomyces, Hanseniaspora, Fabospora, Pachysolen, Saccharomyces, Trichosporon,* and *Torulopsis* (Szczodrak and Targoński 1998). The CCY51-1-1 strain produced 5.6% (w/v) ethanol at 40 and 43 °C from 14% (w/v) glucose. Recently, Avchar et al. (2022) isolated 92 thermotolerant yeast strains from distillery effluent and molasses samples. In the isolated strains, nine species were contained such as *S. cerevisiae, Ogataea polymorpha,* and *Candida ethanolica.* Among the strains, a *K. marxianus* strain MSS6.3 produced maximum concentration of ethanol at 45 °C as well as *Pichia kudriavzevii.* Based on these studies, *K. marxianus* is considered as a good thermotolerant ethanol-producing yeast. Limtong et al. (2007) isolated *K. marxianus* strain DMKU 3-1042 in Thailand and found to produce ethanol well at 40 °C. In the study, strains were enriched in a condition supplemented with 4% ethanol (v/v) as a first step of screening. The ethanol fermentation ability of the DMKU 3-1042 strain was compared with *K. marxianus* strains NCYC587, NCYC1429, and NCYC2791 (Nonklang et al. 2008). Among the four strains, DMKU 3-1042 was the best ethanol producer at 45 °C. The *K. marxianus* strain DBKKUY-103 was also isolated in the presence of 4% ethanol (v/v) at 35 °C and produced 83.5 g/L ethanol at 40 °C using a medium consisting of sweet sorghum juice and $(NH_4)_2HPO_4$ (Nuanpeng et al. 2016; Pilap et al. 2018). Kumar et al. (2009) isolated *Kluyveromyces* sp. IIPE453 as an ethanol-producing yeast at high temperatures from a soil sample in a sugar mill. The strain showed highest ethanol production rates (g/g/h) at 45–50 °C. The strain was reported to produce ethanol from various types of biomass, such as sugarcane juice, molasses, and bagasse hydrolysate at 50 °C (Kumar et al. 2010). Rodrussamee et al. (2011) and Nonklang et al. (2008) reported that *K. marxianus* grows by utilizing various types of sugar, which is useful to utilize various types of biomass for fermentation. However, there was little ethanol production from xylose and none from arabinose as carbon sources.

Several *S. cerevisiae* strains were also isolated for high-temperature fermentation. D'Amore et al. (1989) screened 65 yeast strains containing *S. cerevisiae, Saccharomyces bayanus, K. marxianus, Kluyveromyces lactis, Candida shehatae, Hansenula wingei, Saccharomyces rouxii,* and *Scheffersomyces* (*Pichia*) *stipitis.* The fermentation ability was tested at 40 °C, and a *S. cerevisiae* (former *Saccharomyces diastaticus*) strain showed the highest ethanol production among the strains at 40 °C. Kiran Sree et al. (2000) isolated four *S. cerevisiae* strains from a thermal power plant in India. Among the strains, VS_1 and VS_3 grew at 44 °C. Ethanol production by the strains was highest at 30 and 35 °C but decreased over 40 °C. Nuanpeng et al. (2016) isolated *S. cerevisiae* strain DBKKU Y-53 in Thailand. In the screening process, yeast strains were first enriched at 35 °C in the presence of 4% ethanol. The strain DBKKU Y-53 grew well and produced ethanol at ~42 °C, but maximum production was achieved at 37 °C in sweet sorghum juice. By using the enrichment technique and following tests in molasses at 40 °C, a *S. cerevisiae* strain DMKU 3-S087 was also isolated (Pattanakittivorakul et al. 2019). Ethanol production in a glucose medium at 40 °C reached 72 g/L for 60 h culture.

Pichia kudriavzevii strains were also often isolated and may be another species suitable for high-temperature ethanol fermentation. By using the enrichment culture in the presence of ethanol at 40 °C, *P. kudriavzevii* DMKU 3-ET15 was isolated from traditional fermented pork sausage in Thailand. The strain produced ethanol at approximately 7.7% (w/v) from 18% glucose at 35–42 °C (Yuangsaard et al. 2013). *P. kudriavzevii* RZ8-1 isolated from a plant orchard in Thailand produced a larger amount of ethanol at 40 than 37 °C (Chamnipa et al. 2018). The strain KVMP10 was isolated for orange peel fermentation and was reported to produce 5.4% (w/v) ethanol at 42 °C in a model medium containing ~10% total sugars (Koutinas et al. 2016). In the screening performed by Avchar et al. (2022), several *P. kudriavzevii* strains were isolated as thermotolerant yeasts by colony formation at 40 °C. Four *P. kudriavzevii* strains produced 33–38 g/L ethanol at 45 °C. *P. kudriavzevii* has better stress tolerance to high temperature, ethanol, NaCl, and low pH than *S. cerevisiae* (Li et al. 2021; Pongcharoen 2022). The multi-stress tolerance of *P. kudriavzevii* would be an advantage in fermentation using crude biomass as feedstock.

Several other thermotolerant ethanol-producing strains were also selected from culture collections or isolated from various environments. *Ogataea polymorpha* (former *Hansenula polymorpha*) is a thermotolerant yeast. From six *O. polymorpha* strains tested, several strains produced over 9 g/L ethanol from 2% sucrose at 37 °C (Ryabova et al. 2003). *Candida* species also produced ethanol at relatively high temperatures. *C. tropicalis* strain TH4 was isolated from an Indonesian traditional beverage as an ethanol-producing yeast and produced 4.6% (v/v) ethanol at 42 °C. However, the production was decreased by 30% compared with fermentation at 30 °C (Hermansyah et al. 2015). *Candida acidothermophilum* is reported as a thermotolerant yeast and produced ethanol at 70–80% of the theoretical yield from 5% glucose at 37–45 °C (Kadam and Schmidt 1997). From a hot spring drainage in Japan, a thermotolerant yeast was isolated and produced ethanol 6.6% (w/v) from 15% glucose at 40°C (Ueno et al. 2002). The species was not determined, but rDNA sequence suggested that the strain was similar to *Candida glabrata* and *Kluyveromyces delphensis*. Strains of *Kluyveromyces lactis*, *Cryptococcus neoformans*, *Candida albicans*, and *Debaryomyces hansenii* are reported to be thermotolerant yeasts (Nedeva et al. 1993; Arguelled 1997; Takashima et al. 2009; Prakash et al. 2011). However, these species have little or no ethanol fermentation ability and have not been studied regarding high-temperature ethanol fermentation. However, resources in the world remain to be investigated for the isolation of thermotolerant, ethanol-producing yeasts.

In Table 2.1, optimum ethanol production temperatures of each strain were shaded for comparison between species. The optimum fermentation temperature for the strains is approximately 40 °C. Some *K. marxianus* strains showed the best ethanol production ability or kept fermentation ability at higher temperatures than 40 °C, suggesting that *K. marxianus* is a better species for high-temperature ethanol fermentation.

2.4 Yeast Immobilization for High-Temperature Fermentation

To perform high-temperature fermentation, cell immobilization has been studied. A thermotolerant strain (RA-74-2) of *S. cerevisiae* was immobilized in alginate gel beads. Ethanol production of the immobilized cells was higher than that of free cells at 26–42 °C (Sohn et al. 1997). The immobilized cells also produced ethanol efficiently in three rounds of fermentation at 40 °C. Interestingly, some cells in beads were lost during 40 h fermentation at 43 °C, but the number of cells increased during the following fermentation at 37 °C. By decreasing the fermentation temperature, the fermentation activity of the immobilized cells can be recovered. Ylitervo et al. (2011) encapsulated *S. cerevisiae* CBS 8066 into alginate-chitosan capsules. The encapsulated cells consumed almost 3% of glucose in the medium and converted it to ethanol at more than 85% yield at 42 °C over seven repeated fermentation cycles, although free cells completely lost glucose-consuming activity after two rounds. In addition, the maximum temperature of ethanol fermentation increased from 42 to 45 °C by encapsulation. In another study, *S. cerevisiae* AXAZ-1 was immobilized in brewer's spent grains, and the immobilized cells were used for fermentation 37 times repeatedly at various temperatures (Kopsahelis et al. 2009). For comparison, free cells were also tested in repeated fermentations. In the repeated fermentations, molasses at 12% (w/v) of the initial sugar concentration was used as a substrate. Ethanol concentration and conversion rate in each batch-fermentation were similar between immobilized cells and free cells at 30–35 °C. In contrast, the fermentation time required to consume the sugars with immobilized cells was around half that of free cells. In addition, the effect of immobilization against temperature was also observed. Immobilized cells produced 5.5–6.8% (v/v) ethanol with 14 rounds of fermentation at 38 and 40 °C. In free cells, ethanol production decreased, and fermentation time was extended during repeated fermentations at 38 °C. These reports indicate that immobilization protects yeast cells in high-temperature fermentation.

2.5 High-Temperature SSF

SSF can make ethanol in a simpler way and in a shorter time, but the different optimal temperatures between saccharification and fermentation are a barrier to its industrial application. High-temperature fermentation using a thermotolerant yeast is a possible solution to overcome the barrier because it increases saccharification efficiency with keeping fermentation ability. An attempt of high-temperature SSF is summarized in Table 2.2.

S. cerevisiae NKU5377 produced ethanol from wastepaper in SSF, and ethanol production was higher at 40 °C than at 30 and 37 °C (Park et al. 2010). Liu et al. (2014) demonstrated that ethanol production and fermentation speed in a SSF

Table 2.2 Ethanol production in high-temperature SSF

Strain		Biomass		Temp. (°C)	Ethanol production		Enzyme		Remarks	Ref
Saccharomyces cerevisiae										
	80000012[a]	Corn stover	20%	39	59.8	g/L	30	FPU/g glucan	With pre-hydrolysis, 2% Tween 20	Liu et al. (2014)
	KNU5377	Waste newsprint	250 g/L	50– >40	8.4	% (v/v)	140	FPU/g paper	With pre-hydrolysis	Park et al. (2010)
	VS_3	Sweet Sorghum	100 g	37	8.2	g/g substrate	1300	U/g enzyme		Sree et al. (1999)
Kluyveromyces marxianus										
	L. G.	Solka-floc	10%	42	37.6	g/L	15	FPU/g		Ballesteros et al. (1991)
	NRRL Y-6860	Rice straw	8%	45	11.5	g/L	6.25	FPU	+6.25 U β-glucosidase /g substrate	Castro and Roberto (2014)
	NCIM 3358	Solka-floc	10%	43	3.0	% (w/v)	40	FPU/g		Hari Krishna et al. (2001)
	Y. 01070	Spruce chip	5%	42	7.5	g/L	37	FPU/g	+381 U β-glucosidase/g cellulose	Bollók et al. (2000)
	IMB4	Switchgrass	41 g/L	45	16.6	g/L	15	FPU/g glucan		Suryawati et al. (2008)
	MSS6.3	Rice straw	5%	45	10.9	g/L	15	FPU	Pretreated with NaOH	Avchar et al. (2022)
	CBS6556	Poplar	20%	43	49.0	g/L	15	mg/g glucan	Pretreated with THF and acid	Sengupta et al. (2022)
Candida glabrata										
	NFRI3164	Avicel	5%	42	14	g/L	500	CMC U/g Avicel		Watanabe et al. (2010)

(continued)

Table 2.2 (continued)

Strain		Biomass		Temp. (°C)	Ethanol production		Enzyme		Remarks	Ref
Candida acidothermophilum										
	ATCC 20381	Poplar	6% (3.5% cellulose)	42	18	g/L	25	FPU/g		Kadam and Schmidt (1997)
Pichia kudriavzevii										
	SI	Rice straw	20%	42	33.4	g/L	15	FPU	Pretreated with acid steam	Yuan et al. (2017)
	DSA3.2	Rice straw	5%	45	10.5	g/L	15	FPU	Pretreated with NaOH	Avchar et al. (2022)

[a]Angel Yeast Co. Ltd.

system using corn stover increased on elevating the temperatures from 30 to 39 °C. However, a high concentration of glucose remained at 36 and 39 °C, suggesting that saccharification was enhanced, but glucose consumption by *S. cerevisiae* is not fast enough at high temperatures. In the SSF with *S. cerevisiae* VS_3 strain using starchy materials as feedstock, ethanol production at 42 °C was lower than that at 37 °C, likely due to the lower thermotolerance of *S. cerevisiae* at 42 °C (Sree et al. 1999). Thermotolerance of *S. cerevisiae* may not be sufficient for high-temperature SSF.

High-temperature SSFs with the thermotolerant yeast species are performed at higher temperatures than those in the SSFs with *S. cerevisiae*. SSF with *K. marxianus* L. G. strain at 42 °C produced 37.6 g/L ethanol from 10% of a microcrystalline cellulose in 78 h, but reducing sugars released by enzyme hydrolysis started to accumulate in the latter process (Ballesteros et al. 1991). Castro and Roberto (2014) compared ethanol production (g/L/h) between *K. marxianus* and *S. cerevisiae* in SSF using rice straw as biomass. Maximum ethanol concentration was similar but only in the case of *K. marxianus*; productivity was enhanced at 45 °C than at 30 °C. In addition, glucose remained in the medium of *S. cerevisiae* at 45 °C. Higher temperature is preferred for SSF, but 45 °C is too high for *S. cerevisiae*. Ethanol fermentation in an SSF using poplar solids at 37 °C was also compared using *K. marxianus* and *S. cerevisiae* (Sengupta et al. 2022). In an 18% biomass loading condition, *S. cerevisiae* D5A strain produced 70 g/L ethanol after 8 days. In the same condition, *K. marxianus* CBS6556 produced 62 g/L ethanol in 3 days, but glucose started to accumulate in the culture after 4 days. *K. marxianus* IMB4 strain was used for SSF using switchgrass at 37, 41, and 45 °C. Ethanol production increased on elevating the temperatures, and the highest ethanol production was achieved at 45 °C with an initial pH of 5.5 (Suryawati et al. 2008). *K. marxianus* NCIM 3358 also showed better ethanol production than *S. cerevisiae* NRRL-Y-132 in high-temperature SSF (Hari Krishna et al. 2001). NCIM 3358 produced 3.0% (w/v) ethanol from microcrystalline cellulose at 43 °C in the presence of cellulase, while NRRL-Y-132 produced 2.3% (w/v) ethanol. In addition, ethanol production was enhanced by the addition of β-glucosidase to 3.5% with the *K. marxianus* strain, suggesting that efficient conversion to glucose is required for a higher ethanol yield. Avchar et al. (2022) performed high-temperature SSF using *K. marxianus* and *P. kudriavzevii* strains isolated. *P. kudriavzevii* strains showed ethanol production abilities comparable to a *K. marxianus* strain at 45 °C. The ethanol concentrations after 48 h cultivation were approximately 10 g/L from 5% alkali-treated rice straw. A *P. kudriavzevii* strain SI was isolated at 40 °C from kitchen waste, and ethanol productions in SSF using acid and steam-treated rice straw as feedstock were compared to a *S cerevisiae* strain BCRC20270. At 30 °C, ethanol production by the yeasts gradually increased and reached 25 g/L at 80 h. At 42 °C, ethanol production by SI was accelerated and enhanced, though BCR20270 did not produce ethanol. As a result, over 30 g/L ethanol was produced in 48 h by SI (Yuan et al. 2017).

Candida yeasts were also examined for high-temperature SSF. A *Candida glabrata* wild-type strain produced 14 g/L ethanol in SSF at 42 °C using 5% microcrystalline cellulose as a substrate with agitation at 150 rpm. A respiratory-deficient strain produced 17 g/L ethanol in the same condition (Watanabe et al. 2010). *C. acidothermophilum* produced ethanol faster at 40 °C than 37 °C from acid-treated poplar (Kadam and Schmidt 1997). In the study, it was estimated that cellulase-related cost reduction by increasing temperature in SSF was 16% at 40 °C, 35% at 45 °C, and (if the yeast produces ethanol efficiently) 51% at 50 °C compared with 37 °C.

A xylose-fermenting genetically engineered *S. cerevisiae* (strain RWB222) was used for SSF of paper sludge containing a significant amount of xylan as well as glucan. The ethanol production was 45.2 g/L at 37 °C and 4.01 g/L at 30 °C. The paper sludge used in the study contained 82 g/L glucan and 22 g/L xylan. The final ethanol yield was higher than the theoretical complete conversion of the glucan, indicating that xylose released from xylan was also utilized for ethanol production (Zhang and Lynd 2010).

High-temperature SSF has been studied in many thermotolerant yeast strains. However, still the ethanol production from cellulosic materials does not seem to be sufficient for industrial levels. If the final production level is low, the following distillation cost becomes much higher. Therefore, high performance of yeasts and suitable processes are required for practical high-temperature SSF.

2.6 Pilot-Scale High-Temperature SSF Using *K. marxianus*

These days, cornstarch and sugar cane are major feedstocks for bioethanol production. However, the consumption of these "food" biomass for ethanol production causes an economic problem. Therefore, lignocellulosic material has received much attention as a feedstock because it is nonedible. However, industrially efficient ethanol production from lignocellulose in separated hydrolysis and fermentation as well as SSF is still under development.

Cassava pulp is nonedible waste in tapioca starch factories. It is very cheap and used as a fertilizer and an additive in feed. However, cassava pulp contains approximately 50% starch in dry weight and has the potential to be used for fermentation. In Thailand, up to 2.6 million tons of cassava pulp is produced per year. It accounts for approximately 10% of the total cassava production in Thailand. Cassava is also cultivated in tropical areas, such as Nigeria and Indonesia. Annual production in the world is more than 200 million tons. If 10% remains as cassava pulp by-product, as is the case in Thailand, approximately 10 million tons of starch remains to be used as a feedstock in the world. To produce ethanol from cassava pulp efficiently, we developed a high-temperature SSF process using the thermotolerant yeast *K. marxianus*.

First, we isolated thermotolerant yeast strains (Hoshida et al. 2012). Samples such as fruits, flowers, waste from alcoholic beverages, and a filter cake in a sugar cane factory were collected in Thailand. After enrichment in sugar cane broth at 45 °C, thermotolerant yeasts growing at 50 °C were isolated. Among the isolates, seven strains showed better growth compared with *K. marxianus* DMKU 3-1042, which was previously isolated as a good ethanol-producing yeast at high temperatures (Limtong et al. 2007; Nonklang et al. 2008). Ethanol production abilities of the seven strains at 45 °C in molasses were also comparable with DMKU 3-1042. As a result, strain UBU1-1 was selected. The draft genome sequence of UBU1-1 strongly suggested that it is a *K. marxianus* strain.

In laboratory-scale experiments, minimum amounts of enzymes for efficient ethanol production from cassava pulp in SSF were determined. Cassava pulp was hydrated, treated with α-amylase (0.50 g/kg-pulp of Liquozyme® SC, Novozyme), and autoclaved. The liquefied cassava pulp was mixed with 0.23 g glucoamylase/kg pulp (Spirizyme® Fuel, Novozyme) and 6.67 g cellulase/kg pulp of (Accellerase® DUET, Genecor). Cellulase was added mainly to loosen cellulosic fibers and to improve the access of glucoamylase to starch. The amounts of glucoamylase and cellulase were lower than the levels recommended in the instructions, and increased amounts of glucoamylase and cellulase to two- and tenfold, respectively, did not enhance ethanol production speed or final ethanol concentration. By using *K. marxianus* UBU1-1 strain, cassava pulp (25% w/w-water) was fermented in a 2-L SSF system at 40 °C. As a result, 7.8% ethanol was produced in 42 h. In a 20-L scale, 8.5% (w/v) was produced in 48 h of fermentation at 40 °C. The viscosity of the cassava pulp slurry was very high (24,000–200,000 mPa s) at the fermentation start point. During fermentation, the viscosity of the fermentation broth became lower than we expected. In addition to the degradation of starch and cellulose, polygalacturonase activity produced by *K. marxianus* helped to lower viscosity by hydrolyzing pectin in the cassava pulp. In a larger-scale fermentation, the high viscosity of the fermentation broth in the early phase might have been a problem, causing low heat diffusion and inefficient mixing of enzymes.

To demonstrate that high-temperature SSF of cassava pulp by using *K. marxianus* UBU1-1 is possible in more large-scale fermentation, a 10 kL-size reactor was built in Thailand. The average maximum temperature in the environment is approximately 33 °C in Thailand. Even in such a climate, the temperature of the reactor was controlled via circulating water, which was circulated through a cooling tower but not chilled. Wet cassava pulp is easily utilized for fermentation, but its starch contents were not enough for an economically feasible concentration of ethanol production. Therefore, wet and dry cassava pulp were mixed and used as a substrate. Utilization of dried pulp is preferable for whole-year production because wet pulp is easily spoiled by contamination of microorganisms. Starch contents of cassava pulp used in the pilot tests were sometimes lower than 50% (w/dry weight). In such cases, dried cassava contents were increased. We conducted over 68 rounds of batch fermentations to optimize ethanol fermentation conditions and pretreatment. Optimum fermentation conditions were determined as follows: Dry and wet cassava pulp was pulverized to an average particle size of approximately 200 μm. Pulverized pulp was mixed with water and steamed at 100–105 °C for hydration. After cooling, glucoamylase and cellulase were added, and fermentation was carried out at 40 °C with *K. marxianus* UBU1-1. At the end of fermentation, the viscosity of the broth decreased to 100 mPa s. Typical ethanol production during the high-temperature SSF is shown in Fig. 2.1. More than 8% ethanol, which was the target concentration for this proof-of-concept study, was produced within 60 h.

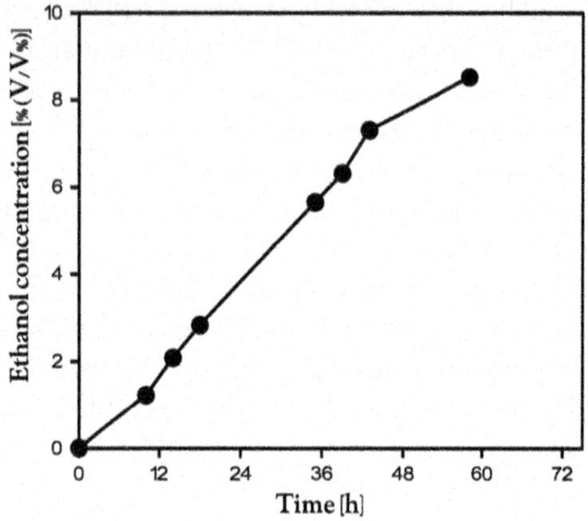

Fig. 2.1 Ethanol production in a 10 kL-scale SSF at 40 °C using *K. marxianus* and cassava pulp as feedstock

2.7 Consolidated Bioprocessing by Natural and Genetically Engineered Yeasts

The cost of enzymes in SSF is a significant consideration, especially for cellulose utilization. The fermentation process, in which polysaccharides are degraded by enzymes secreted by the organism for fermentation and resulting sugars are used for fermentation, is called consolidated bioprocessing (CBP). The concept of CBP may be first used in *K. marxianus* (Margaritis and Bajpai 1982), because *K. marxianus* strains produce large amounts of inulinase (Hoshida et al. 2018), which degrades inulin to fructose. Inulin is a polysaccharide mainly composed of fructose, which binds via β2,1 glycosyl bonds. Inulinase hydrolyzes the β2,1 glycosyl bonds and releases fructose efficiently from inulin. The Jerusalem artichoke is a plant that accumulates inulin in its tuber. Margaritis and Bajpai (1982) reported the production of 46 g/L ethanol from Jerusalem artichoke juice (10% total sugar) using *K. marxianus* at 35 °C in a CBP. This report was published prior to the identification of *K. marxianus* as a thermotolerant yeast for high-temperature fermentation in 1984 by Hacking et al. Charoensopharat et al. (2015) reported high-temperature CBP of Jerusalem artichoke by *K. marxianus* strain DBKKU Y-102. The strain produced 90.6 and 85.4 g/L ethanol from 230 g/L of total sugar of tuber mash at 37 and 40 °C, respectively. The effects of temperatures on the CBPs using *K. marxianus* were also determined (Hu et al. 2012; Charoensopharat et al. 2015). The *K. marxianus* strains used in the studies grew well at 40 °C. However, ethanol productions were significantly decreased at 40 °C compared to 30–38 °C. Activities of inulinase secreted in the medium were decreased at 41 °C (Yuan et al. 2008), while the optimum temperatures of an inulinase for inulin hydrolysis are 40–60 °C (Rouwenhorst et al. 1988). Secretory production of inulinase proteins at high temperatures needs to be increased for efficient ethanol production.

For CBP using starch and cellulosic materials as feedstock, genetic engineering should be applied to produce hydrolyzing enzymes by yeasts. The first example might be the expression of a glucoamylase in *S. cerevisiae*. An engineered strain to express a glucoamylase consumed 93.0 g/L carbohydrate and produced 44.8 g/L ethanol at 32 °C using soluble starch as feedstock (Inlow et al. 1988).

Genetic engineering for high-temperature CBP has often been performed in thermotolerant yeasts. In *K. marxianus*, three genes encoding thermostable secretory cellulases, endo-β-1,4-glucanase, cellobiohydrolase, and β-glucosidase, were introduced. The resulting strain secreted these cellulases but resulted in little ethanol production from carboxymethyl cellulose (CMC). On the other hand, cellobiose was a good sugar for the strain to produce ethanol; a maximum ethanol concentration of 43.4 g/L was achieved from 10% cellobiose at 45 °C (Hong et al. 2007). Yanase et al. (2010) expressed cellulases on the cell membrane of *K. marxianus* by fusing the 3′ region of the α-agglutinin gene. As a result, the resulting K1 strain produced 4.25 g/L ethanol from 10 g/L β-glucan at 48 °C in 12 h by inoculating a large amount of pre-cultured cells. Similar amounts of ethanol were also produced at 40 and 45 °C in 24 h fermentation.

Other thermotolerant yeasts were also engineered for CBP at high temperatures. An α-amylase gene and a glucoamylase gene of yeast *Schwanniomyces occidentalis* were expressed in *O. polymorpha* to utilize starchy material directly. An engineered strain produced 10 g/L ethanol from 3% soluble starch at 47 °C (Voronovsky et al. 2009). A strain of *O. polymorpha* expressing xylan-hydrolyzing enzymes was also constructed. It grew by utilizing xylan as a sole carbon source, but ethanol production was low (Voronovsky et al. 2009). Kitagawa et al. (2010) developed a transformation and heterologous expression system and expressed a β-glucosidase in *Issatchenkia orientalis*. The engineered strain produced 29 g/L ethanol in a CBP at 40 °C using 10% microcrystalline cellulose as a substrate.

S. cerevisiae has also been engineered for CBP, but the fermentation temperatures were not practically high. Engineered strains expressing endo/exoglucanase and β-glucosidase were reported by Cho and Yoo (1999), Den Haan et al. (2007), Wen et al. (2010), and Hu et al. (2016) for CBP using cellulosic substrates. However, CBP temperature of the studies were lower than 38 °C; therefore, we do not describe them in this chapter.

2.8 Strain Improvement for High-Temperature Fermentation

S. cerevisiae is used for industrial ethanol fermentation due to its high ethanol production in conventional fermentation. Taking advantage of this ability in high-temperature fermentation, strain improvements of *S. cerevisiae* strains have been conducted (Table 2.3). Sridhar et al. (2002) treated *S. cerevisiae* VS_1 and VS_3 strains with ultraviolet (UV) radiation and selected mutants growing well at 42 °C. Growth of a mutant named UV-VS_1 100 was improved under high concentrations of glucose, and the strain produced 5.8% (w/v) ethanol from 250 g/L glucose at 42 °C. Another UV mutagenesis of *S. cerevisiae* ATCC 26602 generated a mutant showing approximately 1.3-fold higher ethanol production (72 g/L) compared with

Table 2.3 Strain improvement by mutagenesis and laboratory evolution

Strain names			Methods		Improvement of high-temperature ethanol production			
Sp[a]	Parent	Mutant	Mutagenesis	Screening condition	Parent (g/L)	Mutant (g/L)	Condition	References
Sc	VS_1	UV-VS_1 40	UV	42 °C	50	58	42 °C, 250 g/L glucose	Sridhar et al. (2002)
Sc	ATCC 26602	MT15	UV	45 °C	55[c]	72	40 °C, 150 g/L glucose	Rajoka et al. (2005)
Sc	FAS	MLD10	UV, g-ray	45 °C, deoxy-glucose	48	85	43 °C, 180 g/L sugars	Arshad et al. (2014)
Sc	S288C	Y43	HEPEB[b]	45 °C	0	79	43 °C, 200 g/L sugars	Zhang et al. (2012)
Sc	IR2	IR2-9a	Evolution	42 °C	16	28	40 °C, SSF of 10% substrate	Edgardo et al. (2008)
Sc	p1 (D3)	S1	Evolution, UV, EMS	50 °C, 55 °C	–	46	40 °C, 100 g/L glucose	Balakumar et al. (2001)
Sc	WW	AT	Evolution, mating	42 °C	52	66	41 °C, 150 g/L glucose	Mitsumasu et al. (2014)
Sc	DMKU 3-S087	S087E100-294	EMS	37 °C	60	73	37 °C, molasses	Pattanakittivorakul et al. (2019)
Sc	S288C	TTY23	Evolution	39.5 °C	73	93	40 °C, 200 g/L glucose	Caspeta et al. (2019)
Km	FIM1	KM-100d	Evolution	6% ethanol	20	50	45 °C, 200 g/L glucose	Mo et al. (2019)
Km	DMKU 3-1042	ACT002	Evolution	45 °C[d]	46	56	45 °C, 160 g/L glucose	Pattanakittivorakul et al. (2022)

[a]Species
[b]High-energy pulsed electron beam
[c]Value was read from a graph
[d]Temperatures were gradually increased to 45 °C

the parental strain (55 g/L) at 40 °C (Rajoka et al. 2005). Zhang et al. (2012) used a high-energy-pulse-electron beam for mutagenesis of *S. cerevisiae* S288C. Strain S288C did not grow or produce ethanol at 43 °C, but a mutant strain Y43 produced approximately 8% (w/v) ethanol. Arshad et al. (2014) combined and repeated mutagenesis with UV and γ-ray to obtain thermotolerant and catabolite-resistant mutants. Ethanol production of the most productive mutant strain (MLD10) reached 8.5% (w/v) from 18% fermentable sugars at 42 °C, while the parental strain produced 4.8% (w/v) in the same condition. Ethyl methanesulfonate (EMS) is also used for mutagenesis. A *S. cerevisiae* thermotolerant strain was used for mutagenesis and improved strains were selected among 341 mutants by evaluating ethanol production at 37 °C. One of the strains, S08E100-294, produced 7.3% ethanol in molasses broth at 37 °C, which was 1.3% higher than the wild-type strain (Pattanakittivorakul et al. 2019).

UV mutagenesis was also conducted on *K. marxianus*. The *K. marxianus* strain 8-1 was obtained from NRRLY-1109 after UV-C radiation and produced 2.25% (w/v) ethanol at 46 °C, though the parental strain produced 1.75% (Hughes et al. 2013). Sakanaka et al. (1996) applied protoplast fusion of *K. marxianus* and *S. cerevisiae*. Ethanol production by the strain F-1 obtained was higher than that by the parental *K. marxianus* strain at 43 °C; at this temperature, the parental *S. cerevisiae* did not grow and produce ethanol. These approaches have enhanced the high-temperature ethanol-fermentation ability of yeast strains, but the improvement has not been drastic.

Another approach is laboratory evolution at high temperatures. Edgardo et al. (2008) conducted laboratory evolution by culturing *S. cerevisiae* cells repeatedly from 40 to 48 °C. As a result, a mutant strain was obtained that showed higher ethanol yield at 40 °C in SSF. A *S. cerevisiae* strain evolved at high temperature produced 9.3% (w/v) ethanol at 40 °C, while the parental strain did 7.2%. However, the strain had relatively low ethanol stress tolerance and mitochondria activity compared to the parental strain. To overcome the drawback, fermentation was performed with supplementation of KCl/KOH (Caspeta et al. 2019). Balakumar et al. (2001) combined UV and EMS mutagenesis with laboratory evolution. The finally obtained *S. cerevisiae* mutant S1 produced ethanol 4.6% (w/v) at 40 °C within 36 h from 10% glucose. Mitsumasu et al. (2014) used a *S. cerevisiae* strain for laboratory evolution at high temperature and low pH independently, and evolved strains were mated. The resulting thermo- and acid-tolerant strain (AT) produced 6.58 g/L of ethanol at 41 °C within 24 h, though the ethanol production by the parental WW strain was 5.19 g/L. Laboratory evolution was also applied to genetically engineered strains for SSF and CBP. For example, cellobiose transporter and β-glucosidase genes were introduced into *S. cerevisiae*, and the engineered strain was repeatedly cultured in a cellobiose medium (Hu et al. 2016). As a result, an evolved strain showed faster cellobiose consumption and ethanol production compared to the parental strain. However, the productivity at 38 °C was lower than 34 °C, suggesting that further evolution and selection will be required to take advantage in high temperatures.

In an example of laboratory evolution of *K. marxianus*, a strain was repeatedly cultured at high temperatures with gradual increments (Pattanakittivorakul et al.

2022). Three strains obtained showed clearly increased ethanol production at 40–47 °C. These strains also showed stronger stress tolerances to such as acetic acid and furfural despite the general YPD medium being used during evolution. The strains might have spontaneous mutations to gain a broad range of stress tolerance including thermotolerance. *K. marxianus* cells evolved in the presence of 6% ethanol at 30 °C also produced a higher amount of ethanol than the parental strain at 45 °C. The cells showed enhanced tolerance against temperature, H_2O_2, and NaCl as well as ethanol (Mo et al. 2019). Multiple stress tolerance may be helpful for high-temperature fermentation. Laboratory evolution in a synthetic medium might be also useful to obtain yeast strains suitable for a feedstock with insufficient nutrients in high-temperature fermentation (Ai et al. 2022).

Genetic techniques are also applied for strain improvement for high-temperature fermentation. Overexpression of Hsf1p and Msn2p, respectively, in *S. cerevisiae* enhanced growth and increased ethanol production at 40–43 °C, while the ethanol production at 30 °C did not change. These proteins are stress-related transcription factors, and transcription analysis of the strains suggested that sugar uptake and glucose metabolism were improved at high temperatures, respectively (Li et al. 2017). Khatun et al. (2017) introduced artificial zinc finger proteins in *S. cerevisiae* and selected the strain MNII-AZFP, which showed better growth and ethanol fermentation than a control strain at high temperatures. The MNII-AZFP strain produced 43.1 g/L ethanol from 100 g/L glucose at 42 °C though a control strain did only 7.1 g/L. In the MNII-AZFP strain, expressions of the genes involved in stress tolerance, such as *MSN2*, *HSP30*, and *SOD1*, were enhanced, suggesting that global upregulation of stress-tolerance genes is useful to obtain yeast strains for high-temperature ethanol fermentation to cope with various types of stresses. Wang et al. (2022) compared expressions of a multiple-stress-tolerant strain previously obtained and its parental strain. As a result, *ENA5*, which encodes a sodium pump, was suggested as a key gene to increase stress tolerance. Overexpression of *ENA5* in the multiple-stress-tolerant strain further increased tolerances to ethanol, temperature, and osmolality. The resulting strain produced 43.5 g/L ethanol from 100 g/L glucose at 44 °C. The effect of sodium pump on stress tolerance is not clear but the result also indicates that multiple-stress tolerance is an advantage for yeast strains to be used for high-temperature fermentation.

2.9 Thermotolerant Mechanism of Yeasts

It is well-known that heat-shock proteins protect proteins and cells from stress conditions, including high temperature. Trehalose accumulation and its protective effect against heat is also known in yeast (De Virgilio et al. 1994). These protective mechanisms are applicable for strain improvement for high-temperature fermentation. *S. cerevisiae* overexpressing the trehalose-6-phosphate synthase gene accumulated larger amounts of trehalose and produced a twofold higher concentration of ethanol (60 g/L) at 38 °C compared with non-engineered strains (An et al. 2011). In

O. polymorpha, a significant increase of ethanol production from xylose at 50 °C is achieved by overexpression of *HSP104* and disruption of the trehalase gene *ATH1* (Ishchuk et al. 2009). Expression of a small heat shock protein *HSP12* using a weak promoter also increased thermotolerance of *S. cerevisiae* (Zhang et al. 2023). On the other hand, when we evaluated the growth of diploid non-essential gene knockout strains at 37 °C (high-temperature growth) and viability after exposure to 49 °C for 60 min (heat shock resistance), the identified genes required for high-temperature growth and heat shock resistance did not overlap with each other (unpublished results). For high-temperature growth, genes involved in vesicle and protein transport, RNA metabolism, cell cycle and cell division, and DNA repair account for 63% of the total genes identified. In contrast, mitochondrial genes account for 50% of genes identified in the heat shock condition. These results suggest that high-temperature growth and heat shock resistance should be considered separately.

Thermotolerance is a quantitative trait which is determined by several alleles called quantitative trait locus (QTL). QTL analysis is intensively performed using *S. cerevisiae* strains. A QTL analysis by repeated mating and sporulation of thermotolerant and thermosensitive strains of *S. cerevisiae* found alleles involved in growth at high temperature (Steinmetz et al. 2002). An allele of *END3*, a gene involved in actin cytoskeletal organization, contributes to the thermotolerance. Benjaphokee et al. (2012) revealed that an allele in the *CDC19* gene of a thermotolerant *S. cerevisiae* strain is required for thermotolerance. *CDC19* encodes pyruvate kinase, suggesting that energy metabolism might influence thermotolerance. Recent whole-genome sequencing by next-generation sequencers has become an alternative strategy to identify the genes involved in thermotolerance. Based on genome sequencing of the strains obtained by laboratory evolution of *S. cerevisiae* at high temperature, Caspeta et al. (2014) found ergosterol composition change as a common phenotype in the thermotolerant strains. Wang et al. (2019) performed whole-genome sequencing of two segregant pools of *S. cerevisiae*, superior and inferior to thermotolerance, to identify QTLs. As a result, *VPS34*, *VID24*, and *DAP1* in an industrial thermotolerant strain were identified as major causative alleles for thermotolerance. The authors ascribed the thermotolerance conferred by the alleles to membrane fluidity. In *S. cerevisiae*, CRISPR/Cas-based strategy identified *OLE1* as a gene to enhance thermotolerance through its activated expression (Li et al. 2019). *OLE1* encodes a fatty acid desaturase. A strain evolved in the presence of a sphingolipid inhibitor had different ceramide contents compared to the parental strain and was conferred thermotolerance (Randez-Gil et al. 2020). Chen et al. (2013) analyzed gene expression in high-temperature fermentation by microarray analysis and identified four upregulated and five downregulated genes at 40 °C compared with 30 °C. *ERG9*, which may affect cell membrane structure and integrity, was one of the downregulated genes. These results suggest the importance of membrane properties for thermotolerance. Laboratory evolution and genome sequencing also suggested several genes required for thermotolerance in yeasts. In *S. cerevisiae*, chromatin remodeling, stress signal response including cell wall integrity were suggested to be important (Huang et al. 2018).

Some other approaches are used to identify the genes involved in thermotolerance. Transposon insertion was used to isolate thermotolerant disruptants in *S. cerevisiae*. However, the relation between the transposon-inserted positions and thermotolerance is not clear (Kim et al. 2011). Auesukaree et al. (2009) identified 178 genes required for high-temperature growth using a *S. cerevisiae* knockout collection. In *K. marxianus*, transcription start sites analysis revealed transcriptional differences between growth at 45 and 30 °C. The number of up- and downregulated genes at 45 °C compared with 30 °C were 199 and 508, respectively. Grouping the genes in GO terms suggested that *K. marxianus* copes with high temperature by reducing central metabolism and increasing protein turnover and DNA repair (Lertwattanasakul et al. 2015). These approaches listed a relatively large number of genes as candidates responsible for thermotolerance. Precise confirmation experiments will be required to clearly conclude their importance for thermotolerance.

To develop the strains suitable for high-temperature fermentation via genetic engineering, intensive studies to reveal the genes required for robustness under fermentation conditions at high temperature are desired.

2.10 Conclusion

Many thermotolerant yeast species have been isolated. However, it seems that *K. marxianus* is the practical high-temperature ethanol fermentation species at present. In our high-temperature fermentation using molasses as well as cassava pulp, a *K. marxianus* strain repeatedly produced more than 8% ethanol at 40 °C within 24 h. High-temperature ethanol fermentation has the potential to be used in commercial ethanol production in the near future. For non-conventional and conventional yeast species, fermentation conditions and genetic improvements should be further considered for high-temperature SSF using nonedible biomass. To enhance the performance of ethanol fermentation at high temperatures, understanding of thermotolerant mechanisms is essential. Various genetic approaches must be employed to elucidate thermotolerant mechanisms and to improve thermotolerance of yeasts.

References

Ai Y, Luo T, Yu Y et al (2022) Downregulation of ammonium uptake improves the growth and tolerance of *Kluyveromyces marxianus* at high temperature. Microbiologyopen 11. https://doi.org/10.1002/mbo3.1290

An MZ, Tang YQ, Mitsumasu K et al (2011) Enhanced thermotolerance for ethanol fermentation of *Saccharomyces cerevisiae* strain by overexpression of the gene coding for trehalose-6-phosphate synthase. Biotechnol Lett 33:1367–1374. https://doi.org/10.1007/s10529-011-0576-x

Anderson PJ, McNeil K, Watson K (1986) High-efficiency carbohydrate fermentation to ethanol at temperatures above 40??C by *Kuyveromyces marxianus* var. *marxianus* isolated from sugar mills. Appl Environ Microbiol 51:1314–1320

Arguelled JC (1997) Thermotolerance and trehalose accumulation induced by heat shock in yeast cell of Candida albicans. FEMS Microbiol Lett 146:65–71. https://doi.org/10.1016/S0378-1097(96)00455-7

Arshad M, Ahmed S, Zia MA, Rajoka MI (2014) Kinetics and thermodynamics of ethanol production by *Saccharomyces cerevisiae* MLD10 using molasses. Appl Biochem Biotechnol 172:2455–2464. https://doi.org/10.1007/s12010-013-0689-x

Auesukaree C, Damnernsawad A, Kruatrachue M et al (2009) Genome-wide identification of genes involved in tolerance to various environmental stresses in *Saccharomyces cerevisiae.* J Appl Genet 50:301–310. https://doi.org/10.1007/BF03195688

Auesukaree C, Koedrith P, Saenpayavai P et al (2012) Characterization and gene expression profiles of thermotolerant *Saccharomyces cerevisiae* isolates from Thai fruits. J Biosci Bioeng 114. https://doi.org/10.1016/j.jbiosc.2012.03.012

Avchar R, Lanjekar V, Baghela A (2022) Bioprospecting thermotolerant yeasts from distillery effluent and molasses for high-temperature ethanol production. J Appl Microbiol 132:1134–1151. https://doi.org/10.1111/jam.15288

Balakumar S, Arasaratnam V, Balasubramaniam K (2001) Isolation and improvement of a thermotolerant *Saccharomyces cerevisiae* strain. World J Microbiol Biotechnol 17:739–746. https://doi.org/10.1023/A:1012952809273

Ballesteros I, Ballesteros M, CabaÑas A et al (1991) Selection of thermotolerant yeasts for simultaneous saccharification and fermentation (SSF) of cellulose to ethanol. Appl Biochem Biotechnol 28–29:307–315. https://doi.org/10.1007/BF02922610

Benjaphokee S, Koedrith P, Auesukaree C et al (2012) CDC19 encoding pyruvate kinase is important for high-temperature tolerance in *Saccharomyces cerevisiae.* N Biotechnol 29:166–176. https://doi.org/10.1016/j.nbt.2011.03.007

Bollók M, Réczey K, Zacchi G (2000) Simultaneous saccharification and fermentation of steam-pretreated spruce to ethanol. Appl Biochem Biotechnol 84–86:69–80. https://doi.org/10.1385/ABAB:84-86:1-9:69

Caspeta L, Chen Y, Ghiaci P et al (2014) Altered sterol composition renders yeast thermotolerant. Science 346:75–78. https://doi.org/10.1126/science.1258137

Caspeta L, Coronel J, Montes de Oca A et al (2019) Engineering high-gravity fermentations for ethanol production at elevated temperature with *Saccharomyces cerevisiae.* Biotechnol Bioeng 116:2587–2597. https://doi.org/10.1002/bit.27103

Castro RCA, Roberto IC (2014) Selection of a thermotolerant *Kluyveromyces marxianus* strain with potential application for cellulosic ethanol production by simultaneous saccharification and fermentation. Appl Biochem Biotechnol 172:1553–1564. https://doi.org/10.1007/s12010-013-0612-5

Chamnipa N, Thanonkeo S, Klanrit P, Thanonkeo P (2018) The potential of the newly isolated thermotolerant yeast *Pichia kudriavzevii* RZ8-1 for high-temperature ethanol production. Braz J Microbiol 49:378–391. https://doi.org/10.1016/j.bjm.2017.09.002

Charoensopharat K, Thanonkeo P, Thanonkeo S, Yamada M (2015) Ethanol production from Jerusalem artichoke tubers at high temperature by newly isolated thermotolerant inulin-utilizing yeast *Kluyveromyces marxianus* using consolidated bioprocessing. Antonie van Leeuwenhoek 108:173–190. https://doi.org/10.1007/s10482-015-0476-5

Chen Q, Fang Y, Zhao H et al (2013) Transcriptional analysis of *Saccharomyces cerevisiae* during high-temperature fermentation. Ann Microbiol 63. https://doi.org/10.1007/s13213-013-0606-0

Cho KM, Yoo YJ (1999) Novel SSF process for ethanol production from microcrystalline cellulose using deltaintegrated recombinant yeast, *Saccharomyces cerevisiae* L2612 delta GC. J Microbiol Biotechnol 9:340–345

D'Amore T, Celotto G, Russell I, Stewart GG (1989) Selection and optimization of yeast suitable for ethanol production at 40??C. Enzyme Microb Technol 11:411–416. https://doi.org/10.1016/0141-0229(89)90135-X

Den Haan R, Rose SH, Lynd LR, van Zyl WH (2007) Hydrolysis and fermentation of amorphous cellulose by recombinant *Saccharomyces cerevisiae*. Metab Eng 9:87–94. https://doi.org/10.1016/j.ymben.2006.08.005

De Virgilio C, Hottiger T, Dominguez J et al (1994) The role of trehalose synthesis for the acquisition of thermotolerance in yeast. I. Genetic evidence that trehalose is a thermoprotectant. Eur J Biochem / FEBS 219:179–186. https://doi.org/10.1111/j.1432-1033.1994.tb19928.x

Edgardo A, Carolina P, Manuel R et al (2008) Selection of thermotolerant yeast strains *Saccharomyces cerevisiae* for bioethanol production. Enzyme Microb Technol 43. https://doi.org/10.1016/j.enzmictec.2008.02.007

Hacking AJ, Taylor IWF, Hanas CM (1984) Selection of yeast able to produce ethanol from glucose at 40°C. Appl Microbiol Biotechnol 19:361–363. https://doi.org/10.1007/BF00253786

Hari Krishna S, Janardhan Reddy T, Chowdary GV (2001) Simultaneous saccharification and fermentation of lignocellulosic wastes to ethanol using a thermotolerant yeast. Bioresour Technol 77:193–196. https://doi.org/10.1016/S0960-8524(00)00151-6

Hermansyah N, Sugiyama M, Harashima S (2015) *Candida tropicalis* isolated from Tuak, a North Sumatera-Indonesian traditional beverage, for bioethanol production. Korean J Microbiol Biotechnol 43. https://doi.org/10.4014/mbl.1506.06002

Hong J, Wang Y, Kumagai H, Tamaki H (2007) Construction of thermotolerant yeast expressing thermostable cellulase genes. J Biotechnol 130:114–123. https://doi.org/10.1016/j.jbiotec.2007.03.008

Hoshida H, Kidera K, Takishita R et al (2018) Enhanced production of extracellular inulinase by the yeast *Kluyveromyces marxianus* in xylose catabolic state. J Biosci Bioeng 125:676–681. https://doi.org/10.1016/j.jbiosc.2017.12.024

Hoshida H, Phong TT, Nonklang S, Ano A, Yarimizu T, Tokuda S, Yagi Y, Shite Y, Murashige S, Nakagawa T, Pukahuta C, Sutthikhampa S, Lorroengsil S, Hung LT, Cha-aim K, Akada R (2012) Isolation of thermotolerant ethanologenic yeasts and improvement of fermentation ability by using breeding and genetic engineering, In Summary Book Asia Core Program (2008-2013) on "Capacity building and development of microbial potential and fermentation thechnology twards new era", edt. Mamoru Yamada et al. International exchange core program in Yamaguchi Univeristy (2012), ISBN978-4-9906826-0-6

Huang CJ, Lu MY, Chang YW, Li WH (2018) Experimental Evolution of Yeast for High-Temperature Tolerance. Mol Biol Evol 35:1823–1839. https://doi.org/10.1093/molbev/msy077

Hu N, Yuan B, Sun J et al (2012) Thermotolerant *Kluyveromyces marxianus* and *Saccharomyces cerevisiae* strains representing potentials for bioethanol production from Jerusalem artichoke by consolidated bioprocessing. Appl Microbiol Biotechnol 95:1359–1368. https://doi.org/10.1007/s00253-012-4240-8

Hu ML, Zha J, He LW et al (2016) Enhanced bioconversion of cellobiose by industrial *Saccharomyces cerevisiae* used for cellulose utilization. Front Microbiol 7. https://doi.org/10.3389/fmicb.2016.00241

Hughes DB, Tudroszen NJ, Moye CJ (1984) The effect of temperature on the kinetics of ethanol production by a thermotolerant strain of *Kluveromyces marxianus*. Biotechnol Lett 6:1–6. https://doi.org/10.1007/BF00128221

Hughes SR, Bang SS, Cox EJ et al (2013) Automated UV-C mutagenesis of *Kluyveromyces marxianus* NRRL Y-1109 and selection for microaerophilic growth and ethanol production at elevated temperature on biomass sugars. J Lab Autom 18:276–290. https://doi.org/10.1177/2211068213480037

Inlow D, McRae J, Ben-Bassat A (1988) Fermentation of corn starch to ethanol with genetically engineered yeast. Biotechnol Bioeng 32:227–234. https://doi.org/10.1002/bit.260320213

Ishchuk OP, Voronovsky AY, Abbas CA, Sibirny AA (2009) Construction of *Hansenula polymorpha* strains with improved thermotolerance. Biotechnol Bioeng 104:911–919. https://doi.org/10.1002/bit.22457

Kadam KL, Schmidt SL (1997) Evaluation of *Candida acidothermophilum* in ethanol production from lignocellulosic biomass. Appl Microbiol Biotechnol 48:709–713. https://doi.org/10.1007/s002530051120

Khatun MM, Yu X, Kondo A et al (2017) Improved ethanol production at high temperature by consolidated bioprocessing using *Saccharomyces cerevisiae* strain engineered with artificial zinc finger protein. Bioresour Technol 245:1447–1454. https://doi.org/10.1016/j.biortech.2017.05.088

Kim HS, Kim NR, Yang J, Choi W (2011) Identification of novel genes responsible for ethanol and/or thermotolerance by transposon mutagenesis in *Saccharomyces cerevisiae*. Appl Microbiol Biotechnol 91:1159–1172. https://doi.org/10.1007/s00253-011-3298-z

Kiran Sree N, Sridhar M, Suresh K et al (2000) Isolation of thermotolerant, osmotolerant, flocculating *Saccharomyces cerevisiae* for ethanol production. Bioresour Technol 72. https://doi.org/10.1016/S0960-8524(99)90097-4

Kitagawa T, Tokuhiro K, Sugiyama H et al (2010) Construction of a β-glucosidase expression system using the multistress-tolerant yeast *Issatchenkia orientalis*. Appl Microbiol Biotechnol 87. https://doi.org/10.1007/s00253-010-2629-9

Kopsahelis N, Nisiotou A, Kourkoutas Y et al (2009) Molecular characterization and molasses fermentation performance of a wild yeast strain operating in an extremely wide temperature range. Bioresour Technol 100:4854–4862. https://doi.org/10.1016/j.biortech.2009.05.011

Koutinas M, Patsalou M, Stavrinou S, Vyrides I (2016) High temperature alcoholic fermentation of orange peel by the newly isolated thermotolerant *Pichia kudriavzevii* KVMP10. Lett Appl Microbiol 62. https://doi.org/10.1111/lam.12514

Kumar S, Singh SP, Mishra IM, Adhikari DK (2009) Ethanol and xylitol production from glucose and xylose at high temperature by *Kluyveromyces* sp. IIPE453. J Ind Microbiol Biotechnol 36:1483–1489. https://doi.org/10.1007/s10295-009-0636-6

Kumar S, Singh SP, Mishra IM, Adhikari DK (2010) Feasibility of ethanol production with enhanced sugar concentration in bagasse hydrolysate at high temperature using *Kluyveromyces* sp. IIPE453. Biofuels 1. https://doi.org/10.4155/bfs.10.51

Lertwattanasakul N, Kosaka T, Hosoyama A et al (2015) Genetic basis of the highly efficient yeast *Kluyveromyces marxianus*: complete genome sequence and transcriptome analyses. Biotechnol Biofuels 8:47. https://doi.org/10.1186/s13068-015-0227-x

Li P, Fu X, Zhang L et al (2017) The transcription factors Hsf1 and Msn2 of thermotolerant *Kluyveromyces marxianus* promote cell growth and ethanol fermentation of *Saccharomyces cerevisiae* at high temperatures. Biotechnol Biofuels 10:1–13. https://doi.org/10.1186/s13068-017-0984-9

Li P, Fu X, Zhang L, Li S (2019) CRISPR/Cas-based screening of a gene activation library in *Saccharomyces cerevisiae* identifies a crucial role of *OLE1* in thermotolerance. Microb Biotechnol 12:1154–1163. https://doi.org/10.1111/1751-7915.13333

Li C, Liu Q, Wang Y et al (2021) Salt stress improves thermotolerance and high-temperature bioethanol production of multi-stress-tolerant *Pichia kudriavzevii* by stimulating intracellular metabolism and inhibiting oxidative damage. Biotechnol Biofuels 14:1–17. https://doi.org/10.1186/s13068-021-02071-0

Limtong S, Sringiew C, Yongmanitchai W (2007) Production of fuel ethanol at high temperature from sugar cane juice by a newly isolated *Kluyveromyces marxianus*. Bioresour Technol 98:3367–3374. https://doi.org/10.1016/j.biortech.2006.10.044

Liu Z-H, Qin L, Zhu J-Q et al (2014) Simultaneous saccharification and fermentation of steam-exploded corn stover at high glucan loading and high temperature. Biotechnol Biofuels 7. https://doi.org/10.1186/s13068-014-0167-x

Margaritis A, Bajpai P (1982) Continuous ethanol production from Jerusalem artichoke tubers. I. Use of free cells of *Kluyveromyces marxianus*. Biotechnol Bioeng 24:1483–1493

Matsushita K, Azuma Y, Kosaka T et al (2015) Genomic analyses of thermotolerant microorganisms used for high-temperature fermentations. Biosci Biotechnol Biochem 8451:1–14. https://doi.org/10.1080/09168451.2015.1104235

Mitsumasu K, Liu Z-S, Tang Y-Q et al (2014) Development of industrial yeast strain with improved acid- and thermo-tolerance through evolution under continuous fermentation conditions followed by haploidization and mating. J Biosci Bioeng 118. https://doi.org/10.1016/j.jbiosc.2014.05.012

Mo W, Wang M, Zhan R et al (2019) *Kluyveromyces marxianus* developing ethanol tolerance during adaptive evolution with significant improvements of multiple pathways. Biotechnol Biofuels 12. https://doi.org/10.1186/s13068-019-1393-z

Nedeva TS, Savov VA, Kujumdzieva-Savova AV, Davidov ER (1993) Screening of thermotolerant yeasts as producers of superoxide dismutase. FEMS Microbiol Lett 107:49–52

Nonklang S, Abdel-Banat BMA, Cha-aim K et al (2008) High-temperature ethanol fermentation and transformation with linear DNA in the thermotolerant yeast *Kluyveromyces marxianus* DMKU3-1042. Appl Environ Microbiol 74:7514–7521. https://doi.org/10.1128/AEM.01854-08

Nuanpeng S, Thanonkeo S, Yamada M, Thanonkeo P (2016) Ethanol production from sweet sorghum juice at high temperatures using a newly isolated thermotolerant yeast *Saccharomyces cerevisiae* DBKKU Y-53. Energies (Basel) 9. https://doi.org/10.3390/en9040253

Park I, Kim I, Kang K et al (2010) Cellulose ethanol production from waste newsprint by simultaneous saccharification and fermentation using *Saccharomyces cerevisiae* KNU5377. Process Biochem 45:487–492. https://doi.org/10.1016/j.procbio.2009.11.006

Pattanakittivorakul S, Lertwattanasakul N, Yamada M, Limtong S (2019) Selection of thermotolerant *Saccharomyces cerevisiae* for high temperature ethanol production from molasses and increasing ethanol production by strain improvement. Antonie van Leeuwenhoek 112:975–990. https://doi.org/10.1007/s10482-019-01230-6

Pattanakittivorakul S, Tsuzuno T, Kosaka T et al (2022) Evolutionary adaptation by repetitive long-term cultivation with gradual increase in temperature for acquiring multi-stress tolerance and high ethanol productivity in *Kluyveromyces marxianus* DMKU 3-1042. Microorganisms 10. https://doi.org/10.3390/microorganisms10040798

Pilap W, Thanonkeo S, Klanrit P, Thanonkeo P (2018) The potential of the newly isolated thermotolerant *Kluyveromyces marxianus* for high-temperature ethanol production using sweet sorghum juice. 3 Biotech 8:1–10. https://doi.org/10.1007/s13205-018-1161-y

Pongcharoen P (2022) The ability of *Pichia kudriavzevii* to tolerate multiple stresses makes it promising for developing improved bioethanol production processes. Lett Appl Microbiol 75:36–44. https://doi.org/10.1111/lam.13703

Prakash G, Varma AJ, Prabhune A et al (2011) Microbial production of xylitol from d-xylose and sugarcane bagasse hemicellulose using newly isolated thermotolerant yeast Debaryomyces hansenii. Bioresour Technol 102:3304–3308. https://doi.org/10.1016/j.biortech.2010.10.074

Rajoka MI, Ferhan M, Khalid AM (2005) Kinetics and thermodynamics of ethanol production by a thermotolerant mutant of *Saccharomyces cerevisiae* in a microprocessor-controlled bioreactor. Lett Appl Microbiol 40:316–321. https://doi.org/10.1111/j.1472-765X.2005.01663.x

Randez-Gil F, Prieto JA, Rodríguez-Puchades A et al (2020) Myriocin-induced adaptive laboratory evolution of an industrial strain of *Saccharomyces cerevisiae* reveals its potential to remodel lipid composition and heat tolerance. Microb Biotechnol 13:1066–1081. https://doi.org/10.1111/1751-7915.13555

Rodrussamee N, Lertwattanasakul N, Hirata K et al (2011) Growth and ethanol fermentation ability on hexose and pentose sugars and glucose effect under various conditions in thermotolerant yeast *Kluyveromyces marxianus*. Appl Microbiol Biotechnol 90. https://doi.org/10.1007/s00253-011-3218-2

Rouwenhorst RJ, Visser LE, Van Der Baan AA et al (1988) Production, distribution, and kinetic properties of inulinase in continuous cultures of *Kluyveromyces marxianus* CBS 6556. Appl Environ Microbiol 54(5):1131–1137

Ryabova OB, Chmil OM, Sibirny AA (2003) Xylose and cellobiose fermentation to ethanol by the thermotolerant methylotrophic yeast *Hansenula polymorpha*. FEMS Yeast Res 4:157–164. https://doi.org/10.1016/S1567-1356(03)00146-6

Sakanaka K, Yan W, Kishida M, Sakai T (1996) Breeding a fermentative yeast at high temperature using protoplast fusion. J Ferment Bioeng 81:104–108. https://doi.org/10.1016/0922-338X(96)87585-4

Sengupta P, Mohan R, Wheeldon I et al (2022) Prospects of thermotolerant *Kluyveromyces marxianus* for high solids ethanol fermentation of lignocellulosic biomass. Biotechnol Biofuels Bioproducts 15:1–13. https://doi.org/10.1186/s13068-022-02232-9

Sohn HY, Park W, Jin I, Seu JH (1997) Characterization of ethanol fermentation using alginate immobilized thermotolerant yeast cells. J Microbiol Biotechnol 7:62–67

Sree NK, Sridhar M, Suresh K, Rao LV (1999) High alcohol production by solid substrate fermentation from starchy substrates using thermotolerant *Saccharomyces cerevisiae*. Bioprocess Eng 20:561–563. https://doi.org/10.1007/s004490050629

Sridhar M, Kiran Sree N, Venkateswar Rao L (2002) Effect of UV radiation on thermotolerance, ethanol tolerance and osmotolerance of *Saccharomyces cerevisiae* VS1 and VS3 strains. Bioresour Technol 83:199–202. https://doi.org/10.1016/S0960-8524(01)00221-8

Steinmetz LM, Sinha H, Richards DR et al (2002) Dissecting the architecture of a quantitative trait locus in yeast. Nature 416:326–330. https://doi.org/10.1038/416326a

Suryawati L, Wilkins MR, Bellmer DD et al (2008) Simultaneous saccharification and fermentation of Kanlow switchgrass pretreated by hydrothermolysis using *Kluyveromyces marxianus* IMB4. Biotechnol Bioeng 101:894–902. https://doi.org/10.1002/bit.21965

Szczodrak J, Targoński Z (1998) Selection of thermotolerant yeast strains for simultaneous saccharification and fermentation of cellulose. Biotechnol Bioeng 31:300–303. https://doi.org/10.1002/bit.260310404

Takashima M, Sugita T, Toriumi Y, Nakase T (2009) *Cryptococcus tepidarius* sp. nov., a thermotolerant yeast species isolated from a stream from a hotspring area in Japan. Int J Syst Evol Microbiol 59:181–185. https://doi.org/10.1099/ijs.0.004515-0

Ueno R, Urano N, Kimura S (2002) Effect of temperature and cell density on ethanol fermentation by a thermotolerant aquatic yeast strain isolated from a hot spring environment. Fish Sci 68. https://doi.org/10.1046/j.1444-2906.2002.00463.x

Voronovsky AY, Rohulya OV, Abbas CA, Sibirny AA (2009) Development of strains of the thermotolerant yeast *Hansenula polymorpha* capable of alcoholic fermentation of starch and xylan. Metab Eng 11. https://doi.org/10.1016/j.ymben.2009.04.001

Wang Z, Qi Q, Lin Y et al (2019) QTL analysis reveals genomic variants linked to high-temperature fermentation performance in the industrial yeast. Biotechnol Biofuels 12:1–18. https://doi.org/10.1186/s13068-019-1398-7

Wang L, Li B, Su RR et al (2022) Screening novel genes by a comprehensive strategy to construct multiple stress-tolerant industrial *Saccharomyces cerevisiae* with prominent bioethanol production. Biotechnol Biofuels Bioprod 15:1–19. https://doi.org/10.1186/s13068-022-02109-x

Watanabe I, Nakamura T, Shima J (2010) Strategy for simultaneous saccharification and fermentation using a respiratory-deficient mutant of *Candida glabrata* for bioethanol production. J Biosci Bioeng 110. https://doi.org/10.1016/j.jbiosc.2010.02.003

Wen F, Sun J, Zhao H (2010) Yeast surface display of trifunctional minicellulosomes for simultaneous Saccharification and fermentation of cellulose to ethanol. Appl Environ Microbiol 76:1251–1260. https://doi.org/10.1128/AEM.01687-09

Yanase S, Hasunuma T, Yamada R et al (2010) Direct ethanol production from cellulosic materials at high temperature using the thermotolerant yeast *Kluyveromyces marxianus* displaying cellulolytic enzymes. Appl Microbiol Biotechnol 88. https://doi.org/10.1007/s00253-010-2784-z

Ylitervo P, Franzén CJCJ, Taherzadeh MJMJ (2011) Ethanol production at elevated temperatures using encapsulation of yeast. J Biotechnol 156:22–29. https://doi.org/10.1016/j.jbiotec.2011.07.018

Yuan WJ, Zhao XQ, Ge XM, Bai FW (2008) Ethanol fermentation with *Kluyveromyces marxianus* from Jerusalem artichoke grown in salina and irrigated with a mixture of seawater and freshwater. J Appl Microbiol 105:2076–2083. https://doi.org/10.1111/j.1365-2672.2008.03903.x

Yuan SF, Guo GL, Hwang WS (2017) Ethanol production from dilute-acid steam exploded lignocellulosic feedstocks using an isolated multistress-tolerant *Pichia kudriavzevii* strain. Microb Biotechnol 10:1581–1590. https://doi.org/10.1111/1751-7915.12712

Yuangsaard N, Yongmanitchai W, Yamada M, Limtong S (2013) Selection and characterization of a newly isolated thermotolerant *Pichia kudriavzevii* strain for ethanol production at high temperature from cassava starch hydrolysate. Antonie van Leeuwenhoek 103:577–588. https://doi.org/10.1007/s10482-012-9842-8

Zhang J, Lynd LR (2010) Ethanol production from paper sludge by simultaneous saccharification and co-fermentation using recombinant xylose-fermenting microorganisms. Biotechnol Bioeng 107. https://doi.org/10.1002/bit.22811

Zhang Q, Fu Y, Wang Y et al (2012) Improved ethanol production of a newly isolated thermotolerant *Saccharomyces cerevisiae* strain after high-energy-pulse-electron beam. J Appl Microbiol 112. https://doi.org/10.1111/j.1365-2672.2011.05209.x

Zhang ML, Zhang H, He YX et al (2023) Improving thermo-tolerance of *Saccharomyces cerevisiae* by precise regulation of the expression of small HSP. RSC Adv 13:36254–36260. https://doi.org/10.1039/d3ra05216h

Chapter 3
Development of Yeast *Saccharomyces* Platform for the Production of Biofuels and Bio-based Products

Pattanan Songdech and Nitnipa Soontorngun

Contents

3.1 Carbon Utilization by *Saccharomyces* 78
3.2 Central Carbon Metabolisms 78
3.3 Glycolysis 79
3.4 Crabtree Effect 80
3.5 Biofuels 80
3.6 Bio-ethanol 81
3.7 Biobutanol 81
3.8 Biofuel Production from *S. cerevisiae* 82
3.9 Production of Ethanol in *S. cerevisiae* 82
3.10 Genetic Modifications Biofuels and Bio-based Products 83
3.10.1 Strategies Based on *GPD* Genes 83
3.10.2 Strategies Based on *GLT/GLN/GDH* Genes 84
3.10.3 Strategies Based on *DAK* and *GCY* Genes 84
3.10.4 Strategies Based on *ADH* and *PDC* Genes 85
3.10.5 Strategies Based on Transcription Factors Znf1 and Asg1 86
3.11 Lignocellulose Feedstocks and Lignocellulose-Based Bioproducts 87
3.12 Bioethanol Production from Lignocellulosic Biomass 87
References 88

Abstract *Saccharomyces* stems from *saccharon* (sugar) and *myces* (fungus). The budding yeast *S. cerevisiae* is a eukaryotic cell with a thick layer of cell wall made up of beta glucan network and small proportion of polysaccharide chitin. As a unicellular eukaryotic model organism, it is often used in the field of modern biotechnology. This characteristic feature of yeast's cell wall makes it a good carrier with a

P. Songdech · N. Soontorngun (✉)
Excellent Research Laboratory for Yeast Innovation, Division of Biochemical Technology, School of Bioresources and Technology, King Mongkut's University of Technology Thonburi, Bangkok, Thailand
e-mail: nitnipa.soo@kmutt.ac.th

A. A. Sibirny (ed.), *Biotechnology of Yeasts and Filamentous Fungi*,
Grand Challenges in Biology and Biotechnology,
https://doi.org/10.1007/978-3-031-74726-7_3

long release time. With 12,068 kilobases (kb) over 16 chromosomes, it is the first eukaryotic organism to be sequenced. Because of its comparable cell structure to human cells, *S. cerevisiae* is well-known in the field of medical research applications. *S. cerevisiae* can grow in either haploid or diploid form. It takes basic dietary need for growth and is easy to cultivate in typical laboratory conditions. It comes in the shape of a round to ovoid through budding mechanisms. Furthermore, *S. cerevisiae* has the capacity to develop in both aerobic and anaerobic environments. The yeast *S. cerevisiae* performs the most prevalent type of fermentation (alcoholic fermentation), and it consumes various types of sugars depending on how it grows. Galactose and fructose are the best fermenting sugars for aerobic growth for the yeast *S. cerevisiae*. Owing to such property, *S. cerevisiae* displays high potential for industrial applications such as in biofuel and biochemical production.

3.1 Carbon Utilization by *Saccharomyces*

Glucose is the principal source of energy for most organisms and is the most favored carbon source for *S. cerevisiae*. In *S. cerevisiae*, glucose is the preferred substrate. It is also the starting point for all of the major metabolic pathways. The block of several synthetic enzymes by glucose has been documented. The pathways involved are in gluconeogenesis, the tricarboxylic acid cycle, and respiration are caused by the repression of glucose, which is one of the most important regulating metabolisms (Ma and Liu 2010). The yeast *S. cerevisiae*, on the other hand, undergoes ethanol fermentation during glucose metabolism in aerobic conditions. Furthermore, *S. cerevisiae* has been shown to be involved in the production of several fermenting products. These include wine, beer, and cider; distilled beverages such as rum, vodka, whisky, brandy, and sake; and other alcoholic beverages worldwide as well as soft drinks such as fruit juice, honey-drink, and tea. Likewise, *S. cerevisiae*'s demand in bakery products is also commonly recognized as the baker's yeast (Gombert and van Maris 2015). *S. cerevisiae* has been used for the production of fermented foods such as bread, wine, and beer on an industrial level for decades. Among others, glycolysis is one of the most important pathways in the scale of fermentation for bioethanol production. Furthermore, *S. cerevisiae* was chosen as a model organism due to its unique features among other organisms, such as reduced contaminant chances, ease of growth, and phage resistance.

3.2 Central Carbon Metabolisms

Central carbon metabolisms use a series of enzyme reactions to convert sugar to a metabolic precursor. Acetyl-CoA is a key chemical in central carbon metabolism, needed for essential cellular processes like energy, lipid, and amino acid metabolism. Furthermore, acetyl-CoA is used as a precursor in the production of a variety of industrial chemicals. These include lipids (such as dietary supplements and biodiesels), polyketides (such as antibiotics and anticancer drugs), polyhydroxyalkanoates (biodegradable polymers), and isoprenoids (such as flavors and fragrances, biodiesels, antimicrobials, anticancer drugs, rubber, cosmetic additives, and vitamins).

Acetyl-CoA is synthesized endogenously in *S. cerevisiae* by the four compartments. It starts from the critical intermediary pyruvate in the mitochondria and acetate from the nucleus, peroxisome, and cytoplasm. Acetaldehyde is also an intermediary in the conversion of pyruvate to ethanol. Furthermore, the cytosolic pathway to acetyl-CoA is energetically costly (2 ATP equivalents), which could reduce glucose as product output. At high glucose levels, a considerable percentage of the glucose is converted to reduce the by-products such as ethanol and a variety of other chemicals. Pyruvate is then delivered toward mitochondria at low glucose levels, and once within the mitochondrion, the PDH complex can convert pyruvate to acctyl-CoA (Conrad et al. 2014; Zaman et al. 2008).

3.3 Glycolysis

Glycolysis is a process by which glucose turns into pyruvate; however, glucose is utilized in the glycolysis in order to supply energy in the form of ATP, decreasing equivalent in NADPH form and precursor used other cellular metabolic processes. It is maintained in the Archaea, Bacteria and Eukarya, three domains of life. Glycolysis is also being researched for usage in the baking industry, as a biotransformation catalyst, and in the manufacturing of bioethanol (Conrad et al. 2014; Gancedo 2008; Peeters et al. 2017). In the upper part of glycolysis, two ATP are utilized from glucose to fructose 1,6-bisphosphate and fructose 1,6-bisphosphate. The latter is transformed into two 3-carbon sugars. Pyruvate is the final product converted from 3-carbon sugars releasing 2 ATP and 1 NADH for each. Pyruvate can then be further decomposed via the respiration or fermentation routes. In alcoholic fermentation, pyruvate decarboxylase (Pdc) and alcohol dehydrogenase (Adh) transform pyruvate into ethanol. This results in a net gain of ATP per glucose molecule. ADH converts NADH generated during lower glycolysis into NAD+, allowing alcoholic fermentation to occur without the use of oxygen (Conrad et al. 2014; Gancedo 2008; Peeters et al. 2017).

Pyruvate dehydrogenase (Pdh) functions to reduce pyruvate into acetyl-CoA in the respiration pathway, which is ultimately oxidized to CO_2 in TCA cycle. Other reduced coenzymes are produced, which get oxidized in the mitochondrion in order to provide ATPs. In general, eukaryotes' respiration system can produce up to 38 molecules of ATP through one molecule of glucose, plus two more via glycolysis. The P/O ratio in *S. cerevisiae* is estimated to be around 1.2, with a respiration yield of roughly 18 ATPs per glucose (Peeters et al. 2017).

3.4 Crabtree Effect

Baker's yeast can grow under both anaerobic and aerobic environments. *S. cerevisiae* is noted for its rapid conversion of carbohydrates to ethanol and carbon dioxide. Despite the fact that aerobic respiration with oxygen as the ultimate electron acceptor is possible, *S. cerevisiae* ferments alcohol until the sugar/glucose in the medium is depleted. The Crabtree effect is given to describe this event. Yeasts like *S. cerevisiae* and *Schizosaccharomyces pombe*, for example, can generate ethanol even in aerobic environments. They are termed Crabtree-positive. Crabtree-negative yeast is a yeast that does not have a Crabtree effect. The Crabtree effect has been claimed to be easily proven in chemostat and batch culture. Under glucose-limited chemostat culture and low dilution rates, CO_2 production equals O_2 uptake. Under this situation, biomass production is high at approximately 0.5 g/g, although residual glucose content is minimal. To maintain a steady-state population size in the chemostat, the population is forced to replicate faster as the dilution rates increase. This shows a rise in the concentration of leftover glucose. Both CO_2 generation and O_2 consumption increase as the dilution rate rises. They remain connected until a critical threshold is then achieved. When the dilution rate is increased past this critical point, fermentation thus occurs. CO_2 production accelerates and decouples from O_2 used. The start of fermentation is accompanied by a quick decline in biomass output, which are below 0.2 g/g (Crabtree 1928; Díaz-Ruiz et al. 2008; Peeters et al. 2017).

Batch culture studies can be performed to figure out the glucose content at the fermentation start points. This has been found in *S. cerevisiae* at glucose concentrations of roughly 150 mg/L; in terms of comparison, glucose uptake begins to saturate around 500 mg/L. However, this may vary with species and depend on the specific circumstances. The rapid decrease in biomass yield begins a question on Crabtree-positive yeasts and the "wasteful" fermentation pathway. Despite this, they could theoretically manufacture all ATPs through respiration alone (Peeters et al. 2017).

3.5 Biofuels

Biofuel is a type of alternative fuel made primarily from renewable feedstock and generated by biological processes. The generation of biofuel from renewable resources is taken into consideration because the maximum sustainable opportunity to fossil fuels as it affords high-quality effect toward the environment and economy. Biofuels include bio-alcohol, biodiesel, Fischer-Tropsch diesel, methane, and hydrogen that come in liquid, gaseous, and solid forms. Biofuel has been broadly utilized as a transportation fuel because of its benefits, with biodiesel and bioethanol being the most well-known. Ethanol is currently the most popular automotive fuel due to its various benefits. Butanol, on the other hand, is getting a lot of interest because it has a lot more advantages than ethanol. Unlike ethanol, which has two carbon atoms (C_2H_5OH), butanol has four carbon atoms and the chemical formula $C4H_9OH$. Butanol comes in four different isomers: 1-butanol, 2-butanol, isobutanol, and tert-butanol. 1-Butanol and isobutanol are known to be the most acceptable candidates for transportation fuel among the isomers. *Clostridium acetobutylicum* is the oldest natural producer of 1-butanol via the acetone-butanol-ethanol (ABE) fermentation, whereas yeast can make isobutanol via the Ehrlich route (Van Maris et al. 2007).

3.6 Bio-ethanol

For nearly a century, ethanol has been utilized as a transportation fuel. Ethanol is predicted to be the largest-volume product in industrial biotechnology, with a world production of 100 Mton. Its manufacturing is, presently, primarily based on the fermentation of cane sugar or hydrolyzed corn starch with the yeast *S. cerevisiae* together with the first-generation bioethanol process is distinguished by high ethanol yield on sugar use (>90% of the theoretical maximum yield of 0.51 g ethanol) (g hexose sugar), ethanol titers of up to 21% (w/w), and volumetric productivity of 2–3 $kgm^{-3}\ h^{-1}$ (Verstrepen et al. 2004).

Bioethanol C_2H_5OH, on the other hand, is known as bi-carbon alcohol produced by fermentation from biomass feedstock. It is suggested as a useful biofuel, one of the carbon-neutral products. However, due to the low-energy content and strong hygroscopicity of ethanol, it has been seen as drawbacks in terms of its fuel quality when compared to biobutanol. The latter has been gaining traction as a fuel that is superior to ethanol due to a number of significant advantages such as better energy density, lesser hygroscopicity, and higher octane number (Busti et al. 2010; Conant and Wolfe 2007; Karras et al. 2019).

3.7 Biobutanol

A four-carbon-containing alcohol biobutanol occurs in four isomeric types (e.g., n-butanol, sec-butanol, isobutanol, and ter-butanol). Among all isobutanol is a four-carbon alcohol with a branched structure which is also familiar as 2-methyl-1-propanol. Isobutanol is a valuable chemical raw material. It is a starting material used to make plasticizers, antioxidants, as well as others high-value-added products. Some examples are the plastics, rubber, pharmaceutical, and pesticide industries, as well as a gasoline mix stock and jet fuel upgrade. Isobutanol can also be used as a solvent to make isobutylamine, acrylates, methacrylates, acid-curable lacquers, and baked finishes, as a cleanser for paints and inks, or as a mobile phase in chromatography.

Isobutanol, in contrast, has been recognized as a promising new generation of biofuels. In comparison to ethanol, which is highly corrosive and must be transported in small quantities and mixed with gasoline at the time of distribution. Biobutanol is the more appropriate biofuel due to its better features such as low vapor pressure, thereby allowing it to be openly moved through the pipeline transportation. Also, it can be directly stored in the existing gas station system without any modification. Furthermore, butanol's octane numbers and air fuel ratio are identical with gasoline, making its property more similar to oil. Furthermore, because biobutanol contains less oxygen, it can be mixed with gasoline in any ratio (Pfeiffer and Morley 2014).

3.8 Biofuel Production from *S. cerevisiae*

For a couple of years, yeast, *S. cerevisiae*, has been utilized in the food and biotechnology industries and is considered "generally recognized as save" (GRAS). Yeast, *S. cerevisiae*, is a strong platform for biofuel production, and it is also a good laboratory strain in modern molecular research, for example, gene manipulation, metabolic engineering, and synthetic biology. In the field of genetics and cell biology, *S. cerevisiae* is recognized as the model eukaryotic organism. In order to choose *S. cerevisiae* as a model is the unmatched homologous recombination ability, which aids in the stable integration of foreign DNA fragments with long lengths in vivo, availability of several genetic tools for heterologous overexpression and annoyed knockout of desired genes that have developed. Furthermore, because *S. cerevisiae* is a eukaryote, it has various organelles, which allows for compartmentalized biosynthesis. It also has the ability to tolerate product and toxic inhibitors (Geladé et al. 2003). As the cell factory, *S. cerevisiae* is popular in the field of production of biologically based chemicals and employ in industrial applications such as wine and beer production, but efficient transformation of carbon substrates such as glucose, different types of hexoses and pentoses by yeast hydrolytic enzymes and transporters. The metabolic pathways of yeast produce ethanol, butanol, alkanes, and fatty acids.

3.9 Production of Ethanol in *S. cerevisiae*

In *S. cerevisiae*, the gene family of hexose transporters consists of more than 20 members including 18 genes encoding hexose transporters (*HXT1-HXT17*, *GAL2*) and two genes encoding sensors of glucose signaling (*SNF3, RGT2*) and also both Gpr1 and Hxk2 that sense glucose levels (Gancedo 2008; Rintala et al. 2008). The sugar transporters can be classified into two classes which depend on the glucose affinity: (1) low affinity for glucose and high transport capacity, in which the most important hexose transporters are Hxt1 and Hxt3, and (2) high affinity and low transport capacity, with the key hexose transporters being Hxt2, Hxt4, and Hxt7.

First, glucose enters the glycolytic pathway by hexose transporters, and glucose will be in this pathway to generate pyruvate intermediate. In the first step of glycolysis, glucose is phosphorylated by hexokinases (encoded by *HXK1* and *HXK2*) and glucokinase (encoded by *GLK1*) which requires ATP and converts to glucose-6-phosphate intermediate. The glucose-6-phosphate is isomerized to fructose-6-phosphate catalyzed by phosphoglucose isomerase encoded by *PGI1*. In the next step, the fructose-6-phosphate molecule is converted into fructose 1,6-biphosphate via transferring of inorganic phosphate from ATP which is catalyzed by phosphofructokinase encoded by *PFK1* and *PFK2*. The conversion of fructose 1,6-bisphosphate to glyceraldehyde 3-phosphate and dihydroxyacetone phosphate is catalyzed by fructose 1,6-bisphosphate aldolase encoded by *FBA1*. In the first step fermentative pathway, pyruvate is decarboxylated by pyruvate decarboxylase (Pdc1/5/6), converting pyruvate into acetaldehyde and releasing carbon dioxide. The acetaldehyde will be reduced to the formation of ethanol by alcohol dehydrogenase encoded by *ADH1*, allowing the re-oxidation of NADH to NAD^+. The ethanol is the main product of fermentation, also several other by-products are produced during the process, including cell biomass, glycerol, and some organic acids (van Dijken and Scheffers 1986).

3.10 Genetic Modifications Biofuels and Bio-based Products

3.10.1 Strategies Based on GPD *Genes*

In *S. cerevisiae*, glycerol is synthesized from dihydroxyacetone phosphate that is catalyzed by glycerol-3-phosphate dehydrogenase (encoded by *GPD1* and *GPD2*) and conversion of glycerol-3-phosphate to glycerol catalyzed by glycerol-3-phosphate phosphatase (encoded by *GPP1* and *GPP2*), respectively (Majidian et al. 2018; Nissen et al. 2000). GPD is the key enzyme controlling the overall rate of glycerol synthesis of both reactions, directly or indirectly (Eriksson et al. 1995). Also, the *GPD1* is involved under osmotic stress and *GPD2* is induced under anaerobic conditions important for maintaining redox balance under anaerobic conditions (Ansell et al. 1997). The single deletion of *GPD1* or *GPD2* gene in *S. cerevisiae* is shown. Deletion of *GPD2* produced glycerol at a rate 40% less than the wild-type strain. Ethanol production is increased by 13%, and the biomass production is decreased by 50% in the deletion strain. Deletion of *GPD1* shows no significant differences in terms of growth, glycerol, and ethanol production (Valadi et al. 1998). Also, elimination of *GPD1* decreases glycerol formation by 50% and increases ethanol yield 5.1%, when compared with the wild-type strain (Michnick et al. 1997).

3.10.2 *Strategies Based on* GLT/GLN/GDH *Genes*

The decreasing formation of glycerol led to the increasing production of ethanol in *S. cerevisiae* by reducing the excess NADH formation and increasing ATP consumption by the manipulation of some genes *GLT, GLN,* and *GDH.* They are involved in glutamate metabolism (Nissen et al. 2000). The NADH and/or NADPH are consumed via other pathways rather than the glycerol production pathway which can be targeted and is the pathway of glutamate/glutamine production. In this pathway, ammonium reacts with α-ketoglutarate to form glutamate, and this reaction is catalyzed by an NADPH-dependent glutamate dehydrogenase encoded by *GDH1* and *GDH3* (Kong et al. 2007). However, the reverse reaction is catalyzed by glutamate dehydrogenase 2 encoded by *GDH2* and generates NADH at low ammonium concentrations for amino acids synthesis (Nissen et al. 1997). The other route for glutamate synthesis is through the interaction between glutamine and α-ketoglutarate, catalyzed by glutamate synthase encoded by *GLT1* (Nissen et al. 1997). This reaction is through the reduction of NADH to NAD^+. Glutamine is produced from glutamate that reacts with $NH4^+$, catalyzed by glutamine synthetase (encoded by *GLN1*) using ATP (Kong et al. 2006). Moreover, the overexpression of *GLT1, GLN1,* and *GDH2* genes and/or the deletion of *GDH1/GDH3* genes increase the reduction of NADH to NAD^+ or NADPH to $NADP^+$. Thus, the reaction responsible for decreasing NADH level will lower the generation of glycerol.

The overexpression of *GLT1* and deletion of the *FPS1* gene, which encodes the channel protein responsible for glycerol efflux pump, showed a 14% higher ethanol yield and 30% lower glycerol formation compared with the control strain (Kong et al. 2006). Also, in the overexpression of *GLT1* and deletion of either *GPD1* or *GPD2*, the result of both strains was increased ethanol production by 11% and 13% as well as decreased generation of glycerol by 25.% and 38%, respectively, compared with the wild type (Kong et al. 2007). The individual deletion of *GPD1* or *GDH1* genes decreased the formation of glycerol. However, deletion of both *GPD1* and *GDH1* genes led to significantly decreasing the production of glycerol by 45% compared with the wild-type strain and also increasing the ethanol yield by 2% compared with the control strain (Lee et al. 2002).

3.10.3 *Strategies Based on* DAK *and* GCY *Genes*

The *S. cerevisiae* can ferment glycerol as the sole carbon source via two-branch pathway including in aerobic condition controlled by *GUT1-GUT2* genes and in anaerobic condition controlled by *GCY-DAK* genes. This leads to the formation of dihydroxyacetone phosphate (DHAP), a glycolytic intermediate, and ethanol, which is produced in the final step (Yu et al. 2010). Moreover, the aerobic pathway is catalyzed by glycerol kinase (encoded by *GUT1*) and mitochondrial glycerol 3-phosphate dehydrogenase (encoded by *GUT2*) (Krantz et al. 2009). In the anaerobic route, glycerol is first reduced to dihydroxyacetone by glycerol dehydrogenase (encoded by *GCY1*) and then to dihydroxyacetone phosphate by dihydroxyacetone kinase (encoded by *DAK1*) (Zhang et al. 2013). Thus, the overexpression of *GCY1* and *DAK1* increased the production of ethanol by twofold compared with the wild-type strain. Thus, overexpression of the *GUP1* gene, which encodes glycerol uptake protein, led to an increase in ethanol production by threefold compared with the wild-type strain (Yu et al. 2010). The transcription factor Spt15 is capable of transcribing most of the mRNA genes in the genome while the transcription factor Spt3 activates the transcription of some RNAP II genes involved in ethanol production by interacting with Spt15 led to increase the ethanol yield by 1.8-fold (Yu et al. 2012a).

3.10.4 *Strategies Based on* ADH *and* PDC *Genes*

In alcoholic fermentation, the decarboxylation of pyruvate to acetaldehyde and the subsequent conversion of acetaldehyde to ethanol are critical steps to improving the efficiency of the ethanol yield. These reactions are catalyzed sequentially through pyruvate decarboxylase and alcohol dehydrogenase. There are three isozymes of the Pdc1, Pdc5, and Pdc6 enzymes encoded by *PDC1*, *PDC5*, and *PDC6* genes, respectively. However, *PDC1* gene is the most strongly expressed isozyme in *S. cerevisiae* and a greatly efficient target of genetic manipulation (Yu et al. 2012b). The deletion of *PDC1* strain was dramatically decreased generation of ethanol by 70% compared with the wild-type strain. Moreover, the overexpression of *PDC1* increased decarboxylase activity and also increased ethanol production (Tokuhiro et al. 2009).

The alcohol dehydrogenase isoenzymes, important enzymes involved in ethanol synthesis pathway in *S. cerevisiae*, encoded by *ADH1*, *ADH3*, *ADH4*, and *ADH5*, are responsible for the final step of the alcoholic fermentation and also reduced acetaldehyde to ethanol (Maestre et al. 2008). However, the oxidation of ethanol to acetaldehyde is catalyzed by Adh2 which is repressed by high glucose condition (Ciriacy 1975). The overexpression of the *ADH1* and *PDC1* but lacking *FPS1* and *GPD2* genes increased the production of ethanol by 1.5-fold compared with parental strain. Moreover, ethanol yield increased by approximately 0.6 g ethanol/g glucose in YS-7 (*gpd1*Δ*adh2*Δ*ald4*Δ) strain, but ethanol productivity decreased by 0.2 g/L/h. While the highest increase in ethanol productivity was observed in YS-6 (*adh2*Δ*ald4*Δ) strain around 0.3 (g/L/h), ethanol yield increased by approximately 0.5 g ethanol/g glucose (Liu et al. 2019).

3.10.5 Strategies Based on Transcription Factors Znf1 and Asg1

It appears that one effective strategy to improve strain robustness and raise bioethanol productivity is to overexpress key transcriptional activators of genes involved in stress responses and glycolysis (Songdech et al. 2020). Previously, as shown in this study, the role of transcription factor Znf1 in the control of glucose metabolism was found. When glucose is a sole carbon source, Znf1 activates important genes involved in pyruvate metabolism, glycolysis, and alcoholic fermentation. Overexpression of *ZNF1* accelerated glucose consumption at high glucose; after 24–36 h of fermentation, less than 1% of glucose remains. Significantly, overexpression of *ZNF1* increased ethanol concentrations by nearly 20–25%, with maximum ethanol concentration of approximately 80 g/L with increased ethanol productivity. Znf1 also confers resistance to weak-acid and osmotic stresses, a crucial characteristic in the fermentation of alcohol.

For isobutanol, the native *Saccharomyces cerevisiae* produces only a trace amount of isobutanol under catabolite repression of glucose. Recently, the new functions in nonfermentable carbon transcription factor Znf1 were found, which plays in the synthesis of isobutanol when xylose is utilized through activation of genes involved in iron-sulfur assimilation, valine biosynthesis, and the Ehrlich pathway. Moreover, the engineered ZNXISO strain can produce high levels of isobutanol from xylose. It was created by overexpressing *ZNF1*, xylose-reductase/dehydrogenase (XR-XDH) variants, xylulokinase, a xylose-specific sugar transporter, and enzymes of the isobutanol pathway in the engineered *S. cerevisiae pho13gre3*Δ strain. Thus, transcription factor modification is a crucial strategy toward the more sustainable synthesis of advanced biofuels from xylose, the second most abundant carbon source (Songdech et al. 2024).

Understanding tolerant mechanisms is essential for building robust host strains, which in turn will help develop a cost-effective microbial cell factory for the production of biofuels and biochemicals. Since Znf1 triggers the activation of PPP genes, which are necessary for the production of essential amino acids, cellular metabolites, and preservation of the NADP/NADPH redox balance, it also facilitates *ILV3*, *ILV5*, and *ARO10* genes, linked to the synthesis of essential intermediates for the production of isobutanol. At different points along yeast metabolic pathways, the transcription factor Znf1 keeps redox balance and energy homeostasis intact to ensure swift regulation during stresses (Ali et al. 2024). Another transcription factor Asg1 of stress response genes is shown to regulate the metabolism of fatty acids. Downregulation of Asg1 is correlated with decreased expression of β-oxidation genes and the accumulation of triacylglycerols (TAGs) (Jansuriyakul et al. 2016). Asg1 may play a role in the regulation of biodiesel production in yeasts as shown for Znf1 in bioethanol and isobutanol production.

3.11 Lignocellulose Feedstocks and Lignocellulose-Based Bioproducts

Lignocellulosic feedstocks are the most abundant renewable material from agricultural wastes which mainly consist of the heterogeneous complex of cellulose (30–35%), hemicellulose (25–30%), and lignin (10–20%), respectively. Cellulose is a homopolymer of glucose that is joined together by β-1,4-glycosidic linkages (Klemm et al. 2005). Hemicelluloses are a branched heteropolymer of hexoses (glucose) and pentoses (mainly xylose and arabinose) (Saha 2003). Lignin is derived mainly from three main monomers: coumaryl alcohol, coniferyl alcohol, and sinapyl alcohol. They are coupled via covalent bonds to xylans rendering massiveness and stability to the plant cell wall as well as giving the tensile strength of plant material (Mielenz 2001). However, acid hydrolysis is a popular pretreatment process in the degradation of hemicellulose which leads to the formation of products such as pentose and hexose sugars (such as glucose, xylose, arabinose) and inhibitory compounds including acetic acid, formic acid, levulinic acid, formaldehyde, 5-hydroxymethylfurfural, and furfural. The mixed sugars obtained through the hydrolysis process are fermented by microbial catalysts to make products such as ethanol produced by the yeast *S. cerevisiae*. Moreover, the process of hydrolysis and fermentation can be separated (hydrolysis and fermentation; SHF) which are methods by enzymatic hydrolysis and fermentation or simultaneously (simultaneous saccharification and fermentation; SSF) which is a process that combines the enzymatic hydrolysis with fermentation to produce high-value products. Also, the consolidated bioprocessing (CBP) of lignocellulosic refers to a process in which the microorganism in fermentation also contributes by producing cellulolytic enzymes (van Zyl et al. 2007).

3.12 Bioethanol Production from Lignocellulosic Biomass

After hydrolysis, the sugar molecules are consumed by bacteria or yeasts that can be released by enzymes to produce organic acids, or alcohol in anaerobic conditions (Mussatto et al. 2010). The effect of the fermentation process depends on the effective hydrolysis and selection of appropriate microorganisms. It is to reduce the inhibitory compounds and produce high ethanol yield (Achinas and Euverink 2016). The fermentation process is either saccharification and fermentation (SSF) or separate hydrolysis and fermentation (SHF). In the previous report, accumulation of glucose, cellobiose, and end products of hydrolysis inhibit cellulases activity. In SSF, cellulose hydrolysis and hexose fermentation occur in a single reactor. This is to reduce the inhibitory reactions in cellulose, which overcomes the difficulty of this type of fermentation in SHF. For corn-based and sugar-based biofuel industries, *S. cerevisiae* is one of the mostly used strains for ethanol fermentation and, in the next decade also, will move forward to the economical production of other biofuels and biochemicals (Afdhol et al. 2019).

References

Achinas S, Euverink GJW (2016) Consolidated briefing of biochemical ethanol production from lignocellulosic biomass. Electron J Biotechnol 23:44–53

Afdhol M, Abdurrahman M, Hidayat F, Chong F, Zaid H (2019) Review of solvents based on biomass for mitigation of wax paraffin in Indonesian oilfield. Appl Sci 9:5499

Ali SA, Songdech P, Samakkarn W, Duangphakdee O, Soontorngun N (2024) New regulatory role of Znf1 in transcriptional control of pentose phosphate pathway and ATP synthesis for enhanced isobutanol and acid tolerance. Yeast 41(6):401–417

Ansell R, Granath K, Hohmann S, Thevelein JM, Adler L (1997) The two isoenzymes for yeast NAD^+-dependent glycerol 3-phosphate dehydrogenase encoded by *GPD1* and *GPD2* have distinct roles in osmoadaptation and redox regulation. EMBO J 16(9):2179–2187

Busti S, Coccetti P, Alberghina L, Vanoni M (2010) Glucose signaling-mediated coordination of cell growth and cell cycle in *Saccharomyces cerevisiae*. Sensors (Basel, Switzerland) 10(6):6195–6240

Ciriacy M (1975) Genetics of alcohol dehydrogenase in *Saccharomyces cerevisiae*—II. Two loci controlling synthesis of the glucose-repressible ADH II. MGG. Mol Gen Genet 138(2):157–164

Conant GC, Wolfe KH (2007) Increased glycolytic flux as an outcome of whole-genome duplication in yeast. Mol Syst Biol 3:129

Conrad M, Schothorst J, Kankipati HN, Van Zeebroeck G, Rubio-Texeira M, Thevelein JM (2014) Nutrient sensing and signaling in the yeast *Saccharomyces cerevisiae*. FEMS Microbiol Rev 38(2):254–299

Crabtree HG (1928) The carbohydrate metabolism of certain pathological overgrowths. Biochem J 22(5):1289–1298

Díaz-Ruiz R, Avéret N, Araiza D, Pinson B, Uribe-Carvajal S, Devin A, Rigoulet M (2008) Mitochondrial oxidative phosphorylation is regulated by fructose 1,6-bisphosphate. A possible role in Crabtree effect induction? J Biol Chem 283(40):26948–26955

Eriksson P, André L, Ansell R, Blomberg A, Adler L (1995) Cloning and characterization of *GPD2*, a second gene encoding *sn*-glycerol 3-phosphate dehydrogenase (NAD^+) in *Saccharomyces cerevisiae*, and its comparison with *GPD1*. Mol Microbiol 17(1):95–107

Gancedo JM (2008) The early steps of glucose signalling in yeast. FEMS Microbiol Rev 32(4):673–704

Geladé R, Van de Velde S, Van Dijck P, Thevelein JM (2003) Multi-level response of the yeast genome to glucose. Genome Biol 4(11):233

Gombert AK, van Maris AJ (2015) Improving conversion yield of fermentable sugars into fuel ethanol in 1st generation yeast-based production processes. Curr Opin Biotechnol 33:81–86

Jansuriyakul S, Somboon P, Rodboon N, Kurylenko O, Sibirny A, Soontorngun N (2016) The zinc cluster transcriptional regulator Asg1 transcriptionally coordinates oleate utilization and lipid accumulation in *Saccharomyces cerevisiae*. Appl Microbiol Biotechnol 100(10):4549–4560

Karras C, Smedh M, Förster R, Deschout H, Fernandez-Rodriguez J, Heintzmann R (2019) Successful optimization of reconstruction parameters in structured illumination microscopy—a practical guide. Opt Commun 436:69–75

Klemm D, Heublein B, Fink H-P, Bohn A (2005) Cellulose: fascinating biopolymer and sustainable raw material. Angew Chem Int Ed 44(22):3358–3393

Kong Q-X, Gu J-G, Cao L-M, Zhang A-L, Chen X, Zhao X-M (2006) Improved production of ethanol by deleting *FPS1* and over-expressing *GLT1* in *Saccharomyces cerevisiae*. Biotechnol Lett 28(24):2033–2038

Kong Q-X, Cao L-M, Zhang A-L, Chen X (2007) Overexpressing *GLT1* in *gpd1*Δ mutant to improve the production of ethanol of *Saccharomyces cerevisiae*. Appl Microbiol Biotechnol 73(6):1382–1386

Krantz M, Ahmadpour D, Ottosson L-G, Warringer J, Waltermann C, Nordlander B, Klipp E, Blomberg A, Hohmann S, Kitano H (2009) Robustness and fragility in the yeast high osmolarity glycerol (HOG) signal-transduction pathway. Mol Syst Biol 5(1):281

Lee WJ, Kim MD, Ryu YW, Bisson LF, Seo JH (2002) Kinetic studies on glucose and xylose transport in *Saccharomyces cerevisiae*. Appl Microbiol Biotechnol 60(1-2):186–191

Liu K, Yuan X, Liang L, Fang J, Chen Y, He W, Xue T (2019) Using CRISPR/Cas9 for multiplex genome engineering to optimize the ethanol metabolic pathway in *Saccharomyces cerevisiae*. Biochem Eng J 145:120–126

Ma M, Liu ZL (2010) Mechanisms of ethanol tolerance in *Saccharomyces cerevisiae*. Appl Microbiol Biotechnol 87(3):829–845

Maestre O, García-Martínez T, Peinado RA, Mauricio JC (2008) Effects of *ADH2* overexpression in *Saccharomyces bayanus* during alcoholic fermentation. Appl Environ Microbiol 74(3):702–707

Majidian P, Tabatabaei M, Zeinolabedini M, Naghshbandi MP, Chisti Y (2018) Metabolic engineering of microorganisms for biofuel production. Renew Sust Energ Rev 82:3863–3885

Michnick S, Roustan JL, Remize F, Barre P, Dequin S (1997) Modulation of glycerol and ethanol yields during alcoholic fermentation in *Saccharomyces cerevisiae* strains overexpressed or disrupted for *GPD1* encoding glycerol 3-phosphate dehydrogenase. Yeast 13(9):783–793

Mielenz JR (2001) Ethanol production from biomass: technology and commercialization status. Curr Opin Microbiol 4(3):324–329

Mussatto SI, Dragone G, Guimarães PM, Silva JP, Carneiro LM, Roberto IC, Vicente A, Domingues L, Teixeira JA (2010) Technological trends, global market, and challenges of bio-ethanol production. Biotechnol Adv 28(6):817–830

Nissen TL, Schulze U, Nielsen J, Villadsen J (1997) Flux distributions in anaerobic, glucose-limited continuous cultures of *Saccharomyces Cerevisiae*. Microbiology 143(1):203–218

Nissen TL, Kielland-Brandt MC, Nielsen J, Villadsen J (2000) Optimization of ethanol production in *Saccharomyces cerevisiae* by metabolic engineering of the ammonium assimilation. Metab Eng 2(1):69–77

Peeters K, Van Leemputte F, Fischer B, Bonini BM, Quezada H, Tsytlonok M, Haesen D, Vanthienen W, Bernardes N, Gonzalez-Blas CB, Janssens V, Tompa P, Versées W, Thevelein JM (2017) Fructose-1,6-bisphosphate couples glycolytic flux to activation of Ras. Nat Commun 8(1):922

Pfeiffer T, Morley A (2014) An evolutionary perspective on the Crabtree effect. Front Mol Biosci 1:17

Rintala E, Wiebe MG, Tamminen A, Ruohonen L, Penttilä M (2008) Transcription of hexose transporters of *Saccharomyces cerevisiae* is affected by change in oxygen provision. BMC Microbiol 8:53

Saha BC (2003) Hemicellulose bioconversion. J Ind Microbiol Biotechnol 30(5):279–291

Songdech P, Ruchala J, Semkiv MV, Jensen LT, Sibirny A, Ratanakhanokchai K, Soontorngun N (2020) Overexpression of transcription factor *ZNF1* of glycolysis improves bioethanol productivity under high glucose concentration and enhances acetic acid tolerance of *Saccharomyces cerevisiae*. Biotechnol J 15(7):e1900492

Songdech P, Butkinaree C, Yingchutrakul Y, Promdonkoy P, Runguphan W, Soontorngun N (2024) Increased production of isobutanol from xylose through metabolic engineering of *Saccharomyces cerevisiae* overexpressing transcription factor Znf1 and exogenous genes. FEMS Yeast Res 24

Tokuhiro K, Ishida N, Nagamori E, Saitoh S, Onishi T, Kondo A, Takahashi H (2009) Double mutation of the *PDC1* and *ADH1* genes improves lactate production in the yeast *Saccharomyces cerevisiae* expressing the bovine lactate dehydrogenase gene. Appl Microbiol Biotechnol 82(5):883–890

Valadi H, Larsson C, Gustafsson L (1998) Improved ethanol production by glycerol-3-phosphate dehydrogenase mutants of *Saccharomyces cerevisiae*. Appl Microbiol Biotechnol 50(4):434–439

van Dijken JP, Scheffers WA (1986) Redox balances in the metabolism of sugars by yeasts. FEMS Microbiol Lett 32(3):199–224

Van Maris A, Winkler A, Kuyper M, de laat, W., Dijken, J., Pronk, J. (2007) Development of Efficient Xylose Fermentation in *Saccharomyces cerevisiae*: Xylose Isomerase as a Key Component. Adv Biochem Eng Biotechnol 108:179–204

van Zyl W, Lynd L, Den Haan R, McBride J (2007) Consolidated Bioprocessing for Bioethanol Production Using *Saccharomyces cerevisiae*. Adv Biochem Eng Biotechnol 108:205–235

Verstrepen KJ, Iserentant D, Malcorps P, Derdelinckx G, Van Dijck P, Winderickx J, Pretorius IS, Thevelein JM, Delvaux FR (2004) Glucose and sucrose: hazardous fast-food for industrial yeast? Trends Biotechnol 22(10):531–537

Yu KO, Kim SW, Han SO (2010) Engineering of glycerol utilization pathway for ethanol production by *Saccharomyces cerevisiae*. Bioresour Technol 101(11):4157–4161

Yu KO, Jung J, Ramzi AB, Choe SH, Kim SW, Park C, Han SO (2012a) Increased ethanol production from glycerol by *Saccharomyces cerevisiae* strains with enhanced stress tolerance from the overexpression of SAGA complex components. Enzym Microb Technol 51(4):237–243

Yu KO, Jung J, Ramzi AB, Kim SW, Park C, Han SO (2012b) Improvement of ethanol yield from glycerol via conversion of pyruvate to ethanol in metabolically engineered *Saccharomyces cerevisiae*. Appl Biochem Biotechnol 166(4):856–865

Zaman S, Lippman SI, Zhao X, Broach JR (2008) How Saccharomyces responds to nutrients. Annu Rev Genet 42:27–81

Zhang L, Tang Y, Guo Z, Shi G (2013) Engineering of the glycerol decomposition pathway and cofactor regulation in an industrial yeast improves ethanol production. J Ind Microbiol Biotechnol 40(10):1153–1160

Chapter 4
Glycerol Production and Conversion in Yeasts and Biotechnological Significance of These Processes

Kostyantyn Dmytruk, Marta Semkiv, and Andriy A. Sibirny

Contents

4.1 Introduction 92
4.2 Glycerol Metabolism in Yeast Cells 92
 4.2.1 Glycerol Synthesis 92
 4.2.2 Glycerol Utilization 95
 4.2.3 Glycerol Transport Across Plasma Membrane 98
4.3 Glycerol Role in Yeast Cells 101
 4.3.1 Osmoregulation 101
 4.3.2 Redox Sink During *S. cerevisiae* Growth Under Anaerobic Conditions 104
4.4 Methods of Glycerol Production 105
4.5 Glycerol Production by the Yeast *S. cerevisiae* 107
4.6 Glycerol Production by Osmotolerant Yeasts 109
4.7 Glycerol Application in the Industry 110
4.8 Glycerol Bioconversion to High-Value Chemicals 110
References 113

Abstract Glycerol has many applications in cosmetic, paint, automotive, food, and pharmaceutical industries, as well as in the production of explosives. Currently, crude glycerol is commercially available as a by-product of biodiesel production;

K. Dmytruk (✉)
Department of Molecular Genetics and Biotechnology, Institute of Cell Biology, NAS of Ukraine, Lviv, Ukraine

M. Semkiv
Department of Physiology and Cell Biology, Leibniz Institute of Plant Genetics and Crop Plant Research (IPK), Gatersleben, Germany

A. A. Sibirny
Institute of Biotechnology, University of Rzeszow, Rzeszow, Poland

A. A. Sibirny (ed.), *Biotechnology of Yeasts and Filamentous Fungi*, Grand Challenges in Biology and Biotechnology,
https://doi.org/10.1007/978-3-031-74726-7_4

however, its costs of purification are prohibitive. This chapter contains description of industrial production of glycerol through glucose fermentation by *Candida* sp. and *Saccharomyces cerevisiae*. Besides, it covers glycerol metabolism, its role in yeast cells, and the bioconversion of glycerol by-products to various useful compounds.

4.1 Introduction

Glycerol, also known as propane-1,2,3-triol, glycerine, or glycerin, is a simple polyol compound. It is a colorless, odorless, viscous, sweet, and nontoxic liquid. Glycerol is a constituent of neutral fats (triacylglycerides) and is widely used in the food industry as a sweetener and humectant. It is also employed in the pharmaceutical industry and in the production of cosmetics, toothpaste, paints, drugs, paper, textiles, leather, and explosives (Chilakamarry et al. 2021). Glycerol was first isolated in 1779 by Swiss pharmacist K. W. Scheele, who heated a mixture of litharge (PbO) with olive oil (AlMarzouq and Elnagdi 2019). It can be produced chemically from oils (triacylglycerides) or from propylene, as well as by microbial synthesis, with yeast organisms being promising producers. In recent years, a significant amount of by-product glycerol from the biodiesel industry has entered the market. This glycerol is contaminated with methanol, salts, and other impurities, making its purification quite expensive. This review describes the pathways of glycerol biosynthesis and catabolism in different yeast species, the construction of yeast strains capable of aerobic and anaerobic glycerol overproduction, and strains that effectively convert by-product glycerol into valuable compounds.

4.2 Glycerol Metabolism in Yeast Cells

Glycerol is crucial for living cells as it forms the backbone of phospholipids (common components of cell membranes) and triacylglycerols (widespread storage lipids). Additionally, the free form of glycerol is both utilized and produced by many organisms, particularly yeasts. Glycerol metabolism is most comprehensively described for laboratory strains of the conventional yeast *Saccharomyces cerevisiae* (Fig. 4.1). For other yeasts, the available information is rather fragmentary.

4.2.1 Glycerol Synthesis

In the yeast *S. cerevisiae*, glycerol is synthesized from the glycolysis intermediate dihydroxyacetone phosphate (DHAP) in two consecutive reactions catalyzed by the cytosolic enzymes glycerol-3-phosphate dehydrogenase and glycerol-3-phosphate

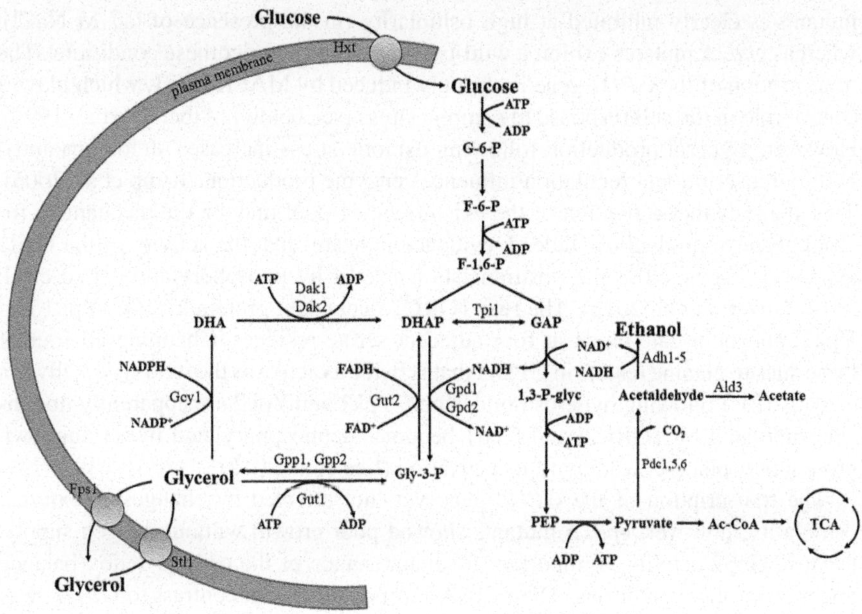

Fig. 4.1 Glycerol metabolism in *S. cerevisiae*. Abbreviations: G-6-P, glucose-6-phosphate; F-6-P, fructose-6-phosphate; F-1,6-P, fructose-1,6-biphosphate; DHAP, dihydroxyacetone phosphate; DHA, dihydroxyacetone; GAP, glyceraldehyde-3-phosphate; Gly-3-P, glycerol-3-phosphate; 1,3-P-glyc, 1,3-phosphoglycerate; PEP, phosphoenolpyruvate; Ac-CoA, acetyl coenzyme A; TCA, tricarboxylic acid cycle; Tpi1, triose phosphate isomerase; Adh1-5, alcohol dehydrogenases; Pdc1,5,6, pyruvate decarboxylases; Ald3, aldehyde dehydrogenase; Gpd1, Gpd2, cytosolic glycerol-3-phosphate dehydrogenases; Gpp1, Gpp2, glycerol-3-phosphate phosphatases; Gut1, glycerol kinase; Gut2, glycerol-3-phosphate dehydrogenase; Gcy1, glycerol dehydrogenase; Dak1, Dak2, dihydroxyacetone kinase; Hxt, Stl1, Fps1, membrane transporters

phosphatase. DHAP originates from the splitting of fructose1,6-bisphosphate into DHAP and glyceraldehyde3-phosphate (GAP). These two compounds serve as starting points for two metabolic pathways: GAP leads to ethanol, while DHAP leads to glycerol. DHAP can be converted into GAP and vice versa by the action of the enzyme triose phosphate isomerase. Cytosolic NAD^+-dependent glycerol-3-phosphate dehydrogenase (GPDH) catalyzes the reduction of DHAP to glycerol-3-phosphate with the simultaneous oxidation of NADH to NAD^+. This is the rate-limiting step in glycerol production (Remize et al. 2001). In *S. cerevisiae*, this enzyme is encoded by two isogenes, *GPD1* and *GPD2*. The *GPD1* gene was first discovered by Larsson et al. (1993) and Albertyn et al. (1994), while the *GPD2* gene was discovered later by Eriksson et al. (1995) and Ansell et al. (1997). The amino acid sequences of Gpd1 and Gpd2 are 69% identical (Eriksson et al. 1995), and the kinetic properties of the two isoenzymes are similar. Despite the similar physical and catalytic properties of their gene products, the *GPD1* and *GPD2* genes are differentially regulated at the transcriptional level. Additionally, mutant phenotypes indicate that Gpd1 and Gpd2 fulfill different roles in metabolism. Growth of *gpd1Δ* mutants is clearly inhibited at high osmolarity (in the presence of 0.7 M NaCl), whereas *gpd2Δ* mutants exhibit a wild-type growth rate under these conditions. The transcription of the *GPD1* gene is strongly induced by MAPK Hog1, which plays a central role in the cell response to osmotic stress (see below) (Albertyn et al. 1994). However, glycerol production following osmotic stress increases almost instantly, before transcriptional regulation influences enzyme production (Klipp et al. 2005). This suggests that activation of the existing Gpd1 pool may be the mechanism for rapid glycerol production. Indeed, evidence indicates that the activity of the Gpd1 enzyme is regulated at the posttranslational level by phosphorylation (Lee et al. 2012; Oliveira et al. 2012). The two TORC2-dependent protein kinases, Ypk1 and Ypk2, phosphorylate Gpd1 at four adjacent serine residues. Changing all four of these sites to alanine results in GPDH that is twice as active as the wild type (Oliveira et al. 2012). Following hyperosmotic shock, Ypk1 and Ypk2 are apparently no longer stimulated by TORC2, and Gpd1 becomes dephosphorylated by an unknown protein phosphatase, leading to its activation (Lee et al. 2012).

The transcription of the *GPD2* gene was not affected by changes in external osmolarity; however, *gpd2Δ* mutants showed poor growth without oxygen supply, because Gpd2 activity is important for maintenance of the cellular redox balance under anaerobic conditions. The mRNA level of *GPD2*, in contrast to *GPD1*, was ninefold upregulated after a shift of *S. cerevisiae* from aerobic to anaerobic conditions (Ansell et al. 1997). During prolonged anaerobic growth, expression of *GPD2* decreased, and in steady-state anaerobic chemostat cultures, the mRNA concentration stabilized at 1.5 times the aerobic level (ter Linde et al. 1999). A *gpd1Δ gpd2Δ* double mutant is totally unable to produce glycerol and cannot grow at all under anaerobic conditions, unless an NADH-oxidizing agent, such as acetaldehyde or acetoin, is added (Bjorkqvist et al. 1997). At high osmolarity, the double mutant does not grow either (Ansell et al. 1997).

Genes encoding NAD^+-dependent GPDH have also been characterized in several other yeast species and filamentous fungi, such as *Candida magnoliae* (Lee et al. 2008), *Candida glycerinogenes* (Chen et al. 2008), *Debaryomyces hansenii* (Thome 2004), and *Aspergillus nidulans* (Fillinger et al. 2001). In addition, $NADP^+$-dependent GPDH activity has been discovered in certain yeast species, such as *Candida versatilis* (Watanabe et al. 2008).

The second step in the glycerol synthesis is the irreversible hydrolysis of the phosphate group from glycerol-3-phosphate, a reaction catalyzed by glycerol-3-phosphatase (G3P) (Norbeck et al. 1996). Similarly, as for the dehydrogenase, two structural genes encoding isoenzymes of the phosphatase have been identified: *GPP1* and *GPP2*. The sequences of these corresponding proteins are 95% identical (Norbeck et al. 1996). Both isoenzymes have a high specificity for DL-glycerol-3-phosphate, and a pH optimum is 6.5. *GPP1* and *GPP2* genes have differential expression regulation which closely resembles that of the *GPD* genes. Mutants lacking Gpp1 activity showed poor growth under anaerobic conditions, and *GPP1* expression seemed to be induced upon switch to anaerobic conditions (Pahlman et al. 2001). The expression of *GPP2* is strongly activated through the High Osmolarity Glycerol (HOG) pathway in the presence of hyperosmotic stress

(Norbeck et al. 1996; Pahlman et al. 2001). Single mutant *gpp1Δ* or *gpp2Δ* did not reveal growth impairment during osmotic stress, indicating that the two paralogs can substitute well for each other (Pahlman et al. 2001). However, the deletion of both genes in the *gpp1Δ gpp2Δ* mutant caused a growth inhibition under anaerobic conditions and increased sensitivity to osmotic stress. The *gpp1Δ gpp2Δ* mutant accumulated increased amounts of glycerol-3-phosphate and produced only minor amounts of glycerol (Pahlman et al. 2001).

In filamentous fungi and fission yeast, another possible pathway for glycerol synthesis through dihydroxyacetone (DHA) has been proposed. The first step in this pathway is assumed to be catalyzed by a so far uncharacterized enzyme that dephosphorylates DHAP to DHA. The second step is catalyzed by an $NADP^+$-dependent glycerol dehydrogenase. This pathway has been suggested based on the fact that $NADP^+$-dependent glycerol dehydrogenase activities converting DHA to glycerol have been identified in fission yeast (Marshall et al. 1989) and several fungi from *Aspergillus* genus (Redkar et al. 1995; Schuurink et al. 1990; Ruijter et al. 2004). More precisely, this pathway was studied on *A. nidulans* (Fillinger et al. 2001; de Vries et al. 2003). Mutants with disruption of gene *gldB* encoding $NADP^+$-dependent glycerol dehydrogenase accumulated significantly reduced glycerol amounts and showed a severe growth defect under hyperosmotic conditions (de Vries et al. 2003), while the deletion of *gfdA*-encoding cytosolic GPDH did not abolish glycerol biosynthesis (Fillinger et al. 2001). Expression of *A. nidulans* paralog *gldB* gene was strongly induced under conditions of hyperosmotic shock. These studies suggested that the glycerol formation through DHA is the major pathway responsible for glycerol accumulation under osmotic stress in *A. nidulans*, whereas the pathway through glycerol-3-phosphate did not play a significant role. However, in order to clearly establish this alternative pathway for glycerol production, a "DHAPase" activity needs to be allocated to certain enzyme, which hasn't been explicitly done yet.

4.2.2 Glycerol Utilization

As glycerol is relatively abundant in nature, it is not surprising that many microorganisms can use it as a source of carbon and energy. Similar to glycerol production, glycerol utilization can occur through glycerol-3-phosphate or dihydroxyacetone. The resulting dihydroxyacetone phosphate can either enter central catabolism by being transformed into glyceraldehyde3-phosphate via triose phosphate isomerase, or it can serve as a substrate for lipid synthesis. The pathway through glycerol-3-phosphate involves a glycerol kinase and an FAD-dependent glycerol 3-phosphate dehydrogenase located at the outer surface of the inner mitochondrial membrane. The latter enzyme directly transfers the electrons via $FADH_2$ to the respiratory chain. In *S. cerevisiae*, this pathway represents main if not explicit route of glycerol utilization. In *S. cerevisiae*, glycerol kinase is encoded by gene *GUT1* (Sprague and Cronan 1977; Pavlik et al. 1993), and glycerol-3-phosphate dehydrogenase is encoded by gene *GUT2* (Ronnow and Kielland-Brandt 1993). *S. cerevisiae* mutants defective in *GUT1* or *GUT2* are unable to utilize glycerol as a sole carbon source (Sprague and Cronan 1977). Mitochondrial glycerol-3-phosphate dehydrogenase Gut2 together with the abovementioned cytosolic glycerol-3-phosphate dehydrogenases Gpd1 and Gpd2 contributes to the "glycerol-3-phosphate shuttle" of yeast, which serves for conversion of mitochondrial NADH into NAD^+ during anaerobic growth of *S. cerevisiae* (Larsson et al. 1998). The route of glycerol dissimilation through glycerol-3-phosphate intermediate has also been observed in a number of other yeasts, e.g., *D. hansenii* (Adler et al. 1985), *Zygosaccharomyces rouxii* (van Zyl et al. 1991), and *C. glycerinogenes* (Wang et al. 2000). A number of yeasts are thought to dissimilate glycerol via the second pathway with dihydroxyacetone as an intermediate. The first step in the pathway is the oxidation of glycerol to dihydroxyacetone by glycerol dehydrogenase followed by the phosphorylation of dihydroxyacetone by the dihydroxyacetone kinase. Mentioned enzymes are encoded by genes *GCY1*, *DAK1*, and *DAK2*. In *S. cerevisiae*, the existence of this pathway was proposed by Norbeck and Blomberg (1997). Jung and coworkers recently confirmed that Gcy1 is the $NADP^+$-dependent glycerol dehydrogenase (Jung et al. 2012). The function of this pathway in *S. cerevisiae* is unclear and most probably does not involve utilization of glycerol as a carbon source, because it cannot substitute nonfunctioning glycerol-3-phosphate catabolic route in *gut1Δ* or *gut2Δ* mutants. The dihydroxyacetone pathway in this yeast might play a role in the regulation of the glycerol concentration during hyperosmotic stress (Blomberg 2000) or redox regulation (Costenoble et al. 2000). However, in other yeasts, this pathway may be more significant. Overall, information about the role of one or another route for glycerol utilization in certain yeast species is ambiguous. Based on the presence of glycerol kinase or NAD^+-dependent glycerol dehydrogenase activity in cell-free extracts, Tani and Yamada (1987) divided the yeast species investigated into three groups: (1) group that was assumed to use exclusively the glycerol-3-phosphate pathway (e.g., *Candida boidinii*), (2) group that use only the dihydroxyacetone pathway (e.g., *Hansenula ofunaensis*), and (3) group where both pathways were considered to be functional (e.g., *Candida valida*). However, evidence for the presence of activities of the enzymes attributive to any of the specific pathway are not sufficient to assess the functionality of a metabolic pathway. More explicit evidence can represent, for example, the fact that deletion of the gene that encodes key enzyme of particular catabolic pathway leads to the reduction or total abolishment of yeast growth on glycerol, as it was shown for *gut1Δ* or *gut2Δ* mutants of *S. cerevisiae* (Sprague and Cronan 1977). In this way, strong evidence for the dihydroxyacetone pathway for glycerol utilization was provided in the yeast *Schizosaccharomyces pombe* (Matsuzawa et al. 2010). *S. pombe* strain with deletion of the gene *gld1* possessed reduced NAD^+-dependent glycerol dehydrogenase activity and was not able to grow in media containing glycerol as the sole carbon source.

Different yeast species and even different strains from the same species show high diversity with regard to the ability for growth on glycerol as the sole source of carbon (Kurtzman et al. 2011). Most strains of conventional yeast *S. cerevisiae* have shown low growth performance on the medium with glycerol as a sole carbon source, but their growth can be improved by the addition of amino acids, nucleobases, or complex supplements to the cultivation medium (Merico et al. 2011; Swinnen et al. 2013). Among 52 *S. cerevisiae* strains, the strain with the fastest biomass accumulation on glycerol was chosen and compared to common laboratory strain in respect of its genetic features (Swinnen et al. 2013). It turned out that better biomass accumulation was caused by three genetic determinants—specific alleles of genes *UBR2*, *SSK1*, and *GUT1* (Swinnen et al. 2016). Whether the role of the last gene for glycerol utilization is obvious, the role of genes *UBR2* and *SSK1* is uncertain, and they encode, respectively, an ubiquitin ligase (Finley et al. 2012) and a component of HOG pathway (Maeda et al. 1994). Two independent studies have reported that wild-type *S. cerevisiae* strains not able to grow in synthetic glycerol medium can be relatively easily evolved by repeated batch cultivations in glycerol medium to reach better growth levels in these conditions (Ochoa-Estopier et al. 2011; Merico et al. 2011). These results may imply that *S. cerevisiae* ancestors were able to utilize glycerol more efficiently, but this capability was lost during evolution. This "loss-of-function" likely occurred because *S. cerevisiae* generally employs a different competitive strategy: it aims to outgrow other microorganisms by rapidly assimilating sugar and producing ethanol, a process driven by the strong Crabtree effect. In this way, potential competitors for the consumption of the fermentation products ethanol and glycerol during the post-diauxic growth phase have already been outcompeted during the sugar-consuming phase.

Heterologous expression of Fps1 homologs from superior glycerol-utilizing yeast species, such as *Pachysolen tannophilus, Komagataella (Pichia) pastoris, Yarrowia lipolytica*, and *Cyberlindnera jadinii*, significantly improves the growth performance on glycerol in the selected glycerol-consuming *S. cerevisiae* wild-type strain (Klein et al. 2016). Heterologous overexpression of the glycerol transporter *STL1* from *Wickerhamomyces anomalus* under the control of the *TEF1* promoter in an *stl1Δ S. cerevisiae* background resulted in a tenfold increase in biomass accumulation on glycerol-containing media compared to the wild type (Da Cunha et al. 2019).

Adaptive laboratory evolution of *S. cerevisiae* toward improved growth on glycerol, followed by transcriptome analysis, revealed that upregulation of genes related to the tricarboxylic acid (TCA) cycle and oxidative phosphorylation contributed to the increased specific growth rate on glycerol. Conversely, genes related to the pentose phosphate pathway were downregulated. Based on these findings, metabolic engineering targets for improving glycerol assimilation were identified. Overexpression of *HAP4* (a transcription factor involved in upregulating TCA cycle genes) and *STL1* (a glycerol/H+-symporter), along with the deletion of *RIM15* (a protein kinase), enhanced the growth of *S. cerevisiae* on glycerol (Kawai et al. 2019). Strains of *S. cerevisiae* with significantly increased glycerol utilization obtained through adaptive evolution possessed a set of hypermutated genes. Transcriptomic, proteomic, and metabolomic analyses of the re-engineered strains confirmed the causality of these mutations. Overexpression of the mutated genes *GUT1*, *KGD1* (α-ketoglutarate dehydrogenase), and *UBC13* (E2 ubiquitin-conjugating enzyme) resulted in improved glycerol utilization in industrial strains, comparable to that of the evolved strain (Strucko et al. 2018).

Some other yeast species exhibit a much better ability to grow on glycerol than *S. cerevisiae*. For example, when the ability of 42 different yeast species to grow on glycerol was tested, *Pichia jadinii* and *Pichia anomala* showed the highest growth rates, approximately three times higher than *S. cerevisiae* (Lages et al. 1999). Current industrial applications employing glycerol-based cultivation media use such yeasts as *Y. lipolytica, K. pastoris,* and *P. tannophilus*. When their growth performance on glycerol was tested, these species showed 1.3–3.3 times higher growth rates than the best glycerol-utilizing *S. cerevisiae* strain (Klein et al. 2016). Overexpression of *GUT1* under the control of the strong constitutive *TEF1* promoter in *Y. lipolytica* led to a 20% increase in glycerol utilization, whereas overexpression of *GUT2* had no effect (Carly et al. 2017). Overexpression of *GUT1* in *Y. lipolytica* resulted in a 13.5% increase in glycerol uptake compared to the parental strain (Ong et al. 2020). Overexpression of sugar alcohol phosphatase (PYP) from *S. cerevisiae*, along with the native glycerol kinase (GK) and transketolase (TKL), in *Y. lipolytica* increased the maximum glycerol utilization rate of the strain from 0.3 to 1.0 g/L/h (Jagtap et al. 2021).

4.2.3 Glycerol Transport Across Plasma Membrane

In order to be used as a carbon source, glycerol must be transported into the cell. On the other hand, excessive amounts of glycerol, e.g., produced for NAD^+ reoxidation (see below), must be exported from the cell. The molecular details of the glycerol flux through a plasma membrane have been best studied in *S. cerevisiae*; for other yeast and filamentous fungi, the available information on this matter is fragmentary. In *S. cerevisiae*, the process of identification of the glycerol transport mechanisms wasn't straightforward as all possible hypotheses have been considered: passive and facilitated diffusion, active transport.

For example, previously, glycerol was generally regarded as a molecule that is able to permeate a plasma membrane by passive diffusion driven by its concentration gradient (Gancedo et al. 1968; Heredia et al. 1968; Romano 1986). However, Oliveira et al. (2003) based on comparison of the glycerol olive oil/water partition coefficient with that of other molecules speculated that glycerol is unlikely to show significant diffusion through biological membranes, although they indicated that some passive diffusion via the lipid bilayer cannot be completely excluded. Since plasma membrane is probably almost impermeable to glycerol, then any detectable passive glycerol diffusion into the cell must be mediated via the channel-like facilitator protein. Several experimental studies demonstrated a saturable glycerol uptake component in *S. cerevisiae*, which was believed to be connected with the protein Fps1, a member of the major intrinsic protein (MIP) family of channel proteins (Luyten et al. 1995; Sutherland et al. 1997). This protein shows high sequence similarity to the glycerol facilitator GlpF from *E. coli* (Heller et al. 1980), the only transporter responsible for glycerol uptake in this bacterium. Subsequent experiments showed, however, that previously detected saturable glycerol uptake kinetics was an artifact caused by considerable Gut1 activity that interfered with the experimental set-up (Oliveira et al. 2003). Deletion of *GUT1* gene eliminated such effect. It has also been shown that *S. cerevisiae fps1Δ* mutants were able to grow on glycerol as sole carbon source (Luyten et al. 1995). Further studies confirmed that Fps1 in *S. cerevisiae* is involved rather in glycerol export during osmoregulation than in glycerol uptake (Tamas et al. 1999).

The existence of a glycerol active transport system in *S. cerevisiae* was predicted based on the observation of simultaneous uptake of glycerol and protons accompanied by intracellular glycerol accumulation (Sutherland et al. 1997). *GUP1* and its paralog *GUP2* had been suggested to encode proteins responsible for glycerol active transport (Holst et al. 2000). These authors showed that the Gup1 protein is essential for the alleviation of the osmotic stress-induced cell death of *gpd1Δ gpd2Δ* mutants: deletion of *GUP1* in a *gpd1Δ gpd2Δ* double deletion mutant suppresses the rescuing effect of glycerol supplementation during hyperosmotic shock (Holst et al. 2000). However, further studies undermined this theory. It has been shown that glycerol uptake activity was enhanced after switching to medium with glycerol or high salt concentrations, whereas *GUP1* and *GUP2* expression was found to be constitutive (Oliveira and Lucas 2004). Also, double deletion mutant *gup1Δ gup2Δ* still shows active glycerol uptake under salt stress (Neves et al. 2004). Therefore, proteins Gup1 and Gup2 probably are involved in glycerol uptake only indirectly.

Ferreira et al. (2005) revealed that a crucial role in glycerol uptake has glycerol/H^+-symporter Stl1. The deletion of *STL1* gene completely abolished active glycerol transport and prevented *S. cerevisiae* growth on glycerol as a sole carbon source. The pattern of *STL1* gene expression is consistent with aforementioned features of glycerol uptake; in particular, transcription of *STL1* is transiently induced after the shift to hyperosmotic medium (Posas et al. 2000; Ferreira et al. 2005) and then started to decrease gradually as glycerol production in stressed cells is activated. This downregulation does not occur in *gpd1Δ gpd2Δ* mutant, which is unable to synthesize intracellular glycerol (Ferreira et al. 2005). Similar H^+- but also Na^+-glycerol symporters have been described for halotolerant yeasts *D. hansenii* (Lucas et al. 1990), *Pichia sorbitophila* (Lages and Lucas 1995), and *Z. rouxii* (van Zyl et al. 1990).

As was mentioned before, *S. cerevisiae* shows relatively low growth rates on glycerol. Gancedo et al. (1968) speculated that glycerol uptake velocity might be a major rate-limiting step for growth on glycerol in this yeast. Homologs of *STL1* and *FPS1* genes have been identified for yeasts with the best growth performance on glycerol, i.e., *Y. lipolytica*, *P. tannophilus*, and *K. pastoris* (Dujon et al. 2004; Mattanovich et al. 2009; Liu et al. 2013). The contribution of Stl1 and Fps1 homologs to glycerol utilization in other yeast species is not yet clear. There are evidence that, in contrast to *S. cerevisiae*, in other yeasts Fps1 homologs might actually play role in glycerol uptake. For example, Liu et al. (2013) showed that expression of *FPS2* gene from *P. tannophilus* in *S. cerevisiae* was able to suppress the growth defect on glycerol of a *stl1Δ* deletion mutant, whereas overexpression of the homologous *S. cerevisiae FPS1* gene was unable to achieve this task. ScFps1 and PtFps2 orthologs are also different with regard to their protein structure. They both have core region with six putative trans-membrane domains (TMDs), but PtFps2 is much shorter as it does not possess long N- and C-terminal extensions. Overexpression in the *S. cerevisiae* strain of *PtFPS2* and similar *FPS1* homologs from different non-conventional yeast species (*C. jadinii*, *P. pastoris*, and *Y. lipolytica*) resulted in significant improvement of its glycerol growth rates. This high growth rates were retained even after *STL1* gene deletion (Klein et al. 2016).

In *S. cerevisiae*, glycerol export from the cells occurs through already mentioned channel protein Fps1 (Luyten et al. 1995). At high osmolarity, this glycerol efflux aquaglyceroporin is closed, and the produced glycerol is retained inside the cells (Luyten et al. 1995; Tamas et al. 1999). In silico and experimental data suggest that Fps1 closure is one of the first events following osmoshock (Petelenz-Kurdziel et al. 2013). After a shift from high to low osmotic strength, the Fps1 channel opens, and the cells release the accumulated glycerol within a few minutes. An obtained *fps1Δ* deletion mutant was extremely sensitive to such a shift from high to low osmolarity, indicating that Fps1 channel is important to protect the cells from a hypo-osmotic shock (Luyten et al. 1995; Tamas et al. 1999). In anaerobic cultures, the *FPS1* expression level is higher than under aerobic conditions, as it is needed to expel an excessive glycerol, produced during NADH reoxidation (ter Linde et al. 1999). Molecular structure of Fps1 has been extensively studied in order to understand the mechanisms of its opening and closure. Fps1 possesses long (230 and 250 amino acids, respectively) N- and C-terminal extensions of the core domain that have been shown to be required for Fps1 closure. Modified Fps1 with deletion of either domain becomes hyperactive, which leads to poor glycerol accumulation, constant excretion of glycerol to the medium and glycerol overproduction in an attempt by the cell to compensate for glycerol loss (Ahmadpour et al. 2014). This effect may also be caused by mutation making Thr231 in the long N-terminal extension of Fps1 unphosphorylatable as well as certain mutations in the well-conserved motif around this residue (Ahmadpour et al. 2014). Lee et al. (2013) demonstrate that the closure mechanism of Fps1 is triggered by the phosphorylation of Rgc2. In its unphosphorylated state, Rgc2 binds to the C-terminal extension of Fps1, keeping the channel open. The fast opening of Fps1 channel after hypo-osmotic shock was suggested to be affected by the Slt2 MAPK, which is stimulated by hypo-osmotic shock, cell wall stress, and morphogenic changes (Levin 2011; Baltanas et al. 2013). It is interesting that this elaborated regulation of Fps1 opening or closure is restricted to yeasts closely related to *S. cerevisiae* (Pettersson et al. 2005). Probably, it was evolved as a means for adaptation to rapid changes in the environmental osmotic pressure. Interestingly, some highly osmotolerant yeasts, such as *D. hansenii* and *Millerozyma farinosa* (formerly *P. sorbitophila*), do not possess Fps1 (Prista et al. 2005; Sabir et al. 2016). The absence of Fps1 in these yeasts, which prevents glycerol leakage, may be linked to their adaptation to naturally high osmolarity conditions.

4.3 Glycerol Role in Yeast Cells

4.3.1 Osmoregulation

Maintenance of the cellular water balance is fundamental for living organisms. Rapid changes in the extracellular osmolarity and water activity can be detrimental for the cells, for example, increased osmolarity of the environment causes water efflux and hence cell shrinkage. The accumulation of compatible solutes to compensate for water loss is a universal strategy for survival in these conditions (Yancey et al. 1982). As implied by their name, such molecules are compatible with intracellular processes and either replace water or/and revert the water concentration gradient and drive water back into cells. The most abundant compatible solutes in microorganisms are small uncharged molecules such as (1) polyols (glycerol, arabitol, trehalose or sucrose), (2) amino acids (proline, glutamate, or glutamine), and (3) ectoines, (ectoine or β-hydroxyectoine) (Grant 2004). In yeast and fungi, glycerol is the most prominent compatible solute to regulate the cell turgor at high extracellular osmolarity (Brown 1978; Blomberg and Adler 1992). For instance, during initial phase of wine fermentation, *S. cerevisiae* produces increased amounts of glycerol as a response to osmotic stress because of the high sugar concentration in the must. Also in *S. cerevisiae,* glycerol production increases in medium containing high concentrations of NaCl (Blomberg and Adler 1989). Moreover, a marked rise in the level of cytoplasmic glycerol-3-phosphate dehydrogenase activity has been observed in such conditions (Blomberg and Adler 1989; Andre et al. 1991). *S. cerevisiae* adaptation to hyperosmotic stress is accomplished via increased glycerol accumulation and its retention inside the cells. Mutants that block the synthesis of glycerol (e.g., *gpd1Δgpd2Δ*) or that cause leakage of glycerol out of the cell (e.g., with hyperactive Fps1) confer an osmo-sensitive phenotype (Hohmann 2002). Other yeasts might employ slightly different mechanisms to increase intracellular glycerol concentration. For example, xerotolerant yeasts such as *Z. rouxii* retains and accumulates glycerol against the concentration gradient (Edgley and Brown 1983; Lages et al. 1999).

In yeast, glycerol accumulation during osmo-adaptation is controlled mainly by the high-osmolarity glycerol (HOG) pathway (Fig. 4.2). Mutations that inactivate the HOG pathway render yeast cells highly sensitive to hyperosmotic stress while mutations causing constantly activated HOG pathway are lethal. The HOG pathway is a typical example of a mitogen-activated protein kinase (MAPK) pathway, regulated by a three-tiered cascade of kinases. Central in the HOG pathway is the MAPK, Hog1 (Brewster et al. 1993), which is activated by the MAP kinase kinase (MAPKK), Pbs2 (Brewster et al. 1993). The two independent upstream osmosensing pathways, the Sln1-branch and Sho1-branch, activate the MAPKK Pbs2 (Hohmann 2002; O'Rourke et al. 2002). These two branches of the HOG pathway appear to function independently and perceive different types of stimuli. After a hyperosmotic shock, Hog1 is activated by phosphorylation in Thr174 and Tyr176 positions. The period of Hog1 phosphorylation correlates with the degree of osmotic

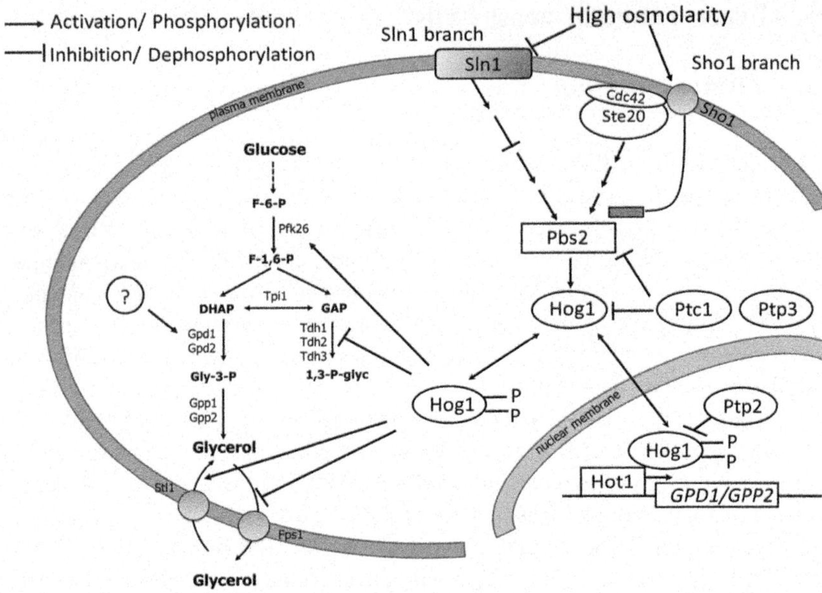

Fig. 4.2 High osmolarity glycerol (HOG) pathway in *Saccharomyces cerevisiae*. Hog1, MAPK; Pbs2, MAPKK; the Sln1-branch and Sho1-branch, the two independent upstream osmo-sensing pathways; Sln1, transmembrane protein; Sho1, transmembrane protein with C-terminal elongation, the SH3 domain, which anchors Pbs2 to the membrane; Ste20, component of the pheromone-response and filamentous MAPK pathways; Hot1, transcriptional activator; Ptp2, Ptp3 and Ptc1, phosphatases that inactivate Hog1; Gpd1, Gpd2, cytosolic glycerol-3-phosphate dehydrogenases; ?, unknown protein phosphatase that dephosphorylates and activates Gpd1; Gpp1, Gpp2, glycerol-3-phosphate phosphatases; Tdh1, Tdh2, Tdh3, glyceraldehyde 3-phosphate dehydrogenases; Pfk26, phosphofructo-2 kinase; Tpi1, triose phosphate isomerase; Stl1, Fps1, membrane glycerol transporters; *GPD1/GPP2*, genes encoding glycerol-3-phosphate dehydrogenase and glycerol-3-phosphate phosphatase; F-6-P, fructose-6-phosphate; F-1,6-P, fructose-1,6-biphosphate; DHAP, dihydroxyacetone phosphate; GAP, glyceraldehyde-3-phosphate; Gly-3-P, glycerol-3-phosphate; 1,3-P-glyc, 1,3-phosphoglycerate

stress and with the time that passes between the initial cell shrinkage and the onset of the volume recovery (Blomberg 2022). Mutants with decreased ability to accumulate glycerol show prolonged HOG pathway activation, and mutants with increased glycerol accumulation show shorter periods of HOG pathway activation (Klipp et al. 2005; Krantz et al. 2004), so pathway downregulation is clearly linked to glycerol accumulation. The inactivation of the HOG pathway occurs through dephosphorylation of Hog1 by the negative regulators Ptp2, Ptp3, and Ptc1 (Warmka et al. 2001; Wurgler-Murphy et al. 1997). The HOG pathway has been in details described in numerous reviews (Chen and Thorner 2007; Saito and Posas 2012; Blomberg 2022).

Activated through dual phosphorylation MAPK Hog1 interacts with cytosolic and nuclear targets to provide cellular adaptation to osmotic stress (Westfall et al.

2004). Westfall et al. (2008) showed that cells deficient in the nuclear import of Hog1 are still capable of adapting to hyperosmotic stress, highlighting the importance of cytosolic targets. This interaction of Hog1 with cytosolic targets prevents glycerol efflux and induces metabolic shift toward glycerol production at the expense of biomass formation (Petelenz-Kurdziel et al. 2013; Bouwman et al. 2011; Westfall et al. 2008).

Glycerol efflux is prevented by closure of Fps1 channel. Although there is clear evidence that Hog1 targets Fps1 to acquire tolerance against arsenite (Thorsen et al. 2006) or acetic acid (Mollapour and Piper 2007), during a hyperosmotic shock, Hog1 appears to regulate Fps1 indirectly through phosphorylation and inhibition of the Fps1-activating proteins Rgc1 and Rgc2 (Beese et al. 2009).

Hog1 also stimulates the increase of glycolytic flux which may help compensating for the flux that is redirected toward glycerol (Petelenz-Kurdziel et al. 2013). It is achieved via upregulation of the key glycolytic enzyme, phosphofructokinase. This enzyme is allosterically controlled by AMP and fructose2,6-bisphosphate, which both increase phosphofructokinase activity. The latter, fructose2,6-bisphosphate is the product of the phosphorylation of fructose6-phosphate by 6-phosphofructo-2-kinase, which is encoded by *PFK26* and *PFK27*. Dihazi et al. (2004) showed that Hog1 activates Pfk26 which in turn leads to higher fructose2,6-bisphosphate production and consequently phosphofructokinase activation. Westfall et al. (2008) postulates that Hog1 triggers decrease in activity of Tdh1 and Tdh3 that catalyze conversion of glyceraldehyde-3-phosphate into 1,3-bisphosphoglycerate. The decrease of these enzymes' activities partially redirects the glycolytic flux to DHAP, the first intermediate of glycerol synthesis.

Rapid activation of glycerol production under hyperosmotic conditions is also caused by increased activity of Gpd1. Gpd1 is activated via dephosphorylation by an unknown protein phosphatase, in Hog1-independent manner (Lee et al. 2012). Also, hyperosmotic stress causes the changes in redox status due to liberation of protein-associated NADH. This, in turn, may trigger shifts in the cellular distribution of Gpd1 and hence its activity (Jung et al. 2010).

In the nucleus, Hog1 stimulates initiation of transcription by interacting with DNA-binding proteins, such as the activators, Hot1 (Alepuz et al. 2003; Rep et al. 2000) and Smp1 (de Nadal et al. 2003), or the repressor Sko1 (Proft et al. 2001; Proft and Struhl 2002), and it is also involved in subsequent steps of transcription (de Nadal et al. 2011). The gene expression changes lead to increased amounts of glycerol biosynthetic enzymes as well as increased capacity for glycerol uptake, especially in the absence of glucose as a carbon source. Hog1 directly or indirectly stimulates the expression of more than 100 genes (O'Rourke et al. 2002), among them are the genes *GPD1* and *GPP2*, to a lesser extent also *GPP1*, as well as *STL1*. Activation of genes *GPD1* and *GPP2* increases cells' capacity to produce glycerol, which may play a role at later stages of the adaptation to hyperosmotic conditions, especially to sustain growth in adapted cells. For a more detailed review on osmoregulation in yeast cells, the reader is referred to Blomberg (2022).

4.3.2 *Redox Sink During* S. cerevisiae *Growth Under Anaerobic Conditions*

The coenzyme nicotinamide adenine dinucleotide (NAD^+) is an important electron transport carrier. Only catalytic amounts of this pyridine nucleotide are present in yeast, and thus it must be continuously cycled between its oxidative and reduced form. The total ratio of NAD^+:NADH is approximately 3 in yeast (Anderson et al. 2002), including the free and bound forms of the coenzyme. A stable intracellular redox state is achieved by accurate balancing of the NAD^+ reduction and NADH oxidation.

In the glucose medium, cytosolic reduction of NAD^+ is mainly catalyzed by the glyceraldehyde3-phosphate dehydrogenase. The mitochondrial NADH originates from the reactions of the TCA cycle. Also, biosynthetic reactions during yeast growth led to a net NADH production. The generated NADH is re-oxidized in *S. cerevisiae* via five main mechanisms, which are (1) ethanol production by the alcohol dehydrogenase, Adh; (2) glycerol production by cytosolic glycerol3-phosphate dehydrogenase, Gpd1/2; (3) respiration of cytosolic NADH via the external mitochondrial NADH dehydrogenases, Nde1/2; (4) respiration of cytosolic NADH via the glycerol3-phosphate shuttle; and (5) oxidation of intra-mitochondrial NADH via a mitochondrial "internal" NADH dehydrogenase, Ndi1 (Bakker et al. 2001).

Assimilation of glucose into biomass involves the formation of "excess" NADH. Most of the NADH generated in biosynthetic reactions originates from the synthesis of amino acids in the cytosol and the mitochondrial matrix (Albers et al. 1998). As redox-neutral ethanol production cannot serve as a redox sink for excess NADH, it has to be re-oxidized by other means. During growth in the presence of oxygen, it can be achieved by the respiratory chain. The external NADH dehydrogenase (Nde) and internal NADH dehydrogenase (Ndi), both localized in the mitochondrial inner membrane, directly transfer the reducing equivalents of their compartment to the respiratory chain. However, under anaerobic conditions, respiration is not possible. So, under these conditions, *S. cerevisiae* re-oxidizes NADH in glycerol-3-phosphate dehydrogenase reaction. Genes *GPD2* and *GPP1* are upregulated after shift to anaerobic conditions, which leads to increased glycerol production. Deletion of these genes results in a strain unable to grow in the absence of oxygen (Ansell et al. 1997; Bjorkqvist et al. 1997). To ensure constant activity of this route for NADH reoxidation, synthesized glycerol is exported from the cells via Fps1 channel.

Since the cytosolic glycerol formation is the exclusive route for NADH reoxidation under anaerobic conditions, NADH that is produced in the mitochondrial matrix during amino acid synthesis has to be somehow transported to the cytosol. However, the mitochondrial inner membrane is virtually impermeable to pyridine nucleotide coenzymes (Von Jagow and Klingenberg 1970). That is why anaerobic reoxidation of intra-mitochondrial NADH requires a shuttle mechanism that exports redox equivalents to the cytosol. Bakker et al. (2001) proposed a possible involvement of the ethanol-acetaldehyde shuttle, in the export of intra-mitochondrial NADH. Another existing glycerol-3-phosphate shuttle serves for the transfer of redox equivalents from cytosolic NADH to the respiratory chain under aerobic conditions. This shuttle consists of two components: cytosolic NAD^+-linked glycerol-3-phosphate dehydrogenase (Gpd1 and Gpd2) and mitochondrial glycerol-3-phosphate:ubiquinone oxidoreductase, often referred to as mitochondrial glycerol-3-phosphate dehydrogenase (Gut2) (Dawson 1979).

It should be mentioned that attempts to increase glycerol production by *S. cerevisiae* (e.g., by overexpression of the gene *GPD1*) violates the redox balance in the cell and causes the production of additional oxidized by-products such as pyruvate, acetate, acetoin, 2,3-butanediol, and succinate (Michnick et al. 1997; Roustan et al. 1999).

Because glycerol production involves dephosphorylation, it may also be involved in replenishment of the cytosolic pool of inorganic phosphate at the switch from gluconeogenic to glycolytic metabolism (Thevelein and Hohmann 1995; van Heerden et al. 2014). It has been shown that overexpression of the first enzyme in glycerol biosynthesis suppresses growth defect in a *tps1Δ* mutant, which suffers from accumulation of sugar phosphatase and depletion of inorganic phosphate pool (Thevelein and Hohmann 1995).

4.4 Methods of Glycerol Production

Glycerol can be produced either by chemical synthesis from petrochemical feedstocks, or by microbial fermentation, or can be recovered as a by-product of soap or biodiesel manufacture (Wang et al. 2001). The oldest among these is the process of soap production, which is known from 2800 BC when soap was obtained for the first time by heating fats in the presence of ash (Hunt 1999). In 1811, French chemist M. E. Chevreul defined the chemical formulas of fatty acids and the formulas of glycerin in vegetable oil and animal fat. His work was patented. It was known as the first industrial method to obtain glycerin soap by reacting fatty material with lime and alkaline material (Gesslein 1999). The process of saponification is now of lesser importance in industrial nations and many developing countries, because of the replacement of soap with detergents (Rehm 1988; Agarwal 1990). Glycerol production by microbial fermentation was discovered by Louis Pasteur (1858). Pasteur showed that glycerol was formed at a quantity of 2.0–3.6% [g/g] from the sugar during alcoholic fermentation. That makes glycerol an important constituent of alcoholic beverages produced by fermentation (wine, beer, and others). For example, increasing glycerol concentrations in wine is considered desirable, as glycerol contributes to the smoothness and overall mouthfeel of wine (Eustace and Thornton 1987). However, glycerol yield during fermentation was too low for industrial application, so alcoholic fermentation was not considered as a source of glycerol until World War I when demand for glycerol in explosive manufacture exceeded the supply from the soap industry (Prescott and Dunn 1959). To meet the increased necessity for glycerol, Neuberg and Reinfurth developed the so-called "second form of fermentation" which implies the addition of steering agents (Neuberg and Reinfurth 1918). Steering agents, such as bi-sulfite ions, were added in the fermentation broth forming a complex with acetaldehyde. Ethanol formation, which consumes glycolytic NADH, was therefore impaired, lacking the intermediate acetaldehyde; consequently, the reoxidation of the cofactor was managed through glycerol production. Neuberg's second form of fermentation increased glycerol synthesis, reaching conversion efficiencies from 23 to 28% depending on the process conditions (Wang et al. 2001), but large amounts of sulfite-acetaldehyde complex were accumulated as a by-product (Agarwal 1990). Another developed at that time process, fermentation at high pH (around 7 or above), also improved the glycerol yield significantly (Connstein and Ludecke 1919). However, these wartime processes were unable to compete with chemically synthesized processes developed after World War II as yields of glycerol from sugar by fermentation were low, and recovery by distillation was inefficient (Vijaikishore and Karanth 1986). Chemical synthesis was responsible for considerable part of world glycerol production during the second half of twentieth century. Since the late 1940s, and following the discovery of synthetic surfactants, glycerol has been produced from epichlorohydrin obtained from propylene (and thus from fossil oil) as large chemical companies forecasted a glycerol shortage and initiated its synthetic production (Frost and Sullivan 2006). A significant amount of glycerol was also synthesized from allyl alcohol. However, the route of glycerol production by the oxidation or chlorination of propylene has declined in relative importance since the early 1970s (Hester 2000) due to environmental concerns and the increase in cost and decrease in the availability of propylene, especially in developing countries. Decline in chemical glycerol synthesis led to a renewed interest in glycerol production by microbial fermentation. To fulfill this task, new strains of *S. cerevisiae* and new species of yeasts were proposed, which will be discussed in greater detail in the next sections.

In recent years, increasing focus on renewable and sustainable energy has boosted the production of biodiesel from rapeseed oil, cooking oil, and animal fats, especially in Europe. Biodiesel is a viable fuel and a fossil diesel additive to replace sulfur, which content is being progressively lowered according to tighter environmental legislation. Glycerol as a by-product of biodiesel production has also seen a dramatic increase in production. Approximately, 10% of the reaction volume in a given biodiesel production process ends up as raw or crude glycerol (Johnson and Taconi 2007). This low-grade glycerol also contains water, salts, and other organic materials, including residual methanol as well as free fatty acids. Each component varies quite a bit in content depending on the raw material used (Thompson and He 2006). Purification of a crude glycerol fraction from biodiesel production to obtain food-grade glycerol is quite expensive (Yang et al. 2012). More preferable is bioconversion of crude glycerol into liquid biofuels, green chemicals, and bioenergy on the basis of fermentation processes, which can provide an efficient solution for sustainable management of glycerol and improve the economics of biodiesel industries (Lynd et al. 2005).

4.5 Glycerol Production by the Yeast *S. cerevisiae*

As has been already mentioned, during both World Wars, wild-type strains of *S. cerevisiae* were used on an industrial scale to produce glycerol from carbohydrate containing feedstocks in sulfite-based process (Agarwal 1990; Rehm 1988). However, this process is inefficient and causes environmental problems. Rapidly developing metabolic engineering tools offer new means for redirecting of carbon flux toward glycerol instead of ethanol production. In general, five directions of *S. cerevisiae* metabolic engineering for improvement of glycerol production can be distinguished: (1) enhancement of glycerol-3-phosphate dehydrogenase activity; (2) improvement of glycerol efflux from the cells; (3) impairment or reduction of triose phosphate isomerase activity; (4) dissection or attenuation of the route that leads to ethanol production; and (5) shifting of NAD^+/NADH ratio in order to increase the amount of NADH available for glycerol-3-phosphate dehydrogenase or changing cofactor-specificity of glycerol-3-phosphate dehydrogenase. Most of the studies combine several or all of these approaches.

To enhance the activity of glycerol-3-phosphate dehydrogenase, the *GPD1* gene was overexpressed, which resulted in a fourfold increase in glycerol production at the expense of ethanol (Michnick et al. 1997). The substitution of all four phosphorylation sites renders glycerol-3-phosphate dehydrogenase twofold more active than that of the wild type (Oliveira et al. 2012). The evolutionary engineering of a bicistronic artificial operon containing both yeast *GPD1* and *GPP2* genes in a heterologous system of *Escherichia coli* resulted in the obtainment of fused gene *GPD1-GPP2* (Salles et al. 2007). The product of this gene, fused protein possessed both catalytic sites of glycerol-3-phosphate dehydrogenase and glycerol-3-phosphate phosphatase and was able to convert DHAP into glycerol more quickly, probably due to substrate channeling between active sites.

An increase of glycerol production could be also achieved by elevation of its outflow from yeast cells. It was shown that a modified form of the Fps1 protein from which 76–230 amino acid residues were removed fails to retain inside the cell the produced glycerol under hyperosmotic stress (Karlgren et al. 2004). The corresponding strain with an engineered Fps1 glycerol channel showed poor growth during fermentation. Further overexpression of the *GPD1* in this strain increased the amount of the produced glycerol; however, a pronounced growth defect retained (Remize et al. 2001).

Decreased ethanol formation was achieved by the deletion of one or several *ADH* genes coding for alcohol dehydrogenases (Drewke et al. 1990; Cordier et al. 2007) or *PDC1* and *PDC5* genes coding for pyruvate decarboxylases (Nevoigt and Stahl 1996). Mutants that are defective in alcohol dehydrogenase, however, accumulate large amounts of acetaldehyde and acetic acid, which are toxic to cells. Acetaldehyde could be partially detoxified due to overexpression *ALD3* coding for aldehyde dehydrogenase (Cordier et al. 2007). The mutants defective in pyruvate decarboxylase cannot grow without adding exogenous ethanol or acetate as the last enzyme is the only source of the cytosolic two-carbon compounds that are required primarily for lipid biosynthesis (Pronk et al. 1996). The most promising results have been reported for mutants of *S. cerevisiae* deleted in the *TPI1* gene coding for triose phosphate isomerase (Overkamp et al. 2002). Unfortunately, such mutants had a severe growth defect, especially in glucose (Compagno et al. 2001). This growth defect was later attributed to an accumulation of DHAP, which has been shown to inactivate *myo*-inositol-3phosphate synthase and render the cells auxotrophic for inositol (Shi et al. 2005). The addition of inositol partly restored the growth of the *tpi1Δ* mutant. The growth defect was also partly overcome by the additional deletion of the two isogenes *NDE1* and *NDE2*, encoding mitochondrial external NADH dehydrogenase, and *GUT2*, encoding the FAD⁺-dependent mitochondrial glycerol dehydrogenase (Overkamp et al. 2002). As these enzymes are primarily responsible for mitochondrial reoxidation of cytosolic NADH, their deletion increased the availability of cytosolic NADH. The *tpi1Δ nde1Δ nde2Δ gut2Δ* quadruple mutant was able to grow on glucose but the specific growth rate very low. Evolutionary engineering of the quadruple mutant in glucose-containing medium partially relieved this growth defect and resulted in a strain with glycerol yield which is very close to the theoretical maximum (Overkamp et al. 2002). Another approach, which was also based on a *tpi1Δ* mutant, involved also the overexpression of *GPD1*, the deletion of *ADH1*, and the overexpression of *ALD3* (Cordier et al. 2007).

An original approach to increase glycerol production, with a simultaneous drop in ethanol titer, was described by the expression of the cytosolic acetolactate synthase Ilv2. The constructed strain has a decrease in intracellular pyruvate concentration and possesses a fourfold increase in glycerol production, with a concomitant 1.8-fold decrease in ethanol production when compared to that of the parental strain (Murashchenko et al. 2016). We constructed strains with a constitutively decreased expression of *TPI1*, overexpression of a fused gene for *GPD1-GPP1*, the modified gene *FPS1*, and truncated gene *ILV2*. The constructed strain combining the above-mentioned genetic manipulations grew on glucose as the sole carbon and energy source under anaerobic conditions and accumulated up to 9 times more glycerol under micro-aerobic conditions and up to 4.7 times more glycerol under anaerobic conditions relative to the parental wild-type strain (Semkiv et al. 2017; Semkiv et al. 2020).

The theoretical maximum for glycerol production in *S. cerevisiae* is limited by the fact that only half of the glucose carbon can be used for NADH-consuming glycerol production, whereas the other half is required for NADH generation via glyceraldehyde 3-phosphate dehydrogenase in the "pyruvate branch" of glycolysis. This problem was addressed by Geertman et al. (2006). They used a *pdc1Δ pdc5Δ pdc6Δ nde1Δ nde2Δ gut2Δ* deletion mutant strain combined with *GPD2* overexpression for overexpression of the gene *FDH1*, encoding a NAD⁺-dependent formate dehydrogenase. Cells supplied with an additional source of cytosolic NADH achieved a glycerol yield exceeding the theoretical maximum; however, this strategy required formate cofeeding. To avoid cofeeding, it has been proposed to switch the redox factor specificity of glycerol formation in *S. cerevisiae* from NADH to NADPH by introducing a NADPH-dependent DHAP reductase and redirecting the main carbon flux through the pentose phosphate pathway by reducing phosphoglucose isomerase activity. In *Bacillus subtilis gpsA* gene, encoding a NADPH-dependent DHAP reductase was successfully introduced into an *S. cerevisiae pgi1* mutant, but that resulted in only 35% increase of glycerol formation, probably due to the low specific activity of the bacterial *gpsA* gene product in the yeast transformant (Nevoigt 2008).

4.6 Glycerol Production by Osmotolerant Yeasts

Glycerol production by osmotolerant yeasts was first described by Nickerson and Carroll (1945) for yeast *Zygosaccharomyces acidifaciens*. This study together with the following reports on glycerol and other polyols (e.g., erythritol, D-arabitol, and mannitol) production by Spencer and Shu (1957) and Onishi et al. (1961) stimulated a comprehensive investigation of osmotolerant or osmophilic yeasts expecting that an organism producing glycerol without a need for a steering agent might be found. However, these investigations did not lead to commercialization because of difficulties in recovering glycerol from fermented broth and the availability of less expensive glycerol from petroleum sources. Later, the development of separation techniques, such as reverse osmosis, ultrafiltration, ion exchange, ion-exclusion, etc. provided means for industrialization of glycerol production by osmotolerant yeast fermentation.

Most osmotolerant yeast species considered for glycerol production belong to the genera *Candida*, *Debaryomyces*, *Hansenula*, *Pichia*, *Saccharomyces*, *Schizosaccharomyces*, *Torulaspora*, and *Zygosaccharomyces* (Onishi 1963; Zhang et al. 1963; Zhuge 1973; Kumar et al. 1989; Petrovska et al. 1999). All of these yeasts need oxygen for growth, which considerably increases the production costs due to additional expenses needed for bioreactor aeration and cooling (Cueto-Rojas et al. 2015).

Glycerol yields as high as 40% have been reached using a strain of *C. magnoliae* (Hajny et al. 1960). *C. magnoliae* is isolated from honeycomb and also known as an erythritol producer (Kim et al. 1996). *C. magnoliae* is able to grow in the presence of up to 50% (w/v) sugars and produces erythritol, mannitol, and glycerol as compatible solutes in response to high sugar concentrations (Yu et al. 2006). When *C. magnoliae* was grown in high concentrations of fructose, it produced a significant amount of glycerol that was almost the same amount as erythritol (Yu et al. 2006). Similarly, intracellular glycerol accumulation is also critical for *C. magnoliae* to maintain osmolarity like in *S. cerevisiae*.

An intensive effort has been made in China since the 1970s to develop a glycerol process based on osmotolerant yeasts. Zhuge (1973) isolated a strain from among 5000 osmotolerant yeast isolates with considerable promise for glycerol production. This strain isolated from glazed fruit in Southern China was later identified as *C. glycerinogenes* (Wang et al. 1999). *C. glycerinogenes* has been commercially exploited to produce glycerol (Zhuge et al. 2001). Compared with other yeasts, *C. glycerinogenes* has several useful properties, such as tolerance to high glucose concentrations, rapid growth, and ability to accumulate high extracellular glycerol concentration. To further enhance glycerol production from bagasse hydrolysate, a homologous modified hexose transporter gene (*HXT4*) and heterologous genes for xylose reductase (*XYL1*), xylitol dehydrogenase (*XYL2*), and xylulose kinase (*XKS1*) from *Scheffersomyces stipitis* were overexpressed in *C. glycerinogenes*. Fermentation of undetoxified bagasse hydrolysate by the constructed strain resulted in a glycerol titer of 40.3 g/L and a yield of 40.4%. This process required only 1 kg of bagasse to produce 93.5 g of glycerol (Jiang et al. 2024). *Z. rouxii* and *Hansenula anomala* are other osmotolerant yeasts that produce a significant amount of glycerol under high osmotic conditions (Groleau et al. 1995; Parekh and Pandey 1985).

4.7 Glycerol Application in the Industry

The applications of glycerol in industry are primarily linked to its physical properties. For instance, due to its humectant property, glycerol is used in the preparation of gels and ointments. It can also be used as a plasticizer to enhance the spread of coatings over beads, granules, and tablets in cosmetics and pharmaceuticals. Glycerol is widely used in pre-processed tobacco to prevent it from crumbling. In the healthcare industry, glycerol is used to maintain pressure in the eyeball and central nervous system. Its hygroscopic nature makes it valuable in glues and adhesives to prevent them from drying too quickly. Additionally, glycerol is used in the textile industry to soften fibers and in flexible foams. At low temperatures, glycerol tends to supercool instead of crystallizing, which is why aqueous glycerol solutions resist freezing and are used as antifreeze in cooling systems. Glycerol is also employed as a stabilizing agent for the storage of cells and proteins.

4.8 Glycerol Bioconversion to High-Value Chemicals

The world production of biodiesel as an alternative to petroleum is dynamically increased (Farouk et al. 2024). Huge amounts of glycerol are generated as a by-product of biodiesel production (about 10% by weight) through transesterification with methanol catalyzed by NaOH or KOH (Garlapati et al. 2016). As the utilization of this glycerol is an important component of the economics of the biodiesel industry, there is a worldwide demand for the efficient use of glycerol (Willke and Vorlop 2008). The chemical properties of glycerol are also valuable. There is a growing interest in using glycerol as a central raw material for producing fine chemicals such as glyceric acid, dihydroxyacetone, ketomalonic (or mesoxalic) acid, poly(ketomalonate), 1,2,3-tri-tert-butyl glycerol, propylene glycol (1,2-propanediol), acrolein, epichlorohydrin, and others (Pagliaro et al. 2007). However, chemical synthesis often involves the use of hazardous compounds and can lead to environmental pollution. In contrast, bioconversion of glycerol, especially the inexpensive by-product of the biodiesel industry, into value-added compounds by certain microorganisms is more environmentally friendly. Below are some existing or potential biotechnological processes that utilize glycerol as a substrate for conversion or as a carbon and energy source for microbial growth:

1. Production of dihydroxyacetone (DHA). DHA is predominantly used as coloring agent in sunless tanning creams and lotions (Levy 1992). Its chemical synthesis is expensive and involves dealing with hazardous reactants (Hekmat et al. 2003). Biotechnological production of DHA employs the bacterium *Gluconobacter oxydans*, which has membrane-bound pyrroloquinoline quinone-dependent glycerol dehydrogenase that catalyzes the oxidation of the secondary hydroxy group of glycerol to DHA (Matsushita et al. 1994). Bacteria *Gluconobacter melanogenus* (Flickinger and Perlman 1977) and *Acetobacter xylinum* (Nabe et al. 1979) have also been reported to synthesize DHA from glycerol.
2. Production of 1,3-propanediol. 1,3-Propanediol (1,3-PDO) is used for production of polyesters, polyethers, polyurethanes, adhesives, composites, laminates, coatings, and moldings. In addition, 1,3-PDO together with terephthalic acid is used for the production of valuable biodegradable polymer polytrimethylene terephthalate. Chemical synthesis of 1,3-PDO has many issues, such as high energy consumption, toxic intermediates, and expensive catalysts. Fortunately, 1,3-PDO can be biosynthesized from glycerol in two sequential reactions: first, glycerol dehydratase or diol dehydratase converts glycerol into 3-hydroxypropionaldehyde (3-HPA) (Schneider and Pawelkiewicz 1966; Toraya et al. 1978); in the second step, 3-HPA is reduced to 1,3-PDO by NADPH- or NADH-dependent 1,3-PDO dehydrogenase. Production of 1,3-PDO has been shown for strains of *Klebsiella pneumoniae*, *Clostridium butyricum*, *Clostridium acetobutylicum*, etc. (Forage and Foster 1982; Biebl 1991).
3. Production of ethanol. Fuel ethanol can be used as an addition to gasoline in the car engines. Exponential growth in the production of fuel ethanol was observed over the past decade (OECD/FAO 2021). Fuel ethanol is mainly gained from sugarcane, corn, or sugar beets (Da Silva et al. 2009). Considerable efforts have been made toward the development of profitable technology for ethanol production from lignocellulosic feedstocks (Dellomonaco et al. 2010). Raw glycerol (by-product of biodiesel production) is also considered as an inexpensive feedstock for fuel ethanol production (Licht 2010). Unfortunately, the most frequently used ethanol producer *S. cerevisiae* does not exhibit acceptable growth rates in synthetic glycerol medium if no growth-supporting medium supplements (such as amino acids and nucleic bases) are added (Swinnen et al. 2013). The growth of *S. cerevisiae* on glycerol may be considerably improved, by selection of glycerol-adapted *S. cerevisiae* strains (Ochoa-Estopier et al. 2011), expression of heterologous glycerol transporters (Klein et al. 2016), etc. A new metabolic engineering strategy to reroute glycerol to ethanol production in *S. cerevisiae* involved the expression of four genes in the dihydroxyacetone pathway (*Candida utilis* glycerol facilitator, *Ogataea polymorpha* glycerol dehydro-

genase, endogenous dihydroxyacetone kinase, and triosephosphate isomerase). This strategy also included eliminating the first step of both glycerol biosynthesis and glycerol catabolism via G3P by knocking out the *GPD1* and *GUT1* genes. Additionally, the expression of the NADH oxidase gene *noxE* from *Lactococcus lactis* compensated for the lack of NAD^+. The engineered strain converted 1.28 times more glycerol to ethanol than the parental strain (Rezk Khattab and Watanabe 2021). Also, raw glycerol can be converted to ethanol by engineered strains of methylotrophic yeast *O. polymorpha* (Hong et al. 2010; Kata et al. 2016). Elementary mode analysis and metabolic evolution of *E. coli* mutants led to substantial improvement of ethanol production from glycerol (Trinh and Srienc 2009). Also, some new microorganisms with good ability to grow and produce ethanol on crude glycerol have been isolated, e. g., bacterium *Kluyvera cryocrescens* (Choi et al. 2011) and yeast *P. tannophilus* (Liu et al. 2012).

4. Production of 2,3-butanediol. 2,3-Butanediol (2,3-BDO) is used as a fuel, solvent, and for the production of polymers (Perego et al. 2003; Saha and Bothast 1999). Biosynthesis of 2,3-BDO occurs from pyruvate in three steps. First, acetolactate synthase catalyzes the condensation of two pyruvate molecules to form acetolactate and CO2. Then, acetolactate is decarboxylated to acetoin by acetolactate decarboxylase. Finally, acetoin is reduced to 2,3-butanediol by 2,3-BDO dehydrogenase or acetoin reductase (Juni 1952). Thus, for 2,3-BDO production all substrates have to be converted to pyruvate. Many substrates have been used for the production of 2,3-BDO, e.g., starch hydrolysates, raw and decolored molasses, and corncob hydrolysates (Perego et al. 2000; Ma et al. 2010). Glycerol might also be a good carbon source for 2,3-BDO production, because 2,3-BDO is a known by-product of 1,3-propanediol production from glycerol by *K. pneumoniae* (Biebl et al. 1998). Production of 2,3-BDO from glycerol by *K. pneumonia* G31 resulted in high product yields under intense aeration and alkaline pH conditions (Petrov and Petrova 2010).
5. Production of glyceric acid. Glyceric acid is a promising and expensive chemical used for pharmaceutical purposes. It is a known by-product of dihydroxyacetone production from glycerol with *G. oxydans*. The biosynthesis of glyceric acid from glycerol involves two dehydrogenases: glycerol is first oxidized to glyceraldehyde, which is then further oxidized to glyceric acid (Habe et al. 2009b). Among various acetic acid bacteria, *Acetobacter tropicalis* has proven to be the most efficient organism for producing glyceric acid (Habe et al. 2009a).
6. Production of biosurfactants. Surfactants are used as cleaners, emulsifiers, and laundry detergents or in paints. Biosurfactants consist of a hydrophilic part (consist of a sugar, peptide, or protein) and a hydrophobic/lipophilic part (contains fatty acids or fatty alcohols). The biosurfactants have advantages over chemically produced detergents because they are biodegradable and less toxic, and they can be produced from renewable resources. Natural biosurfactant producers are bacteria, yeast, and fungi (Mulligan 2005). Pure or raw glycerol is often used in biosurfactant production, as it constitutes the backbone of the lipid component of biosurfactants. Among microorganisms that can carry out biosurfactants pro-

duction using glycerol and other supplies, there are yeast *Pseudozyma antarctica* (Morita et al. 2007), *Candida bombicola* (Ashby and Solaiman 2010), bacterium *Pseudomonas aeruginosa* (Wu et al. 2008), and others.

Thus, glycerol is a valuable compound which finds its application in different branches of the industry.

Acknowledgment This work was supported by grants of National Science Centre of Poland (NCN) UMO-2021/41/B/NZ1/01224. This work was also supported by "Presidential Discretionary-Ukraine Support Grants" from Simons Foundation, Award No 1030281 and No 1290613.

References

Adler L, Blomberg A, Nilsson A (1985) Glycerol metabolism and osmoregulation in the salt-tolerant yeast *Debaryomyces hansenii*. J Bacteriol 162:300–306

Agarwal GP (1990) Glycerol. In: Fiechter A (ed) Advances in biochemical engineering/biotechnology 41: microbial bioproducts. Springer, Berlin, pp 95–128

Ahmadpour D, Geijer C, Tamas MJ, Lindkvist-Petersson K, Hohmann S (2014) Yeast reveals unexpected roles and regulatory features of aquaporins and aquaglyceroporins. Biochim Biophys Acta 1840:1482–1491

Albers E, Liden G, Larsson C, Gustafsson L (1998) Anaerobic redox balance and nitrogen metabolism in *Saccharomyces cerevisiae*. Recent Res Dev Microbiol 2:253–279

Albertyn J, Hohmann S, Thevelein JM, Prior BA (1994) *GPD1*, which encodes glycerol-3-phosphate dehydrogenase is essential for growth under osmotic stress in *Saccharomyces cerevisiae* and its expression is regulated by the high-osmolarity glycerol response pathway. Mol Cell Biol 14:4135–4144

Alepuz PM, de Nadal E, Zapater M, Ammerer G, Posas F (2003) Osmostress-induced transcription by Hot1 depends on a Hog1-mediated recruitment of the RNA Pol II. EMBO J 22:2433–2442

AlMarzouq DS, Elnagdi NMH (2019) Glycerol and Q-tubes: green catalyst and technique for synthesis of polyfunctionally substituted heteroaromatics and anilines. Molecules 24(9):1806

Anderson RM, Bitterman KJ, Wood JG, Medvedik O, Cohen H, Lin SS, Manchester JK, Gordon JI, Sinclair DA (2002) Manipulation of a nuclear NAD^+ salvage pathway delays aging without altering steady-state NAD^+ levels. J Biol Chem 277:18881–18890

Andre L, Hemming A, Adler L (1991) Osmoregulation in *Saccharomyces cerevisiae*: studies on the osmotic induction of glycerol production and glycerol-3-phosphate dehydrogenase (NAD^+). FEBS Lett 286:13–17

Ansell R, Granath K, Hohmann S, Thevelein JM, Adler L (1997) The two isoenzymes for yeast NAD^+-dependent glycerol 3-phosphate dehydrogenase encoded by *GPD1* and *GPD2* have distinct roles in osmoadaptation and redox regulation. EMBO J 16:2179–2187

Ashby RD, Solaiman DK (2010) The influence of increasing media methanol concentration on sophorolipid biosynthesis from glycerol-based feedstocks. Biotechnol Lett 32:1429–1437

Bakker BM, Overkamp KM, van Maris AJ, Kotter P, Luttik MA, van Dijken JP, Pronk JT (2001) Stoichiometry and compartmentation of NADH metabolism in *Saccharomyces cerevisiae*. FEMS Microbiol Rev 25:15–37

Baltanas R, Bush A, Couto A, Durrieu L, Hohmann S, Colman-Lerner A (2013) Pheromone-induced morphogenesis improves osmoadaptation capacity by activating the HOG MAPK pathway. Sci Signal 6:ra26

Beese SE, Negishi T, Levin DE (2009) Identification of positive regulators of the yeast fps1 glycerol channel. PLoS Genet 5(11):e1000738

Biebl H (1991) Glycerol fermentation of 1,3-propanediol by *Clostridium butyricum*—measurement of product inhibition by use of a Ph-Auxostat. Appl Microbiol Biotechnol 35:701–705

Biebl H, Zeng AP, Menzel K, Deckwer WD (1998) Fermentation of glycerol to 1,3-propanediol and 2,3-butanediol by Klebsiella pneumoniae. Appl Microbiol Biotechnol 50:24–29

Bjorkqvist S, Ansell R, Adler L, Liden G (1997) Physiological response to anaerobicity of glycerol 3-phosphate dehydrogenase mutants of *Saccharomyces cerevisiae*. Appl Environ Microbiol 63:128–132

Blomberg A (2000) Metabolic surprises in Saccharomyces cerevisiae during adaptation to saline conditions: questions, some answers and a model. FEMS Microbiol Lett 182:1–8

Blomberg A (2022) Yeast osmoregulation—glycerol still in pole position. *FEMS Yeast Res* 22(1):foac035

Blomberg A, Adler L (1989) Roles of glycerol and glycerol-3-phosphate dehydrogenase (NAD^+) in acquired osmotolerance of *Saccharomyces cerevisiae*. J Bacteriol 171:1087–1092

Blomberg A, Adler L (1992) Physiology of osmotolerance in fungi. Adv Microb Physiol 33:145–212

Bouwman J, Kiewiet J, Lindenbergh A, van Eunen K, Siderius M, Bakker BM (2011) Metabolic regulation rather than de novo enzyme synthesis dominates the osmo-adaptation of yeast. Yeast 28:43–53

Brewster JL, de Valoir T, Dwyer ND, Winter E, Gustin MC (1993) An osmosensing signal transduction pathway in yeast. Science 259:1760–1763

Brown AD (1978) Compatible solutes and extreme water stress in eukaryotic micro-organisms. Adv Microb Physiol 17:181–242

Carly F, Vandermies M, Telek S, Steels S, Thomas S, Nicaud JM, Fickers P (2017) Enhancing erythritol productivity in *Yarrowia lipolytica* using metabolic engineering. Metab Eng 42:19–24

Chen RE, Thorner J (2007) Function and regulation in MAPK signaling pathways: lessons learned from the yeast *Saccharomyces cerevisiae*. Biochim Biophys Acta 1773:1311–1340

Chen X, Fang H, Rao Z, Shen W, Zhuge B, Wang Z, Zhuge J (2008) Cloning and characterization of a NAD+-dependent glycerol-3-phosphate dehydrogenase gene from *Candida glycerinogenes*, an industrial glycerol producer. FEMS Yeast Res 8:725–734

Chilakamarry CR, Sakinah AMM, Zularisam AW, Pandey A (2021) Glycerol waste to value added products and its potential applications. Syst Microbiol Biomanuf 1(4):378–396

Choi WJ, Hartono MR, Chan WH, Yeo SS (2011) Ethanol production from biodiesel-derived crude glycerol by newly isolated *Kluyvera cryocrescens*. Appl Microbiol Biotechnol 89:1255–1264

Compagno C, Brambilla L, Capitanio D, Boschi F, Ranzi BM, Porro D (2001) Alterations of the glucose metabolism in a triose phosphate isomerase-negative *Saccharomyces cerevisiae* mutant. Yeast 18(7):663–670

Connstein W, Ludecke K (1919) Uber glycerin-Gewinnung durch Garung. Ber deut chem Ges 52:1385–1391

Cordier H, Mendes F, Vasconcelos I, Francois JM (2007) A metabolic and genomic study of engineered *Saccharomyces cerevisiae* strains for high glycerol production. Metab Eng 9:364–378

Costenoble R, Valadi H, Gustafsson L, Niklasson C, Franzen CJ (2000) Microaerobic glycerol formation in Saccharomyces cerevisiae. Yeast 16:1483–1495

Cueto-Rojas HF, van Maris AJ, Wahl SA, Heijnen JJ (2015) Thermodynamics-based design of microbial cell factories for anaerobic product formation. Trends Biotechnol 33(9):534–546

Da Cunha AC, Gomes LS, Godoy-Santos F, Faria-Oliveira F, Teixeira JA, Sampaio GMS, Trópia MJM, Castro IM, Lucas C, Brandão RL (2019) High-affinity transport, cyanide-resistant respiration, and ethanol production under aerobiosis underlying efficient high glycerol consumption by *Wickerhamomyces anomalus*. J Ind Microbiol Biotechnol 46:709–723

Da Silva GP, Mack M, Contiero J (2009) Glycerol: a promising and abundant carbon source for industrial microbiology. Biotechnol Adv 27:30–39

Dawson AG (1979) Oxidation of cytosolic NADH formed during aerobic metabolism in mammalian cells. Trends Biochem Sci 4:171–176 de Nadal E, Casadome L, Posas F (2003) Targeting the MEF2-like transcription factor Smp1 by the stress-activated Hog1 mitogen-activated protein kinase. Mol Cell Biol 23:229–237

de Nadal E, Ammerer G, Posas F (2011) Controlling gene expression in response to stress. Nat Rev Genet 12:833–845

de Vries RP, Flitter SJ, van de Vondervoort PJ, Chaveroche MK, Fontaine T, Fillinger S, Ruijter GJ, d'Enfert C, Visser J (2003) Glycerol dehydrogenase, encoded by *gldB* is essential for osmotolerance in *Aspergillus nidulans*. Mol Microbiol 49:131–141

Dellomonaco C, Fava F, Gonzalez R (2010) The path to next generation biofuels: successes and challenges in the era of synthetic biology. Microb Cell Fact 9:3

Dihazi H, Kessler R, Eschrich K (2004) High osmolarity glycerol (HOG) pathway-induced phosphorylation and activation of 6-phosphofructo-2-kinase are essential for glycerol accumulation and yeast cell proliferation under hyperosmotic stress. J Biol Chem 279:23961–23968

Drewke C, Thielen J, Ciriacy M (1990) Ethanol formation in adh0 mutants reveals the existence of a novel acetaldehyde-reducing activity in *Saccharomyces cerevisiae*. J Bacteriol 172(7):3909–3917

Dujon B, Sherman D, Fischer G, Durrens P, Casaregola S, Lafontaine I, De Montigny J, Marck C, Neuvéglise C, Talla E, Goffard N, Frangeul L, Aigle M, Anthouard V, Babour A, Barbe V, Barnay S, Blanchin S, Beckerich JM, Beyne E, Bleykasten C, Boisramé A, Boyer J, Cattolico L, Confanioleri F, De Daruvar A, Despons L, Fabre E, Fairhead C, Ferry-Dumazet H, Groppi A, Hantraye F, Hennequin C, Jauniaux N, Joyet P, Kachouri R, Kerrest A, Koszul R, Lemaire M, Lesur I, Ma L, Muller H, Nicaud JM, Nikolski M, Oztas S, Ozier-Kalogeropoulos O, Pellenz S, Potier S, Richard GF, Straub ML, Suleau A, Swennen D, Tekaia F, Wésolowski-Louvel M, Westhof E, Wirth B, Zeniou-Meyer M, Zivanovic I, Bolotin-Fukuhara M, Thierry A, Bouchier C, Caudron B, Scarpelli C, Gaillardin C, Weissenbach J, Wincker P, Souciet JL (2004) Genome evolution in yeasts. Nature 430:35–44

Edgley M, Brown AD (1983) Yeast water relations: physiological changes induced by solute stress in *Saccharomyces cerevisiae* and *Saccharomyces rouxii*. J Gen Microbiol 129:3453–3463

Eriksson P, Andre L, Ansell R, Blomberg A, Adler L (1995) Characterization of *GPD2*, a second gene encoding *sn*-glycerol 3-phosphate dehydrogenase (NAD^+) in *Saccharomyces cerevisiae*, and its comparison with *GPD1*. Mol Microbiol 17:95–107

Eustace R, Thornton RJ (1987) Selective hybridization of wine yeasts for higher yields of glycerol. Can J Microbiol 33:112–117

Farouk SM, Tayeb AM, Abdel-Hamid SMS, Osman RM (2024) Recent advances in transesterification for sustainable biodiesel production, challenges, and prospects: a comprehensive review. Environ Sci Pollut Res 31:12722–12747

Ferreira C, van Voorst F, Martins A, Neves L, Oliveira R, Kielland-Brandt MC, Lucas C, Brandt A (2005) A member of the sugar transporter family, Stl1p is the glycerol/H+ symporter in *Saccharomyces cerevisiae*. Mol Biol Cell 16:2068–2076

Fillinger S, Ruijter G, Tamas MJ, Visser J, Thevelein JM, d'Enfert C (2001) Molecular and physiological characterization of the NAD-dependent glycerol 3-phosphate dehydrogenase in the filamentous fungus *Aspergillus nidulans*. Mol Microbiol 39:145–157

Finley D, Ulrich HD, Sommer T, Kaiser P (2012) The ubiquitin-proteasome system of *Saccharomyces cerevisiae*. Genetics 192:319–360

Flickinger MC, Perlman D (1977) Application of Oxygen-Enriched Aeration in the Conversion of Glycerol to Dihydroxyacetone by *Gluconobacter melanogenus* IFO 3293. Appl Environ Microbiol 33:706–712

Forage RG, Foster MA (1982) Glycerol fermentation in *Klebsiella pneumoniae*: functions of the coenzyme B12-dependent glycerol and diol dehydratases. J Bacteriol 149:413–419

Frost and Sullivan (2006) R&D creating new avenues for glycerol (August 4, 2006). Available online at https://www.frost.com/prod/servlet/market-insight-top.pag?docid=77264824

Gancedo C, Gancedo JM, Sols A (1968) Glycerol metabolism in yeasts. Pathways of utilization and production. Eur J Biochem 5:165–172

Garlapati VK, Shankar U, Budhiraja A (2016) Bioconversion technologies of crude glycerol to value added industrial products. Biotechnol Rep 9:9–14

Geertman JM, van Maris AJ, van Dijken JP, Pronk JT (2006) Physiological and genetic engineering of cytosolic redox metabolism in *Saccharomyces cerevisiae* for improved glycerol production. Metab Eng 8:532–542

Gesslein BW (1999) Humectants in personal care formulation: a practical guide. In: Schueller R, Romanowski P (eds) Conditioning agents for hair and skin. New York, Marcel Dekker, pp 95–96

Grant WD (2004) Life at low water activity. Philos Trans R Soc Lond Ser B Biol Sci 359:1249–1266. Discussion 1266–1247

Groleau D, Chevalier P, Yuen TH, TLS. (1995) Production of polyols and ethanol by osmophilic yeast *Zygosaccharomyces rouxii*. Biotechnol Lett 17:315–320

Habe H, Fukuoka T, Kitamoto D, Sakaki K (2009a) Biotransformation of glycerol to D-glyceric acid by *Acetobacter tropicalis*. Appl Microbiol Biotechnol 81:1033–1039

Habe H, Shimada Y, Yakushi T, Hattori H, Ano Y, Fukuoka T, Kitamoto D, Itagaki M, Watanabe K, Yanagishita H, Matsushita K, Sakaki K (2009b) Microbial production of glyceric acid, an organic acid that can be mass produced from glycerol. Appl Environ Microbiol 75:7760–7766

Hajny GJ, Hendershot WF, Peterson WH (1960) Factors affecting glycerol production by a newly isolated osmophilic yeast. Appl Microbiol 8:5–11

Hekmat D, Bauer R, Fricke J (2003) Optimization of the microbial synthesis of dihydroxyacetone from glycerol with *Gluconobacter oxydans*. Bioprocess Biosyst Eng 26:109–116

Heller KB, Lin EC, Wilson TH (1980) Substrate specificity and transport properties of the glycerol facilitator of *Escherichia coli*. J Bacteriol 144:274–278

Heredia CF, Sols A, DelaFuente G (1968) Specificity of the constitutive hexose transport in yeast. Eur J Biochem 5:321–329

Hester A (2000) Microbial glycerol. Ind Bioprocess 22(4):3–5

Hohmann S (2002) Osmotic stress signaling and osmoadaptation in yeasts. Microbiol Mol Biol Rev 66: 300–372

Holst B, Lunde C, Lages F, Oliveira R, Lucas C, Kielland-Brandt MC (2000) *GUP1* and its close homologue *GUP2*, encoding multimembrane-spanning proteins involved in active glycerol uptake in *Saccharomyces cerevisiae*. Mol Microbiol 37:108–124

Hong WK, Kim CH, Heo SY, Luo LH, Oh BR, Seo JW (2010) Enhanced production of ethanol from glycerol by engineered *Hansenula polymorpha* expressing pyruvate decarboxylase and aldehyde dehydrogenase genes from *Zymomonas mobilis*. Biotechnol Lett 32:1077–1082

Hunt JA (1999) A short history of soap. Pharm J 263(7076):985–989

Jagtap SS, Bedekar AA, Singh V, Jin YS, Rao CV (2021) Metabolic engineering of the oleaginous yeast *Yarrowia lipolytica* PO1f for production of erythritol from glycerol. Biotechnol Biofuels 14:188

Jiang D, Wang M, Zhao X, Lu X, Zong H, Zhuge B (2024) Glycerol production from undetoxified lignocellulose hydrolysate by a multiresistant engineered *Candida glycerinogenes*. J Agric Food Chem 72(3):1630–1639

Johnson DT, Taconi KA (2007) The glycerin glut: options for the value-added conversion of crude glycerol resulting from biodiesel production. Environ Prog 26:338–348

Jung S, Marelli M, Rachubinski RA, Goodlett DR, Aitchison JD (2010) Dynamic changes in the subcellular distribution of Gpd1p in response to cell stress. J Biol Chem 285:6739–6749

Jung JY, Kim TY, Ng CY, Oh MK (2012) Characterization of *GCY1* in *Saccharomyces cerevisiae* by metabolic profiling. J Appl Microbiol 113:1468–1478

Juni E (1952) Mechanisms of Formation of Acetoin by Bacteria. J Biol Chem 195:715–726

Karlgren S, Filipsson C, Mullins JGL, Bill RM, Tamas MJ, Hohmann S (2004) Identification of residues controlling transport through the yeast aquaglyceroporin Fps1 using a genetic screen. Eur J Biochem 271:771–779

Kata I, Semkiv MV, Ruchala J, Dmytruk KV, Sibirny AA (2016) Overexpression of the genes *PDC1* and *ADH1* activates glycerol conversion to ethanol in the thermotolerant yeast *Ogataea (Hansenula) polymorpha*. Yeast 33(8):471–478

Kawai K, Kanesaki Y, Yoshikawa H, Hirasawa T (2019) Identification of metabolic engineering targets for improving glycerol assimilation ability of *Saccharomyces cerevisiae* based on adaptive laboratory evolution and transcriptome analysis. J Biosci Bioeng 128:162–169

Kim SY, Park SS, Jeon YJ, Seo JH (1996) Analysis of fermentation characteristics for production of erythritol by *Candida sp*. Kor J Food Sci Technol 28:935–939

Klein M, Islam Z-u, Knudsen PB, Carrillo M, Swinnen S, Workman M, Nevoigt E (2016) The expression of glycerol facilitators from various yeast species improves growth on glycerol of *Saccharomyces cerevisiae*. Metab Eng Commun 3:252–257

Klipp E, Nordlander B, Kruger R, Gennemark P, Hohmann S (2005) Integrative model of the response of yeast to osmotic shock. Nat Biotechnol 23:975–982

Krantz M, Nordlander B, Valadi H, Johansson M, Gustafsson L, Hohmann S (2004) Anaerobicity prepares *Saccharomyces cerevisiae* cells for faster adaptation to osmotic shock. Eukaryot Cell 3:1381–1390

Kumar J, Agarwal SC, Basu SK, Spivastava OP (1989) Isolation, evaluation and taxonomic characterization of some glycerol producing osmotolerant yeasts. Biol Membr 15:129–135

Kurtzman CP, Fell JW, Boekhout T (eds) (2011) The yeasts—a taxonomic study. Elsevier, Amsterdam

Lages F, Lucas C (1995) Characterization of a glycerol/H+ symport in the halotolerant yeast *Pichia sorbitophila*. Yeast 11:111–119

Lages F, Silva-Graca M, Lucas C (1999) Active glycerol uptake is a mechanism underlying halotolerance in yeasts: a study of 42 species. Microbiology 145(9):2577–2585

Larsson K, Ansell R, Eriksson P, Adler L (1993) A gene encoding sn-glycerol-3-phosphate dehydrogenase (NAD^+) complements an osmosensitive mutant of *Saccharomyces cerevisiae*. Mol Microbiol 10:1101–1111

Larsson C, Pahlman IL, Ansell R, Rigoulet M, Adler L, Gustafsson L (1998) The importance of the glycerol 3-phosphate shuttle during aerobic growth of *Saccharomyces cerevisiae*. Yeast 14:347–357

Lee DH, Kim MD, Ryu YW, Seo JH (2008) Cloning and characterization of *CmGPD1*, the *Candida magnoliae* homologue of glycerol-3-phosphate dehydrogenase. FEMS Yeast Res 8:1324–1333

Lee YJ, Jeschke GR, Roelants FM, Thorner J, Turk BE (2012) Reciprocal phosphorylation of yeast glycerol-3-phosphate dehydrogenases in adaptation to distinct types of stress. Mol Cell Biol 32:4705–4717

Lee J, Reiter W, Dohnal I, Gregori C, Beese-Sims S, Kuchler K, Ammerer G, Levin DE (2013) MAPK Hogl closes the *S. cerevisiae* glycerol channel Fps1 by phosphorylating and displacing its positive regulators. Genes Dev 27:2590–2601

Levin DE (2011) Regulation of cell wall biogenesis in *Saccharomyces cerevisiae*: the cell wall integrity signaling pathway. Genetics 189:1145–1175

Levy SB (1992) Dihydroxyacetone-Containing Sunless or Self-Tanning Lotions. J Am Acad Dermatol 27:989–993

Licht FO (2010) World ethanol and biofuels. Report 8

Liu X, Jensen PR, Workman M (2012) Bioconversion of crude glycerol feedstocks into ethanol by *Pachysolen tannophilus*. Bioresour Technol 104:579–586

Liu X, Mortensen UH, Workman M (2013) Expression and functional studies of genes involved in transport and metabolism of glycerol in *Pachysolen tannophilus*. Microb Cell Fact 12:27

Lucas C, Da Costa M, van Uden N (1990) Osmoregulatory active sodium-glycerol cotransport in the halotolerant yeast *Debaryomyces hansenii*. Yeast 6:187–191

Luyten K, Albertyn J, Skibbe WF, Prior BA, Ramos J, Thevelein JM, Hohmann S (1995) Fps1, a yeast member of the MIP family of channel proteins, is a facilitator for glycerol uptake and efflux and is inactive under osmotic stress. EMBO J 14:1360–1371

Lynd LR, van Zyl WH, McBride JE, Laser M (2005) Consolidated bioprocessing of cellulosic biomass: an update. Curr Opin Biotechnol 16:577–583

Ma CQ, Wang AL, Wang Y, Jiang TY, Li LX, Xu P (2010) Production of 2,3-butanediol from corncob molasses, a waste by-product in xylitol production. Appl Microbiol Biotechnol 87:965–970

Maeda T, Wurgler-Murphy SM, Saito H (1994) A two-component system that regulates an osmosensing MAP kinase cascade in yeast. Nature 369:242–245

Marshall JH, Kong YC, Sloan J, May JW (1989) Purification and properties of glycerol: $NADP^+$ 2-oxidoreductase from *Schizosaccharomyces pombe*. J Gen Microbiol 135:697–701

Matsushita K, Toyama H, Adachi O (1994) Respiratory chains and bioenergetics of acetic acid bacteria. Adv Microb Physiol 36:247–301

Matsuzawa T, Ohashi T, Hosomi A, Tanaka N, Tohda H, Takegawa K (2010) The *gld1* gene encoding glycerol dehydrogenase is required for glycerol metabolism in *Schizosaccharomyces pombe*. Appl Microbiol Biotechnol 87:715–727

Mattanovich D, Graf A, Stadlmann J, Dragosits M, Redl A, Maurer M, Kleinheinz M, Sauer M, Altmann F, Gasser B (2009) Genome, secretome and glucose transport highlight unique features of the protein production host *Pichia pastoris*. Microb Cell Fact 8:29

Merico A, Ragni E, Galafassi S, Popolo L, Compagno C (2011) Generation of an evolved *Saccharomyces cerevisiae* strain with a high freeze tolerance and an improved ability to grow on glycerol. J Ind Microbiol Biotechnol 38:1037–1044

Michnick S, Roustan JL, Remize F, Barre P, Dequin S (1997) Modulation of glycerol and ethanol yields during alcoholic fermentation in *Saccharomyces cerevisiae* strains overexpressed or disrupted for *GPD1* encoding glycerol 3-phosphate dehydrogenase. Yeast 13:783–793

Mollapour M, Piper PW (2007) Hog1 mitogen-activated protein kinase phosphorylation targets the yeast Fps1 aquaglyceroporin for endocytosis, thereby rendering cells resistant to acetic acid. Mol Cell Biol 27:6446–6456

Morita T, Konishi M, Fukuoka T, Imura T, Kitamoto D (2007) Microbial conversion of glycerol into glycolipid biosurfactants, mannosylerythritol lipids, by a basidiomycete yeast, *Pseudozyma antarctica* JCM 10317(T). J Biosci Bioeng 104:78–81

Mulligan CN (2005) Environmental applications for biosurfactants. Environ Pollut 133:183–198

Murashchenko L, Abbas C, Dmytruk K, Sibirny A (2016) Overexpression of the truncated version of ILV2 enhances glycerol production in Saccharomyces cerevisiae. Yeast 33(8):463–469

Nabe K, Izuo N, Yamada S, Chibata I (1979) Conversion of glycerol to dihydroxyacetone by immobilized whole cells of *Acetobacter xylinum*. Appl Environ Microbiol 38:1056–1060

Neuberg C, Reinfurth E (1918) Natürliche und erzwungene Glycerinbildung bei der alkoholischen Gärung. BiochemZ 92:32

Neves L, Lages F, Lucas C (2004) New insights on glycerol transport in *Saccharomyces cerevisiae*. FEBS Lett 565:160–162

Nevoigt E (2008) Progress in metabolic engineering of *Saccharomyces cerevisiae*. Microbiol Mol Biol Rev 72:379–412

Nevoigt E, Stahl U (1996) Reduced pyruvate decarboxylase and increased glycerol-3-phosphate dehydrogenase [NAD^+] levels enhance glycerol production in *Saccharomyces cerevisiae*. Yeast 12:1331–1337

Nickerson WJ, Carroll WR (1945) On the metabolism of *Zygosaccharomyces*. Arch Biochem 7:257–271

Norbeck J, Blomberg A (1997) Metabolic and regulatory changes associated with growth of *Saccharomyces cerevisiae* in 1.4 M NaCl. Evidence for osmotic induction of glycerol dissimilation via the dihydroxyacetone pathway. J Biol Chem 272:5544–5554

Norbeck J, Pehlman AK, Akhtar N, Blomberg A, Adler L (1996) Purification and characterization of two isoenzymes of dl-glycerol-3-phosphatase from *Saccharomyces cerevisiae*. Identification of the corresponding *GPP1* and *GPP2* genes and evidence for osmotic regulation of Gpp2p expression by the osmosensing mitogen-activated protein kinase signal transduction pathway. J Biol Chem 271:13875–13881

Ochoa-Estopier A, Lesage J, Gorret N, Guillouet SE (2011) Kinetic analysis of a *Saccharomyces cerevisiae* strain adapted for improved growth on glycerol: Implications for the development of yeast bioprocesses on glycerol. Bioresour Technol 102(2):1521–1527

OECD/FAO (2021) OECD-FAO agricultural outlook 2021-2030. OECD, Paris. https://doi.org/10.1787/19428846-en

Oliveira R, Lucas C (2004) Expression studies of *GUP1* and *GUP2*, genes involved in glycerol active transport in *Saccharomyces cerevisiae*, using semi-quantitative RT-PCR. Curr Genet 46:140–146

Oliveira R, Lages F, Silva-Graca M, Lucas C (2003) Fps1p channel is the mediator of the major part of glycerol passive diffusion in *Saccharomyces cerevisiae*: artefacts and re-definitions. Biochim Biophys Acta 1613:57–71

Oliveira AP, Ludwig C, Picotti P, Kogadeeva M, Aebersold R, Sauer U (2012) Regulation of yeast central metabolism by enzyme phosphorylation. Mol Syst Biol 8:623

Ong KL, Fickers P, Lin CSK (2020) Enhancing succinic acid productivity in the yeast *Yarrowia lipolytica* with improved glycerol uptake rate. Sci Total Environ 702:134911

Onishi H (1963) Osmophilic yeasts. Adv Food Res 12:53–94

Onishi H, Saito N, Koshiyama I (1961) Studies on osmoplilic yeasts. Part XI. Various factors affecting on Polyalcohol production by *Pichia miso*. Agric Biol Chem 25:124–130

O'Rourke SM, Herskowitz I, O'Shea EK (2002) Yeast go the whole HOG for the hyperosmotic response. Trends Genet 18:405–412

Overkamp KM, Bakker BM, Kotter P, Luttik MA, Van Dijken JP, Pronk JT (2002) Metabolic engineering of glycerol production in *Saccharomyces cerevisiae*. Appl Environ Microbiol 68:2814–2821

Pagliaro M, Ciriminna R, Kimura H, Rossi M, Della PC (2007) From glycerol to value-added products. Angew Chem Int Ed Engl 46:4434–4440

Pahlman AK, Granath K, Ansell R, Hohmann S, Adler L (2001) The yeast glycerol 3-phosphatases Gpp1p and Gpp2p are required for glycerol biosynthesis and differentially involved in the cellular responses to osmotic, anaerobic, and oxidative stress. J Biol Chem 276:3555–3563

Parekh SR, Pandey NK (1985) Production of glycerol by *Hansenula anomala*. Biotechnol Bioeng 27:1081–1091

Pasteur ML (1858) Production constante de glycérine dans la fermentation alcoolique. C R Acad Sci 46:857

Pavlik P, Simon M, Schuster T, Ruis H (1993) The glycerol kinase (*GUT1*) gene of *Saccharomyces cerevisiae*: cloning and characterization. Curr Genet 24:21–25

Perego P, Converti A, Del Borghi A, Canepa P (2000) 2,3-butanediol production by *Enterobacter aerogenes*: selection of the optimal conditions and application to food industry residues. Bioprocess Eng 23:613–620

Perego P, Converti A, Del Borghi M (2003) Effects of temperature, inoculum size and starch hydrolyzate concentration on butanediol production by *Bacillus licheniformis*. Bioresour Technol 89:125–131

Petelenz-Kurdziel E, Kuehn C, Nordlander B, Klein D, Hong KK, Jacobson T, Dahl P, Schaber J, Nielsen J, Hohmann S, Klipp E (2013) Quantitative analysis of glycerol accumulation, glycolysis and growth under hyper osmotic stress. PLoS Comput Biol 9:e1003084

Petrov K, Petrova P (2010) Enhanced production of 2,3-butanediol from glycerol by forced pH fluctuations. Appl Microbiol Biotechnol 87:943–949

Petrovska B, Winkelhausen E, Kuzmanova S (1999) Glycerol production by yeasts under osmotic and sulfite stress. Can J Microbiol 45:695–699

Pettersson N, Filipsson C, Becit E, Brive L, Hohmann S (2005) Aquaporins in yeasts and filamentous fungi. Biol Cell 97:487–500

Posas F, Chambers JR, Heyman JA, Hoeffler JP, de Nadal E, Arino J (2000) The transcriptional response of yeast to saline stress. J Biol Chem 275:17249–17255

Prescott SC, Dunn CG (1959) Industrial microbiology, 3rd edn. McGraw-Hill, New York, pp 208–217

Prista C, Loureiro-Dias MC, Montiel V, García R, Ramos J (2005) Mechanisms underlying the halotolerant way of *Debaryomyces hansenii*. FEMS Yeast Res 5(8):693–701
Proft M, Struhl K (2002) Hog1 kinase converts the Sko1-Cyc8-Tup1 repressor complex into an activator that recruits SAGA and SWI/SNF in response to osmotic stress. Mol Cell 9:1307–1317
Proft M, Pascual-Ahuir A, de Nadal E, Arino J, Serrano R, Posas F (2001) Regulation of the Sko1 transcriptional repressor by the Hog1 MAP kinase in response to osmotic stress. EMBO J 20:1123–1133
Pronk JT, Yde Steensma H, van Dijken JP (1996) Pyruvate metabolism in *Saccharomyces cerevisiae*. Yeast 12(16):1607–1633
Redkar RJ, Locy RD, Singh NK (1995) Biosynthetic pathways of glycerol accumulation under salt stress in *Aspergillus nidulans*. Exp Mycol 19:241–246
Rehm H-J (1988) Microbial production of glycerol and other polyols. In: Rehm H-J, Reed HJ (eds) Biotechnology, vol. 6B. Special microbial processes. VCH, Verlagsgesellschaft, Weinheim, pp 51–69
Remize F, Barnavon L, Dequin S (2001) Glycerol export and glycerol-3-phosphate dehydrogenase, but not glycerol phosphatase, are rate limiting for glycerol production in *Saccharomyces cerevisiae*. Metab Eng 3:301–312
Rep M, Krantz M, Thevelein JM, Hohmann S (2000) The transcriptional response of *Saccharomyces cerevisiae* to osmotic shock. Hot1p and Msn2p/Msn4p are required for the induction of subsets of high osmolarity glycerol pathway-dependent genes. J Biol Chem 275:8290–8300
Rezk Khattab SM, Watanabe T (2021) Efficient conversion of glycerol to ethanol by an engineered *Saccharomyces cerevisiae* strain. Appl Environ Microbiol 87(23):e0026821
Romano AH (1986) Microbial sugar transport systems and their importance in biotechnology. Trends Biotechnol 4:207–213
Ronnow B, Kielland-Brandt MC (1993) Gut2, a gene for mitochondrial glycerol-3-phosphate dehydrogenase of *Saccharomyces cerevisiae*. Yeast 9:1121–1130
Roustan JL, Salblayrolles JM, Barre P, Dequin S (1999) Glycerol overproduction by engineered *Saccharomyces cerevisiae* wine yeast strains leads to substantial changes in by-product formation and to a stimulation of fermentation rate in stationary phase. Appl Environ Microbiol 65:143–149
Ruijter GJ, Visser J, Rinzema A (2004) Polyol accumulation by *Aspergillus oryzae* at low water activity in solid-state fermentation. Microbiology 150:1095–1101
Sabir F, Loureiro-Dias MC, Prista C (2016) Comparative analysis of sequences, polymorphisms and topology of yeasts aquaporins and aquaglyceroporins. FEMS Yeast Res 16(3):fow025
Saha BC, Bothast RJ (1999) Production of 2,3-butanediol by newly isolated *Enterobacter cloacae*. Appl Microbiol Biotechnol 52:321–326
Saito H, Posas F (2012) Response to hyperosmotic stress. Genetics 192:289–318
Salles IM, Forchhammer N, Croux C, Girbal L, Soucaille P (2007) Evolution of a *Saccharomyces cerevisiae* metabolic pathway in *Escherichia coli*. Metab Eng 9:152–159
Schneider Z, Pawelkiewicz J (1966) The properties of glycerol dehydratase isolated from *Aerobacter aerogenes*, and the properties of the apoenzyme subunits. Acta Biochim Pol 13:311–328
Schuurink R, Busink R, Hondmann DH, Witteveen CF, Visser J (1990) Purification and properties of NADP$^{(+)}$-dependent glycerol dehydrogenases from *Aspergillus nidulans* and *A. niger*. J Gen Microbiol 136:1043–1050
Semkiv MV, Dmytruk KV, Abbas CA, Sibirny AA (2017) Metabolic engineering for high glycerol production by the anaerobic cultures of *Saccharomyces cerevisiae*. Appl Microbiol Biotechnol 101(11):4403–4416
Semkiv MV, Ruchala J, Dmytruk KV, Sibirny AA (2020) 100 years later, what is new in glycerol bioproduction? Trends Biotechnol 38(8):907–916
Shi Y, Vaden DL, Ju S, Ding D, Geiger JH, Greenberg ML (2005) Genetic perturbation of glycolysis results in inhibition of de novo inositol biosynthesis. J Biol Chem 280:41805–41810

Spencer JFT, Shu P (1957) Polyhydric alcohol production by osmophilic yeasts: effect of oxygen tension and inorganic phosphate concentration. Can J Microbiol 3:559

Sprague GF, Cronan JE (1977) Isolation and characterization of *Saccharomyces cerevisiae* mutants defective in glycerol catabolism. J Bacteriol 129:1335–1342

Strucko T, Zirngibl K, Pereira F, Kafkia E, Mohamed ET, Rettel M, Stein F, Feist AM, Jouhten P, Patil KR, Forster J (2018) Laboratory evolution reveals regulatory and metabolic trade-offs of glycerol utilization in Saccharomyces cerevisiae. Metab Eng 47:73–82

Sutherland FC, Lages F, Lucas C, Luyten K, Albertyn J, Hohmann S, Prior BA, Kilian SG (1997) Characteristics of Fps1-dependent and -independent glycerol transport in *Saccharomyces cerevisiae*. J Bacteriol 179:7790–7795

Swinnen S, Klein M, Carrillo M, McInnes J, Nguyen HT, Nevoigt E (2013) Re-evaluation of glycerol utilization in *Saccharomyces cerevisiae*: characterization of an isolate that grows on glycerol without supporting supplements. Biotechnol Biofuels 6:157

Swinnen S, Ho PW, Klein M, Nevoigt E (2016) Genetic determinants for enhanced glycerol growth of *Saccharomyces cerevisiae*. Metab Eng 36:68–79

Tamas MJ, Luyten K, Sutherland FC, Hernandez A, Albertyn J, Valadi H, Li H, Prior BA, Kilian SG, Ramos J, Gustafsson L, Thevelein JM, Hohmann S (1999) Fps1p controls the accumulation and release of the compatible solute glycerol in yeast osmoregulation. Mol Microbiol 31:1087–1104

Tani T, Yamada K (1987) Glycerol metabolism in methylotrophic yeasts. Agric Biol Chem 51:1927–1933

ter Linde JJ, Liang H, Davis RW, Steensma HY, van Dijken JP, Pronk JT (1999) Genome-wide transcriptional analysis of aerobic and anaerobic chemostat cultures of Saccharomyces cerevisiae. J Bacteriol 181:7409–7413

Thevelein JM, Hohmann S (1995) Trehalose synthase, guard to the gate of glycolysis in yeast? Trends Biochem Sci 20:3–10

Thome PE (2004) Isolation of a *GPD* gene from *Debaryomyces hansenii* encoding a glycerol 3-phosphate dehydrogenase (NAD+). Yeast 21:119–126

Thompson JC, He BB (2006) Characterization of crude glycerol from biodiesel production from multiple feedstocks. Appl Eng Agric 22:261–265

Thorsen M, Di Y, Tängemo C, Morillas M, Ahmadpour D, Van der Does C, Wagner A, Johansson E, Boman J, Posas F, Wysocki R, Tamás MJ (2006) The MAPK Hog1p modulates Fps1p-dependent arsenite uptake and tolerance in yeast. Mol Biol Cell 17:4400–4410

Toraya T, Honda S, Kuno S, Fukui S (1978) Coenzyme B12-dependent diol dehydratase: regulation of apoenzyme synthesis in *Klebsiella pneumoniae* (*Aerobacter aerogenes*) ATCC 8724. J Bacteriol 135:726–729

Trinh CT, Srienc F (2009) Metabolic engineering of *Escherichia coli* for efficient conversion of glycerol to ethanol. Appl Environ Microbiol 75:6696–6705

van Heerden JH, Wortel MT, Bruggeman FJ, Heijnen JJ, Bollen YJ, Planqué R, Hulshof J, O'Toole TG, Wahl SA, Teusink B (2014) Lost in transition: start-up of glycolysis yields subpopulations of nongrowing cells. Science 343:1245114

van Zyl PJ, Kilian SG, Prior BA (1990) The role of an active transport mechanism in glycerol accumulation during osmoregulation by *Zygosaccharomyces rouxii*. Appl Microbiol Biotechnol 34:231–235

van Zyl PJ, Prior BA, Kilian SG (1991) Regulation of glycerol metabolism in *Zygosaccharomyces rouxii* to osmotic stress. Appl Microbiol Biotechnol 36:369–374

Vijaikishore P, Karanth NG (1986) Glycerol production by fermentation—a review. Proc Biochem 21:54–57

Von Jagow G, Klingenberg M (1970) Pathways of hydrogen in mitochondria of *Saccharomyces carlsbergensis*. Eur J Biochem 12:583–592

Wang Z-X, Zhuge J, Fang H-Y (1999) A new osmotolerant glycerol-highly-producing species of *Candida sp.–Candida glycerolgenesis* Zhuge sp. nov. Acta Microbiol Sin 39:68–74

Wang Z-X, Zhuge J, Cao Y, Chen J, Fang H-Y (2000) The key enzymes of metabolism of glycerol in *Candida glycerolgenesis*. Acta Microbiol Sin 40:180–187

Wang Z-X, Zhuge J, Fang H, Prior BA (2001) Glycerol production by microbial fermentation: a review. Biotechnol Adv 19:201–223

Warmka J, Hanneman J, Lee J, Amin D, Ota I (2001) Ptc1, a type 2C Ser/Thr phosphatase, inactivates the HOG pathway by dephosphorylating the mitogen-activated protein kinase Hog1. Mol Cell Biol 21:51–60

Watanabe Y, Nagayama K, Tamai Y (2008) Expression of glycerol 3-phosphate dehydrogenase gene (*CvGPD1*) in salt-tolerant yeast *Candida versatilis* is stimulated by high concentrations of NaCl. Yeast 25:107–116

Westfall PJ, Ballon DR, Thorner J (2004) When the stress of your environment makes you go HOG wild. Science 306:1511–1512

Westfall PJ, Patterson JC, Chen RE, Thorner J (2008) Stress resistance and signal fidelity independent of nuclear MAPK function. Proc Natl Acad Sci USA 105:12212–12217

Willke T, Vorlop K (2008) Biotransformation of glycerol into 1,3-propanediol. Eur J Lipid Sci 110:831–840

Wu JY, Yeh KL, Lu WB, Lin CL, Chang JS (2008) Rhamnolipid production with indigenous *Pseudomonas aeruginosa* EM1 isolated from oil-contaminated site. Bioresour Technol 99:1157–1164

Wurgler-Murphy SM, Maeda T, Witten EA, Saito H (1997) Regulation of the Saccharomyces cerevisiae HOG1 mitogen-activated protein kinase by the PTP2 and PTP3 protein tyrosine phosphatases. Mol Cell Biol 17:1289–1297

Yancey PH, Clark ME, Hand SC, Bowlus RD, Somero GN (1982) Living with water stress: evolution of osmolyte systems. Science 217:1214–1222

Yang F, Hanna MA, Sun R (2012) Value-added uses for crude glycerol—a byproduct of biodiesel production. Biotechnol Biofuel 5:13

Yu JH, Lee DH, Oh YJ, Han KC, Ryu YW, Seo JH (2006) Selective utilization of fructose to glucose by *Candida magnoliae*, an erythritol producer. Appl Biochem Biotechnol 129–132:870–879

Zhang S-Z, Yang X-W, Wang H-L (1963) A study on production of glycerol and arabitol by osmophilic yeasts: II. The condition for production of arabitol by *Hansenula arabitolgenes*. Fang Acta Microbiol Sin 9:134–139

Zhuge J (1973) A study on glycerol production by fermentation with a new strain of osmophilic yeast. Kexue Tongbao (Sci Bull Sin) 18:188–189

Zhuge J, Fang HY, Wang ZX, Chen DZ, Jin HR, Gu HL (2001) Glycerol production by a novel osmotolerant yeast *Candida glycerinogenes*. Appl Microbiol Biotechnol 55:686–692

Chapter 5
Production of Industrially Relevant Organic Acids by Yeasts and Filamentous Fungi

Riccardo Milanesi, Danilo Porro, Paola Branduardi, and Immacolata Serra

Contents

5.1 Introduction 124
5.2 Citric Acid (2.9 Mtons/Year) 126
5.3 Lactic Acid (400 kTons/Year) 128
5.4 Fumaric Acid (90 kTons/Year) 129
5.5 Succinic Acid (70 kTons/Year) 131
5.6 Itaconic Acid (41.4 kTons/Year) 132
5.7 Renewable Organic Acids from Yeast and Fungi 134
5.8 Conclusions and Perspectives 135
References 138

Abstract Organic acids are broadly distributed in nature, and humans have used them in their natural sources since the early ages, essentially for food and drink preparations. Organic acids are low-molecular-weight organic compounds with one or more acidic groups such as carboxyl, sulfonic, alcohol, and thiol groups. These functional groups make organic acids key building block chemicals, which have huge markets, and some of them are still commonly refined from petroleum. Indeed, during the first half of the twentieth century, advances in chemical synthesis offered competitive manufacturing processes. This scenario changed starting from the 1990s with the development of the industrial biotechnology field and the integration of biomass processing industries; fermentation and downstream procedures improved the economics for the biological synthesis. In this chapter, we mainly review the aspects of current or piloting large-scale successful industrial processes for yeast or fungal organic acid productions. Perspectives are also reported.

R. Milanesi · D. Porro (✉) · P. Branduardi (✉) · I. Serra
Department of Biotechnology and Biosciences, University of Milano-Bicocca, Milano, Italy
e-mail: danilo.porro@unimib.it; paola.branduardi@unimib.it

A. A. Sibirny (ed.), *Biotechnology of Yeasts and Filamentous Fungi*,
Grand Challenges in Biology and Biotechnology,
https://doi.org/10.1007/978-3-031-74726-7_5

5.1 Introduction

Building block chemicals are molecules with multiple functional groups potentially allowing for the transformation into new families of useful high-value chemicals or materials, which have huge markets and are commonly refined from fossil resources. Twenty years ago, the US Department of Energy (DOE) (Bozell and Petersen 2010) identified 12 building block chemicals that can be produced from sugars via biological or chemical conversions (Table 5.1). Since this first report, many studies have been carried out to develop microbial cell factories and biorefineries for their industrial productions. Biorefinery development has two strategic goals: the displacement of petroleum in favor of renewable raw materials as feedstock and the establishment of a robust bio-based industry (Bozell and Petersen 2010).

About 70% of the first top 12 building blocks suggested in 2004 are organic acids, low-molecular-weight molecules with one or more acidic groups. All of them, including succinic acid, 2,5-furan dicarboxylic acid, 3-hydroxypropionic acid (3-HPA), aspartic acid, glutamic acid, itaconic acid, and levulinic acid, have been already or are being commercialized. Table 5.2 summarizes molecules, companies, and processes by which these building blocks are produced nowadays.

About 36% (succinic, fumaric, itaconic, and lactic acids) of the 2004 DOE's derived top organic chemicals can be produced by yeast or fungi, while the remaining are essentially still produced by chemical synthesis (malic, 2,5 furan dicarboxylic, aspartic, glucaric, and levulinic acids) or from bacterial cells (3-hydroxypropionic acid and glutamic acids), as shown in Table 5.2.

A general criterion to be met for the production of commodity chemicals is to develop a process for the (bio)synthesis of the desired molecule at a price lower than 2 EUR *per* kg (Van Dien 2013). When coming to microbial-based processes, this usually implies that the cell factory/biocatalyst should accumulate natural or

Table 5.1 The 2004 DOE's and revised 2010 top sugar-derived building blocks (Bozell and Petersen 2010)

2004	2010
1,4-Diacids (succinic, fumaric, and malic acids)	3-Hydroxypropionic acid/aldehyde
2,5-Furan dicarboxylic	Bio-hydrocarbons
3-Hydroxybutyrolactone	Ethanol
3-Hydroxy propionic acid	Furans
Aspartic acid	Glycerol and derivative
Glucaric acid	Lactic acid
Xylitol/arabitol	Levulinic acid
Glutamic acid	Succinic acid
Itaconic acid	Xylitol
Levulinic acid	Sorbitol
Glycerol	
Sorbitol	

Table 5.2 Production of the 2004 and 2010 revised DOE's top organic acids: overview of the commercialization or demonstration: pilot scale successes (Bozell and Petersen 2010)

Molecule of interest	Process type	Company(ies)
Furan dicarboxylic acid	Chemical (dehydration and oxidation)	ADM (USA) Ava Biochem (CH) Avantium (NL) Corbion (NL)
3-Hydroxy propionic acid	Fermentation (*Escherichia coli*)	Cargill (USA) Novozymes (DK) Perstorp (S)
Aspartic acid	Enzymatic conversion from fumaric acid	Ajinomoto (J) Grace Fine Chemical Manufacturing (USA)
Fumaric acid	Chemical (oxidation and isomerization) Fermentation (*Rhizopus oryzae*)	MBI international (USA) Myriant Corporation (USA)
Glucaric acid	Chemical (oxidation)	Cargill (USA) Rennovia, Inc. (USA) Rivertop Renewables (USA)
Glutamic acid	Fermentation (*Corynebacterium glutammicum*)	Many companies
Itaconic acid	Fermentation (*Aspergillus terreus*)	Cargill (USA) Itaconix Corporation (USA) Qingdao Kehai Biochemistry Co., Ltd (CHN)
Lactic acid	Fermentation (lactic acid bacteria, yeast)	Blue Marble Biomaterials (USA) Calysta (USA) Cargill (USA) Corbion (NL) Galactic (USA) Glycosbio (USA) Myriant Corporation (USA)
Levulinic acid	Chemical (acid treatment—dehydration, hydration)	Arzeda (USA) Avantium (NL) Biofine Technology LLC (USA) GF Biochemicals (IT) Segetis (USA)
Malic acid	Chemical and fermentation	Different companies
Succinic acid	Fermentation (yeast)	Bioamber (USA) Myriant (USA) Reverdia (NL) Succinity (D)

non-natural chemical products to a non-physiological titer and often start from a non-preferred substrate. Moreover, this must be scalable from research scale to industrial one and be robust toward chemical-physical oscillation and genetic perturbations. In spite of all the criteria to be met, an increasing number of microbial processes are matching scale-up parameters providing tangible alternatives to

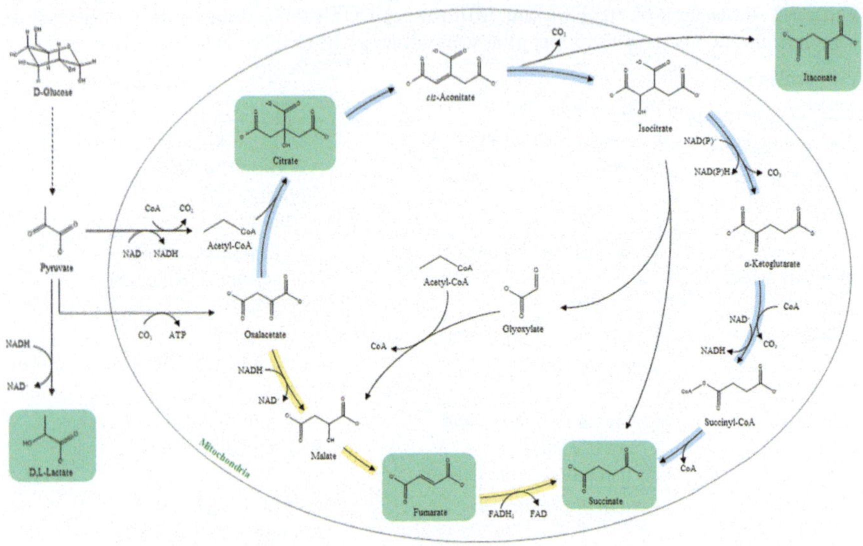

Fig. 5.1 Metabolic pathways for the production of citric, fumaric, itaconic, lactic, and succinic acid. Forward and reverse TCA cycle reactions are highlighted in blue and yellow, respectively

chemical synthesis at industrial level. From a chronological point of view, citric acid has been the first organic acid industrially produced by wild-type/recombinant yeast/fungi, followed by lactic acid, fumaric acid, succinic acid, and itaconic acid (Fig. 5.1).

In this chapter, we will present the microbial processes that reached relevant scales (pilot or commercial plant) and, more in general, the readiness level of these technologies.

5.2 Citric Acid (2.9 Mtons/Year)

(2-Hydroxy-propane-1,2,3-tricarboxylic acid; $C_6H_8O_7$; 192.12 g/mol; pKa1 = 2.63; pKa2 = 4.11; pKa3 = 5.34 at 25 °C)

Citric acid (CA) is a key intermediate metabolite in the TCA cycle and derives its name from the Latin word *citrus*, as it was originally obtained from lemons, the fruit of the citrus tree. The first report dates back 1784 by Carl Wilhelm Scheele, a Swedish chemist (1742–1786). Citric acid is a universal intermediate of biochemical pathways, and its traces are found in virtually all plants and animals (Fig. 5.1). It is extensively used in the food and beverage industry as it combines a pleasant taste with low toxicity and palatability. It serves several functions in the food formulation, like preservation, flavor fixation and enhancement, bacterial stabilization, and standardization of acidity levels. Citric acid was first produced commercially in

England around 1826 from imported Italian lemons (Papagianni 2007). In around 1880, the firm Charles Pfizer developed a chemical synthesis route. Different synthetic routes using different starting materials have since been published. In the early twentieth century, surface culture methods using *Aspergillus niger* were implemented and in the middle of the last century further developed toward a submerged process. Global citric acid production has reached 2.9 million tons with an expected CAGR of 2.6% for the period by 2032 (Citric Acid Market Size, Share and Trends Analysis 2032). The vast majority of this is produced via fermentation using *A. niger*, although many other microorganisms can be used to produce citric acid (*A. wentii*, *A. awamori*, *A. foetidus*, *A. fenicis*, *A. fonsecalus*, *A. fumaricus*, *A. luchensis*, *A. saitoi* and *A. usumii*, *Candida lipolytica*, *C. tropicalis*, *C. guilliermondii*, *C. intermedia*, *C. parapsilosis*, *C. zeylanoides*, *C. fibriae*, *C. subtropicalix*, *C. oleophila* from the genus *Candida*, among others) (Książek 2023). Specific strains that are capable of overproducing citric acid have been developed for various types of fermentation processes. The yield of citric acid from these strains often exceeds 70% of the theoretical yield on the carbon source (Papagianni 2007). In order to successfully produce citric acid, some critical parameters must be monitored and maintained: high substrate concentration, low and finite content of nitrogen, defined concentration of trace metals, high dissolved oxygen, and low pH of the fermentation media (Książek 2023). Biochemically, CA is produced through the oxidation of glucose via glycolysis and TCA cycle, while it is exported from the mitochondria by the tricarboxylate transporter (Fig. 5.1) (Oke Teleky and Cristian Vodnar 2019). Since aconitase affinity for CA is lower than that of tricarboxylate transporter, this results in the mitochondrial export of citrate and import of malate as counterion (Karaffa and Kubicek 2003). A high glycolytic flux promotes CA production, guaranteeing a high carbon flux into the TCA cycle. Nevertheless, it has been shown that, in *A. niger*, citric acid itself is a potent inhibitor of glycolysis acting on phosphofructokinase I (PfkI). This inhibition is counteracted by a posttranslational modification of PfkI triggered by a drop of the intracellular pH and a cAMP peak that takes place before the onset of citric acid accumulation (Mlakar and Legiša 2006). The enzyme is cleaved and phosphorylated, leading to a fragment that is highly active and highly inducible by ammonium, but less susceptible to citrate inhibition, which explains why glycolysis is not repressed. *A. niger* cellular morphology is also a crucial prerequisite for the efficient production of organic acid (X. Sun et al. 2018). Growth in the form of pellets of <1 mm in diameter is associated with high production rates and yields (Książek 2023). To date, the molecular mechanism underlying this regulation remains unclear; nevertheless, the conditions for obtaining such pellets have been determined empirically, with copper concentration and pH of the growth media being the most influential parameters (Książek 2023).

The use of yeast strains capable of reaching high yield, such as *Yarrowia lipolytica* and different *Candida* species, allows to increase the concentration of carbon source and reduce the oxygenation of the growth media. Many other yeast species have been proven to accumulate citric acid; among these are *Candida tropicalis*, *Candida oleophila*, and *Hansenula anomala* (Di Lorenzo et al. 2022). Nevertheless, none of these microorganisms have provided better performances than *A. niger*, which is still the preferred cell factory for CA production, taking advantage of either stirred tank or bubble column reactors for submerged fermentation. Among the main players in citric acid production, it is possible to find Archer Daniels Midland Company; Cargill, Inc.; Tate & Lyle PLC; Jungbunzlauer Suisse AG; Cofco Biochemical (Anhui) Co., Ltd.; Huangshi Xinghua Biochemical Co. Ltd.; RZBC Group Co. Ltd.; Weifang Ensign Industry Co., Ltd.; Gadot Biochemical Industries Ltd.; and S.A. Citrique Belge N.V.

5.3 Lactic Acid (400 kTons/Year)

(2-Hydroxypropanoic acid; $C_3H_6O_3$; 90.07948 g/mol; pKa1 = 3.78 at 25 °C)

Lactic acid (LA) was first identified from sour milk by the chemist Carl Wilhelm Scheele in 1780. It can occur in two isomeric forms: L(+) or D(-) or in the racemic form DL-lactic acid. LA production can be obtained chemically or through fermentation processes, which have the advantages to provide the pure forms (L(+) or D(-)). Fermentation processes that involve the production of lactic acid have been used in food preparation since centuries. The 90% of nowadays production of LA is obtained by fermentative processes with several players involved in the business (Ajala et al. 2020). Lactic acid has versatile properties as a precursor for various chemicals and materials (Fig. 5.2). Nevertheless, the major application of LA is in the manufacturing of polylactic acid (PLA), which is considered the most promising polyester for bioplastic production, as the properties of the PLA-based materials can find applications in packaging, agriculture, transport, electronics, and textiles and are also suitable for 3D printing (Naser et al. 2021). PLA is made through

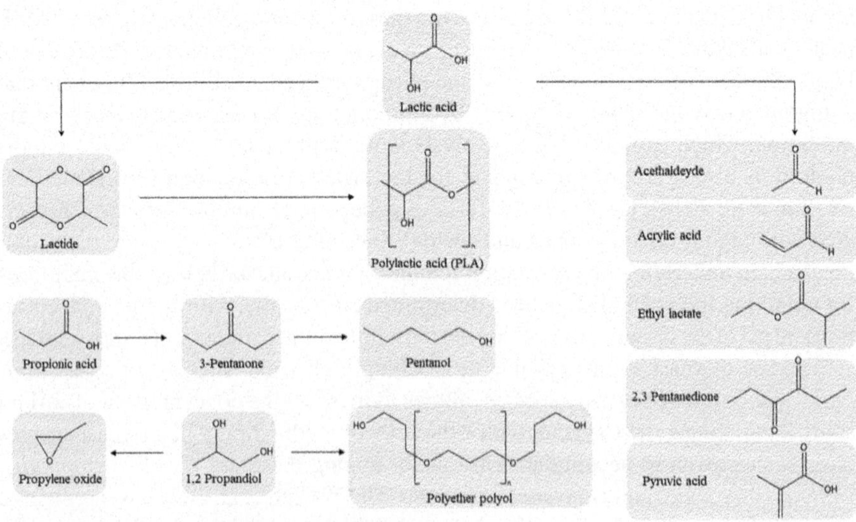

Fig. 5.2 Lactic acid and industrially relevant derivatives

sequential chemical reactions often occurring through the cyclization of two LA molecules into lactide, followed by polymerization. At lab scale, it has been very elegantly proven that PLA can be directly produced as non-naturally occurring biomaterial at intracellular level from engineered bacteria (Jung and Lee 2011) (WO2017108577A1 2008). The biggest LA producers at industrial scale are Corbion and Cargill, which produce around 240,000 and 180,000 tons, respectively (Jem and Tan 2020), while the main suppliers of PLA are Cargill (Minneapolis, MN, USA) Galactic (Escanaffles, Belgium), Purac (Amsterdam, The Netherlands), FKuR (Willich, Germany), and Shimadzu (Kyoto, Japan).

Downstream and purification contribute for 50% of the overall production costs of LA, and this relates directly to the choice of the production organism. Several microorganisms can produce pure L-LA, including bacteria, fungi, cyanobacteria, algae, and yeasts, while D-LA production has been proven for *Lactobacillus delbrueckii*, *Sporolactobacillus*, *Escherichia coli*, and *Bacillus coagulans* (Di Lorenzo et al. 2022).

Lactic acid bacteria (LABs) are naturally producers of LA through the reduction of glycolytic pyruvate, and they are still considered the best performers (Di Lorenzo et al. 2022). LABs produce LA in anaerobic and micro-aerobic conditions, and it is possible to distinguish homofermentative LABs (such as *Streptococcus*, *Lactococcus*, *Enterococcus*, *Pediococcus*, and *Lactobacillus* genera) that produce LA as main fermentation product from heterofermentative LABs (such as *L. brevis*, *L. fermentum*, and *L. reuteri*), which produce LA alongside with ethanol, acetic acid, and carbon dioxide. Homofermentative LABs are considered the best performing ones, due to their ability to reach yield values very close to the maximum theoretical one (1 g/g through the Embden-Meyerhof-Parnas pathway) (Di Lorenzo et al. 2022).

Since low pH negatively affects the metabolic activity of LABs, the buffering of the cultivation media is pivotal to maximize LA production. This is achieved by the addition of either $Ca(OH)_2$ or $CaCO_3$ causing the formation of calcium lactate that must be treated with H_2SO_4 to obtain the desired lactic acid generating calcium sulfate, gypsum, a waste product with serious disposal concerns (Porro et al. 1999). In light of this, different yeasts have been engineered with diverse heterologous lactate dehydrogenase encoding genes, to produce lactic acid: yeasts can grow and be metabolically active at pH values lower than the pKa of lactic acid, therefore ensuring the production of the desired undissociated form of the organic acid. To maximize yield, alcoholic fermentation has been partially or fully abolished (Sauer et al. 2010) (Suominen 2006), while to maximize titer, adaptive laboratory evolution or screening protocols (Valli et al.) have been applied (Valli et al. 2006).

5.4 Fumaric Acid (90 kTons/Year)

(*trans*-Butenedioic acid; $C_4H_4O_4$; 116.07 g/mol; pKa1 = 3.03; pKa2 = 4.54 at 25 °C)

Fumaric acid is a key intermediate metabolite in the TCA cycle firstly isolated from the plant *Fumaria officinalis*, from which it derives its name (Fig. 5.1). FA market size by 2023 is reported to be around 700M USD with a CAGR of 4% by

2032 (Fumaric Acid Market Size, Price Trends | Forecast 2032). The majority of global fumaric acid production is used for the production of resins and unsaturated polyesters (UPRs), but its application is also spreading to food and beverages and ruminal feed (Li et al. 2018). The use of fumaric acid in food applications is based upon its acidulent, preservative, and flavor enhancing and dough improvement characteristics, which differentiate fumaric acid from other more general purpose acidulates, such as citric acid and phosphoric acid. Although the current predominant route to fumaric acid is petrochemical, via maleic anhydride (using either benzene or butane as a feedstock) at high temperatures (90–100 °C), fermentation processes have been optimized over the years. Some high yield processes for the bio-based production of fumaric acid have been reported in the literature, mainly using *Rhizopus* species (Roa Engel et al. 2011) as cell factories. Few alternative processes based on the enzymatic conversion (maleate isomerases from *Bacillus*, *Pseudomonas*, or *Arthrobacter*) of maleic acid have been developed, but none of them reached industrial applications yet (Di Lorenzo et al. 2022).

In 1940 Pfizer reported the first industrial bioprocess for FA production using *Rhizopus arrhizus*, reaching a production of 4000 tons/year. This fungus, alongside with *R. oryzae*, *R. nigricans*, and *R. formosa*, can accumulate and secrete FA in aerobic and nitrogen-limited conditions through the reductive TCA cycle (rTCA) (Fig. 5.1). FA biosynthesis relies on the cytosolic production of oxaloacetate through the condensation of pyruvate and CO_2 (with ATP consumption) mediated by the pyruvate carboxylase. Oxaloacetate is then converted into malate (malate dehydrogenase) and fumarate (fumarase) (Roa Engel et al. 2011). Nitrogen limitation impairs cellular growth, while the simultaneous occurrence of TCA and rTCA results in the consumption of carbon source, ATP production, and CO_2 fixation into fumarate. This flux distribution justifies the suboptimal yields (lower than theoretical 2 mol *per* mol of glucose) of FA production in glucose-fed, aerobic, and nitrogen-limited bioprocesses exploiting *Rhizopus* species (Fig. 5.1). Nevertheless, a *R. arrhizus* strain able to produce more than 130 g/L of FA was patented by Du Pont as the best performing cell factory (Ling and Ng 1988). Generally speaking, these processes are limited by the fungi morphology and by the pH of the production medium. Pellet morphology is the best performing one, allowing for efficient mass transfer and preventing the increase in medium viscosity (Zhou et al. 2011). pH also plays a crucial role in preserving the cell factory performances, and $CaCO_3$ is the most commonly neutralizing agent (the process is conducted at close to neutral pH) used in the process. However, being the calcium fumarate salt insoluble, it accumulates during the fermentation, increasing the viscosity of the media, limiting oxygen mass transfer, and increasing the power input required to keep the culture broth well mixed. The formation of an insoluble fraction also complicates the product recovery, and a broth heating step is required to solubilize the calcium fumarate and separate it from the biomass and other solids in the fermentation broth. The use of $CaCO_3$ also results in the generation of gypsum ($CaSO_4 \bullet H_2O$) during the separation/purification of the fumaric acid. As for lactic acid, this by-product constitutes an industrial waste, and the associated cost of disposal negatively impacts the overall process economics (Gangl et al. 2024).

A possible work around for this issue may be to release the pH control in the late phase of the batch. This has been shown not to affect the fermentation performance and strongly reduce the amount of $CaCO_3$ required to buffer the media for the development of a one-step recovery fermentation process characterized by titers > 80 g/L with a productivity of 1.8 g/L h and 0.57 (g/g glucose) yield. The process was scaled to 2000 L pilot scale, and the techno-economic analysis projected interesting production costs for an appropriate use in the food and beverage application. Meanwhile, recycling and reuse of the base were demonstrated and incorporated into the process, as well as the ability of *R. oryzae* to produce fumaric acid from other carbon sources, including biomass hydrolysate (Di Lorenzo et al. 2022).

5.5 Succinic Acid (70 kTons/Year)

(Butanedioic acid; $C_4H_6O_4$; 118.09 g/mol; pKa1 = 4.2; pKa2 = 5.6 at 25 °C)

Succinic acid (SA) is a chemical building block that replaces petroleum-based maleic anhydride in polyesters, alkyd resins, polyurethanes, plasticizers, and solvents: it is also used in the manufacture of food and pharmaceuticals, among other products (Fig. 5.3). Succinic acid demand is predicted to expand to commodity chemical

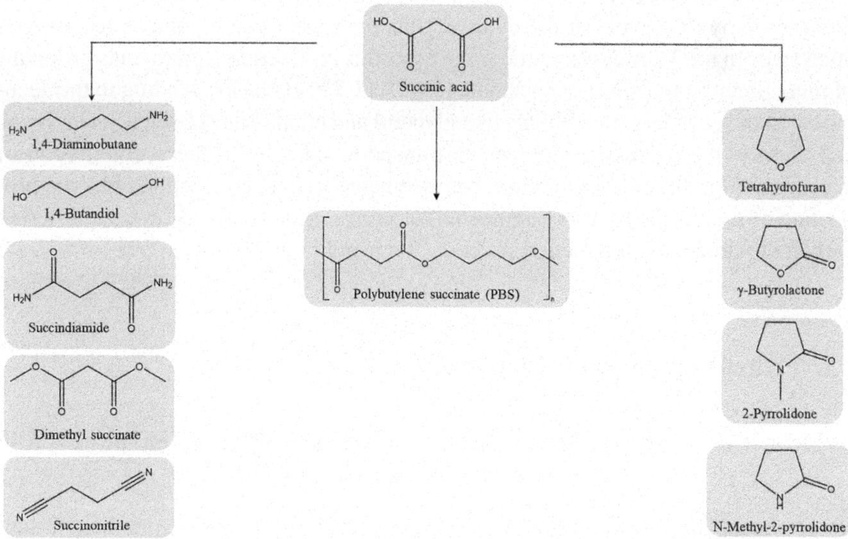

Fig. 5.3 Succinic acid and industrially relevant derivatives

status with 70,000 tons/year and a CAGR of 6.4% by 2032 (Succinic Acid Market Size and Share Report 2022-2030). SA represents the key precursor for the production of several bulk chemicals including 1,4-butanediol, gamma-butyrolactone, tetrahydrofuran, adipic acid, *N*-methylpyrrolidone, linear aliphatic esters, or biodegradable polymers such as polybutylene succinic acid (PBS) (Fig. 5.3). Succinic acid is a key intermediate metabolite in the central carbon flux and can be produced via the reductive or oxidative pathway of TCA cycle or via the glyoxylate shunt (Fig. 5.1). Production of succinic acid has been actively studied using diverse microorganisms including rumen bacteria such as *Actinobacillus succinogenes*, *Mannheimia succiniciproducens*, *Basfia succiniciproducens*, and other microorganisms such as *Escherichia coli*, *Anaerobiospirillum succiniciproducens*, *Corynebacterium glutamicum*, and *Saccharomyces cerevisiae* (Choi et al. 2015). Nowadays, production of succinic acid by microbial fermentation is still less remunerative than petrochemical processes (Dienst and Onderzoek 2015). Nevertheless, several companies are commercializing bio-based succinic acid, with Myriant (Quincy, MA, USA), Reverdia (Cassano Spinola, Italy), and Succinity (Montmelò, Spain) among the main players. The company Reverdia developed a process based on engineered *S. cerevisiae* strain capable of producing 100 g/L of succinic acid by fermentation. The proprietary yeast-based fermentation process, which operates at a much lower pH than competing processes, allows succinic acid to be produced with significantly higher energy efficiencies compared to the bacteria or chemical traditional method. Data for industrially achieved productivities are not available; however, the engineering strategy of the *S. cerevisiae* cell factory includes a combination of the reductive and oxidative TCA cycle (Raab and Lang 2011). In brief, a yeast was constructed with deletions for the genes of succinate and isocitrate dehydrogenase, which redirected carbon flux into the glyoxylate cycle in the cytosol, thus bypassing two oxidative decarboxylation steps in the TCA cycle; at the same time, this redirection circumvents problems of mitochondrial transport of succinate (Fig. 5.1). The pathways leading to the accumulation of secondary metabolites like glycerol and ethanol have been deleted. In the cell factory, the overexpression of the transporter leading to the accumulation of succinic acid in the culture medium has also been considered. Remarkably, the production of succinic acid from engineered *S. cerevisiae* cells also allows sequestering carbon dioxide in the production process (Raab and Lang 2011).

5.6 Itaconic Acid (41.4 kTons/Year)

(2-Methylene butanedioic acid; $C_5H_6O_4$; 130.10 g/mol; pKa1 = 3.85; pKa2 = 5.45 at 25 °C)

Itaconic acid (IA) is an unsaturated carboxylic acid also known as methylene succinic acid. It is used exclusively in nonfood applications as precursor in styrene-butadiene rubber, synthetic latex, methyl methacrylate (MMA), and unsaturated polyester resins (UPRs) (Fig. 5.4) (Oke Teleky and Cristian Vodnar 2019). Itaconic

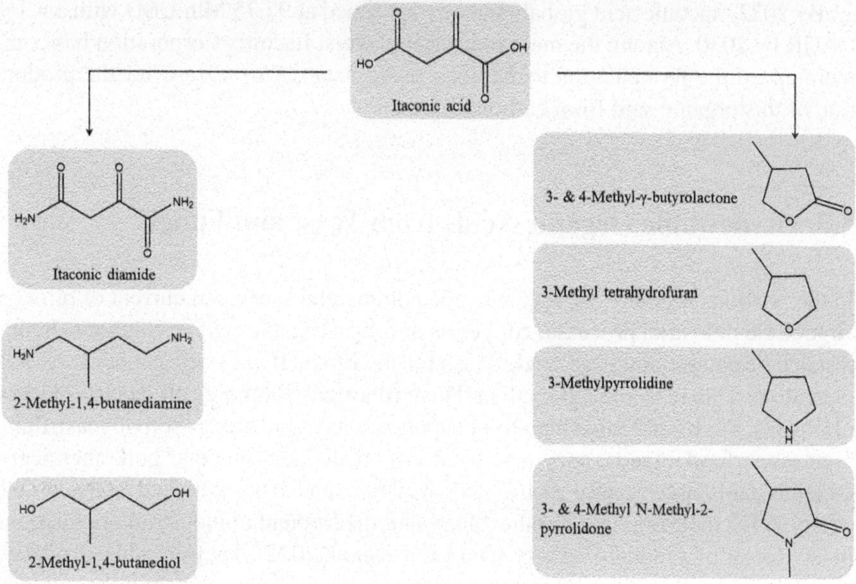

Fig. 5.4 Itaconic acid and industrially relevant derivatives

acid is not a primary metabolite, so both the anabolism and catabolism of this acid are relatively rare metabolic attributes. This acid was selected as a top building block chemical in the 2004 report, but it was excluded in the 2010 report because the market size was not expanded as expected. However, itaconic acid is also a building block for adhesives and sealants, finishing agents, paint and coating additives, detergents and cleaners, and absorbents and dispersants and can replace banned chemical phosphates in detergents.

Fermentation processes for itaconic acid production are generally preferred to chemical synthesis as the last is time demanding and expensive and results in lower yields. The biomanufacturing of itaconic acid mainly relies on *Aspergillus terreus* due to its high production yield and titer (Cunha da Cruz et al. 2018). Nevertheless, the filamentous fungus *Ustilago maydis* has also been recently explored for industrial production of itaconic acid (Cunha da Cruz et al. 2018).

In *A. terreus*, mitochondrial acetyl-CoA is condensed with oxaloacetate into *cis*-aconitate that, in turn, is decarboxylated to itaconate by *cis*-aconitate decarboxylase (Fig. 5.1). Differently, *U. maydis* lacks *cis*-aconitate decarboxylase and requires the isomerization of *cis*-aconitate into its *trans* form, followed by the decarboxylation of *trans*-aconitate in itaconic acid (Zhao et al. 2018). To push itaconic acid production, both *U. maydis* and low pH tolerant species such as *Pichia kudriavzevii* and *A. niger* have been engineered with *A. terreus* production pathways. Nevertheless, the natural producer *A. terreus* remains by now the best performing cell factory (W. Sun et al. 2020; Xie et al. 2020).

By 2022, itaconic acid global market was valued at 97.75 Mln USD with a 4.1% CAGR by 2030. Among the main industrial players, Itaconix Corporation has commercialized the fermentation technology using *Aspergillus terreus* for the production of this organic acid from carbohydrates.

5.7 Renewable Organic Acids from Yeast and Fungi

In the previous sections, we reviewed the commercial aspects of current or piloting large-scale industrial processes for yeasts or fungal organic acid productions. In this respect, yeast and fungi are workhorses for the production of organic acids by fermentation. This is facilitated by the relative robustness of these cell factories to low pH values. Yeasts and fungi also found applications in the production of yeast-fungi biomasses, foods, feeds, enzymes, pharmaceuticals, and fine and bulk chemicals. Organic acids are broadly distributed in nature and have attracted considerable attention for their role in natural ecology and their potential industrial applications in any sector of a modern society (Di Lorenzo et al. 2022). They could be produced at industrial level using wild-type or engineered cells. Yeast and fungi are already producers, even if not at economic level, of other important organic acids like acetic acid, ascorbic acid, butyric acid, epoxy succinic acid, formic acid, gallic acid, gluconic acid, isobutyric acid, isocitric acid, kojic acid, malic acid, oxalic acid, oxaloacetic acid, propionic acid, pyruvic acid, tartaric acid, and α-ketoglutaric acid, to mention the most relevant. Even prokaryotic cell factories can be used for the production of organic acids. However, they often require neutral pH growth conditions and consequently accumulate the salt forms of organic acids, which require cost-intensive acidification and precipitation to retrieve the desired products. Advances and improvements in metabolic engineering and bioprocess development will make yeasts and fungi better candidates for environmentally responsible and economically feasible fermentation processes for the production of organic acids. Furthermore, the arising of synthetic biology, prompted by the development of novel cloning tools and faster and cheaper synthesis of DNA fragments, opens to the design of specific metabolic and regulatory circuits using orthogonal biological components. The application of this strategy to metabolic engineering will facilitate and expand the construction of novel cell factories also taking advantage of novel AI tools (Nielsen et al. 2022). The combination of this paradigm to newly discovered enzymes, pathways, and species will allow to leverage the available biodiversity to foster strain development.

5.8 Conclusions and Perspectives

Analysts predict a rapid expansion of renewable chemicals production in the near future based on planned capacity expansion or new construction. In March 2024, the European Commission published a communication to the EU Parliament and Council focusing on the boosting of biotechnology and biomanufacturing in the EU area (Committee and the Committee of the Regions Building the Future with Nature: Boosting Biotechnology and Biomanufacturing in the EU 1 2024). While the EU biotechnology strategy will be updated by the end of 2025, the cited communication describes biotechnology as "a critical technology for the economic security perspective." By 2021, the global biotechnology market size reached EUR 720 billion, of which the USA shared 60%, followed by EU (12%) and China (11%). From 2008 to 2018, the EU biotechnology sector increased more than twice as fast as the overall economy of the area making it one of the fastest growing industries in the EU. The impact of these technologies on agriculture, forestry, energy, food, and feed sectors can make the EU more competitive and resilient. Nevertheless, tangible actions are required to overcome several limitations still present in the EU panorama. The technological transfer of research output to the market will require better access to financing and the streamlining of regulatory procedures. Moreover, the financial education and research must be incremented, also focusing on the public acceptance of novel technologies. To pursue this goal, the EU Commission financed, among other actions, the developing distributed research infrastructure IBISBA (Industrial Biotechnology Innovation and Synthetic Biology Accelerator) developed to provide academics and industry with services and expertise to facilitate the technological transfer of early-stage research to industry (Committee and the Committee of the Regions Building the Future with Nature: Boosting Biotechnology and Biomanufacturing in the EU 1 2024). In this framework and according to its roadmap, synthetic biology will promote the commercial development of new renewable chemicals via faster and cheaper methodologies that take advantage from an appropriate design of microorganisms (Nielsen et al. 2022). In this context, the constraints for creating a viable bio-business with the production of commodities with a low selling price starting from renewable resources are still tight. In order to establish an economically viable industrial process, the classical parameters to fulfil are titer, yield, and productivity (Porro et al. 2014). Titer (product concentration in the fermentation broth) is important to keep the purification costs as low as possible. Yield (product *per* substrate ratio) is important to gain sufficient revenue from the invested substrate costs. Productivity (product *per* bioreactor volume and time) relates to the size of the bioreactors and therefore to the investment costs. Remarkable microbial successful commercial examples outside the yeast and fungi world are the production of octadecane dioic acid (ODDA) (this C18 diacid is now commercially available from two pathways, one using microbial fermentation—no more data are available—and one using metathesis of natural oils; Elevance Renewable Sciences, USA), acetone (Green Biologics, USA), artemisinin (Amyris, USA), biofuels (many companies in the USA, EU, Asia), bioisoprene (DuPont Ind. Biosciences, Goodyear, USA, and Ajinomoto, J), hydrocortisone (Sanofi, F), isoprene (GlycosBio, USA), polyhydroxyalkanoate (PHA, Metabolix's, USA), polytetrahydrofuran (from renewable 1,4-BDO, BASF, D), resveratrol (Fluxome, DK), 1,2-propanediol (ADM, USA), 1,3 propanediol (1,3-PDO™, DuPont Tate & Lyle, USA), and 1,4-butanediol (1,4-BDO) (BASF, D) (Novamont, It), with many other new compounds at the pilot plant phase (Choi et al. 2015).

Analyzing all these bioprocesses, the rational engineering was first focused on overexpressing genes for enzymes involved in the pathway and knocking out genes responsible for competing reactions (to abolish by-product formation). However, in most cases, this is not enough to come even close to industrially relevant titer, yield, and productivity. Aspects related to the metabolic balance including (1) stoichiometry, (2) redox, and (3) energy households need to be balanced. Furthermore, (4) flux coupling, (5) irreversible reactions and (6) process, and cell factory robustness add to the efficiency of the pathway when the balances are fulfilled (Table 5.3).

Cellular robustness depicts the ability of a living organism to withstand internal and external perturbation maintaining the overall functionality of the system. In industrial biotechnological applications, this property is declined as the ability to preserve the cell factory performance moving from lab scale to industrial scale processes and for the entire duration of the production campaign (Olsson et al. 2022). Among the several stressors that can negatively affect the cell factory performances may be found the seasonal or territorial variability of the substrates; the spatial heterogeneity of the fermentation parameters such as temperature, oxygen supply, concentration of nutrients, and toxic end products; the accumulation of mutations; and

Table 5.3 Metabolic features constituting successful microbial production processes

Metabolic feature	Significance
Thermodynamic feasibility	Thermodynamics is ultimately decisive if a chemical reaction or metabolic pathway is possible
Stoichiometric balance	The theoretical stoichiometric yield—especially of carbon atoms—defines the potential economic feasibility of a production process
Redox balance	If substrate and product have a different degree of reduction, re-equilibration of the electron balance leads either to by-product formation or energy overflow
Energy balance	If production requires metabolic energy, part of the substrate is lost for its supply
Flux coupling	Coupling the product flux to the central carbon flux by a shared reaction may create a strong driving force toward product formation
Irreversible reactions	Including an irreversible step, like decarboxylation, creates a metabolic driving force pulling flux toward the product
Transport	High production must be associated to an efficient accumulation of the final compound at extracellular level. Uptake of the substrate and transport among different organelles and the cytoplasm are included
Microbial and bioprocess robustness	Including tolerance to the final product and to adverse bioprocess conditions (i.e. low pH, high T, low RT, low OTR, high substrate concentration, etc.)

cellular damage (Olsson et al. 2022). Modularity, redundancy, and dynamic control of gene expression are the three main identified strategies by which microbial populations robustly withstand perturbations. Modularity defines the cellular architecture by which specific pathways, such as metabolic or signal transduction ones, work independently from the rest of the system. This strategy allows to reduce the propagation of the perturbation effect (such as gene deletion or enzymatic inhibition) to the entire biological system. Redundancy defines the existence of alternative molecular strategies to achieve the same biological function (such as the availability of multiple copies for essential genes or enzymes with overlapping specificity). In turn, dynamical control of gene expression allows for the activation of cellular processes upon specific perturbation. This robustness strategy is, among others, the more specialized one requiring a dynamic evolution of receptors, activators, and signaling molecules enabling the direct response to environmental and genetic conditions (Olsson et al. 2022).

Interestingly, an increasing amount of scientific evidence supports the pivotal role of the population variability of a cell culture toward environmental and genetic perturbations. For instance, it has been described that an isogenic *S. cerevisiae* culture growing on glucose presents a subpopulation of slow growing cells characterized by the overexpression of *TSL1*, a gene involved in trehalose production and heat shock protection. Upon heat shock, *TSL1* overexpressing subpopulation resulted in a higher fitness, representing the majority of the active yeast cells (Jing et al. 2018).

To engineer and exploit microbial robustness, it is necessary to develop a theoretical framework for its quantitative description. Kitano defined cell robustness as the integration of the product of the probability function of occurrence of a specific perturbation and the function evaluating the impact of this perturbation on a specific cellular function (Olsson et al. 2022; Kitano 2007). Furthermore, Becker and coworker recently developed a mathematical framework to analyze the impact of the variation in some process parameters on the process outputs allowing for the identification of upper and lower limits of the key process variables (Becker et al. 2023).

A last and important consideration should be dedicated to the transport of the final product. Organic acids in their salt-charged status cannot diffuse across the cellular membranes, while the lipophilic undissociated form of the acid (RCOOH) may permeate the plasma membrane by simple diffusion, even if with a very small efficiency (this is the key base principle of a weak organic acid as preservatives). Consequently, high organic acid productions must be associated with the secretion and accumulation of the final compound at extracellular level (see Table 5.3). Surprisingly, of all the industrial organic acid productions described in this chapter, only for succinic acid the transport of the final compound at extracellular level has been taken into consideration during the rational design of the cell factory.

On the contrary, to produce lactic acid from recombinant yeast cells, the secretion of the organic acid is still based on the activity of one or more endogenous, probably nonspecific, organic acid transporter(s). For all the other organic acids, the accumulation at extracellular level is obtained throughout the endogenous transporter(s), even if in most cases it is still unknown/uncharacterized. It can be easily anticipated that engineering of the plasma membranes, as well as the development of synthetic membranes, for improving the efficiency of the transport and reducing the passive diffusion of the undissociated form back inside the cells, will lead to great developments (Ferraz et al. 2021).

Finally, it might be worthwhile to rethink organic acid chemistry considering which compounds can be produced efficiently by microbial activity. Classical chemistry developed to find applications for the abundant substrates retrievable from petrol, often introducing novel functional groups. On the contrary, nowadays biotechnology is dedicated to the replacement of current platform chemicals (developed and used by petrochemistry) starting from renewable biomasses and generally involving a reduction in the number of functional groups. Since most of the classical building block molecules are not efficiently produced naturally, we strongly suggest a paradigm shift toward bio-based platform chemicals, going beyond the substitution of existing production processes and rethinking them for the large-scale valorization of biomanufacturable compounds.

Acknowledgments PB, DP, and IS acknowledge the National Center 5 "National Biodiversity Future Center" (Award Number: Project code CN_00000033, Concession Decree No. 1034 of 17 June 2022 adopted by the Italian Ministry of University and Research, CUP H43C22000530001, Project title "National Biodiversity Future Center – NBFC. Funder: Project funded under the National Recovery and Resilience Plan (NRRP), Mission 4 Component 2 Investment 1.4—Call for tender No. 3138 of 16 December 2021, rectified by Decree n.3175 of 18 December 2021 of Italian Ministry of University and Research funded by the European Union – NextGenerationEU).

PB and RM acknowledge MUSA—Multilayered Urban Sustainability Action—project, funded by the European Union, NextGenerationEU, under the National Recovery and Resilience Plan (NRRP) Mission 4 Component 2 Investment Line 1.5: Strengthening of Research Structures and Creation of R&D "Innovation Ecosystems," set up of "territorial leaders in R&D."

References

Ajala EO, Olonade YO, Ajala MA, Akinpelu GS (2020) Lactic acid production from lignocellulose – a review of major challenges and selected solutions. ChemBioEng Reviews 7(2):38–49. https://doi.org/10.1002/CBEN.201900018

Becker L, Sturm J, Eiden F, Holtmann D (2023) Analyzing and understanding the robustness of bioprocesses. Trends Biotechnol 41(8):1013–1026. https://doi.org/10.1016/J.TIBTECH.2023.03.002

Bozell JJ, Petersen GR (2010) Technology development for the production of biobased products from biorefinery carbohydrates-the US Department of Energy's "top 10" revisited. Green Chem 12:539. https://doi.org/10.1039/b922014c

Choi S, Song CW, Shin JH, Lee SY (2015) Biorefineries for the production of top building block chemicals and their derivatives. Metab Eng 28:223–239. https://doi.org/10.1016/J.YMBEN.2014.12.007

Citric Acid Market Size, Share and Trends Analysis 2032. Retrieved 9 July 2024, from https://www.imarcgroup.com/citric-acid-manufacturing-plant

Committee and the Committee of the Regions (2024) Building the future with nature: Boosting Biotechnology and Biomanufacturing in the EU. Retrieved 9 July 2024, from https://www.statista.com/statistics/1246614/top-countries-share-of-global-biotech-value/

Cunha da Cruz J, Machado de Castro A, Camporese Sérvulo EF (2018) World market and biotechnological production of itaconic acid. 3Biotech 8(3):1–15. https://doi.org/10.1007/S13205-018-1151-0/FIGURES/5

Di Lorenzo RD, Serra I, Porro D, Branduardi P (2022) State of the art on the microbial production of industrially relevant organic acids. Catalysts 12:234. https://doi.org/10.3390/CATAL12020234

Dienst S, Onderzoek L (2015) From the sugar platform to biofuels and biochemicals. Final report for the European Commission Directorate-General Energy Consorzio per la Ricerca e la Dimostrazione sulle Energie Rinnovabili (RE-CORD)

Ferraz L, Sauer M, Sousa MJ, Branduardi P (2021) The plasma membrane at the cornerstone between flexibility and adaptability: implications for saccharomyces cerevisiae as a cell factory. Front Microbiol 12:715891. https://doi.org/10.3389/fmicb.2021.715891

Fumaric Acid Market Size, Price Trends | Forecast 2032. Retrieved 9 July 2024 from https://www.imarcgroup.com/fumaric-acid-market

Gangl IC, Weigand WA, Keller FA (2024) Economic comparison of calcium fumarate and sodium fumarate production by *Rhizopus arrhizus*. Appl Biochem Biotechnol 24:663–677

Jem KJ, Tan B (2020) The development and challenges of poly (lactic acid) and poly (glycolic acid). Adv Ind Eng Polym Res 3(2):60–70. https://doi.org/10.1016/j.aiepr.2020.01.002

Jing W, Camellato B, Roney IJ, Kaern M, Godin M (2018) Measuring single-cell phenotypic growth heterogeneity using a microfluidic cell volume sensor. Sci Rep 8:17809. https://doi.org/10.1038/s41598-018-36000-3

Jung YK, Lee SY (2011) Efficient production of polylactic acid and its copolymers by metabolically engineered *Escherichia coli*. J Biotechnol 151(1):94–101. https://doi.org/10.1016/J.JBIOTEC.2010.11.009

Karaffa L, Kubicek CP (2003) *Aspergillus niger* citric acid accumulation: do we understand this well working black box? Appl Microbiol Biotechnol 61:189–196. https://doi.org/10.1007/s00253-002-1201-7

Kitano H (2007) Towards a theory of biological robustness. Mol Syst Biol 3:137. https://doi.org/10.1038/msb4100179

Książek E (2023) Citric acid: properties, microbial production, and applications in industries. Molecules (Basel, Switzerland) 29(1):22. https://doi.org/10.3390/MOLECULES29010022

Li Z, Liu N, Cao Y, Jin C, Li F, Cai C, Yao J (2018) Effects of fumaric acid supplementation on methane production and rumen fermentation in goats fed diets varying in forage and concentrate particle size. J Anim Sci Biotechnol 9:21. https://doi.org/10.1186/s40104-018-0235-3

Ling LB, Ng TK (1988) Fermentation process for carboxylic acids. Retrieved 14 July 2024 from https://patents.google.com/patent/US4877731A/en. US4877731A

Mlakar T, Legiša M (2006) Citrate inhibition-resistant form of 6-Phosphofructo-1-kinase from *Aspergillus niger*. Appl Environ Microbiol 72(7):4515. https://doi.org/10.1128/AEM.00539-06

Naser AZ, Deiab I, Darras BM (2021) Poly(lactic acid) (PLA) and polyhydroxyalkanoates (PHAs), green alternatives to petroleum-based plastics: a review. RSC Adv 11:17151–17196. https://doi.org/10.1039/d1ra02390j

Nielsen J, Tillegreen CB, Petranovic D (2022) Innovation trends in industrial biotechnology. Trends Biotechnol 40(10):1160–1172. https://doi.org/10.1016/J.TIBTECH.2022.03.007

Oke Teleky B-E, Cristian Vodnar D (2019) Biomass-derived production of Itaconic acid as a building block in specialty polymers. Polymers 11(6):1035. https://doi.org/10.3390/polym11061035

Olsson L, Rugbjerg P, Torello Pianale L, Trivellin C (2022) Robustness: linking strain design to viable bioprocesses. Trends Biotechnol 40(8):918–931. https://doi.org/10.1016/J.TIBTECH.2022.01.004

Papagianni M (2007) Advances in citric acid fermentation by *Aspergillus niger*: biochemical aspects, membrane transport and modeling. Biotechnol Adv 25(3):244–263. https://doi.org/10.1016/J.BIOTECHADV.2007.01.002

Porro D, Bianchi MM, Brambilla L, Menghini R, Bolzani D, Carrera V, Lievense J, Liu CL, Ranzi BM, Frontali L, Alberghina L (1999) Replacement of a metabolic pathway for large-scale production of lactic acid from engineered yeasts. Appl Environ Microbiol 65(9):4211–4215. https://doi.org/10.1128/AEM.65.9.4211-4215.1999

Porro D, Branduardi P, Sauer M, Mattanovich D (2014) Old obstacles and new horizons for microbial chemical production. Curr Opin Biotechnol 30:101–106. https://doi.org/10.1016/J.COPBIO.2014.06.009

Raab AM, Lang C (2011) Oxidative versus reductive succinic acid production in the yeast *Saccharomyces cerevisiae*. Bioeng Bugs 2(2):120–123. https://doi.org/10.4161/bbug.2.2.14549

Roa Engel CA, Van Gulik WM, Marang L, Van der Wielen LAM, Straathof AJJ (2011) Development of a low pH fermentation strategy for fumaric acid production by *Rhizopus oryzae*. Enzym Microb Technol 48(1):39–47. https://doi.org/10.1016/J.ENZMICTEC.2010.09.001

Sauer M, Porro D, Mattanovich D, Branduardi P, Sauer. (2010) 16 years research on lactic acid production with yeast – ready for the market? Biotechnol Genet Eng Rev 27(1):229–256. https://doi.org/10.1080/02648725.2010.10648152

Succinic Acid Market Size & Share Report, 2022–2030. Retrieved 9 July 2024 from https://www.grandviewresearch.com/industry-analysis/succinic-acid-market

Sun X, Wu H, Zhao G, Li Z, Wu X, Liu H, Zheng Z (2018) Morphological regulation of *Aspergillus niger* to improve citric acid production by chsC gene silencing. Bioprocess Biosyst Eng 41:1029–1038. https://doi.org/10.1007/s00449-018-1932-1

Sun W, Vila-Santa A, Liu N, Prozorov T, Xie D, Faria NT, Ferreira FC, Mira NP, Shao Z (2020) Metabolic engineering of an acid-tolerant yeast strain *Pichia kudriavzevii* for itaconic acid production. Metab Eng Commun 10:e00124. https://doi.org/10.1016/J.MEC.2020.E00124

Suominen (2006) Genetically modified yeast of the species *Issatchenkia orientalis* and closely relates species, and fermentation processes using same

Valli M, Sauer M, Branduardi P, Borth N, Porro D, Mattanovich D (2006) Improvement of lactic acid production in *Saccharomyces cerevisiae* by cell sorting for high intracellular pH. Appl Environ Microbiol 72(8):5492. https://doi.org/10.1128/AEM.00683-06

Van Dien S (2013) From the first drop to the first truckload: commercialization of microbial processes for renewable chemicals. Curr Opin Biotechnol 24(6):1061–1068. https://doi.org/10.1016/J.COPBIO.2013.03.002

WO2017108577A1 - Recombinant yeast cells producing polylactic acid and uses thereof - Google Patents. (2008). Retrieved 14 July 2024 from https://patents.google.com/patent/WO2017108577A1/en

Xie H, Ma Q, Wei D, Wang F (2020) Metabolic engineering of an industrial *Aspergillus niger* strain for itaconic acid production. *3*. Biotech 10:113. https://doi.org/10.1007/s13205-020-2080-2

Zhao M, Lu X, Zong H, Li J, Zhuge B (2018) Itaconic acid production in microorganisms. Biotechnol Lett 40(3):455–464. https://doi.org/10.1007/S10529-017-2500-5/FIGURES/1

Zhou Z, Du G, Hua Z, Zhou J, Chen J (2011) Optimization of fumaric acid production by *Rhizopus delemar* based on the morphology formation. Bioresour Technol 102(20):9345–9349. https://doi.org/10.1016/J.BIORTECH.2011.07.120

Chapter 6
Functional Amino Acid Engineering: A New Breeding Technology for Brewer's Yeasts

Hiroshi Takagi

Contents

6.1 Introduction 142
6.2 Amino Acid Metabolism in Yeast 142
6.3 What Are Sake and Awamori? 144
6.4 How to Isolate or Construct Brewer's Yeasts with Higher Amino Acid Content 145
6.5 Enhanced Proline Synthesis with Lower Succinate and Its Application to "Light-Bodied" Sake Brewing 146
6.6 Enhanced Ornithine Synthesis via Proline/Arginine Metabolism and Its Application to "Healthy Image" Sake Brewing 148
6.7 Enhanced Volatile Compound Synthesis via Ehrlich Pathway and Its Application to "Flavor-Rich" Sake Brewing 150
6.8 Enhanced Isoamyl Acetate Synthesis via Leucine Metabolism and Its Application to "Aromatic" Awamori Brewing 153
6.9 Regulatory Mechanism of Proline Utilization and Its Application to "Proline-Less" Wine and Beer Brewing 155
6.10 Conclusions and Future Perspective 157
References 158

Abstract Amino acids are components of proteins in living organisms and exist not only as nutrients, energy sources, and taste/flavor components but also in free state within cells and plasma, where they play a variety of physiological functions. In recent years, their functions have been attracting attention, and many foods, beverages, supplements, pharmaceuticals, cosmetics, etc. with enhanced functionality by adding amino acids have been widely commercialized in the world. The budding yeast *Saccharomyces cerevisiae* is an important microorganism not only in basic

H. Takagi (✉)
Institute for Research Initiatives, Nara Institute of Science and Technology, Ikoma, Nara, Japan
e-mail: hiro@bs.naist.jp

A. A. Sibirny (ed.), *Biotechnology of Yeasts and Filamentous Fungi*, Grand Challenges in Biology and Biotechnology,
https://doi.org/10.1007/978-3-031-74726-7_6

biology but also in the fermentation industry. Amino acid metabolism and its regulatory mechanisms in yeast are diverse, and in basic science, the elucidation of these molecular mechanisms can deepen our understanding of biological phenomena and cellular functions. In the field of applied research, the artificial control of amino acid content inside and outside the cell is expected to lead to the production of amino acids themselves and metabolism-related substances, as well as fermented foods and alcoholic beverages with enhanced taste, flavor, health, and nutritional functions. We have established a breeding technology called "functional amino acid engineering" that artificially controls the content of "functional" amino acids, which can improve fermentation ability, diversify taste and flavor, impart a healthy image, and increase nutritional value, thereby developing high functionality of microorganisms and improving the productivity of useful substances.

6.1 Introduction

From the perspective of biotechnology applications, industrial yeast breeding has two major objectives: improving fermentation ability with enhanced stress tolerance and diversifying product taste and flavor by modifying metabolic pathways. Amino acids are one of the important factors affecting yeast fermentation. In yeast, amino acid metabolism and its regulatory mechanisms vary under different growth environments by controlling the processes of assimilation and catabolism, including uptake and export. Furthermore, these are complex processes that form robust networks, and cross talk occurs between the regulatory systems of various metabolites and signaling molecules within the cell. Controlling the composition and content of amino acids is expected to contribute to improving productivity and adding value to alcoholic beverages and fermented foods. This chapter mainly introduces our research outcomes through industry-government-academia collaboration, including the construction of highly functional brewer's yeast strains focusing on the metabolic regulatory mechanisms and physiological roles of "functional amino acids" and their applications to brewing of alcoholic beverages such as unique Japanese *sake* and *awamori*.

6.2 Amino Acid Metabolism in Yeast

Amino acids are the building blocks of proteins in living organisms and exist not only as nutrients, energy sources, and flavor components but also in a free state within the cell and plasma, where they play a variety of biological roles. In recent years, their physiological functions have attracted attention, and many foods, beverages, supplements, pharmaceuticals, cosmetics, and other products with enhanced functionality by adding amino acids have appeared on the worldwide market.

The yeast *Saccharomyces cerevisiae* is a model organism for higher eukaryote and an extremely important microorganism in the fermentation chemical industry. The amino acid metabolism and its regulatory mechanisms in yeast are diverse, and in basic science, the elucidation of these molecular mechanisms can deepen our understanding of biological phenomena and cellular functions. In the field of applied research, the artificial control of amino acid content inside and outside the cells is expected to lead to the production of amino acids themselves and metabolism-related substances, as well as fermented and brewed foods with enhanced taste, flavor, health, and nutritional functions.

Yeast can synthesize all amino acids intracellularly, given a carbon and nitrogen source (Ljungdahl and Daignan-Fornier 2012). Amino acids are classified into the glutamate, aromatic, serine, aspartate, and pyruvate families according to the precursors of their synthetic pathways (histidine is related to nucleotide synthesis). In particular, glutamate and glutamine are important amino acids involved in nitrogen metabolism and aminotransferase reactions. Recent studies have shown that the target of rapamycin (TOR) and general control non-repressible (GCN) pathways are important for the regulation of amino acid synthesis (Conrad et al. 2014). Yeast also divides nitrogen sources (amino acids, ammonia, urea, etc.) into those that are easy to utilize (preferred; good) and those that are difficult to utilize (non-preferred; poor). The transcription of metabolism-related genes is regulated by nitrogen catabolite repression (NCR) (Magasanik and Kaiser 2002). When an easily available nitrogen source is insufficient, NCR is released and a less available nitrogen source becomes available. This regulation depends on the genetic background of the yeast and differs depending on the strain. Genes involved in amino acid synthesis are regulated by general amino acid control (GAAC), and multiple synthetic enzymes are induced during amino acid deficiency (Hinnebusch 2005). GAAC is a transcription factor: Gcn4 binds to the promoter, and Gcn2 kinase phosphorylates translation initiation factor eIF2. Regulation of Gcn4 expression is also involved in the release of the NCR and reduction of proteasomal degradation.

Yeast senses extracellular nutritional status using cell membrane sensors. In particular, the SPS (Ssy1-Ptr3-Ssy5) sensor responds to external amino acids and regulates gene expression (Zhang et al. 2018). Ssy1 does not uptake amino acids but instead senses concentration differences through changes in local structure. Ssy5 has protease activity and is involved in the processing of Stp1 and Stp2. In addition, because the subcellular localization and transport of amino acids are important for their metabolism, transporters have been extensively studied (Bianchi et al. 2019). Yeast has a variety of transport systems and uptakes in amino acids in the right amount at the right time in response to environmental changes. At the cell membrane, Pma1 cotransports with H^+, while in the mitochondrial inner membrane, the MCP family is responsible for transport based on concentration differences (Zhao et al. 2021). The vacuole is the main storage site for amino acids, and transporters in the vacuolar membrane maintain amino acids at appropriate levels by V_0V_1 type H^+-ATPase (Kawano-Kawada et al. 2018).

6.3 What Are Sake and Awamori?

There are two types of alcoholic beverages, non-distilled alcoholic beverages, such as beer and wine, and distilled spirits, such as whisky and brandy. Among them, *sake* and *shochu* are unique Japanese alcoholic beverages that are the result of collaboration between fungi and yeasts (Fig. 6.1). Since sake, Japanese rice wine, is an integral part of the Japanese lifestyle and culture, it has been approved as the national alcoholic beverage, *Kokushu*. In addition, Japanese cuisine becomes more popular internationally, so the number of people who enjoy sake with Japanese food increases all over the world. Sake is made from steamed rice, which produces saccharification enzymes and ethanol from glucose by multiple parallel fermentations (simultaneous saccharification and fermentation) with the fungus *Aspergillus oryzae*, as *ki-koji* mold, and the yeast *S. cerevisiae*, respectively. The yeast strain used in sake brewing (sake yeast) can produce ethanol at a concentration of about 20% (v/v). During sake fermentation, yeast cells are exposed to high concentrations of ethanol, which inhibits the growth, viability, and fermentation of yeast cells (Kodama 1993). Therefore, increasing the ethanol tolerance of sake yeast strains could improve ethanol productivity and shorten the fermentation time. We expected that sake yeast cells would possess specific genes or mutations that enhance ethanol stress tolerance. To our surprise, however, sake yeast cells seem to be workaholics, losing self-protective genes, such as *MSN4* and *RIM15*, that inhibit ethanol production (Watanabe et al. 2012, 2016, 2019).

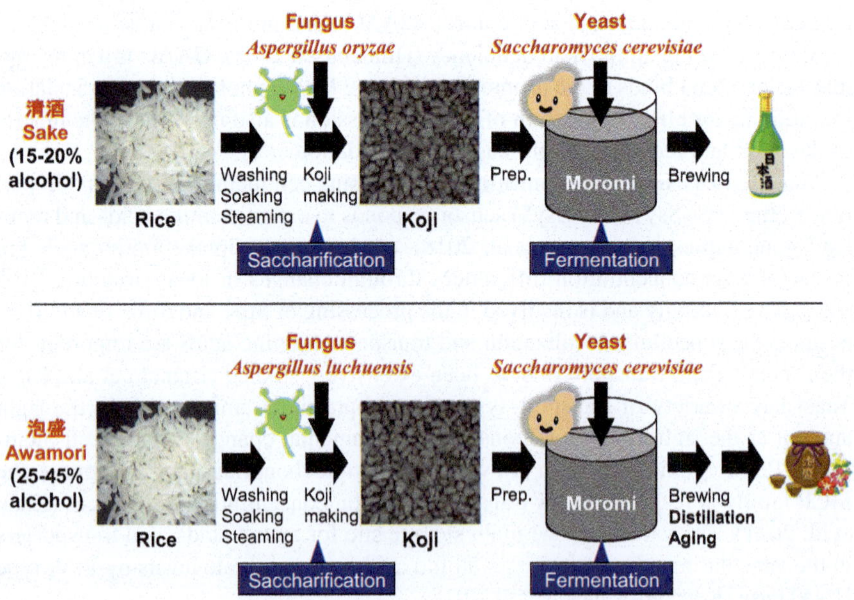

Fig. 6.1 How to brew unique Japanese alcoholic beverages, *sake* and *awamori*, through collaborations with fungus and yeast

Awamori is one type of shochu, a traditional spirit (distilled alcoholic beverage) made from steamed rice in the subtropical Okinawa region of Japan. With a history spanning more than 600 years, this traditional spirit represents Okinawa's culture and industry. Its brewing method is quite different from other distilled alcoholic beverages such as whiskey, which is made from grain mash, in which malt enzymes and yeast convert starch into sugar and alcohol. In awamori brewing, the fungus *Aspergillus luchuensis*, as *kuro-koji* mold, and the yeast *S. cerevisiae* are widely used to make fermented mash (*moromi*) and alcohol, respectively, through multiple parallel fermentation (simultaneous saccharification and fermentation). After the completion of fermentation for about 2 weeks, the alcohol is mainly distilled under atmospheric pressure and aged in barrels, resulting in a rich and strong flavor (Taira et al. 2012). Therefore, awamori has a unique aroma that is clearly different from other types of shochu, and the reduced-pressure distillation gives it a clean and light taste. Since flavors of awamori are mainly affected by the metabolism of yeast cells, yeast breeding technology and natural isolation could contribute to the awamori industry.

6.4 How to Isolate or Construct Brewer's Yeasts with Higher Amino Acid Content

We focused on the metabolic regulatory mechanisms and physiological roles of amino acids, which are important factors that affect the fermentation ability of yeast and the flavor of yeast products such as sake and awamori. How can we isolate yeast mutants that accumulate specific amino acids? In general, microorganisms including yeast that overproduce various amino acids have been obtained by isolating mutants resistant to the toxic analogs of those amino acids. For instance, Fig. 6.2 shows the resistance mechanism to L-azetidine-2-carboxylate (AZC), a toxic analog of L-proline (Pro). It is said that amino acid analogs are transported into the cell via amino acid transporters on the cell membrane. Some amino acid analogs (e.g., AZC) cause misfolding of proteins that are competitively incorporated with the corresponding amino acid (e.g., Pro), thereby inhibiting cell growth. However, cells that accumulate large amounts of Pro are resistant to AZC. In this way, toxic amino acid analog resistant mutants derived from the parent yeast strain can be isolated. Basically, the parent cells cannot grow in the presence of high concentrations of AZC due to its toxicity. However, when ethyl methanesulfonate (EMS) treatment or ultraviolet (UV) irradiation was used to introduce random mutations into the genome of many cells, several colonies appeared on agar plates containing AZC. AZC-resistant mutants were then collected, and the intracellular Pro content was measured.

As an application of recombinant yeast, self-cloning (SC) yeast is more readily accepted by consumers than genetically modified (GM) yeast. According to Japanese government guidelines, SC yeast does not need to be treated as GM yeast

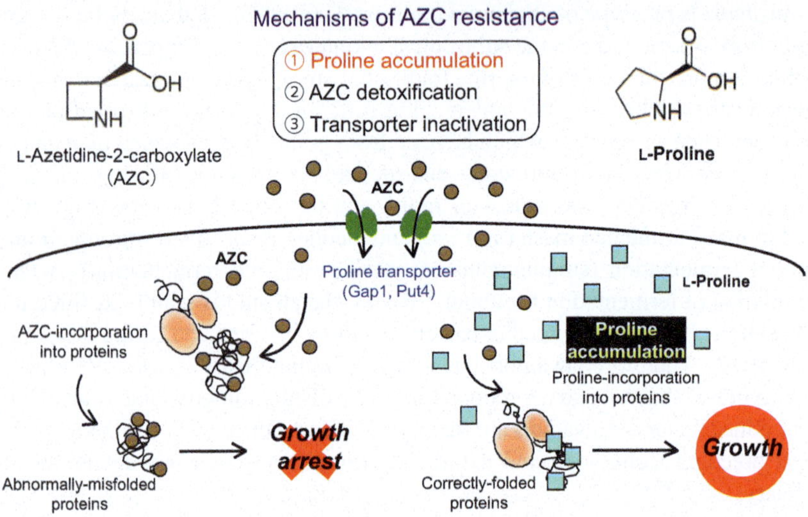

Fig. 6.2 How to isolate yeast mutants with Pro accumulation. The toxic Pro analog L-azetidine-2-carboxylate (AZC) causes misfolding of the proteins into which it is incorporated competitively with Pro and thereby inhibits cell growth. However, Pro-accumulating cells are resistant to AZC

because the SC process is considered to be the same as naturally occurring gene conversion. However, the food industry has not yet accepted SC yeast. In Japan, since 2019, genome-editing technologies, such as CRISPR-Cas9 and TALENs, have begun to be applied to breeding of crops and microorganisms for making foods and beverages, as well as conventional mutagenesis with EMS or UV (Tsuda et al. 2019). To apply genome-editing technology, the removal of foreign DNAs (or RNAs) derived from the plasmid to introduce Cas9 and gRNA must be proven, but to date, the Japanese government has not clarified the method for proving the removal of foreign DNA or the criteria for removal. Therefore, it is currently difficult to construct desirable brewer's yeast strains using genome-editing technologies. To overcome these obstacles, conventional mutagenesis using EMS or UV is considered to be the best method for breeding brewer's yeast strains.

6.5 Enhanced Proline Synthesis with Lower Succinate and Its Application to "Light-Bodied" Sake Brewing

The taste of sake is determined by the combination of many compounds, including sugars, organic acids, amino acids, nucleotides, and inorganic salts. In particular, since sugars have a sweet taste and organic acids have a sour taste, a variety of sake flavors can be produced by changing the balance between sugars and organic acids.

On the other hand, amino acids, which have a sweet, bitter, and umami taste, affect the sensory properties of sake, but it is often thought that a high amino acid content impairs the taste of sake. Amino acids in sake mash come mainly from the digestion of rice proteins by sake malt enzymes, but they are also synthesized by yeast cells during fermentation. In sake, amino acids and other taste components are mainly derived from *S. cerevisiae* cells, and various yeast strains have been constructed to develop differentiated sake products. Recently, Gekkeikan Sake Brewery isolated AZC-resistant mutants from a diploid sake yeast strain. Some of these mutants had high amounts of Pro in the brewed sake. Among them, one mutant strain (K-9-AZC) was found to produce higher Pro and lower succinate in sake mash than the parent strain Kyokai no. 9 (K-9) and is expected to brew low-carbohydrate sake (Kotaka et al. 2019). However, the molecular mechanisms of increased Pro and decreased succinic acid production in strain K-9-AZC have not yet been elucidated. Thus, we identified a novel gene mutation that increases intracellular Pro levels in strain K-9-AZC. We also analyzed the fermentation and metabolic profiles of strain K-9-AZC during sake brewing (Murakami et al. 2020).

Several AZC-resistant mutants were shown to accumulate higher amounts of Pro in the cells than the parent strain. Pro-accumulating strains of *S. cerevisiae* usually have a mutation in the *PRO1* gene, which encodes γ-glutamyl kinase (GK; Pro1). GK is the rate-limiting enzyme in Pro biosynthesis from L-glutamate (Glu), and its activity is allosterically controlled by the end product, Pro (Sekine et al. 2007) (Fig. 6.3). Most of the Pro-accumulating strains have been obtained by expressing Pro1 variants such as Asp154Asn and Ile150Thr, which are less sensitive to feed-back inhibition by Pro (Morita et al. 2003; Sekine et al. 2007) (Fig. 6.3). We found that strain K-9-AZC harbors a novel heteroallelic mutation in the *PRO1* gene, which encodes the Gln79His variant of Pro1 (Fig. 6.3). This mutation caused extreme desensitization to feedback inhibition by Pro, leading to overproduction of Pro. The amino acid residue at position 79 is highly conserved in other microbial GKs. Biochemical and in silico analyses showed that the amino acid residue at position 79 is involved in the stabilization of the Pro-binding pocket in Pro1 via a hydrogen-bonding network, which plays an important role in feedback inhibition (Nishimura et al. 2021). Therefore, we proposed a possible mechanism underlying the feedback inhibition of GK activity. Interestingly, sake brewed with strain K-9-AZC contained 3.7 times more Pro, but only 25% less succinate than sake brewed with strain K-9. There were no significant differences in sake meter value, alcohol, or amino acid content between strains K-9 and K-9-AZC, but the acidity was shown to be reduced by approximately 70% compared to strain K-9. Furthermore, sensory test showed that sake brewed with strain K-9-AZC had suppressed acidity. Thus, based on both analytical data and sensory evaluation, light-bodied sake with a lower acidity could be brewed. Metabolomic analysis suggests that the reduction in succinate is due to a decrease in the levels of 2-oxoglutarate, which is converted to Glu. The approach here may be a practical method for breeding yeast strains involved in the diversity of sake taste.

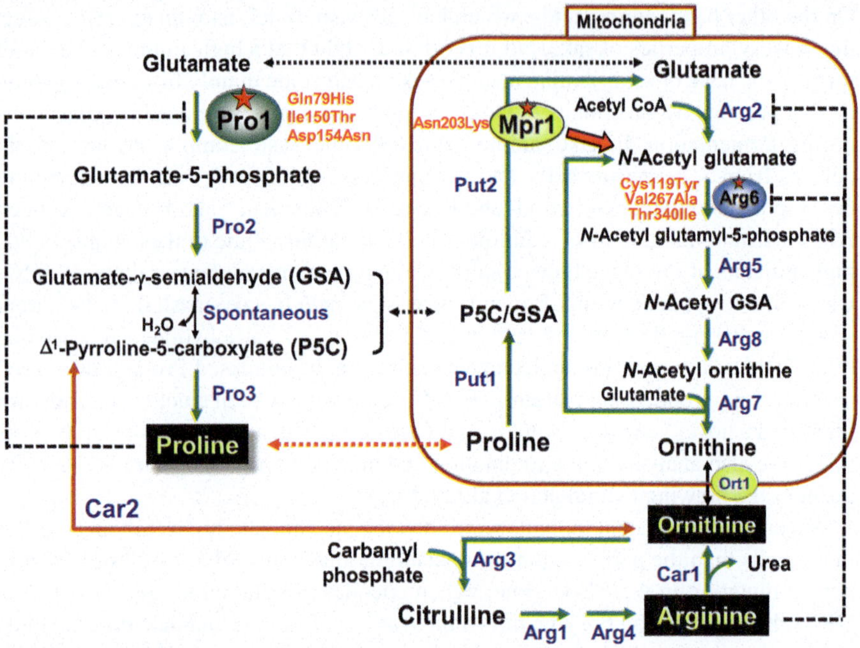

Fig. 6.3 Metabolisms of Pro/Orn/Arg in *S. cerevisiae*. γ-Glutamyl kinase (Pro1) and *N*-acetyl glutamate kinase (Arg6) are key enzymes that regulate the synthesis of Pro and Arg, and their enzymatic activities are subject to feedback inhibition by Pro and Arg, respectively. *N*-Acetyltransferase Mpr1 may contribute to Arg synthesis by supplying *N*-acetyl glutamate

6.6 Enhanced Ornithine Synthesis via Proline/Arginine Metabolism and Its Application to "Healthy Image" Sake Brewing

Increasing the ethanol tolerance of sake yeast strains may improve ethanol productivity and reduce the time required for fermentation. *N*-Acetyltransferase Mpr1 was originally discovered as an enzyme that detoxifies AZC through its *N*-acetylation in the laboratory yeast *S. cerevisiae* Σ1278b (Nasuno et al. 2013; Shichiri et al. 2001; Takagi et al. 2000). Mpr1 protects yeast cells from oxidative stress, such as H_2O_2, heat shock, freezing, or ethanol treatment, by reducing reactive oxygen species (ROS) level (Du and Takagi 2005, 2007; Ishchuk et al. 2010; Nomura and Takagi 2004; Sasano et al. 2010). Interestingly, Mpr1 may contribute to L-arginine (Arg) synthesis by supplying *N*-acetyl glutamate, an intermediate metabolite of Arg synthesis (Nasuno et al. 2016; Nishimura et al. 2010) (Fig. 6.3). Arg as the sole nitrogen source or as the additive in the medium increased the fermentation rate of yeast (Gutiérrez et al. 2012; Thomas and Ingledew 1992), suggesting that the construction of new yeast cells expressing Mpr1 could lead to improved fermentation rates due to the Mpr1-mediated enhancement of Arg synthesis. Recently, we constructed a diploid sake yeast strain expressing Mpr1 and found that a stable Asn203Lys variant of Mpr1 increases the fermentation rate in yeast cells under laboratory-scale sake brewing conditions (Ohashi et al. 2019) (Fig. 6.3).

We previously reported that Pro improves ethanol tolerance in yeast cells possibly by reducing ROS levels (Takagi et al. 2005, 2007, 2016). To improve the ethanol productivity of sake, we succeeded in isolating several Pro-accumulating mutants from a diploid sake yeast by conventional mutagenesis methods. Interestingly, one of them, strain A902-4, produced more than ten times amount of L-ornithine (Orn) and Pro compared to the parent strain (K901). Orn is an important precursor of Arg and is biosynthesized from Glu in the mitochondria and then transported to the cytoplasm (Fig. 6.3). However, the molecular mechanisms involved in the high production of Pro and Orn in strain A902-4 remain unclear. Recently, a novel genetic variant that increases intracellular Pro and Orn levels was identified and analyzed (Ohashi et al. 2020).

Orn is a non-proteinogenic amino acid that functions in the urea cycle. The urea cycle detoxifies ammonia generated by the degradation of amino acids in protein to urea. Orn has several physiological functions, including ameliorating negative states, such as fatigue and lassitude, and improving sleep quality (Kokubo et al. 2013). Thus, Orn-rich sake would be expected to relieve feelings of fatigue after drinking. We identified a homoallelic mutation in the *ARG6* gene encoding Thr340Ile variant *N*-acetyl glutamate kinase (NAGK; Arg6) in strain A902-4 (Fig. 6.3). The NAGK activity of the Thr340Ile variant was highly insensitive to feedback inhibition by Arg and caused intracellular Orn accumulation. Although the removal of feedback inhibition by amino acid substitution in NAGK has been reported in not only the laboratory *S. cerevisiae* strain (Pauwels et al. 2003) but also some bacteria, such as *Corynebacterium glutamicum* (Huang et al. 2015) and *C. crenatum* (Zhang et al. 2017), this is the first report in the industrial yeast strains including sake yeast. Based on the structure of the yeast NAGK, Arg interacts with these amino acid residues. Thr340 is conserved in NAGKs, which are sensitive to feedback inhibition by Arg, suggesting that this residue is involved in Arg recognition. In addition, the local structure around Thr340, which forms the Arg-binding cavity, is highly conserved among NAGKs that are sensitive to feedback inhibition by Arg. In contrast, NAGKs that are insensitive to Arg showed a different local folding. This difference in this local structure is one of the most important mechanisms explaining the Arg sensitivity of NAGKs. Therefore, we concluded that the replacement of Thr340 with Ile disrupts the local conformation via intramolecular interactions and removes Arg-feedback inhibition. The variant is not inhibited by Arg, but the transfer of an acetyl group from *N*-acetyl Orn to Glu in the cyclic pathway regenerates *N*-acetyl glutamate with Orn production, resulting in overproduction of Orn in the mutant. In addition, we newly identified two mutations in the *ARG6* gene encoding the Cys119Tyr or Val267Ala variant of Arg6 in sake yeast mutants with intracellular Orn accumulation (Ohashi et al. 2022) (Fig. 6.3). Although it is unlikely that Cys119 and Val267 in wild-type enzyme are directly involved in Arg recognition, we found here that two variants of Arg6 were insensitive to Arg-feedback inhibition and contributed to high-level production of Orn. Structural analysis of NAGK suggests that these amino acid substitutions influence the sensitivity to Arg-feedback inhibition through alterations in local conformation around each residue. Several GM microorganisms, such as *C. crenatum* (Shu et al. 2018) and *S. cerevisiae* (Liu et al. 2018), have been developed to overproduce Orn, using metabolic engineering or modular pathway rewiring. However, these strains have not been directly applied in food production because the food industry and costumers still have not accepted GM microorganisms. In contrast, the Orn-overproducing sake yeast mutants could be readily applied to sake brewing.

On the other hand, there are no mutations in the *PRO1* gene that are associated with Pro accumulation in this mutant. However, because yeast cells can synthesize Pro from Orn via the Orn transaminase Car2, this mutant may accumulate Pro from excess Orn via Car2 and the Δ^1-pyrroline-5-carboxylate reductase Pro3. Finally, we brewed laboratory-scale sake and analyzed the fermentation profiles of the parent and mutant strains. After fermentation, strain A902-4 produced slightly more ethanol (20.4%) than the parent strain (19.5%), and sake and sake cake brewed with strain A902-4 contained four to five times more Orn and Pro than those brewed with the parent strain. With the aid of Nara Prefecture Institute of Industrial Development, three local breweries in Nara have commercialized this "healthy image" sake containing a high level of Orn (2017, 2019, 2020). This approach could be a practical method to develop superior yeast strains for Orn overproduction.

6.7 Enhanced Volatile Compound Synthesis via Ehrlich Pathway and Its Application to "Flavor-Rich" Sake Brewing

Since sake has become popular all around the world, the development of yeast strains for the production of unique and distinctive sakes has attracted a great deal of attention. During the sake fermentation process by *S. cerevisiae*, branched-chain amino acids (BCAAs; L-isoleucine (Ile), L-valine (Val), and L-leucine (Leu)) are catabolized by the Ehrlich pathway to form volatile compounds, known as fusel alcohols: active amyl alcohol (2-methyl-1-butanol) from Ile, isobutanol (2-methyl-1-propanol) from Val, and isoamyl alcohol (3-methyl-1-butanol) from Leu (Hazelwood et al. 2008) (Fig. 6.4). The composition of fusel alcohols and their esters have a great impact on the flavor and taste of alcoholic beverages owing to their fruit-like flavors (Lilly et al. 2006; Pires et al. 2014).

Ile-derived volatile compounds, active amyl alcohol (a fruit-like flavor) and its acetate ester (2-methylbutyl acetate) (an apple-like flavor), contribute to a variety of the flavors and tastes of alcoholic beverages. Threonine deaminase (TD; Ilv1) catalyzes the first step in Ile biosynthesis, and its activity is subject to feedback inhibition by Ile (Holmberg and Petersen 1988) (Fig. 6.4). We recently isolated an Ile-accumulating mutant (K9-I48) from diploid sake yeast mutants resistant to the Ile toxic analog *O*-methyl threonine by conventional mutagenesis (Isogai et al.

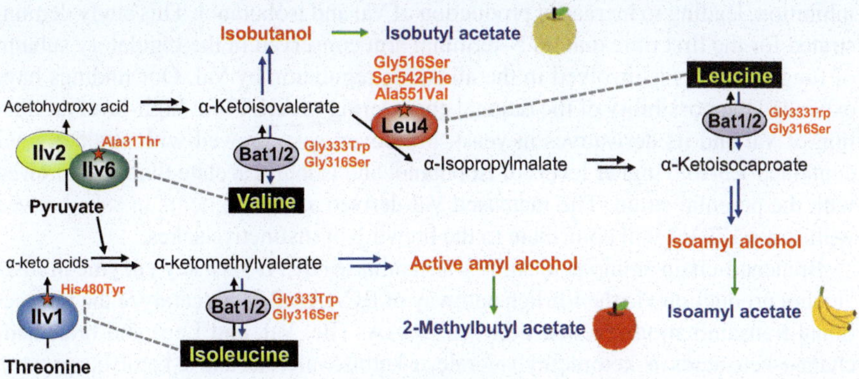

Fig. 6.4 Metabolism of BCAAs and fusel alcohols via the Ehrlich pathway in *S. cerevisiae*. Threonine deaminase (Ilv1), acetohydroxy acid synthase (Ilv2, Ilv6), and α-isopropylmalate synthase (Leu4) are key enzymes that regulate the synthesis of Ile, Val, and Leu, and their enzymatic activities are subject to feedback inhibition by Ile, Val, and Leu, respectively. Branched-chain amino acid aminotransaminases (Bat1 and Bat2) catalyze the bidirectional transamination reaction between BCAAs (Ile, Val, and Leu) and branched-chain α-keto acids (α-ketomethylvalerate, α-ketoisovalerate, and α-ketoisocaproate), respectively

2022). Strain K9-I48 carries a novel heteroallelic mutation in the *ILV1* gene encoding the His480Tyr variant of Ilv1 (Fig. 6.4). Interestingly, the TD activity of the His480Tyr variant was markedly insensitive to feedback inhibition by Ile, but was not upregulated by Val, leading to intracellular accumulation of Ile and extracellular overproduction of active amyl alcohol derived from Ile, in *S. cerevisiae* cells. We demonstrated for the first time that the conserved histidine residue located in a linker region between two regulatory domains is involved in allosteric regulation of Ilv1. It is possible that the inter-monomer hydrogen-bond formation in the His480Tyr variant contributes to the stabilization of the Ilv1 homodimer and prevents conformational alteration by Ile binding. Moreover, sake brewed with strain K9-I48 contained two to three times more active amyl alcohol and 2-methylbutyl acetate than sake brewed with the parent strain. These findings are valuable for the engineering of Ilv1 to increase the productivity of Ile and its derived fusel alcohols.

Val is catabolized into fruit-like aroma compounds, isobutanol and its acetate ester, isobutyl acetate (a melon- or banana-like flavor), during fermentation. We recently isolated a Val-accumulating sake yeast mutant (K7-V7), which is resistant to the Val toxic analog norvaline, from a diploid sake yeast by conventional mutagenesis (Isogai et al. 2023). Strain K7-V7 carries a novel heteroallelic mutation in the *ILV6* gene encoding the Ala31Thr variant of the regulatory subunit for acetohydroxy acid synthase (AHAS) (Fig. 6.4). AHAS catalyzes the first step in the Val biosynthesis, and its activity is subject to feedback inhibition by Val (Pang and Duggleby 1999). AHAS consists of the catalytic subunit Ilv2 that catalyzes decarboxylation and the regulatory subunit Ilv6 that activates the catalytic activity, which are encoded by the *ILV2* and *ILV6* genes, respectively (Fig. 6.4). The Ala31Thr substitution in Ilv6 partially desensitized the AHAS activity to Val-feedback inhibition, leading to increased production of Val and isobutanol. This study demonstrated for the first time that an N-terminal arm conserved in the regulatory subunit of fungal AHASs is involved in the allosteric regulation by Val. Our findings have expanded the possibility of the rational engineering of AHAS for high-level production of Val and its derivatives in yeast. Moreover, sake brewed with strain K7-V7 contained 1.5-fold higher levels of isobutanol and isobutyl acetate than sake brewed with the parental strain. The increased Val-derived aroma contents in sake brewed with strain K7-V7 will contribute to the brewing of distinctive sakes.

Branched-chain amino acid aminotransaminase (BCAT) is a key enzyme in fusel alcohol production via the Ehrlich pathway of BCAAs. BCATs catalyze the bidirectional transamination reaction between BCAAs (Ile, Val, and Leu) and branched-chain α-keto acids (α-ketomethylvalerate, α-ketoisovalerate, and α-ketoisocaproate), respectively. In *S. cerevisiae*, there are two BCATs isoforms, the mitochondrial Bat1 and the cytosolic Bat2, encoded by the *BAT1* and *BAT2* genes, respectively (Toyokawa et al. 2021) (Fig. 6.4). Many studies have shown the effects of deletion, mutation, or overexpression of *BAT1* and *BAT2* on fusel alcohol production (Kingsbury et al. 2015; Koonthongkaew et al. 2020; Takpho et al. 2018), but enhancement of fusel alcohol production by functional variants of BCAT has not been reported. To improve fusel alcohol productivity, we designed variants of Bat1 and Bat2 with altered enzyme activity by in silico computational analysis, namely, the Gly333Ser and Gly333Trp Bat1 and corresponding the Gly316Ser and Gly316Trp Bat2 variants, respectively (Koonthongkaew et al. 2022) (Fig. 6.4). Most of these variants caused growth defects in minimal medium when expressed in *S. cerevisiae* cells. Interestingly, the Gly333Trp Bat1 and Gly316Ser Bat2 variants achieved 18.7-fold and 17.4-fold increases in isobutanol production, respectively, compared to the wild-type enzyme. Enzyme assay revealed that the catalytic activities of all four BCAT variants were lower than that of the wild-type enzyme. Our results indicate that the reduced BCAT activity promoted fusel alcohol production by reducing BCAA synthesis, which occurs via a pathway that directly competes with fusel alcohol production. Such approach provides new insight into the functions of BCAT and will be useful in the construction of enzymes optimized for high-level production of fusel alcohols.

2-Phenylethanol, a high-value aromatic alcohol with a roselike flavor, is synthesized from L-phenylalanine (Phe) via the Ehrlich pathway (Fig. 6.5). The Ehrlich pathway consists of three steps: conversion of Phe to phenylpyruvate by aromatic transaminases (Aro8 and Aro9), decarboxylation of phenylpyruvate to 2-phenylacetaldehyde by phenylpyruvate decarboxylase (Aro10), and reduction of 2-phenylacetaldehyde to 2-phenylethanol by alcohol dehydrogenase (Qian et al. 2019). The transcription of *ARO9* and *ARO10* is regulated by the transcriptional activator Aro80. Therefore, adjusting the quantitative balance between Phe and 2-phenylethanol may give additional value to sake. Recently, we isolated a sake yeast mutant (K9-F39) with intracellular Phe accumulation from the Phe toxic analog (*p*-fluoro-DL-phenylalanine)-resistant colonies and found a missense mutation in the *ARO80* gene encoding the His309Gln variant of Aro80 (Nishimura et al. 2022b) (Fig. 6.5). The DNA-binding form of Aro80 may directly bind with Phe,

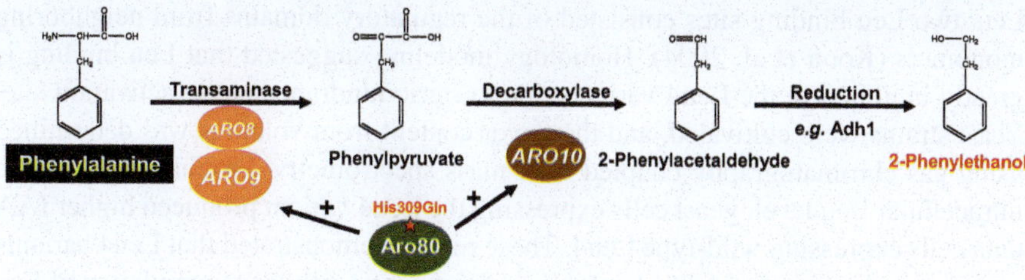

Fig. 6.5 Metabolism of Phe and 2-phenylethanol via the Ehrlich pathway in *S. cerevisiae*. The transcriptional factor Aro80 induces the transcription of the *ARO9* and *ARO10* genes, encoding aromatic transaminase and phenylpyruvate decarboxylase, respectively

resulting in conformational changes of Aro80 that correlates with transcriptional activity. We speculated that the *ARO80* mutation reduces transcriptional activity, inhibits Phe catabolism, and increases intracellular Phe. Indeed, sake brewed with strain K9-F39 had 60% more Phe than sake brewed with the parent strain, but only 10% less 2-phenylethanol. These results indicated that the use of the *ARO80* mutants is appropriate for controlling the content of Phe and 2-phenylethanol. Sake brewing with the *ARO80* mutants could be promising for the productions of a distinctive sake.

6.8 Enhanced Isoamyl Acetate Synthesis via Leucine Metabolism and Its Application to "Aromatic" Awamori Brewing

Isoamyl acetate (IAA; 3-methylbutyl acetate) is known for its banana-like flavor and *ginjoko* aroma in sake (Ashida et al. 1987; Watanabe et al. 1990). In yeast, IAA is converted from α-ketoisocaproate, an intermediate of the L-leucine (Leu) pathway, in two steps catalyzed by α-keto acid decarboxylase and alcohol dehydrogenase. α-Isopropylmalate synthase (IPMS; Leu4) is the key enzyme controlling intracellular Leu levels, and its activity is subject to feedback inhibition by Leu (Fig. 6.4). To increase the content of fruity flavors such as IAA in awamori, we attempted to isolate an awamori yeast mutant that accumulates Leu intracellularly by conventional mutagenesis (Takagi et al. 2015). Therefore, the Leu toxic analog 5,5,5-trifluoro-DL-Leu (TFL)-resistant mutants were isolated from the diploid awamori yeast strain (101-18). Among many TFL-resistant mutants, we succeeded in obtaining mutant strain 18-T55 with increased intracellular Leu content and found heteroallelic mutations in the *LEU4* gene encoding Ser542Phe and Ala551Val variant Leu4 (Fig. 6.4). The Leu content in cells expressing the Leu4 variant was higher than in cells expressing wild-type Leu4. The IPMS activity of wild-type cells was markedly inhibited by Leu, whereas the IPMS activity in the mutant was not affected by Leu-feedback inhibition, leading to excessive synthesis of Leu. In the crystal structure of IPMS from *Mycobacterium tuberculosis* bound with inhibitor

Leu, two Leu binding sites consisted of the regulatory domains from neighboring monomers (Koon et al. 2004). Homology modeling suggested that Leu binding is greatly inhibited in the Leu4 variants due to steric hindrance at the activation site. Yeast strains were cultivated, and the flavor content from volatiles was determined using gas chromatography coupled with mass spectrometry. Proportionally to the intracellular Leu level, yeast cells expressing the Leu4 variant produced higher IAA than cells expressing wild-type Leu4. These results demonstrated that Leu4 variants with desensitization to feedback inhibition caused intracellular accumulation of Leu and extracellular hyperproduction of IAA in awamori yeast. This study is the first to report the isolation of a diploid awamori yeast mutant that overproduces IAA. The approach described here could be a practical method for breeding novel awamori yeasts to expand the taste and flavor diversity of awamori. Finally, awamori brewed with strain 18-T55 was successfully commercialized by Shinzato Shuzo, the oldest brewery in Okinawa, as the brand name HYPER YEAST 101 (2016).

We recently isolated two new strains of *S. cerevisiae* from hibiscus flowers (HC02-5-2; HB yeast) (Abe et al. 2019) and banana stems (35a14; BA yeast) (Tsukahara et al. 2023) in Okinawa. Both HB and BA yeast strains produce high amounts of ethanol and 4-vinylguaiacol (4-VG), a precursor of vanillin, known as the key flavor of aged awamori. However, the IAA concentrations observed in both strains were lower than those observed in awamori yeast (101-18). Whole-genome information revealed that HB yeast is adjacent to wine yeast and is a different clade with classical awamori yeast in the phylogenetic tree, suggesting that HB yeast does not share any ancestry with sake or shochu yeast strains. Conventional awamori yeast (101-18) carrying the *FDC1* pseudogene does not produce 4-VG, but HB yeast with intact *PAD1* and *FDC1* genes, which encode phenylacrylate decarboxylase and ferulic acid decarboxylase involved in the decarboxylation of aromatic carboxylic acids, is advantageous to use in a novel kind of awamori (Mukai et al. 2014). To increase IAA content, mutant strain T25 with Leu accumulation was obtained from TFL-resistant mutants by conventional mutagenesis. In this mutant, we found a novel heteroallelic mutation in the *LEU4* gene encoding the Gly516Ser variant of Leu4, which is less sensitive to feedback inhibition by Leu, leading to intracellular Leu accumulation (Fig. 6.4). Homology modeling of Leu4 also suggested that Leu binding is dramatically inhibited in this mutant due to steric hindrance at the activation site. In laboratory-scale tests, awamori brewed with strain T25 showed higher concentrations of IAA than those brewed with HB yeast (Abe et al. 2019).

Interestingly, a phylogenic analysis using whole-genome information showed that BA yeast, in which the intact *PAD1* and *FDC1* genes are present, did not belong to any domesticated industrial yeast clades, such as sake or bread (Marsit et al. 2017), but was located near to the wild yeasts isolated from Malaysia, the USA, and Japan, suggesting that BA yeast is an endemic strain from Okinawa that evolved independently from other yeasts. We also selected strain BNNL-80 with a higher content of intracellular Leu than the parent strain (35a14) from TFL-resistant mutants. In a laboratory-scale fermentation test, awamori brewed with BNNL80 significantly increased IAA compared to 35a14. In this mutant, we found a homoallelic mutation in the *LEU4* gene encoding the Asp578Asn variant of Leu4, leading to overproduction of Leu thereby and increase in isoamyl acetate (Tsukahara et al. 2023) (Fig. 6.4). This mutation is suggested to reduce sensitivity to feedback inhibition by Leu by altering the local structure of Leu4 similar to the previously studied Asp578Tyr substitution found in sake yeast (Oba et al. 2005).

Finally, these "aromatic" awamori containing high amounts of IAA and 4-VG brewed with strains T25 and BNNL80 have been commercialized by Kamimura Shuzo (2019) and Higa Shuzo (2024) in Okinawa under the brand name of SHO KAMIMURA and ZANPA Shima-Banana Yeast, respectively. This combined approach to the isolation of yeast from nature and its breeding can be applied to the variation of alcoholic beverage quality in the fermentation industry.

6.9 Regulatory Mechanism of Proline Utilization and Its Application to "Proline-Less" Wine and Beer Brewing

Pro is a predominant amino acid in grape must and wort and represents a potentially significant and effective source of nitrogen (Valero et al. 2003). However, *S. cerevisiae* cells cannot utilize Por during wine and beer fermentation, leading to the high accumulation of Pro in wine and beer. This sometimes leads to a nitrogen deficiency during fermentation and Pro accumulation in wine and beer. Although the presence of other nitrogen sources under fermentation conditions is likely to interfere with Pro utilization, the inhibitory mechanisms of Pro utilization remain unclear. We first constructed a Pro auxotrophic yeast strain and identified an inhibitory factor by observing the growth of cells when Pro was present as a sole nitrogen source. Intriguingly, Arg, which is the second-most abundant amino acid in grapes, not ammonium ion, was found to clearly inhibit the growth of Pro auxotrophic cells. In addition, Arg prevented the Pro consumption of wild-type and Pro auxotrophic cells, indicating that Arg is an inhibitory factor of Pro utilization in *S. cerevisiae*. Next, quantitative PCR analysis showed that Arg partially repressed the expression of genes involved in Pro degradation and uptake. Arg also induced the endocytosis of the Pro transporters Put4 and Gap1, whereas ammonium induced the endocytosis of only Gap1 (Nishimura et al. 2020b). Furthermore, we found that the activity of the ubiquitin ligase Rsp5 was essential for the Arg-induced endocytosis of Put4. Because Put4 contains no Rsp5-binding motif, the adaptor protein that plays a role in the Arg-induced endocytosis of Put4 was screened. Genetic and biochemical analyses clearly revealed that the *ART3* gene-disrupted cells were defective in Put4 endocytosis, indicating that the α-arrestin protein Art3 is a key regulator for Put4 endocytosis. More importantly, deletion of *ART3* remarkably canceled the inhibitory effects of Arg on Pro utilization (Nishimura et al. 2020a). In our proposed model, Arg induces the ubiquitination of Put4 dependent on Rsp5 and Art3, causing the rapid endocytosis of Put4 (Fig. 6.6).

We recently reported a possible role of Arg sensing in the inhibition of Pro utilization. First, it was found that two basic amino acids, Orn and L-lysine, inhibit Pro

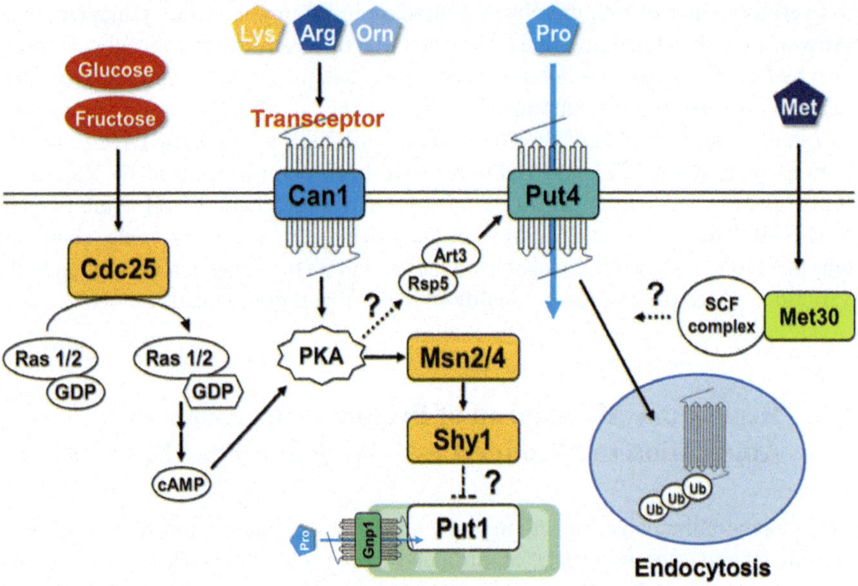

Fig. 6.6 Inhibitory mechanism of Pro utilization in *S. cerevisiae*. Basic amino acids, Arg, Lys, and Orn, are inhibitors of Pro utilization. Can1 is a transceptor that possesses the activity of both transporter and receptor of Arg. Yeast strains that can efficiently utilize the abundant Pro in grape must and wort are promising for making wine and brewing beer with new taste and flavor

utilization by inducing Put4 endocytosis in a manner similar to Arg, but citrulline does not. Genetic screening revealed that the Arg transporter Can1 is involved in the inhibition of Pro utilization by Arg (Fig. 6.6). Intriguingly, the Arg uptake activity of Can1 was not required for the Arg-dependent inhibition of Pro utilization, suggesting that Can1 has a function beyond its commonly known function of transporting Arg. More importantly, biochemical analyses revealed that Can1 activates signaling cascades of protein kinase A in response to extracellular Arg, without increasing cAMP levels. In recent years, transporters that have receptor activity for external factors in addition to transport activity have attracted attention as "transceptors." These proteins are known to activate protein kinase A signaling in a cAMP-independent manner (Van Zeebroeck et al. 2020). We proposed that Can1 regulates Pro utilization by functioning as a transceptor possessing the activity of both a transporter and receptor of Arg (Tanahashi et al. 2023a) (Fig. 6.6). *S. cerevisiae* strains that can assimilate Pro during fermentation are a promising tool in wine and beer production.

In addition, yeast strains with mutations in the genes related to these (*CDC25*, *MSN2/4*, *SHY1*) were able to consume small amounts of Pro (Nishimura et al. 2022a, 2023) (Fig. 6.6). Furthermore, analysis of another mutant strain revealed that Met30 (Skp1/Cullin/F-box ubiquitin ligase) controls Pro utilization independently of Can1 (Fig. 6.6). On the other hand, wild yeasts with different amino acid assimilation abilities exist in nature, and from a large yeast collection, many strains that utilize Pro on model fermentation media for wine and beer were obtained (Tanahashi et al. 2023b, c). We are currently analyzing the Pro utilization mechanism of these strains and attempting to brew wine and beer with reduced Pro content. These findings will contribute to the development of novel yeast strains that can solve fermentation problems and improve the quality and productivity of wine and beer.

6.10 Conclusions and Future Perspective

As described in this chapter, we have established a breeding technology to artificially control the intracellular and extracellular content of "functional" amino acids that have functions such as improving fermentation ability (Pro, Arg), diversifying taste and flavor (BCAA, Phe), adding healthy image (Orn), and increasing nutritional value to develop high functionality in microorganisms and improve the productivity of useful substances, which we have named "functional amino acid engineering" (Fig. 6.7) (Takagi 2019, 2021). Also, by isolating yeast strains from nature with whole-genome analysis and metabolism-focused mutations, we propose a combinatorial approach for industrial yeast breeding that has great potential for improving the quality and adding value of alcoholic beverages. Because *S. cerevisiae* is highly reliable and safe in both food and beverage productions, the development of novel strains that overproduce functional amino acids will contribute greatly to food-related industries.

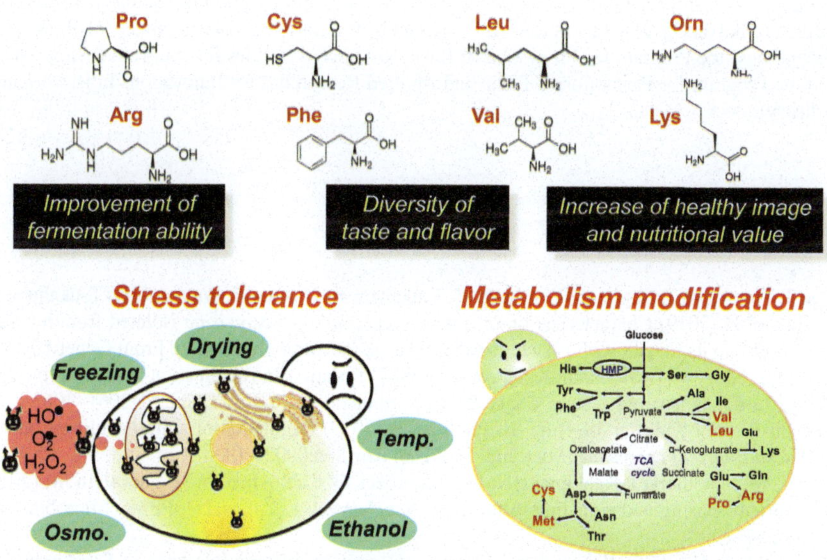

Fig. 6.7 Concept of "functional amino acids engineering." This is a new breeding technology for industrial yeast strains which contributes to improvement of fermentation ability, diversity of product taste and flavor, addition of a healthy image, or improvement of nutritional value

Since amino acids are common to all living organisms, "functional amino acid engineering" is a technology that can be applied not only to microorganisms but also to plants and insects and has a large ripple effect in terms of both basic and applied aspects. Additionally, our results can be said to have led to academic review of the methods for obtaining mutant strains used in research on fermentation production of amino acids in the 1960s and 1970s. By selecting amino acid-overproducing strains from yeast mutants that are resistant to the analog corresponding to the target amino acid, and using databases such as whole-genome analysis with next-generation sequencers, transcriptional profile analysis, comprehensive analysis of metabolites, and structural information on enzymes, the mechanism of high production can be clarified and a new breeding strategy can be developed. The method of constructing a new breeding strategy has introduced new perspectives and ideas to amino acid research. In addition, research focusing on physiological functions found in intracellular amino acids, such as the stress tolerance mechanism by Pro/Arg metabolism, is expected to lead to a new paradigm in amino acid science (environmental adaptation by amino acid metabolism) common to all organisms, not just microorganisms.

Acknowledgments I am grateful to Daisuke Watanabe, Ryo Nasuno, Akira Nishimura, Yoichi Toyokawa, Shota Isogai, Keiko Ashida, Keisuke Hashida, and Yukiko Sugimoto (Nara Institute of Science and Technology, Nara, Japan); Masataka Ohashi (Nara Prefecture Institute of Industrial Development, Nara, Japan); Atsushi Kotaka, Naoyuki Murakami, Natsuki Hotta, Kengo Matsumura, Yoji Hata, and Hiroki Ishida (Gekkeikan Sake Co., Ltd., Kyoto, Japan); and Masatoshi Tsukahara, Takayuki Abe, Maiko Nezuo, Tomoya Iha, Haruna Azuma, and Keiko Tsukahara (BioJet Co., Ltd., Okinawa, Japan) for their contributions and collaborations. This research was mainly supported by grants from the projects of the NARO Bio-oriented Technology Research Advancement Institution (the Research Program on Development of Innovative Technology, the Program for Promotion of Basic Research Activities for Innovative Biosciences, and the Program for Promotion of Basic and Applied Researches for Innovations in Bio-oriented industry).

References

Abe T, Toyokawa Y, Sugimoto Y, Azuma H, Tsukahara K, Nasuno R, Watanabe D, Tsukahara M, Takagi H (2019) Characterization of a new *Saccharomyces cerevisiae* isolated from hibiscus flower and its mutant with L-leucine accumulation for awamori brewing. Front Genet 10:490

Ashida S, Ichikawa E, Suginami K, Imayasu S (1987) Isolation and application of mutants producing sufficient isoamyl acetate, a sake flavor component. Agric Biol Chem 51:2061–2065

Bianchi F, van't Klooster JS, Ruiz SJ, Poolman B (2019) Regulation of amino acid transport in *Saccharomyces cerevisiae*. Microbiol Mol Biol Rev 83:e00024-19

Conrad M, Schothorst J, Kankipati HN, Van Zeebroeck G, Rubio-Texeira M, Thevelein JM (2014) Nutrient sensing and signaling in the yeast *Saccharomyces cerevisiae*. FEMS Microbiol Rev 38:254–299

Du X, Takagi H (2005) *N*-Acetyltransferase Mpr1 conferred freeze tolerance in *Saccharomyces cerevisiae* by reducing reactive oxygen species. J Biochem 138:391–397

Du X, Takagi H (2007) *N*-Acetyltransferase Mpr1 confers ethanol tolerance on *Saccharomyces cerevisiae* by reducing reactive oxygen species. Appl Microbiol Biotechnol 75:1343–1351

Gutiérrez A, Chiva R, Sancho M, Beltran G, Arroyo-López FN, Guillamon JM (2012) Nitrogen requirements of commercial wine yeast strains during fermentation of a synthetic grape must. Food Microbiol 31:25–32

Hazelwood LA, Daran JM, Van Maris AJA, Pronk JT, Dickinson JR (2008) The Ehrlich pathway for fusel alcohol production: A century of research on *Saccharomyces cerevisiae* metabolism. Appl Environ Microbiol 74:2259–2266

Hinnebusch AG (2005) Translational regulation of *GCN4* and the general amino acid control of yeast. Annu Rev Microbiol 59:407–450

Holmberg S, Petersen JG (1988) Regulation of isoleucine-valine biosynthesis in *Saccharomyces cerevisiae*. Curr Genet 13:207–217

Huang Y, Zhang H, Tian H, Li C, Han S, Lin Y, Zheng S (2015) Mutational analysis to identify the residues essential for the inhibition of *N*-acetyl glutamate kinase of *Corynebacterium glutamicum*. Appl Microbiol Biotechnol 99:7527–7537

Ishchuk OP, Abbas CA, Sibirny AA (2010) Heterologous expression of *Saccharomyces cerevisiae MPR1* gene confers tolerance to ethanol and L-azetidine-2-carboxylic acid in *Hansenula polymorpha*. J Ind Microbiol Biotechnol 37:213–218

Isogai S, Nishimura A, Kotaka A, Murakami N, Hotta N, Ishida H, Takagi H (2022) High-level production of isoleucine and fusel alcohol by expression of the feedback inhibition-insensitive threonine deaminase in *Saccharomyces cerevisiae*. Appl Environ Microbiol 88:e0213021

Isogai S, Nishimura A, Murakami N, Hotta N, Kotaka A, Toyokawa Y, Ishida H, Takagi H (2023) Improvement of valine and isobutanol production in sake yeast by Ala31Thr substitution in the regulatory subunit of acetohydroxy acid synthase. FEMS Yeast Res 23:1–9

Kawano-Kawada M, Kakinuma Y, Sekito T (2018) Transport of amino acids across the vacuolar membrane of yeast: Its mechanism and physiological role. Biol Pharm Bull 41:1496–1501

Kingsbury JM, Sen ND, Cardenas ME (2015) Branched-chain aminotransferases control TORC1 signaling in *Saccharomyces cerevisiae*. PLoS Genet 11:e1005714

Kodama K (1993) Sake-brewing yeasts. In: Rose AH, Harrison JS (eds) The yeasts, vol 3. Academic, London, pp 129–168

Kokubo T, Ikeshima E, Kirisako T, Miura Y, Horiuchi M, Tsuda A (2013) A randomized, double-masked, placebo-controlled crossover trial on the effects of L-ornithine on salivary cortisol and feelings of fatigue of flushers the morning after alcohol consumption. Biopsychosoc Med 7:6

Koon N, Squire CJ, Baker EN (2004) Crystal structure of LeuA from *Mycobacterium tuberculosis*, a key enzyme in leucine biosynthesis. Proc Natl Acad Sci USA 101:8295–8300

Koonthongkaew J, Toyokawa Y, Ohashi M, Large CRL, Dunham MJ, Takagi H (2020) Effect of the Ala234Asp replacement in mitochondrial branched-chain amino acid aminotransferase on the production of BCAAs and fusel alcohols in yeast. Appl Microbiol Biotechnol 104:7915–7925

Koonthongkaew J, Ploysongsri N, Toyokawa Y, Ruangpornvisuti V, Takagi H (2022) Improvement of fusel alcohol production by engineering of the yeast branched-chain amino acid aminotransaminase. Appl Environ Microbiol 88:e0055722

Kotaka A, Nakamura Y, Kasai H, Watanabe Y, Morimoto-Sakakibara M, Matsumura K, Hata Y (2019) Development of amino-acid-analogue-resistant and low acid-producing sake yeast for commercial scale sake brewing. J Brew Soc Jpn 114:787–794. (in Japanese)

Lilly M, Bauer FF, Styger G, Lambrechts MG, Pretorius IS (2006) The effect of increased branched-chain amino acid transaminase activity in yeast on the production of higher alcohols and on the flavour profiles of wine and distillates. FEMS Yeast Res 6:726–743

Liu Q, Yu T, Campbell K, Nielsen J, Chen Y (2018) Modular pathway rewiring of yeast for amino acid production. Methods Enzymol 608:417–439

Ljungdahl PO, Daignan-Fornier B (2012) Regulation of amino acid, nucleotide, and phosphate metabolism in *Saccharomyces cerevisiae*. Genetics 190:885–929

Magasanik B, Kaiser CA (2002) Nitrogen regulation in *Saccharomyces cerevisiae*. Gene 290:1–18

Marsit S, Leducq J-B, Durand E, Marchant A, Filteau M, Landry CR (2017) Evolutionary biology through the lens of budding yeast comparative genomics. Nature 18:581–598

Morita Y, Nakamori S, Takagi H (2003) L-Proline accumulation and freeze tolerance in *Saccharomyces cerevisiae* are caused by a mutation in the *PRO1* gene encoding γ-glutamyl kinase. Appl Eviron Microbiol 69:212–219

Mukai N, Masaki K, Fujii T, Iefuji H (2014) Single nucleotide polymorphisms of *PAD1* and *FDC1* show a positive relationship with ferulic acid decarboxylation ability among industrial yeasts used in alcoholic beverage production. J Biosci Bioeng 118:50–55

Murakami N, Kotaka A, Isogai S, Ashida K, Nishimura A, Matsumura K, Hata Y, Ishida H, Takagi H (2020) Effects of a novel variant of the yeast γ-glutamyl kinase Pro1 on its enzymatic activity and sake brewing. J Ind Microbiol Biotechnol 47:715–723

Nasuno R, Hirano Y, Itoh T, Hakoshima T, Hibi T, Takagi H (2013) Structural and functional analysis of the yeast *N*-acetyltransferase Mpr1 involved in oxidative stress tolerance via proline metabolism. Proc Natl Acad Sci USA 110:11821–11826

Nasuno R, Hirase S, Norifune S, Takagi H (2016) Structure-based molecular design for thermostabilization of *N*-acetyltransferase Mpr1 involved in a novel pathway of L-arginine synthesis in yeast. J Biochem 159:271–277

Nishimura A, Kotani T, Sasano Y, Takagi H (2010) An antioxidative mechanism mediated by the yeast *N*-acetyltransferase Mpr1: oxidative stress-induced arginine synthesis and its physiological role. FEMS Yeast Res 10:687–698

Nishimura A, Tanahashi R, Takagi H (2020a) The yeast α-arrestin Art3 is a key regulator for arginine-induced endocytosis of the high-affinity proline transporter Put4. Biochem Biophys Res Commun 531:416–421

Nishimura A, Tanikawa T, Takagi H (2020b) Inhibitory effect of arginine on proline utilization in *Saccharomyces cerevisiae*. Yeast 37:531–540

Nishimura A, Takasaki Y, Isogai S, Toyokawa Y, Tanahashi R, Takagi H (2021) Role of Gln79 in feedback inhibition of the yeast γ-glutamyl kinase by proline. Microorganisms 9:1902

Nishimura A, Ichikawa K, Nakazawa H, Tanahashi R, Morita F, Sitepu I, Boundy-Mills K, Fox G, Takagi H (2022a) The Cdc25/Ras/cAMP-dependent protein kinase A signaling pathway regulates proline utilization in wine yeast *Saccharomyces cerevisiae* under a wine fermentation model. Biosci Biotechnol Biochem 86:1318–1326

Nishimura A, Isogai S, Murakami N, Hotta N, Kotaka A, Matsumura K, Hata Y, Ishida H, Takagi H (2022b) Isolation and analysis of a sake yeast mutant with phenylalanine accumulation. J Ind Microbiol Biotechnol 49:kuab08

Nishimura A, Tanahashi R, Nakazawa H, Oi T, Mima M, Takagi H (2023) PKA-Msn2/4-Shy1 cascade controls inhibition of proline utilization under wine fermentation model. J Biosci Bioeng 136:438–442

Nomura M, Takagi H (2004) Role of the yeast acetyltransferase Mpr1 in oxidative stress: Regulation of oxygen reactive species caused by a toxic proline catabolism intermediate. Proc Natl Acad Sci USA 101:12616–12621

Oba T, Nomiyama S, Hirakawa H, Tashiro K, Kuhara S (2005) Asp578 in LEU4p is one of the key residues for leucine feedback inhibition release in sake yeast. Biosci Biotechnol Biochem 69:1270–1273

Ohashi M, Nasuno R, Watanabe D, Takagi H (2019) Stable *N*-acetyltransferase Mpr1 improves ethanol productivity in the sake yeast *Saccharomyces cerevisiae*. J Ind Microbiol Biotechnol 46:1039–1045

Ohashi M, Nasuno R, Isogai S, Takagi H (2020) High-level production of ornithine by expression of the feedback inhibition-insensitive *N*-acetyl glutamate kinase in the sake yeast *Saccharomyces cerevisiae*. Metab Eng 62:1–9

Ohashi M, Isogai S, Takagi H (2022) Functional analysis of feedback inhibition-insensitive variants of *N*-acetyl glutamate kinase found in sake yeast mutants with ornithine overproduction. Microbiol Spectr 10:e00822–e00822

Pang SS, Duggleby RG (1999) Expression, purification, characterization, and reconstitution of the large and small subunits of yeast acetohydroxyacid synthase. Biochemistry 38:5222–5231

Pauwels K, Abadjieva A, Hilven P, Stankiewicz A, Crabeel M (2003) The *N*-acetylglutamate synthase/*N*-acetylglutamate kinase metabolon of *Saccharomyces cerevisiae* allows coordinated feedback regulation of the first two steps in arginine biosynthesis. Eur J Biochem 270:1014–1024

Pires EJ, Teixeira JA, Brányik T, Vicente AA (2014) Yeast: The soul of beer's aroma—a review of flavour-active esters and higher alcohols produced by the brewing yeast. Appl Microbiol Biotechnol 98:1937–1949

Qian X, Yan W, Zhang W, Dong W, Ma J, Ochsenreither K, Jiang M, Xin F (2019) Current status and perspectives of 2-phenylethanol production through biological processes. Crit Rev Biotechnol 39:235–248

Sasano Y, Takahashi S, Shima J, Takagi H (2010) Antioxidant *N*-acetyltransferase Mpr1/2 of industrial baker's yeast enhances fermentation ability after air-drying stress in bread dough. Int J Food Microbiol 138:181–185

Sekine T, Kawaguchi A, Hamano Y, Takagi H (2007) Desensitization of feedback inhibition of the *Saccharomyces cerevisiae* γ-glutamyl kinase enhances proline accumulation and freezing tolerance. Appl Eviron Microbiol 73:4011–4019

Shichiri M, Hoshikawa C, Nakamori S, Takagi H (2001) A novel acetyltransferase found in *Saccharomyces cerevisiae* Σ1278b that detoxifies a proline analogue, azetidine-2-carboxylic acid. J Biol Chem 276:41998–42002

Shu Q, Xu M, Li J, Yang T, Zhang X, Xu Z, Rao Z (2018) Improved L-ornithine production in *Corynebacterium crenatum* by introducing an artificial linear transacetylation pathway. J Ind Microbiol Biotechnol 45:393–404

Taira J, Tsuchiya A, Furudate H (2012) Initial volatile aroma profiles of young and aged awamori shochu determined by GC/MS/pulsed FPD. Food Sci Technol Res 18:177–181

Takagi H (2019) Metabolic regulatory mechanisms and physiological roles of functional amino acids and their applications in yeast. Biosci Biotechnol Biochem 83:1449–1462

Takagi H (2021) Adventures in brewing exotic Japanese alcoholic beverages with "Amino acid-rich yeast". SIMB News 71:8–17

Takagi H, Shichiri M, Takemura M, Mohri M, Nakamori S (2000) *Saccharomyces cerevisiae* Σ1278b has novel genes of the *N*-acetyltransferase gene superfamily required for L-proline analogue resistance. J Bacteriol 182:4249–4256

Takagi H, Takaoka M, Kawaguchi A, Kubo Y (2005) Effect of L-proline on sake brewing and ethanol stress in *Saccharomyces cerevisiae*. Appl Environ Microbiol 71:8656–8662

Takagi H, Matsui F, Kawaguchi A, Wu H, Shimoi H, Kubo Y (2007) Construction and analysis of self-cloning sake yeasts that accumulate proline. J Biosci Bioeng 103:377–380

Takagi H, Hashida K, Watanabe D, Nasuno R, Ohashi M, Iha T, Nezuo M, Tsukahara M (2015) Isolation and characterization of awamori yeast mutants with L-leucine accumulation that overproduce isoamyl alcohol. J Biosci Bioeng 119:140–147

Takagi H, Taguchi J, Kaino T (2016) Proline accumulation protects *Saccharomyces cerevisiae* cells in the stationary phase from ethanol stress by reducing reactive oxygen species levels. Yeast 33:355–363

Takpho N, Watanabe D, Takagi H (2018) Valine biosynthesis in *Saccharomyces cerevisiae* is regulated by the mitochondrial branched-chain amino acid aminotransferase Bat1. Microb Cell 5:293–299

Tanahashi R, Nishimura A, Morita F, Nakazawa H, Taniguchi A, Ichikawa K, Nakagami K, Boundy-Mills K, Takagi H (2023a) The arginine transporter Can1 acts as a transceptor for regulation of proline utilization in the yeast *Saccharomyces cerevisiae*. Yeast 40:333–348

Tanahashi R, Nishimura A, Nguyen M, Sitepu I, Fox G, Boundy-Mills K, Takagi H (2023b) Large-scale screening of yeast strains that can utilize proline. Biosci Biotechnol Biochem 87:358–362

Tanahashi R, Nishimura A, Nguyen M, Sitepu I, Fox G, Boundy-Mills K, Takagi H (2023c) Isolation of yeast strains with higher proline uptake and their applications to beer fermentation. J Fungi 9:1137

Thomas KC, Ingledew WM (1992) Relationship of low lysine and high arginine concentrations to efficient ethanolic fermentation of wheat mash. Can J Microbiol 38:626–634

Toyokawa Y, Koonthongkaew J, Takagi H (2021) An overview of branched-chain amino acid aminotransferases: functional differences between mitochondrial and cytosolic isozymes in yeast and human. Appl Microbiol Biotechnol 105:8059–8072

Tsuda M, Watanabe KN, Ohsawa R (2019) Regulatory status of genome-edited organisms under the Japanese Cartagena Act. Front Bioeng Biotechnol 7:387

Tsukahara M, Isogai S, Azuma H, Tsukahara K, Toyokawa Y, Takagi H (2023) Characterization of a new *Saccharomyces cerevisiae* isolated from banana stems and its mutant with L-leucine accumulation for awamori brewing. Biosci Biotechnol Biochem 87:240–244

Valero E, Millán C, Ortega JM, Mauricio JC (2003) Concentration of amino acids in wine after the end of fermentation by *Saccharomyces cerevisiae* strains. J Sci Food Agric 83:830–835

Van Zeebroeck G, Demuyser L, Zhang Z, Cottignie L, Thevelein JM (2020) Nutrient sensing and cAMP signaling in yeast: G-protein coupled receptor versus transceptor activation of PKA. Microb Cell 8:17–27

Watanabe M, Fukuda K, Asano K, Ohta S (1990) Mutants of baker's yeasts producing a large amount of isobutyl alcohol or isoamyl alcohol, flavor components of bread. Appl Microbiol Biotechnol 34:154–159

Watanabe D, Araki Y, Zhou Y, Maeya N, Akao T, Shimoi H (2012) A loss-of-function mutation in the PAS kinase Rim15p is related to defective quiescence entry and high fermentation rates of *Saccharomyces cerevisiae* sake yeast strains. Appl Environ Microbiol 78:4008–4016

Watanabe D, Zhou Y, Hirata A, Sugimoto Y, Takagi K, Akao T, Ohya Y, Takagi H, Shimoi H (2016) Inhibitory role of Greatwall-like protein kinase Rim15p in alcoholic fermentation via upregulating the UDP-glucose synthesis pathway in *Saccharomyces cerevisiae*. Appl Environ Microbiol 82:340–351

Watanabe D, Kajihara T, Sugimoto Y, Takagi K, Mizuno M, Zhou Y, Chen J, Takeda K, Tatebe H, Shiozaki K, Nakazawa N, Izawa S, Akao T, Shimoi H, Maeda T, Takagi H (2019) Nutrient signaling via the TORC1-Greatwall-PP2A$^{B55\delta}$ pathway responsible for the high initial rates of alcoholic fermentation in sake yeast strains of *Saccharomyces cerevisiae*. Appl Environ Microbiol 85:e02083-18

Zhang J, Xu M, Ge X, Zhang X, Yang T, Xu Z, Rao Z (2017) Reengineering of the feedback-inhibition enzyme *N*-acetyl-L-glutamate kinase to enhance L-arginine production in *Corynebacterium crenatum*. J Ind Microbiol Biotechnol 44:271–283

Zhang W, Du G, Zhou J, Chen J (2018) Regulation of sensing, transportation, and catabolism of nitrogen sources in *Saccharomyces cerevisiae*. Microbiol Mol Biol Rev 82:e00040–e00017

Zhao P, Zhao C, Chen D, Yun C, Li H, Bai L (2021) Structure and activation mechanism of the hexameric plasma membrane H^+-ATPase. Nat Commun 12:6439

Chapter 7
Lipids of Yeasts and Filamentous Fungi and Their Importance to Biotechnology

Volkmar Passoth and Bettina Müller

Contents

7.1 Introduction 164
7.2 Oleaginous Fungi 165
7.3 Biochemistry of Lipid Synthesis 168
7.4 Lipid Containing Biomolccules 176
7.4.1 Fungal Glycolipids as Biosurfactants 176
7.4.2 Lipid Modification of Proteins 183
7.5 Lipid Production by Fungi for Food, Feed, Biofuels, and Other Applications 185
7.5.1 Lipid Production from Lignocellulose and Crude Glycerol 185
7.5.2 Manipulating Yeasts for Improving Lipid Production 190
7.6 Production of Specific Chemicals and Fuels Derived from Lipid Metabolism 194
7.6.1 Fatty Alcohols 194
7.6.2 Free Fatty Acids 197
7.6.3 Synthesis of Polyunsaturated Fatty Acids (PUFAs) by Fungi 200
7.6.4 Hydrocarbons 202
7.6.5 Isoprenoids 204
7.7 Lipid Extraction 209
7.8 Conclusions 211
References 212

Abstract Lipids are essential for any organism. They are hydrophobic compounds and include, for instance, acylglycerides, free fatty acids, sterols, and isoprenoids. Their synthesis starts from acetyl-CoA, which is converted to lipids through a variety of complex biochemical reactions. Lipids consist of highly reduced molecules,

V. Passoth (✉) · B. Müller
Department of Molecular Sciences, Uppsala Biocentrum, Swedish University of Agricultural Sciences (SLU), Uppsala, Sweden
e-mail: volkmar.passoth@slu.se

A. A. Sibirny (ed.), *Biotechnology of Yeasts and Filamentous Fungi*,
Grand Challenges in Biology and Biotechnology,
https://doi.org/10.1007/978-3-031-74726-7_7

and the redox factor providing electrons for their synthesis is NADPH, mainly produced by the pentose-phosphate pathway and the malic enzyme. Various fungi can accumulate lipids, with oleaginous fungi capable of accumulating 20 to more than 80% of their biomass as lipids. Many of these fungi can convert carbon sources derived from second-generation substrates or low-value residues, such as lignocellulose hydrolysate, volatile fatty acids, or crude glycerol from biodiesel production. There are also substantial efforts to genetically manipulate fungi to produce lipids or lipid-derived compounds efficiently from low-value substrates.

Lipids have broad potential in biotechnological applications. Polyunsaturated fatty acids are essential for human nutrition and can be produced in high amounts by certain zygomycetous species or genetically modified yeasts. Acylglycerides are not only the primary components of edible oils but can also be converted into biodiesel and high-value chemicals. It is possible to obtain biodiesel, jet fuel, specific chemicals and carotenes as chemicals and food and feed additives from isoprenoids. Lipids can be combined with other biomolecules. Glycolipids have a high biotechnological potential as biosurfactants. Lipoproteins are essential for the viability of any eukaryotic cell and can be targets for developing novel antifungal drugs. As most lipids are stored intracellularly, there is a high demand for process development towards sustainable production of lipid-derived fuels and chemicals.

The lipid metabolism of fungi and its biotechnological utilisation has recently gained significant attention and is a rapidly developing field.

7.1 Introduction

Lipids are regarded as one of the three macronutrients, apart from carbohydrates and proteins. They are important as energy storage substances and essential for cell integrity since the cell membrane and all membranes of organelles are mainly formed by lipids. Usually, lipids are defined by their insolubility in water and their solubility in organic solvents such as chloroform, hydrocarbons, and alcohols, etc. Lipids comprise an extremely heterogeneous group of compounds, both from a structural and functional aspect. Several attempts have been made to classify lipids. In 2005, the International Lipid Classification and Nomenclature Committee established a comprehensive classification system of lipids, based on (bio)chemical principles. Eight categories of lipids have been defined: fatty acyls, glycerolipids, glycerophospholipids, sphingolipids, saccharolipids, polyketides, sterols, and prenol lipids (Fahy et al. 2011). In terms of potential biotechnological applications of microbial lipids, the term "microbial oils" is frequently used, which refers to intracellular storage lipids mainly comprising triacylglycerols (Ochsenreither et al. 2016). A variety of fungi, so-called "oleaginous fungi" have the ability to accumulate lipids to large amounts, i.e. more than 20% of the total biomass (Abeln and Chuck 2021). The ability to accumulate lipids is attracting growing attention, as these lipids can be used in a broad range of biotechnological applications, such as generating biofuels, chemicals, or food and feed additives. Specific chemicals such as long-chain polyunsaturated fatty acids, fatty alcohols, alkanes, or carotenes can be derived from lipid metabolism (Béligon et al. 2016; Mata-Gómez et al. 2014; Ochsenreither et al. 2016; Passoth 2017a; Passoth et al. 2023; Sandmann 2022; Sitepu et al. 2014). Lipids can be bound to other biomolecules, such as carbohydrates or proteins, forming glycolipids and lipoproteins, respectively. Lipoproteins are essential for cell viability of all eukaryotes and may represent novel targets for antifungal drugs (Santiago-Tirado and Doering 2016). Glycolipids form biosurfactants, which have a broad potential for applications, for instance in bioremediation of polluted soils and as antimicrobial agents (Fernandes et al. 2023).

Oleaginous organisms have been isolated to enable production of lipid-derived compounds from a variety of substrates. Methods have been established to genetically modify lipid-accumulating fungi to improve lipid accumulation from different substrates or to convert non-oleaginous fungi to oil-accumulating organisms (Beopoulos et al. 2009; Chen et al. 2013). Production of high-value compounds from low-value residues, such as lignocellulose hydrolysate, residual glycerol, or municipal waste (via the carboxylic acid platform), from biodiesel production would be an important step towards a sustainable society that is largely independent of fossil fuels (Passoth et al. 2023; Tomás-Pejó et al. 2023). This chapter aims to summarise recent progress in understanding fungal lipid metabolism and its biotechnological application.

7.2 Oleaginous Fungi

Intracellular lipids are often synthesised by organisms to store energy and carbon. In fungi, especially yeasts, oleaginous organisms are usually defined as those that can accumulate lipids to more than 20% of their biomass. A number of fungi have been identified that meet this criterion. These fungi are quite diverse and belong to at least three fungal phyla: asco- and basidiomycetes (mainly yeasts) and zygomycetes (within *Mortierellomycotina* and *Mucoromcotina*) (Abeln and Chuck 2021; Zhao et al. 2021). Among yeasts, about 160 species have been identified as oleaginous (Abeln and Chuck 2021). *Lipomyces starkeyi*, *Lipomyces lipofer*, and *Yarrowia lipolytica* are examples of ascomycetous oleaginous yeast species. Basidiomycetous oleaginous yeast species belong, for instance, to the genera *Cutaneotrichosporon* or *Rhodotorula* (Flores-Cotera et al. 2001; Jin et al. 2015; Ochsenreither et al. 2016). Note that in the course of the principle "one fungus, one name" many genus and species names have been revised, the genus *Rhodosporidium* is outdated (Wang et al. 2015) and consequently, in this chapter species previously included in this genus will be named *Rhodotorula* spp, such as *R. toruloides* or *R. babjevae*. Similarly, the names *Cutaneotrichosporon curvatum* or *Cutaneotrichosporon oleaginosus* will be used (Liu et al. 2015c). There are on-going activities to identify novel oleaginous fungi. A variety of oleaginous yeasts have been found among cryophilic yeasts, e.g. *Leucosporidium creatinivorum*, which accumulated lipids to more than 70% of the cell dry weight with a yield of 0.2 g lipids per g consumed glucose (Filippucci et al. 2016). In the course of the search for oleaginous species, it has become more and more apparent that oleaginicity is rather related to strains than to species (Salvador López et al. 2022). For instance, not all strains of *Yarrowia lipolytica* reach a lipid content of 20% and should thus not be regarded as oleaginous (Carsanba et al. 2020). Moreover, oleaginous strains of *Saccharomyces cerevisiae* have been isolated (He et al. 2018; Watsuntorn et al. 2021). Similarly, in zygomycetes, strains of some species were found to be oleaginous, while others of the same species were not (Zhao et al. 2021). Salvador Lopez et al. initiated a discussion about a modification of the definition of being oleaginous (Salvador López et al. 2022). The current limit of at least 20% lipid content is arbitrarily chosen. It is also not defined which class of lipids is included. As mentioned above, lipids have a quite diverse chemical structure. Several detection methods identify total lipids, while others only detect fatty acid-derived lipids. Moreover, the definition usually only refers to intracellularly accumulated lipids. However, yeasts can also produce extracellular lipid compounds, such as *Starmerella bombicola*, which produces glycolipids. The amount of glycolipid produced can sometimes exceed the biomass, i.e. lipid accumulation is above 100% of the cell dry weight. Thus, one may discuss whether organisms producing considerable amounts of extracellular lipid compounds should also be regarded as oleaginous (Salvador López et al. 2022). Zhao et al. claimed that a genus should be regarded as oleaginous, when one of its members is identified as oleaginous (Zhao et al. 2021). However, taking into account the diversity of oleaginous behaviour within species, such a generalisation seems not to be appropriate. There are, moreover, also examples of engineered strains of species that are usually not regarded as oleaginous such as *S. cerevisiae*, which can accumulate more than 20% of lipids per cell dry matter. A distinction of "natural" and "artificial" oleaginicity is not necessarily meaningful, since even processes of metabolic engineering are based on processes that are found in nature. Oleaginous *S. cerevisiae* strains were even achieved by sexual crossing (Arhar et al. 2021). Therefore, the term oleaginous should primarily be associated with strains. If associated with a species (or a genus), a significant majority of its members should have been proven to be oleaginous.

Many oleaginous organisms have a broad diversity in utilising substrates for lipid accumulation, including glucose, xylose, glycerol, organic acids and polymers such as starch and hydrophobic compounds such as alkanes (Abeln and Chuck 2021; Passoth et al. 2023; Thevenieau et al. 2018; Tomás-Pejó et al. 2023). Filamentous fungi are usually able to produce longer polyunsaturated fatty acids (Dzurendova et al. 2022), whereas yeasts can accumulate more oil in their cells; in fact, oleaginous yeasts are amongst the highest known producers of lipids and can reach lipid proportions of more than 85% of their dry weight within a few days (Ochsenreither et al. 2016; Passoth et al. 2023). Table 7.1 provides some examples of oleaginous fungi.

Table 7.1 Examples of oleaginous fungal species. If no other reference is given, information was compiled from Salvador López et al. (2022) and Sitepu et al. (2014) for oleaginous yeasts and from Béligon et al. (2016) and Bellou et al. (2016) for oleaginous zygomycetes

Phylum	Taxonomic placement	Species	Lipid content [% DM]	Remarks[a]
Zygomycetes	Order Mucorales	*Cunninghamella echinulata*	24–54	17–22% of total FA GLA
		Cunninghamella blakesleeana	41	22% of total FA PUFA
		Mucor circinelloides	15–36	15–26% of total FA PUFA
		Thamnidium elegans	71	Chatzifragkou et al. (2011)
	Order Mortierellales	*Mortierella elongata*	24	58% of total FA ARA, 13% EPA
		Mortierella alpina	50	43% of total FA ARA
	Order Umbelopsidales	*Umbelopsis isabellina* (syn. *Mortierella isabellina*)	71	11% of total FA ALA, 2.5 GLA
Basidiomycetes	Subphylum Pucciniomycotina Class Cystobasidiomycetes Order Cystobasidiales	*Rhodotorula minuta*	25	
	Class Microbotryomycetes Order Leucosporidiales	*Leucosporidium creatinivorum*	49–61, 70	The latter value was obtained at 20 °C in a test of cryophilic yeasts (Filippucci et al. 2016)
	Order Sporodiobolales	*Rhodotorula colostri*	27	
		Rhodotorula mucilaginosa	28–33	
		Rhodotorula paludigenum	25.7–36.3	
		Rhodotorula sphaerocarpum	15–43	
		Rhodotorula babjevae	22.6–65	
		Rhodotorula diobovatum	20–41	
		Rhodotorula toruloides	20–76.1	67.5% in high-cell density culture (Li et al. 2007)
		Rhodotorula glutinis	35–57	
		Rhodotorula graminis	40	
		Rhodotorula fluviale	25	
		Rhodotorula glacialis	42–68	
	Order Uncertain	*Rhodotorula terpenoidalis*	27	
	Subphylum Agaricomycotina Class Tremellomycetes Order Filobasidiales	*Solicoccozyma terricola* (syn. *Cryptococcus terricolus*)	40.7–69	
		Solicoccozyma terrea (syn. *Cryptococcus terreus*)	25.4–51.7	
		Naganishia adeliensis	32.1–59.1	Obtained at 20°C in a test of cryophilic yeasts (Filippucci et al. 2016; Tasselli et al. 2018)
	Order Cystofilobasidiales	*Phaffia rhodozyma*	37	Flores-Cotera et al. (2001)
	Order Trichosporonales	*Cutaneotrichosporon cutaneum*	20–45	Liu et al. (2013b)

(continued)

Table 7.1 (continued)

Phylum	Taxonomic placement	Species	Lipid content [% DM]	Remarks[a]
Ascomycetes	Subphylum Saccharomycotina Class Saccharomycetes Order Saccharomycetales Family Lipomycetaceae	*Myxozyma melibiosi*	23	
		Lipomyces lipofer	54.8–66.7	
		Lipomyces tetrasporus	27.6–68.6	
		Lipomyces starkeyi	20.4–68.1	
		Lipomyces doorenjongii	39.6–72.3	
		Lipomyces kockii	54.8–78	
	Family Metschnikowiaceae	*Kodamaea ohmeri*	53	
		Metschnikowia pulcherrima	30	
	Family Wickerhamomycetacaea	*Cyberlindnera jadinii*	20–30	
		Cyberlindnera saturnus	25	
	Family Debaryomycetaceae	*Schwanniomyces occidentalis*	23	
		Candida diddensiae	37	
		Kurtzmaniella cleridarum	33	
	Family Dipodascaceae	*Geotrichum histeridarum*	34	
		Galactomyces candidus	50	
		Yarrowia lipolytica	20–70.7	In engineered strains, more than 90% was achieved
	Family Saccharomycetaceae	*Torulaspora delbrueckii*	40	
		Saccharomyces cerevisiae	45	A strain engineered in multiple genes of lipid metabolism (Kamisaka et al. 2013)

[a]Abbreviations: *FA* fatty acids; *ARA* arachidonic acid; *EPA* eicosapentaenoic acid; *ALA* α-linolenic acid, *GLA* γ-linolenic acid; *PUFA* polyunsaturated fatty acids

7.3 Biochemistry of Lipid Synthesis

Fatty acids for microbial oils of oleaginous fungi are synthesised in the cytoplasm from acetyl-CoA, following a reversed β-oxidation. Acetyl-CoA carboxylase synthesises ATP-dependent malonyl-CoA from acetyl-CoA, which is the substrate of the fatty acid synthase (FAS)-enzyme complex. FAS elongates the acyl-CoA chain by two carbons in each step, consuming NADPH. Usually, synthesis starts with acetyl-CoA, resulting in even-numbered fatty acids. However, it is also possible that propionyl-CoA is used in the first condensation, which results in odd-numbered fatty acids (Tehlivets et al. 2007). Odd-numbered fatty acids have been observed in some yeast strains, frequently in relatively low proportions but in few cases up to a proportion of more than 20% of the total fatty acids (Brandenburg et al. 2018; Olstorpe et al. 2014). The proportion of C17:1 could be enhanced in several oleaginous yeast strains by adding propionate as carbon source (Kolouchová et al. 2015; Park et al. 2021). An enhanced proportion of odd chain fatty acids (mainly C17) has also been observed in *Komagataella pastoris* during methylotrophic growth. In this case, odd chain fatty acid synthesis could be driven by α-oxidation, as a gene potentially involved in this process was upregulated (Ma et al. 2021).

Two NADPH are consumed per elongation step to reduce the acetyl groups; thus, for the synthesis of one mole of C18 saturated fatty acid (stearic acid), 8 moles ATP and 16 moles NADPH are required (Ratledge 2014). When glucose is the carbon source, the acetyl-CoA is derived from citric acid, which is exported out of the mitochondria. This happens when there is a surplus of carbon and energy for the cell, which implies that there is a limitation of another essential compound, such as nitrogen, phosphorus, or sulphur (Ratledge and Wynn 2002; Wu et al. 2010, 2011). Under those circumstances, AMP, which is an essential cofactor of isocitrate dehydrogenase (ICDH), is degraded and thus ICDH is no longer able to perform the conversion of isocitrate to α-keto-glutarate. As a result, isocitrate accumulates and levels out with citrate, which is then exported, usually in exchange with malate, to the cytoplasm. In the cytoplasm, citrate is converted to oxaloacetate and acetyl-CoA by ATP:citrate lyase (ACL), consuming ATP and Coenzyme A (Tang et al. 2015). ACL has been seen as a common feature of all oleaginous fungi (Ratledge and Wynn 2002). However, in the light of the discovery of novel oleaginous strains of species that have not been regarded as oleaginous, this view may be challenged. The oleaginous *S. cerevisiae* strain D5a lacks an ACL. Transcriptional responses of genes involved in pathways of N- and energy metabolism were regulated in a way that fluxes were redirected towards providing higher amounts of cytoplasmic acetyl-CoA and NADPH than in non-oleaginous strains. Nevertheless, further investigations are required to fully understand the mechanism of lipid accumulation in non-ACL containing strains (He et al. 2018).

Supply of NADPH has been seen as the limiting factor in lipid synthesis. The major pathways for NADPH-generation are the pentose-phosphate pathway (PPP) with the two NADPH-generating enzymes, glucose-6-phosphate dehydrogenase (G6PDH) and 6-phospho-gluconate dehydrogenase (6PGDH), and the pyruvate-oxaloacetate-malate (POM) or transhydrogenase cycle. In this cycle, pyruvate is carboxylated to oxaloacetate by pyruvate carboxylase (PC), then the oxaloacetate is reduced to malate, which consumes one mole NADH per mole of oxaloacetate, and then malic enzyme (ME) converts the malate to pyruvate, CO_2, and NADPH. There is also a cytoplasmic, $NADP^+$-dependent ICDH in many eukaryotic organisms, which may play a role in generating NADPH during lipid synthesis (Ratledge 2014).

Malic enzyme has been suggested to be the major provider of NADPH for lipid production. A physical interaction between malic enzyme and other lipid synthesis enzymes has been proposed (Ratledge 2014; Ratledge and Wynn 2002). However, in many oleaginous yeasts, no cytoplasmic malic enzyme has been found. *Lipomyces starkeyi* possesses a putative cytoplasmic malic enzyme, but it has a preference for NAD^+ instead of $NADP^+$ (Tang et al. 2010; Vorapreeda et al. 2013). *Y. lipolytica* contains only a mitochondrial ME, and its overexpression did not affect lipid accumulation (Beopoulos et al. 2011). Similarly, the overexpression of a $NADP^+$-dependent ME from *Mortierella alpina* did not result in increased lipid production in *Y. lipolytica* (Zhang et al. 2013). Metabolic flux analysis showed that there is no NADPH-production by ME in this yeast, and that the PPP, although not significantly upregulated under lipid accumulation conditions, was obviously sufficient for the production of reducing cofactors for lipid synthesis. This might also be due to the fact that *Y. lipolytica* accumulates rather low amounts of lipids compared to other oleaginous fungi (Zhang et al. 2016a). On the other hand, by overexpressing genes of acetyl-CoA carboxylase (*ACC1*) and diacylglycerol acyltransferase (*DGA1*), the final step of triacyl-glyceride synthesis, the lipid content in *Y. lipolytica* could be increased to more than 60% of cell dry weight, without further manipulation of NADPH-generating enzymes (Tai and Stephanopoulos 2013).

In other fungi, a clear impact of NADPH-generating enzymes on lipid synthesis has been shown. In *C. cutaneum*, ME was upregulated under lipid accumulation conditions (Liu et al. 2013b). Overexpression of a ME from *Mucor circinelloides* resulted in substantially increased lipid contents both in *M. circinelloides* (Zhang et al. 2007) and in *Rhodotorula glutinis* (Della-Bianca et al. 2013). On the other hand, overexpressing a native ME-gene in *R. toruloides* did not have a large effect on lipid accumulation (Zhang et al. 2016b). In *M. circinelloides*, both PPP and ME contribute to NADPH-formation during lipid synthesis (Tang et al. 2015; Zhao et al. 2015). *M. alpina*, which accumulates significantly more lipids than *M. circinelloides*, has a substantially higher and more constant ME-activity than *M. circinelloides*. Most probably, oxaloacetate, which is formed in the cytoplasm as a result of the ACL-reaction, is NADH-dependently reduced to malate, which is then converted to pyruvate, CO_2, and NADPH by ME. Pyruvate is subsequently transported into the mitochondrion by the pyruvate/malate/citrate shuttle system. In *M. circinelloides*, a substantial part of the malate is also transported back to the mitochondria, since there is an accumulation of cytoplasmic malate due to the low activity of ME (Zhou et al. 2016). Conversion of cytoplasmic malate to pyruvate by ME and transport of pyruvate back to mitochondria would, however, result in a lack of oxaloacetate in the mitochondria of filamentous fungi, as there is obviously no mitochondrial PC (Wynn et al. 2001) and thus pyruvate can only be converted to acetyl-CoA by pyruvate dehydrogenase (PDH). Additional OA can, however, be generated by the glyoxylate cycle, where isocitrate is first converted to glyoxylate and succinate by isocitrate lyase (ICL). Glyoxylate can then be condensed with acetyl-CoA, resulting in malate, which can be transported out from the mitochondria, serving as additional substrate for ME, generating more NADPH (Vongsangnak et al. 2013). Indeed, increased activities of the glyoxylate cycle have been demonstrated in *M. circinelloides* under lipid accumulation conditions (Zhao et al. 2015) as well as in the oleaginous *S. cerevisiae* strain D5a (He et al. 2018). Yeasts have both, cytoplasmic and mitochondrial PC activities, and in *C. cutaneum*, an increased amount of mitochondrial oxaloacetate synthesised by mitochondrial PC has been observed, while cytoplasmic oxaloacetate was mainly produced by ACL (Liu et al. 2013b).

When all NADPH has to be provided by the PPP, as it is possibly the case in yeasts lacking $NADP^+$-dependent cytoplasmic ME-activity, the maximum lipid yield per g glucose would be 0.276 g lipids (Ratledge 2014), and indeed, in one study a yield of 0.27 g lipids was reported in engineered *Y. lipolytica* (Tai and Stephanopoulos 2013). The simultaneous overexpression of a glyceraldhyde-3-phosphate dehydrogenase with a preference for $NADP^+$ instead of NAD^+, and a ME from *M. circinelloides*, and overexpression of a NAD^+/NADH kinase resulted in a maximum lipid yield of 0.28 g/g glucose (Qiao et al. 2017). This is, however, still not reaching the maximum theoretical lipid yield from glucose, when NADPH is not exclusively generated by PPP, which was calculated to be 0.316 g/g consumed sugar (Ratledge 2014). Still, it is considerably higher than the yields that have been obtained until now, where values higher than 0.22 g/g seldomly were obtained (Jin et al. 2015; Ratledge 2014; Robles-Iglesias et al. 2023).

However, it has been pointed out that NADPH-availability is not necessarily the only limiting factor in lipid synthesis. Rodríguez-Frómeta and colleagues discovered that a *M. circinelloides* strain overexpressing a gene encoding cytosolic ME produced more lipids than the original mutant strain but not more than a control strain, which was only transformed with the selection marker LeuA, encoding α-isopropylmalate isomerase, a step in leucine biosynthesis. Adding external leucine to the growth medium did not have a positive effect on lipid synthesis, indicating that the leucine-synthesis pathway itself is activating lipid synthesis (Rodríguez-Frómeta et al. 2013). A similar phenomenon has been observed in *S. cerevisiae*, where complementing of a *LEU2*-mutation resulted in a higher lipid content of the cells (Kamisaka et al. 2007). The mechanism behind this activation of lipid production by an active leucine-synthesis pathway is not completely clear. It is possible that acetyl-CoA is released from leucine degradation, which can serve as substrate for lipid synthesis. The substrate of α-isopropylmalate isomerase, α-isopropylmalate, which accumulates in *leuA*-mutants, plays a pivotal role in leucine metabolism in yeasts. It binds to Leu3p, which turns this protein from a transcriptional repressor to an activator of leucine biosynthesis. One may speculate that similar regulatory phenomena also occur in *M. circinelloides* and that accumulation of α-isopropylmalate in the *leuA*-mutants activates synthesis and prevents degradation of leucine and by this supply of acetyl-CoA from this pathway (Rodríguez-Frómeta et al. 2013).

Other substrates than glucose have also been considered for single cell oil production. This includes, for instance, xylose, which is the major component of hemicellulose and the second most abundant sugar in nature. Xylose is present in lignocellulose hydrolysates in considerable amounts. Substantial efforts have been made to engineer strains of *Saccharomyces cerevisiae* to perform ethanol production from xylose; however, only recently strains capable of surviving the harsh industrial conditions have been obtained. As an alternative to ethanol production, the natural ability of many oleaginous fungi could be harnessed to generate lipids from xylose (Passoth 2014). Xylose is metabolised via the PPP. To achieve this, the sugar is first converted to xylulose, which in most fungi occurs in a two-step reaction. At first, xylose is reduced to xylitol by xylose reductase (XR). In most fungi, this reaction is NADPH-dependent; only few yeasts, those that can form ethanol from xylose, have a dual NADPH/ NADH-dependent activity. In the second step, xylitol is re-oxidised by xylitol dehydrogenase (XDH) to xylulose, which is then phosphorylated and enters the PPP. In the oleaginous yeast *R. toruloides*, an alternative pathway of xylose utilisation has been identified. The xylulose formed from XR and XDH is re-oxidised to ribulose, which is then phosphorylated and further metabolised by the PPP (Jagtap et al. 2021; Jagtap and Rao 2018). Transcription studies demonstrated that arabitol-dehydrogenase, catalysing the conversion of xylulose to arabitol and of arabitol to ribulose, is upregulated in the presence of xylose, while XK was poorly expressed (Jagtap et al. 2021; Martín-Hernández et al. 2023). Due to the demand for NADPH in the XR step, cells grown on xylose would need more NADPH, and thus, more hexose generated in the PPP would have been recycled through the PPP to generate this NADPH. Under those circumstances, 1 mole acetyl-CoA would be formed out of 100 g xylose (about 0.66 moles, thus, about 1.5 moles would be formed per mole xylose). Therefore, an alternative pathway of xylose utilisation has been postulated, involving phosphorolytic cleavage of xylulose-5-P by phosphoketolase. This enzyme generates acetyl phosphate and glyceraldehyde-3-phosphate. Indeed, the corresponding enzyme has been found in a variety of yeast species grown on xylose medium without glucose, including oleaginous species such as *Lipomyces starkeyi* and *Rhodotorula toruloides* (Evans and Ratledge 1984). Moreover, it has been shown that transcription of this gene is enhanced in *R. toruloides*, when growing on crude glycerol in the presence of some hemicellulose hydrolysate, containing xylose, acetic acid, and glucose to a little extend (Martín-Hernández et al. 2023). Via this pathway, lipid yield on weight basis from xylose would be slightly higher than from glucose. For xylose, 1.2 moles acetyl-CoA would be formed out of 100 g xylose, corresponding to 0.34 g lipids per g sugar. For glucose, 1.1 moles acetyl-CoA would be formed out of 100 g (0.56 moles) of glucose, corresponding to 0.32 g lipids per g sugar (Evans and Ratledge 1984; Jin et al. 2015; Papanikolaou and Aggelis 2011). There is varying information about the real yields obtained from xylose; however, there are examples where higher lipid yields have been obtained from xylose compared to those from glucose (Fakas et al. 2009). Glycerol is another interesting substrate for lipid production, as it is generated as a side product of transesterification in biodiesel production. Glycerol is phosphorylated by glycero-kinase and then oxidised to dihydroxyacetone phosphate or glyceraldehyde phosphate, which are further assimilated following the glycolytic pathway. The theoretical lipid yield for glycerol has been estimated with 0.3 g/g, but in practice, usually values of 0.1–0.17 g/g are obtained (Chmielarz et al. 2021; Papanikolaou and Aggelis 2011). In two cases, for *Aspergillus* sp. and *Thamnidium elegans* lipid yields of about 0.2 g/g were obtained (André et al. 2010; Chatzifragkou et al. 2011). Interestingly, when crude glycerol was mixed with a small amount of hemicellulose hydrolysate, yields up to 0.25 g/g consumed carbon source were obtained (Chmielarz et al. 2021; Papanikolaou and Aggelis 2011). Organic acids, especially acetic acid, have also been considered as potential substrates for lipid production (Jagtap et al. 2021). Acetic acid is a side product of a variety of processes and also frequently found in lignocellulosic hydrolysate (Papanikolaou and Aggelis 2011; Passoth and Sandgren 2019; Tomás-Pejó et al. 2023). Acetic acid is converted to acetyl-CoA by acetyl-CoA synthase, encoded by *ACS1* and is further assimilated via the assimilatory glyoxylate cycle (Barnett and Kornberg 1960). Upregulation of *ACS1* and genes of the carboxylate cycle has been observed in *R. toruloides* during growth on acetic acid as the sole carbon source. *ACL1* was downregulated when *ACS1* was highly transcribed (Jagtap et al. 2021). A similar observation has been made when *R. toruloides* was cultivated on a mixture of crude glycerol and hemicellulose hydrolysate, containing initially a mixture of glycerol, xylose, acetic acid, and glucose as carbon sources (Martín-Hernández et al. 2023). The pathways of lipid formation from different carbon sources are summarised in Fig. 7.1.

From the initially saturated, mainly C16 and C18 fatty acids, unsaturated fatty acids are synthesised at the endoplasmic reticulum by the action of desaturases and elongases (Bellou et al. 2016). Eukaryotic systems use an aerobic pathway of desaturation (Fig. 7.2). This pathway consists of three membrane-bound proteins, NAD(P)H-cytochrome b_5 reductase, cytochrome b_5, and the terminal cyanide-sensitive desaturase. The desaturase complex is bound to the endoplasmic reticulum (Certik and Shimizu 1999). It has been stated that in fungi the desaturases are dependent on NADPH (Kendrick and Ratledge 1992; Ratledge 2014). However, for the yeast *Y. lipolytica* and the filamentous fungus *M. circinelloides*, a dual cofactor dependency of NADH and NADPH has been found, for the latter even a higher activity with NADH (Jackson et al. 1998; Pugh and Kates 1979).

Desaturation starts with palmitic (C16) and stearic (C18) acids and the first double bond is generated at the Δ9 position, i.e. palmitoleic acid (C16:1 *cis* 9) and oleic acid (C18:1 *cis* 9) are generated. From oleic acid, further desaturations may occur (Fig. 7.3), at Δ12 position to generate linoleic acid (C18:2), which may be further desaturated at Δ15 position, generating the Ω-3 fatty acid α-linolenic acid (ALA, C18:3, *cis* 9,12,15). Oleic, linoleic, and linolenic acid can be converted further by Δ6 desaturase, leading to the Ω-9, Ω-6, and Ω-3 series of fatty acids (Certik and Shimizu 1999). Yeasts usually have a limited ability to form a great variety of fatty acids. The major fatty acids are C16:0 and C18:1. Fatty acids longer than C18 are only produced in traces. Usually, low amounts of C18:2 and C18:3 are produced (depending on culture conditions, up to 10% of the total fatty acids), in basidiomycetes such as *R. toruloides* usually a bit more than in ascomycetes; still, C18:2 and C18:3 have also been found in ascomycetous yeasts (Brandenburg et al. 2016; Olstorpe et al. 2014; Ratledge 2004; Sigtryggsson et al. 2023). However, some filamentous zygomycetes can also synthesise longer polyunsaturated fatty acids (PUFA). In the Ω-9-series of fatty acids, after the Δ6 desaturase-step, one elongase- and Δ5 desaturase reaction result in the production of mead acid (MA, 20:3 *cis* 9,12,15). In the Ω-6 and Ω-3 series of fatty acids, elongase- and Δ5 desaturase reactions generate arachidonic acid (AA, 20:4 *cis* 5,8,11,14) and eicosapentaenoic acid (EPA, 20:5 *cis* 5,8,11,14,17), respectively. EPA can also be synthesised from AA by a Δ17 desaturase, which especially occurs at low temperatures (Certik and Shimizu 1999). To date, no fungi are known to naturally elongate fatty acids further to produce the longer chain Ω-3 fatty acid docosahexaenoic acid (DHA, 22:6 *cis* 4,7,10, 13,16,19). Thraustochytrids are commercially used to produce DHA as a food additive and produce fatty acids by using a different pathway, based on the polyketide

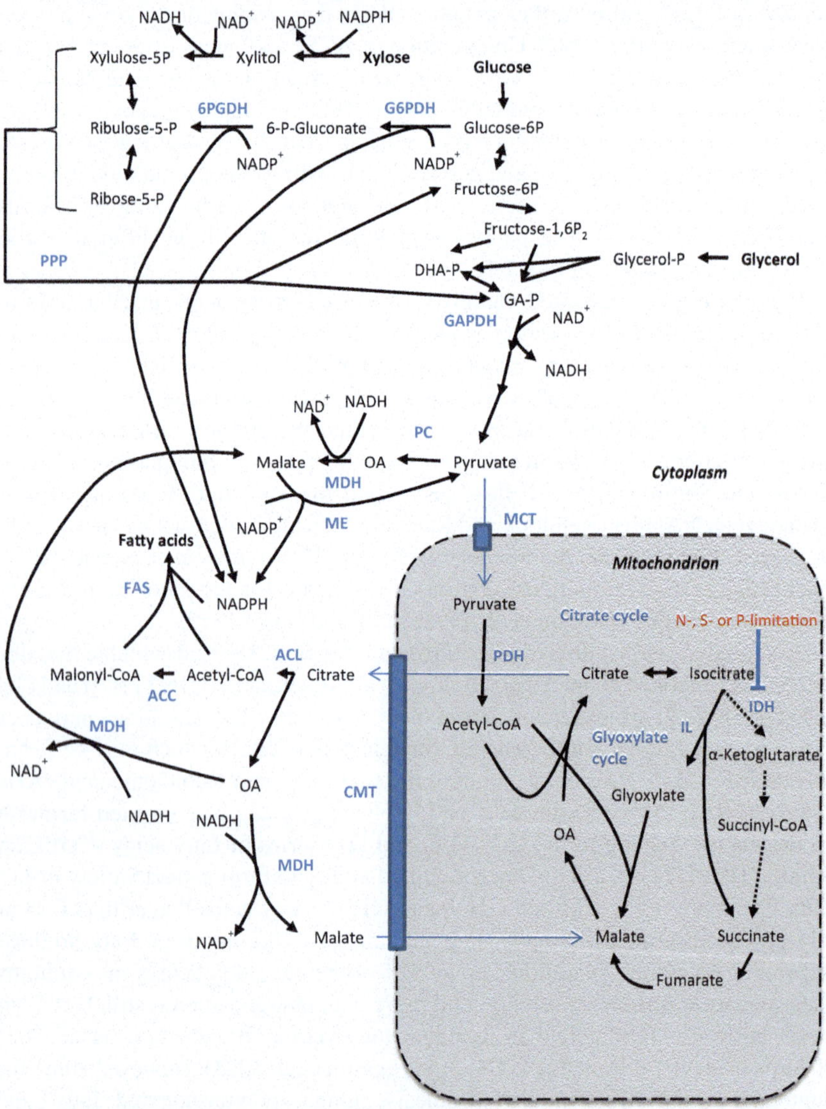

Fig. 7.1 Simplified scheme of fatty acid biosynthesis from glucose, xylose, and glycerol. Acetyl-CoA is provided from the ATP-citrate lyase (ACL) reaction, NADPH from malic enzyme (ME), and the pentose-phosphate pathway (PPP)-enzymes, glucose-6-P dehydrogenase (G6PDH) and 6-P-gluconat dehydrogenase (6PGDH). Citrate-malate can be transported by citrate-malate translocase (CMT) or specific transporters, pyruvate by monocarboxylic acid transporter (MCT). *DHA-P* dihydroxyacetone-P; *GA-P* glycerol aldehyde-P; *GAPDH* glycerol aldehyde-P dehydrogenase; *PC* pyruvate carboxylase; *OA* oxaloacetate; *MDH* malate dehydrogenase; *FAS* fatty acid synthetase; *ACC* acetyl-CoA carboxylase; *PDH* pyruvate dehydrogenase; *IL* isocitrate lyase; *IDH* isocitrate dehydrogenase. From Passoth (2017b)

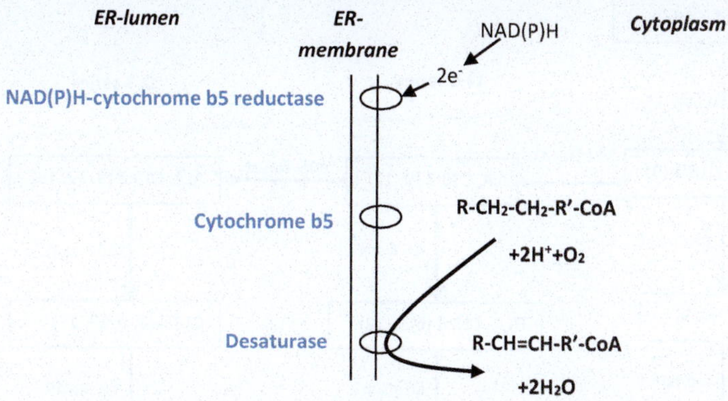

Fig. 7.2 Aerobic desaturation of fatty acids in fungi. Electrons from NAD(P)H are transported along a membrane-bound electron transport chain at the membrane of the endoplasmic reticulum (ER) and finally, together with two electrons from the saturated binding, transferred to O_2 as final electron acceptor, generating an unsaturated fatty acid and H_2O. From Passoth (2017b)

synthase route (Du et al. 2021). Originally classified as fungi, these organisms belong to a group of heterotrophic marine protists within the *Labyrinthulomycota,* aligned with Stramenopila/ Heterokonta (Cavalier-Smith et al. 1994; Leyland et al. 2017; Wang et al. 2024).

Free fatty acids are then converted to tri-acyl-glycerides (TAG) in a series of enzymatic reactions starting from glycerol phosphate or dihydroxyacetone phosphate, which are in several steps converted to phosphatidic acid. Phosphatidic acid is then dephosphorylated to diacylglycerol and subsequently acylated to triacylglycerol (Sorger and Daum 2003). Storage lipids are localised in lipid bodies in the cell. Lipid-body formation is a highly organised process. Lipid bodies are generated between the two leaflets of the lipid-bilayer of the endoplasmic reticulum (ER). Neutral lipids—TAG and sterol esters—accumulate between the leaflets, due to their hydrophobicity. When they reach a critical concentration, they phase-separate from phospholipids, forming lens-like structures. Multiple smaller lenses fuse into larger ones. Finally, the lipid body buds from the cytosolic side of ER. A variety of proteins are associated with lipid bodies (see Renne et al. 2020 for a comprehensive review). Most of the proteins are involved in lipid metabolism, many of them are also associated with the ER. Among them are the enzymes for TAG synthesis, which are essential for lipid-body formation (Renne et al. 2020; Wang 2015; Zhu et al. 2015), seipins, conserved membrane proteins, and proteins interacting with them play an essential role in lipid-body formation and their interaction with the ER (Klug et al. 2021; Renne et al. 2020). Perilipins constitute a major group of lipid-body proteins and are thought to act as coat proteins of the lipid body (Renne et al. 2020). A lipid-body specific protein containing a putative perilipin domain has been identified in *R. toruloides*, and its overexpression in *S. cerevisiae* resulted in giant lipid bodies (Zhu et al. 2015). Lipid bodies have a variety of important cellular functions, including protecting cells from lipid toxicity and allowing coordinated lipid mobilisation (Renne et al. 2020).

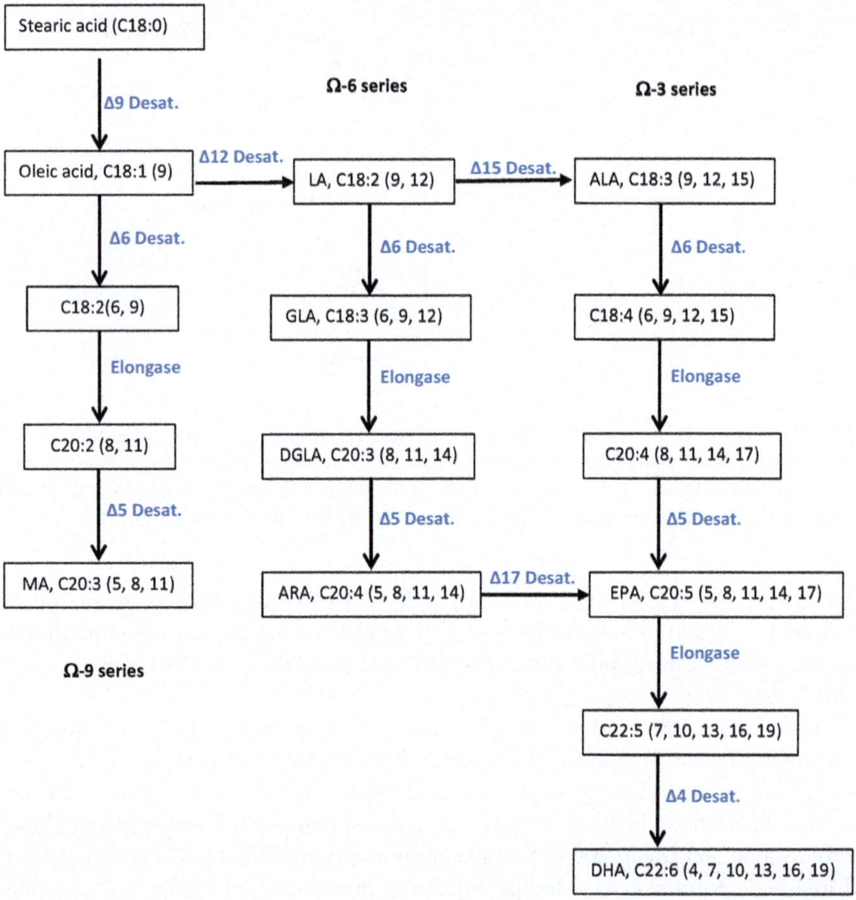

Fig. 7.3 Biosynthesis of unsaturated fatty acids. Production of C18 and longer unsaturated fatty acids starts by Δ9 desaturation of stearic acid by an according desaturase (Desat.). Δ17-desaturase is activated by low temperatures. By a series of desaturations and elongations, fatty acids of Ω9, Ω6, and Ω3-series are generated. *MA* mead acid; *LA* linoleic acid; *GLA* γ-linolenic acid; *DGLA* dihomo-γ-linolenic acid; *ARA* arachidonic acid; *ALA* α-linolenic acid; *EPA* eicosapentaenoic acid; *DHA* docosahexaenoic acid. Currently, no fungi are known that can naturally produce DHA. From Passoth (2017b)

7.4 Lipid Containing Biomolecules

7.4.1 Fungal Glycolipids as Biosurfactants

Glycolipids are derivatives from the lipid metabolism and have biotechnological potential as biosurfactants. They consist of a long-chain fatty acid, representing the hydrophobic part, and a carbohydrate moiety, which is hydrophilic. The carbohydrate component can be glucose, trehalose, mannose, mannosylerythritol, polyols,

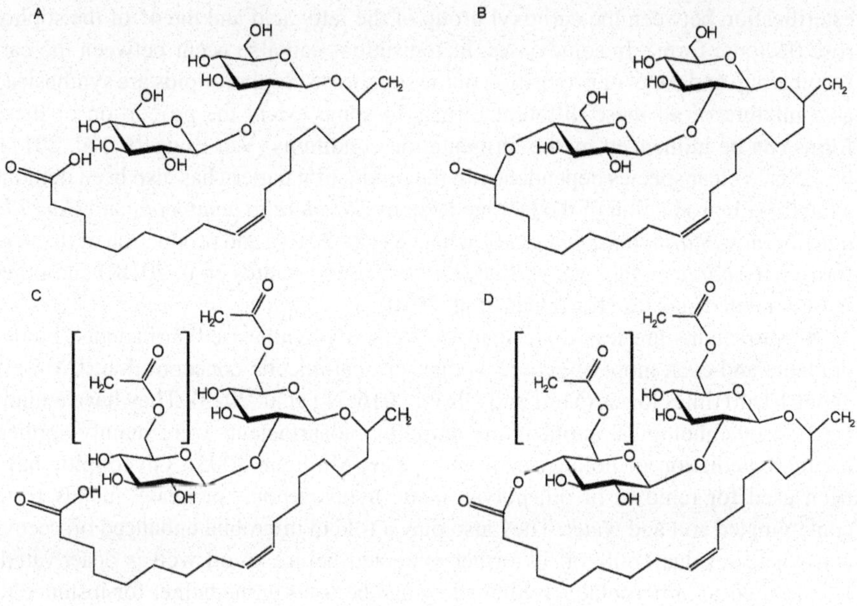

Fig. 7.4 Sophorolipids are formed from sophorose (hydrophilic part), a β-1,2 diglucose, and a terminal or subterminal hydroxylated fatty acid (hydrophobic part), often hydroxyoleate or hydroxystearate. The fatty acid is β-glycosically linked to the sophorose via its hydroxy-group. Sophorolipids can have an acidic form (**a**, **c**) or a lactonic form (**b**, **d**), both these forms can be mono- or diacetylated (**c**, **d**). From Passoth (2017b)

galactose, sophorose, cellobiose, or rhamnose. Mainly rhamno- and sophorolipids are produced commercially by a variety of companies; however, rhamnolipids are produced by the opportunistically pathogenic bacterium *Pseudomonas aeruginosa*, whereas sophorolipids are produced by non-pathogenic yeast strains belonging to a broad range of species (see below) (Jezierska et al. 2018; Jimoh and Lin 2019).

Sophorolipids (Fig. 7.4) are synthesised from C16 or C18 fatty acids, with a single or a double desaturation. Initially, the fatty acid is hydroxylated at Ω-1 or Ω-2 position by cytochrome P450 monooxygenase. Then, by two glycosyltransferase steps, glucose is connected to the molecule, consuming one UDP glucose per step. These glycosylations are performed by two different glycosyltransferases, UgtA1 and UgtB1 (Saerens et al. 2011a, c). The first glucose is β-glycosidic bound to the hydroxyl group. The second glucose is bound β-1,2 glycosidically to the first glucose, forming sophorose. The resulting molecule can be acetylated by an acetyltransferase (Saerens et al. 2011b). Acetylation usually occurs at the 6′ and/ or the 6″-position of the sophorose (Van Bogaert et al. 2011). The sophorolipid is then transported out of the cell by an ABC transporter (Mdr) into the extracellular space (Van Bogaert et al. 2013), where it can undergo further lactonisation by a cell wall-bound lactonesterase. Interestingly, this enzyme shows homology to a biotechnologically interesting lipase of *Pseudozyma antarctica* (Ciesielska et al. 2014). Sophorolipids can exist with a free fatty acid tail (acid form) or undergo internal esterification between the carboxyl group of the fatty acid and the 4″ of the sophorose (lactonic form). In some cases, lactonisations can also occur between the carboxyl group and the 6′ or 6″ position of the sophorose. Sophorolipids are synthesised as a mixture of all these different forms. To some extent the proportion of these forms can be influenced by the fermentation conditions (Van Bogaert et al. 2011); however, a clear species dependency in the production pattern has also been demonstrated, with production of the lactone form by *Starmerella bombicola* and *Candida apicola* (now *Starmerella apicola* (Lachance et al. 2010)) and production of the acid form by *Candida stellata* (now *Starmerella stellata* (Santos et al. 2018)) and other closely related species (Kurtzman et al. 2010).

Biosurfactants are less toxic than synthetically synthesised surfactants, biodegradable and even more effective, as their critical micelle concentration (CMC) is about 10–40 times lower (Mnif and Ghribi 2016; Ng et al. 2023). They have a number of biotechnological applications. Initially, biosurfactants were mainly applied for bioremediation of polluted soil and water (Mulligan 2005). Glycolipids have been used for removal of oil, polyaromatic hydrocarbons, or heavy metals from contaminated soil and water. They also play a role in microbial enhanced oil recovery, where residue crude oil is further extracted before an oil well is deactivated. Due to their antimicrobial activities, they can be used in medicine, for instance to develop antimicrobial films for coatings and wound healing, or to disturb biofilms (Ng et al. 2023). Sophorolipids have been shown to inhibit cancer cells ((Paulino et al. 2016; Van Bogaert et al. 2011), and references therein); non-acetylated sophorolipid forms have antiviral activity (Shah et al. 2005). Sophorolipids can also be used as antimicrobial finish for textile materials. Novel applications include their use in food production or cosmetics, where glycolipids can serve as emulsifiers, for foaming, wetting, and as solubilisers. In food, glycolipids can also act as biopreservative agents, due to their antimicrobial activities (Ng et al. 2023). They are also used in ecological cleaning or dishwasher products (Paulino et al. 2016).

A variety of yeast species are able to produce sophorolipids. The current major producer of sophorolipids is *S. bombicola*, (syn. *Candida bombicola*). There are a variety of other yeast species within the genus *Starmerella*, which are able to produce sophorolipids, including *S. stellata*, *S. floricola*, *S. riodocensis*, *S. apicola,* and *S. batistae*. However, sophorolipids have also been found to be produced by phylogenetically more distant species, including *Wickerhamomyces anomalus* and *Wickerhamiella domercqiae* and even by the basidiomycetous yeast *Pseudohyphozyma bogoriensis* (previous name *Rhodotorula bogoriensis*, (Wang et al. 2015). *S. apicola* was the first yeast species in which sophorolipid production was discovered, and similar final concentrations and product yields have been obtained compared to *S. bombicola*; however, research and commercialisation efforts have mainly focused on the latter species during the last years (Paulino et al. 2016; Van Bogaert et al. 2011).

S. bombicola generates lipid concentrations of more than 400 g/l, and 70% of the substrate was converted to sophorolipids. Accordingly, *S. bombicola* has been commercialised and a range of patents have been filed (Van Bogaert et al. 2011).

For the production of sophorolipids, a sugar and a hydrophobic compound are usually used as carbon sources. Standard substrates consist of 100 g/l glucose, considering that *S. bombicola* tolerates quite high sugar concentrations, along with a fatty acid or a triglyceride (Van Bogaert et al. 2011). Alkanes have also successfully been used, as this yeast can assimilate alkanes, which triggers the production of sophorolipids (Inoue and Ito 1982). Glucose can be kept in excess, while the hydrophobic substrate is usually added in a controlled fed-batch mode (Davila et al. 1992). Alternative substrates have also been considered. When non-glucose sugars are used, sophorolipids are still produced, though in smaller quantities compared to glucose. Using fatty acids or alkanes with the right chain length (C16–C18) significantly increases surfactant production. Free fatty acids are more efficiently converted than their respective oils, while methylesters of free fatty acids are better converted than the corresponding free fatty acids (Asmer et al. 1988; Van Bogaert et al. 2011). Cheaper, alternative substrates can also be used for sophorolipid production, including molasses to replace the sugar, or glycerol as hydrophobic substrate; however, sophorolipids production from these sources is lower (Van Bogaert et al. 2011). Waste products such as biodiesel waste streams, used frying oil, or restaurant waste have also been tested to replace the standard hydrophobic compounds. The results were difficult to evaluate, as comparison to standard substrates were missing; nevertheless, a better yield has been obtained compared to fermentations without adding the hydrophobic substrates, and thus, a potential valorisation of these waste products seems possible (Nitschke and Costa 2007; Paulino et al. 2016; Van Bogaert et al. 2011).

Nitrogen sources usually include yeast extract or corn steep liquor; however, other N-sources such as ammonium or urea are also possible. Some minerals should also be present. Similar to lipid accumulation, sophorolipid production starts when the cells enter stationary phase due to nitrogen limitation. The stationary phase can be maintained for about 20 days, but typical fermentations last for 10 days. The process tolerates a broad temperature-range; typical fermentations are running at 25–30 °C, although 21 °C has been determined as optimal. In contrast, C/N ratio, pH, and oxygenation must be carefully controlled. At low C/N ratio, the formed sophorolipids are metabolised, indicating that they are formed as a kind of external storage material (Van Bogaert et al. 2011). Indeed, the genes encoding sophorolipid-synthesis are induced by nitrogen limitation. This is also true for gene clusters encoding mannosylerythritol and cellobioselipids (Roelants et al. 2014). The pH decreases during fermentation to 3.5, and should thereafter be held at this value. Oxygen is required by the cytochrome P450 monooxygenase, and the oxygen transfer rate should be kept between 50 and 80 mM O_2 per l and h (Van Bogaert et al. 2011). A comparatively high concentration of sophorolipids (422 g/l) was obtained in a two-step process, where whey was initially converted to oily biomass by the oleaginous yeast *Cutaneotrichosporon curvatum*, and in the second step, the disintegrated yeast biomass was used as nutrient source for *S. bombicola* (Daniel et al. 1999). This represents also a way of introducing residues into the production process, which are usually not metabolised by *S. bombicola* (such as lactose from whey). The reached concentration is still superior compared to other studies (Daniel et al. 1999; Venkataraman et al. 2022). Downstream processing of biosurfactants after fermentation represents another challenge in biosurfactant production. The required degree of purification depends on the respective application. Purification techniques can include precipitation, phase separation, solvent extraction, ultrafiltration, crystallisation, and chromatography for ultra-purification (Venkataraman et al. 2022).

The genes involved in sophorolipid production in *S. bombicola* have been identified. Similar to the synthesis pathways of many other secondary metabolites, they are organised in a subtelomeric cluster (Van Bogaert et al. 2013), while the gene for lactonisation is located elsewhere in the genome (Ciesielska et al. 2014). Based on developed tools for genetic engineering of *S. bombicola* (Van Bogaert et al. 2008a, b), production of tailored sophorolipids is possible. For instance, cells can be manipulated to produce non-acetylated sophorolipids by deleting the acetyltransferase gene (Saerens et al. 2011b). Non-acetylated sophorolipid forms have, for instance, been shown to have antiviral activity (Shah et al. 2005). The majority of patents on yeast-based glycolipids are on sophorolipids, probably due to a highly efficient production system. Further experiments have shown that *S. bombicola* can be manipulated to produce other compounds, such as a cellobioselipid and the bioplastic polyhydroxyalkanoate. Although there are still many factors to optimise, *S. bombicola* clearly has great potential for the production of high-value, lipid-related compounds (Jezierska et al. 2018; Ng et al. 2023).

Apart from sophorolipids, other glycolipids are also produced by fungi, including mannosylerythritol lipids (MEL) and cellobioselipids (CBL) (Jezierska et al. 2018; Münßinger et al. 2024). MEL consist of a mannosylerythritol disaccharide, which is acetylated at positions R4 and R6, and acylated with short- chain (C2–C8) and medium-chain (C10–C16) fatty acids at positions R2 and R3 (Fig. 7.5). A variety of different MELs have been originally identified, called "ustilipids" as they are produced by fungi belonging to Ustilaginales (Kurz et al. 2003). Four different MELs have been distinguished, MEL-A (fully acetylated), MEL-B (acetylated at R6), MEL-C (acetylated at R4), and MEL-D (completely deacetylated). However, during the last years, novel variants have been identified, monoacylated and triacylated (with a fatty acid residue even on the erythritol-subunit) (Münßinger et al. 2024). A new nomenclature for MEL has been proposed, reflecting the degree of acylation and the differences in the chain lengths of the fatty acids. The designation

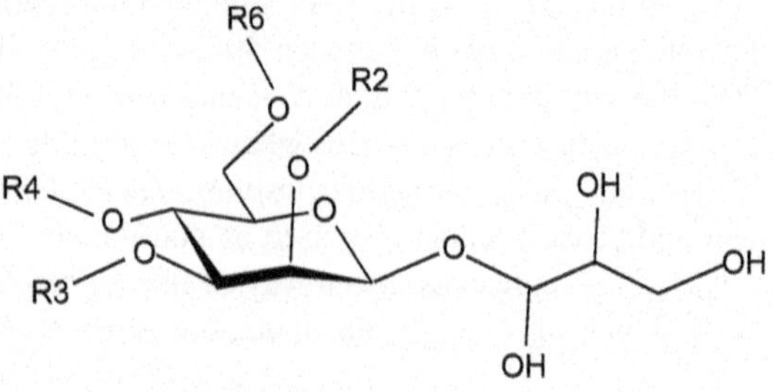

Fig. 7.5 Structure of mannosylerythritol lipids (MEL). R2, R3-fatty acid (C2–C16); MEL-A: R4, R6 –$COCH_3$; MEL-B: R4 –H, R6 –$COCH_3$; MEL-C: R4 –$COCH_3$, R6 –H; MEL-D no acetylation in R4 and R6. From Passoth (2017b)

would then be MEL-X-C_n-C_m, where X stands for the degree of acylation, and m and n for the length of the respective fatty acid (Beck et al. 2019). In *Moesziomyces antarcticus*, novel glycolipids have even been identified where the erythrose-moiety is replaced by mannitol, arabitol, or ribitol (Morita et al. 2012; Morita et al. 2009).

A variety of fungi can produce MEL, including several *Pseudozyma*-species, the dimorphic fungus *Ustilago maydis* and relatives, *Moesziomyces antarcticus* and relatives and the ascomycete *Geotrichum candidum* (Münßinger et al. 2024). Various factors influence the production of MEL. Since most cultivations have been done in shake flasks, it is difficult to provide a general pattern to maximise production. Temperature, pH, kind of carbon, and nitrogen source have been found to affect the final production. Addition of a hydrophobic substrate, such as vegetable oil, combined with a hydrophilic carbon source (sugars, sugar alcohols), appears essential for obtaining high MEL productivities (Münßinger et al. 2024). Rau et al. obtained a final concentration of 165 g/l after 283 h of cultivation of *Moesziomyces aphidis* strains in a 72 l bioreactor in a two-stage process using glucose and soybean oil as carbon source, nitrate, and yeast extract as nitrogen source (Rau et al. 2005). This corresponded to a volumetric productivity of 13.9 g/l and day and a yield of 0.92 g MEL per consumed carbon source. These values still present benchmarks for MEL -production. In the first description of a pilot-scale experiment in a 1000 l bioreactor, a MEL concentration of 76 g/l was reached, a productivity of 0.403 g/l and h, and a yield of 0.69 (Yang et al. 2023).

The genes involved in biosynthesis of MEL have been identified in *U. maydis*; it was actually the first gene cluster responsible for the production of glycolipids identified in fungi. Five genes encode the whole pathway. The first gene (*emt1*) encodes a glucosyltransferase, which performs a stereospecific mannosylation from GDP-mannose of meso-erythritol at the C4-position. The second gene (*mat1*) is an acetyltransferase gene, which results in the acetylation at the R4 and R6-positions of mannose. Acetylation at R4 occurs after acylation at R2 with a short-chain fatty acid (performed by an acyltransferase encoded by *mac1*) or with a medium to long (C10–C16) chain fatty acid (catalysed by an acyltransferase encoded by *mac2*). Both acyltransferases are highly specific, both in terms of the regioselectivity and of the fatty acid chain length. Acylation was obviously essential for MEL secretion. The fifth gene (*mmf1*) encodes a protein from the major facilitators' family, essential for glycolipid secretion (Hewald et al. 2006). Gene clusters in different species of *Ustilaginales* encoding for MEL seem to be very similar, as a MEL-biosynthesis gene cluster with a high degree of identity to the *U. maydis*-cluster has also been found in various other species (Münßinger et al. 2024).

MELs are highly valued in a range of applications, from the treatment of schizophrenia and/or diseases caused by dopamine metabolic dysfunction, to the utilisation as chemical tools for the purification of proteins or as anti-agglomeration agents of ice slurry (Im et al. 2003; Kitamoto et al. 2000, 2001; Roelants et al. 2014). Various forms with different degrees of acetylation or acylation with uncommon fatty acids have been generated, opening up additional potential applications (Roelants et al. 2014 and references therein). MELs with antifungal activities have also been identified (Kitamoto et al. 1993). A specific type of MEL is now

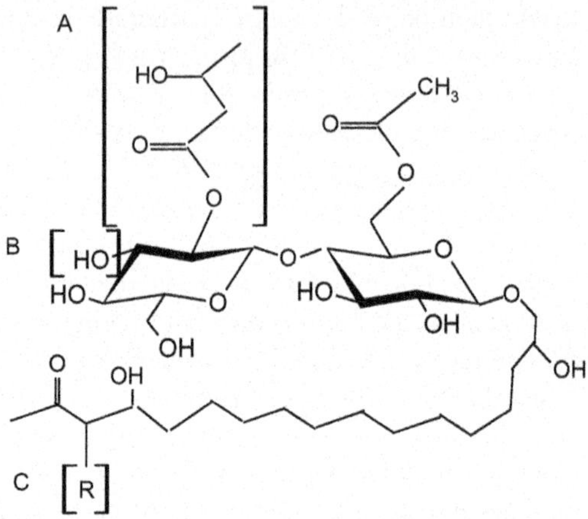

Fig. 7.6 Structure of the cellobioselipids (CBL) ustilagic acid and flocculosin. Ustilagic acid: A: β-hydroxy-hexanoic or octanoic acid, B-OH, C-H or –OH. Flocculosin: A: β-hydroxy-octanoic acid, B-acetyl, C-H. From Passoth (2017b)

commercially available from the Japanese company Toyobo Co., Ltd. under the brand name Ceramela, marketed as a moisturising and skin-barrier enhancing cosmetic ingredient (Münßinger et al. 2024).

Cellobioselipids (CBL) are produced by a small number of basidiomycetous yeasts, mainly belonging to *Ustilaginales*, a few also to the *Trichosporonales*. Ustilagic acid (Fig. 7.6) contains a cellobiose moiety, esterified with an acetyl group (at 6′) and a short-chain (C6 or C8) β-hydroxy fatty acid (at 2″). The cellobiose is O-glycosidically linked to the terminal hydroxy-group of (2, 15, 16) tri- or (15, 16) dihydroxy palmitic acid (Kulakovskaya et al. 2010; Roelants et al. 2014). Decoration of the sugar and hydroxylation of the fatty acid may vary between different species. *Pseudozyma flocculosa* produces flocculosin, a cellobioselipid with antifungal activity. Flocculosin (Fig. 7.6) consists of (3, 15, 16) trihydroxy palmitic acid as lipid component. The cellobiose is acetylated at the 6′ and 3″ positions, and the short-chain fatty acid is exclusively C8 β-hydroxy fatty acid (Teichmann et al. 2011). Similar to the MEL-synthesis genes, the genes for CBL are organised in clusters, and the clusters in *U. maydis* and *P. flocculosa* are highly similar: the former is 40 kb in size, while the latter is 60 kb (Teichmann et al. 2007, 2011). They contain two genes responsible for double hydroxylation of palmitic acid, and a glycosyltransferase gene for glycosylation of the terminal hydroxyl group of the hydroxy-fatty acid by two subsequent additions of glucose from UDP glucose. It is also possible that one cellobiose moiety is added in a single step, but this has not been proven. There are genes encoding acetyl- and acyltransferases, and genes for the synthesis of the short-chain fatty acid and its β-hydroxylation. There is an additional acetyltransferase in *P. flocculosa* that lacks a homologue in *U. maydis*. This gene is responsible for the additional acetylation in flocculosin. In *U. maydis*, α-hydroxylation of the fatty acid is the last step of glycolipid synthesis. In *P. flocculosa*, α-hydroxylation occurs before other hydroxylations, and the gene responsible for this step is still unknown. Both clusters also contain an ABC transporter (in contrast to the MEL clusters, which contain a major facilitator) necessary for glycolipid secretion (Roelants et al. 2014; Teichmann et al. 2007, 2011).

CBL have pronounced antifungal activities and are regarded as alternative fungicides. Many species of the genus *Pseudozyma* have been shown to have antifungal activities (Passoth and Schnürer 2003), and a biocontrol product based on *P. flocculosa* for the treatment of powdery mildew has been commercialised. The involvement of a CBL in antifungal activity of *P. flocculosa* has been demonstrated (Cheng et al. 2003). Fermentation processes to obtain high amounts of CBL are relatively poorly investigated (Münßinger et al. 2024). The best result so far was a final concentration of 17.6 g/l in a 1 l bioreactor, using glucose as carbon source in a repeated batch fermentation with *Sporisorium scitamineum*. The purity was 85–93%, with MEL as side product (Oraby et al. 2020).

The field of glycolipids is an excellent example of the huge biotechnological potential of fungal lipid metabolism. It is rapidly evolving towards novel potential applications in cleaning, pharmaceuticals, cosmetics, food and agricultural industry; novel strains with high potential for glycolipid production are isolated and novel glycolipids are identified and synthesised (Jezierska et al. 2018; Münßinger et al. 2024; Ng et al. 2023).

7.4.2 *Lipid Modification of Proteins*

Lipid modification of proteins plays an essential role in all kingdoms of life. Adding hydrophobic groups to proteins can influence their folding, function, and localisation. Four main classes of lipid modification have been found in fungi, including myristoylation, isoprenylation, S-acylation, and glycosylphosphatidylinositol (GPI)-modification (Jiang et al. 2018; Santiago-Tirado and Doering 2016).

N α-myristoylation is the binding of myristate, a saturated C14-fatty acid, to an N-terminal glycine, which is exposed to the N-terminus after the removal of the initial N-terminal methionine. Binding occurs via an amide bond to the α-amino group of the glycine and is irreversible. It is found in a variety of fungal proteins, for instance at the Rpt2-subunit of the 26S proteasome of *S. cerevisiae*. When the glycine is removed or replaced by another amino acid, protease activity is not influenced, but localisation is disturbed. Corresponding mutants showed phenotypes consistent with reduced degradation of misfolded proteins. In the case of the *S. cerevisiae* proteasome, the probable role of myristoylation is nuclear anchoring (Hirano et al. 2016). Myristoylation is catalysed by N-myristoyltransferase, which is present from yeast to humans and has been shown to be essential in a variety of fungi (Santiago-Tirado and Doering 2016).

Isoprenylation (also called prenylation) is the addition of isoprenoids (farnesyl or geranylgeranyl, see below) to a cysteine residue in a C-terminal CAAX consensus sequence, where C stands for cysteine, A for any aliphatic amino acid, and X for any amino acid. The AAX is removed after prenylation. Prenylation is performed by protein prenyltransferases and the reaction is irreversible. The protein is additionally capped by methylation (Nguyen et al. 2010). Isoprenylation is an essential process and a variety of proteins have been found to be prenylated. Farnesylated proteins were firstly discovered when studying mating pheromones in the basidiomycetous yeast *R. toruloides* (Kamiya et al. 1978). Prenylated pheromones are common among both basidio- and ascomycetes. The *S. cerevisiae* a-factor is the best-studied prenylated fungal pheromone. Prenylation makes the resulting molecule highly hydrophobic, requiring a specific secretion mechanism involving an ABC transporter (encoded by *STE6*). Secretion of the a-factor is thus completely different from that of the α-factor, which is hydrophilic and exported using the "classical" protein secretion pathway via endoplasmic reticulum (ER) and Golgi apparatus. Basidiomycetes produce exclusively hydrophobic pheromones, indicating that their mating mechanism is different from that of ascomycetes (Michaelis and Barrowman 2012). GTP-binding proteins of the Ras-family are another example of isoprenylated proteins, indicating that isoprenylation plays an essential role in cell signalling, stress response, and signal transduction (Omer and Gibbs 1994).

Protein S-acylation is the reversible binding of a fatty acid residue, typically palmitate (palmitoylation), to a cysteine residue of the polypeptide. There is no known recognition sequence for S-acylation, and both soluble and membrane-bound proteins can be palmitoylated. S-acylation regulates various characteristics of a protein, including stability, localisation, interaction with other molecules, and its function. Palmitoylation of soluble proteins can mediate interaction with plasma membranes, sometimes in connection with other lipid modification of the protein. For membrane-bound proteins, it regulates the protein stability and the intracellular trafficking. The reversibility of S-acylation has long been neglected, but it is now recognised as an important factor in regulating protein function (Santiago-Tirado and Doering 2016; Zaballa and van der Goot 2018).

GPI-modification, or GPI-anchoring, is the transfer of a glycolipid to a protein at its C-terminus by transamidation. The glycolipid is synthesised at the membrane of the ER and consists of a phosphatidylinositol linked to a linear glycan composed of glucosamine, three mannose-residues, and a phosphoethanolamine, the latter being required for binding to the protein. Species-specific modifications with additional ethanolamine, sugars, or lipid groups may occur (Muñiz and Zurzolo 2014; Santiago-Tirado and Doering 2016). GPI-modification is evolutionarily conserved and essential in all eukaryotes. GPI-anchored proteins are placed on the outer surface of the cell. These proteins possess two signal sequences: one at the N-terminus targeting it to the ER and one GPI-attachment site at the C-terminus (designated the ω-site). Both signal sequences are removed during processing (Orlean and Menon 2007). GPI-anchoring is irreversible, but under certain circumstances, the protein can be released by the action of a phosphatidylinositol-specific phospholipase. GPI proteins may also be transferred from the membrane to the cell wall, which may be determined by additional sequences surrounding the GPI-anchor site (Frieman and Cormack 2004). Pathogenic fungi utilise GPI-anchored proteins to adhere to tissues and resist host immune response (Santiago-Tirado and Doering 2016).

As lipid modification is essential for eukaryotic cells, it is regarded as potential target for developing antifungal drugs. Prenylation is essential for the activation of small G-proteins, including several oncogenic Ras-proteins. Therefore, a variety of prenylation inhibitors, especially farnesyltransferases have been developed, originally for anti-cancer treatment. The structure of the farnesyltransferase of the pathogenic dimorphic fungus *Cryptococcus neoformans* has been studied in detail. For instance, a substrate-induced conformational change has been observed, that does not occur in human farnesyltransferases and may thus represent a potential therapeutic target (Hast et al. 2011). Palmitoylation is also a potential therapeutic target. Many of the proteins potentially involved in pathogenesis are palmitoylated by the protein *S*-acyltransferase encoded by *PFA4*, including Ras1, which enables high temperature growth. Pfa4 thus seems to be a promising target for novel anti-*C. neoformans* drugs, especially from the background that the closest human homologue, DHHC6, is to only 21% identical to *C. neoformans* Pfa4 on the amino acid level (Nichols et al. 2015; Santiago-Tirado and Doering 2016). N-myristoylation is another potential target for antifungal drugs. In *C. neoformans*, there is one N-myristoyltransferase, and the deletion of its gene is lethal. N-myristoyltransferases from different origins have been shown to be conserved regarding their acyl-CoA substrates; however, they have very specific peptide substrates. Based on this, N-myristoyltransferase inhibitors have been developed, mimicking the N-terminal structure of one of the N-myristoyltransferase targets, the ADP-ribosylation factor Arf1 (Lodge et al. 1998). As mentioned above, the GPI-core structure is conserved among eukaryotes, but there are species-specific modifications including additional sugars, lipids, or phosphoethanolamines, or acylation of the inositol ring. Accordingly, specific inhibitors of fungal GPI-anchoring have been identified that may have a great potential in antifungal therapy (Mann et al. 2015; McLellan et al. 2012; Watanabe et al. 2012). Another potential application of the GPI-anchoring system is to display proteins on the surface of fungal cells. *S. cerevisiae* was manipulated to display cellulose- and hemicellulose degrading enzymes on its surface by connecting those enzymes to a GPI-anchor. Such so-called "armed" strains were able to directly produce ethanol from lignocellulose (Hasunuma and Kondo 2012). Various further applications have now been developed using surface display, including the developments of vaccines by exposing antigens on the cell surface, such as vaccines against candidiasis (Shibasaki et al. 2013) or *Toxoplasma gondii* (Wang et al. 2018) in *S. cerevisiae*, as well as novel food ingredients, biocontrol agents, and more (see Teymennet-Ramírez et al. 2021 for a comprehensive review).

7.5 Lipid Production by Fungi for Food, Feed, Biofuels, and Other Applications

7.5.1 Lipid Production from Lignocellulose and Crude Glycerol

Biodiesel and green diesel are currently the second most abundant biofuels in the world, after ethanol. They are mainly produced from edible vegetable oil (VO) obtained from oil plants such as soy, oil palms, or rapeseed. Waste cooking oils or oil of animal origin are also used to a lesser extent. For generating biodiesel, the triglycerides of the VO undergo a transesterification reaction where they are esterified with methanol, replacing the glycerol and generating fatty acid methyl esters (FAME). Transesterification with ethanol instead of methanol is also possible, but is currently less applied. Glycerol is thus a side product of the biodiesel process. Green diesel is produced by direct hydrogenation of the above-mentioned oils. In this reaction, oxygen is removed and hydrocarbons are formed, which are fully compatible to petrol diesel (Lucantonio et al. 2023). Moreover, VO is also an important resource for food and feed production, in fact, it is among the fastest growing food commodities in the world, and used for the production of various chemicals (Fridrihsone et al. 2020; OECD and Food and Agriculture Organization of the United Nations 2016). However, oil plants are grown on arable land and their energy yield per hectare is relatively low. Thus, a substantial area of arable land is required for food production and replacement of large proportions of mineral-oil derived diesel and chemicals. Oil plant production has been associated with deforestation of tropical forests, increased greenhouse gas emissions, and loss of biodiversity (Beyer and Rademacher 2021; Schmidt 2015).

Oils from oleaginous fungi, especially yeasts, may represent an alternative to vegetable oil. Most oleaginous yeasts can convert a variety of substrates to lipids, including sugars and acids from lignocellulose hydrolysates and crude glycerol, the residue from biodiesel production (Passoth et al. 2023; Sitepu et al. 2014; Tomás-Pejó et al. 2023). On glucose, fairly high lipid concentrations have been reached. In high-cell density cultures of *R. toruloides* on glucose, a final lipid content of 72 g/l was obtained (Li et al. 2007). Metabolically engineered *R. toruloides* reached 89 g/l (Zhang et al. 2016b) and in engineered *Y. lipolytica*, even 99 g/l was reached (Qiao et al. 2017). These values are close to theoretical limits required for economically viable biodiesel production by oleaginous yeasts, according to the National Renewable Energy Laboratory (NREL) (http://www.nrel.gov/docs/fy15osti/62498.pdf). However, sustainable biodiesel production should preferably use residual substrates such as lignocellulose or crude glycerol from biodiesel production. Apart from biodiesel production, yeast lipids can also be used for other purposes. Yeast oils are quite similar to vegetable oils in their composition. Oleaginous yeasts grown on residues from the paper industry (spent sulphite liquor) were used to produce fat for food purposes (fat for baking or as butter replacement) in Germany during both world wars. However, these processes were too expensive and were stopped immediately when vegetable lipids could be imported (Abeln and Chuck 2021; Lundin 1950). Oil from *L. starkeyi* and *R. toruloides* cultivated on wheat straw hydrolysate has been tested as a component in fish feed (arctic char, *Salvelinus alpinus*). These experiments showed no negative impact on fish performance, with no significant differences found in taste and flesh quality compared to the control. The content of potential pollutants such as heavy metals and organic pollutants was below the limits defined by the regulations of the European Union (Blomqvist et al. 2018; Brunel et al. 2022, 2024). This demonstrates that yeast oil can be used for food and feed purposes.

The C/N ratio of lignocellulose is high (Hyvönen et al. 2000; Reinertsen et al. 1984) which triggers lipid accumulation in fungi (see above). However, lignocellulose hydrolysate contains inhibitors toxic to microbes (Jönsson et al. 2013). Hydroxymethylfurfural (HMF) had a slight impact on oleaginous yeasts, influencing lipid accumulation only at higher concentrations and only in some strains. Nevertheless, a prolonged lag-phase upon exposure to HMF has been observed in several strains. Furfural and vanillin showed a strong inhibition of oleaginous yeasts (Chen et al. 2009; Hu et al. 2009). Acetic acid at a concentration of 5 g/l completely inhibited growth of strains belonging to *L. starkeyi*, *R. glutinis,* and *R. toruloides. C. cutaneum* showed some growth at this concentration (Chen et al. 2009). However, the effect of acetic acid and other weak organic acids depends on the pH of the medium. Different degrees of inhibition have been observed in cultures of *L. starkeyi* in hemicellulose hydrolysate at pH 5 and 6 (Brandenburg et al. 2016).

Due to the inhibitors in lignocellulose hydrolysate, final lipid concentrations, yields, and productivities are usually lower with this substrate. On sugarcane bagasse, 15.7 g/l lipids were obtained using *Dipodascus fermentans* (Huang et al. 2012), on rice straw 11.5 g/l (Huang et al. 2009). On corn cob hydrolysate, 12.3 g/l was achieved by *C. cutaneum* (Gao et al. 2014). Corn stover hydrolysate (hydrolysed by ammonia fibre expansion (AFEX)) and switchgrass hydrolysate (hydrolysed by dilute acid hydrolysis) were tested as substrate for oleaginous yeasts. Switchgrass hydrolysate was more inhibitory and only a few of the tested strains could grow on it. Three strains, belonging to *Lipomyces kononenkoae*, *Lipomyces tetrasporus,* and *Rhodotorula toruloides* were able to convert the acid hydrolysate to lipids (at 15–20% solid load at pH 6–7 and some amendments with nitrogen), yielding final lipid concentrations of 25–30 g/l. Similar values were obtained by these strains and by a *Saitoella coloradoensis* strain on corn stover hydrolysate. The achieved yields were 0.13–0.15 g lipids per consumed carbon source, and the production rates 0.12–0.22 g/l and h (Slininger et al. 2016). On wheat straw hydrolysate, a final lipid concentration of 18.1 g/l was obtained with a *R. babjevae*-strain, corresponding to a lipid yield of 0.24 g/g consumed carbon source and a lipid production rate of 0.175 g/h (Brandenburg et al. 2021).

These values are still far below those identified as minimum to obtain competitive fuel production (final lipid concentration 90 g/l, yield 0.28 g lipid/g carbon source, production rate 1.3 g/l h, http://www.nrel.gov/docs/fy15osti/62498.pdf). System analysis calculated a minimum selling price of $ 9.55 per gasoline gallon equivalent (about $ 2.50 per litre gasoline equivalent), when only producing fuel. In this scenario, a lipid yield of 0.24 g/g available sugar was assumed, a lipid productivity of 0.34 g/l h, and a cellular lipid content of 60%. Conversion of the pentose-fraction of the lignocellulose to succinate had a great impact on the total costs and could decrease the minimum selling price to $ 5.28 per gasoline gallon equivalent (about $ 1.40 per litre gasoline equivalent). Increasing lipid yield, decreasing extraction costs, and obtaining higher lipid contents of the biomass were some critical factors to improve the process (Biddy et al. 2016). On the other hand, conversion of the hexose fraction to ethanol is an established process, thus, lipid production from only the hemicellulose fraction may be another strategy to make biofuel production more efficient. Acid pretreatment generates a liquid and a solid fraction, and the hemicellulose is almost completely present in the liquid fraction. Thus, it can be separated relatively easily from cellulose/lignin. Since the liquid fraction also contains most of the inhibitors released during pretreatment, its conversion to lipids can be complicated. This problem might be circumvented by establishing fed-batch cultivation. The hemicellulose fraction of birch wood hydrolysate was converted to lipids using *L. starkeyi*. The yeast co-consumed acetic acid and xylose, enabling establishment of a pH-regulated fed-batch fermentation. A final concentration of 8 g/l and a yield of 0.1 g/g carbon source were obtained, which are currently the highest values achieved with the hemicellulose fraction of lignocellulose (Brandenburg et al. 2016). If one takes into account biogas production from the residues, i.e. connects biolipid production to another biofuel production process, lipid yield and proportion of the yeast biomass become less critical for the total energy efficiency of the process. However, since an aerobic fermentation process requires more energy than, for instance, the more or less anaerobic production of ethanol, rapid fermentation of the sugars is one of the crucial issues to improve the efficiency of microbial lipid production (Karlsson et al. 2016). Isolation of rapidly lipid-accumulating strains and identification of fermentation conditions allowing rapid fermentation are, therefore, important tasks of current research on oleaginous organisms. Interestingly, mixing crude glycerol with hemicellulose hydrolysate resulted in a substantial acceleration of the conversion of the substrate to lipids (Chmielarz et al. 2021). This correlated, among others, with an enhanced transcription of genes involved in the respiratory chain, i.e., energy supply for the cell (Martín-Hernández et al. 2023). Another approach may be represented by remodelling the metabolism towards producing more NADPH at the cost of NADH, which, apart from providing redox-equivalents for fatty acid synthesis, decreases the demands for oxygen in respiration (Qiao et al. 2017).

Utilising cheap, locally available substrates can also reduce the price of biodiesel. Sugarcane juice was tested as substrate for *R. toruloides*-based biodiesel production in Brazil. Final production costs were estimated to about $ 0.50 per litre biodiesel. Interestingly, the amount of biodiesel produced per ha when converting sugarcane juice using the oleaginous yeast was more than six times higher than the biodiesel production per ha obtained from soybean oil (Soccol et al. 2017).

Another approach for improving the energy efficiency of biodiesel production would be utilisation of the crude glycerol produced during transesterification (Karlsson et al. 2016). Although pure glycerol is a high-value compound, used, for instance, in the production of foods, beverages, pharmaceutics, or cosmetics, crude glycerol is highly problematic: it contains salt, methanol, and other toxic compounds that prevent its direct use as a platform chemical and are inhibitory to any fermentation process. Crude glycerol is produced at about 10% (w/w) of the generated biodiesel, and the increase in biodiesel volumes has saturated the marked for crude glycerol, making expensive cleaning procedures for this side product economically unfavourable (Haas et al. 2006; Johnson and Taconi 2007). Consequently, various attempts have been made to utilise crude glycerol as carbon source for biolipid production. Cultivation of *C. curvatum* was optimised in terms of temperature, pH, and crude glycerol concentration. Based on these parameters, fed-batch cultivation was designed and a final lipid concentration of 22 g/l was obtained after 12 days of cultivation (Cui et al. 2012). In another attempt, glycerol from recycled restaurant oil (yellow grease) was tested as substrate for *C. curvatus* and a final lipid concentration of 17.1 g/l was obtained after 12 days of fed-batch cultivation (Liang et al. 2010). Uçkun Kiran et al. tested crude glycerol and rapeseed meal, another problematic residue from biodiesel production, as nitrogen source. The rapeseed meal was pre-fermented with the proteolytic fungus *Aspergillus oryzae*, to make the nitrogen available for the oleaginous yeast in the subsequent fermentation. Lipid production was tested with *R. toruloides* Y4, and after 120 h of fed-batch cultivation, a maximum lipid concentration of 13 g/l was obtained, corresponding to a lipid yield of 0.19 g/g glycerol. When yeast extract was used as nitrogen source, final lipid concentration was only 9.4 g/l. This may indicate that rapeseed meal is a better nitrogen source, or that it provides some additional carbon source. Interestingly, higher final lipid contents and yields were obtained with 50 g/l glycerol compared to 100 g/l glycerol, potentially due to some inhibitory effects of the glycerol on the yeast (Uçkun Kiran et al. 2013). Another study showed that thin stillage has superior characteristics as nitrogen source compared to yeast extract. Lipid production by *R. glutinis* resulted in 5.4 g lipids per litre, compared to 2.3 g/l on yeast extract (Yen et al. 2012). A similar activating effect was found when mixing crude glycerol with hemicellulose hydrolysate. A maximum lipid concentration of 12.5 g/l was obtained with *R. toruloides* on crude glycerol mixed with hemicellulose hydrolysate within 96 h. This corresponded to a yield of 0.25 g/g. When crude glycerol without additions was used, a maximum lipid concentration of 0.17 g/l was obtained after more than 120 h cultivation, corresponding to a yield of 0.17 (Chmielarz et al. 2021). Tchakouteu et al. tested a variety of strains belonging to *C. curvatum*, *Y. lipolytica*, *R. toruloides,* and *L. starkeyi*. They found both *R. toruloides* and *L. starkeyi* to be best for lipid production, generating more than 12 g/l lipids in shake flask cultures. This was reached after 395 and 470 h, respectively, at an initial glycerol concentration of 120 g/l. The lipid yields were about 0.1 g/g consumed glycerol. At a glycerol concentration of 50 g/l, *R. toruloides* reached a final lipid concentration of 7.9 g/l after 195 h, and a lipid yield of 0.17 g/g. *L. starkeyi* did not have an increased lipid yield at lower initial glycerol concentrations (Tchakouteu et al. 2015). The potential of *R. glutinis* to co-produce lipids and carotenoids on glycerol was investigated. In an optimised fermentation, in terms of C/N ratio, initial glycerol concentration, and Tween-20 addition, a maximum lipid concentrations of 6 g/l and carotene concentration of 135 mg/l (yields of 0.033 g/g and 0.7 mg/g, respectively) were obtained after a fermentation time of 72 h (Saenge et al. 2011). Recently, crude glycerol conversion to lipids was tested using *R. toruloides* ATCC 10788. In these experiments, a lipid yield of 0.44 g/g consumed glycerol was obtained. This is beyond the theoretical maximum, which was explained by the consumption of additional carbon sources from the crude glycerol, such as fatty acid methyl esters, soap, diglycerides, monoglycerides, and residual triglycerides (Uprety et al. 2017). The results obtained so far indicate that a variety of strains are able to metabolise crude glycerol from biodiesel production. However, huge differences regarding growth rate, lipid production rate and yield indicate that not all factors of lipid accumulation from crude glycerol are well understood (Chmielarz et al. 2021). Yields and production rates are far from those that would enable commercially viable lipid production, thus, there is an urgent demand for further research in this field.

7.5.2 Manipulating Yeasts for Improving Lipid Production

For most of the above-mentioned oleaginous yeast species, methods for genetic engineering are either poorly developed or unavailable. For some oleaginous yeasts, tools for genetic manipulation have been developed. The most advanced yeast in this regard is the ascomycete *Y. lipolytica* (Abeln and Chuck 2021). Multiple strategies have been developed to engineer strains with enhanced lipid production. These include expanding the range of substrates, maximising the flux towards lipid accumulation, minimising the flux through competing pathways, engineering alternative acetyl-CoA pathways to decouple nitrogen limitation, engineering cytosolic NADPH pathways to provide more NADPH, engineering oxidative stress defence to mitigate lipotoxicity and improve cell morphology, evolutionary metabolic engineering, and screening metabolic modification targets based on predictions from genome-scale models (Wang et al. 2020a).

Y. lipolytica can convert a variety of carbon sources to lipids, including glucose and glycerol, but not xylose (or only to a small extent by certain strains), which is the major sugar of hemicellulose. However, efforts have been directed towards generating xylose-assimilating strains. By complementing the corresponding *Escherichia coli* mutants, Rodriguez et al. identified a cryptic xylose assimilation pathway in *Y. lipolytica* (Rodriguez et al. 2016). Obviously, the genes for xylose assimilation are present in the yeast, but expressed only at a very low level. This obstacle could be overcome by expressing the genes from a strong promoter. In another attempt, xylose reductase and xylitol dehydrogenase from the xylose-fermenting yeast *S. stipitis* and the native xylulokinase were overexpressed in *Y. lipolytica*. The resulting strains were able to grow and produce lipids on xylose; in a bioreactor fed-batch cultivation, a final lipid concentration of 20.1 g/l was reached, with a lipid yield of 0.08 g/g and a production rate of 0.19 g/l h. On mixed carbon source (xylose and glycerol), a final lipid concentration of 50.5 g/l was achieved, the lipid yield was 0.12 g/g consumed carbon source and the production rate 0.23 g/l h. Apart from lipids, citric acid and some xylitol were also produced (Ledesma-Amaro et al. 2016b). Various genetic engineering efforts to improve growth on further carbon sources such as glycerol, disaccharides, starch or inulin, and their conversion to lipids and other valuable compounds by *Y. lipolytica* have been performed (summarised in Ledesma-Amaro and Nicaud 2016).

Overexpression of *DGA1*, encoding diacylglycerol acyltransferase, and *ACC1*, encoding acetyl-CoA carboxylase, resulted in a lipid content of 62% of biomass dry weight, a lipid yield of 0.27, and a production rate of 0.253 g/l h during the lipid production phase. Overexpression was achieved by using an intron-containing TEF promoter, which ensured a 17-fold overexpression compared to the intron-less promoter (Tai and Stephanopoulos 2013). Comparison of gene expression in lipid-accumulating human tissues identified Δ9 stearoyl-CoA desaturase (SCD) as a potential target for manipulation towards higher lipid concentrations. Indeed, overexpressing this protein in *Y. lipolytica* (in a background overexpressing *ACC1* and *DGA1*) resulted in a final lipid concentration of 55 g/l and a lipid yield of 0.266 g/g glucose. Sugar uptake rate was enhanced in the transformant strain. The metabolic background of this effect was probably the removal of the saturated fatty acids palmitic and stearic acid, which can act as inhibitors of ACS (Qiao et al. 2015). A similar effect was discovered in a mutant with a modified regulator protein *mga2-G643R*, where glycine at amino acid residue 643 was replaced by arginine. The resulting mutant had an increased production of unsaturated fatty acid on the expense of the corresponding saturated fatty acids. *MGA2* is a transcriptional activator of the Δ9-desaturase, and it was demonstrated that Δ9-desaturase was upregulated in a strain expressing the *mga2-G643R*-variant. Additionally, glycolytic genes were upregulated, but not those of the citrate cycle, likely generating an imbalance between glycolysis and citric acid cycle, resulting in the higher production of precursors for lipid production. When expression of the *mga2-G643R*-variant was combined with the overexpression of *DGA1*, a final lipid titre of 25 g/l, a productivity of 0.145 g/l h, a yield of 0.213 g lipids per g glucose, and an intracellular lipid content of about 85% could be reached (Liu et al. 2015a).

Glycogen can contribute up to 16% of the total biomass of *Y. lipolytica*. Deleting glycogen synthesis resulted in a 60% increase in TAG synthesis on both glucose and glycerol (Bhutada et al. 2017). Overexpression of *DGA2* and *GPD1*, involved in production of glycerol phosphate, resulted in reduction of the competing pathway of citrate secretion, and thus an increase in lipid production (Sagnak et al. 2018).

Overexpression of peroxisomal carnitine acetyltransferase (*perCAT*2) from *S. cerevisiae* together with *ACC1* and *DGA1* overexpression, as well as overexpression of the phosphoketolase-phosphotransacetylase pathway to couple NADPH regeneration with acetyl-CoA production, resulted in a threefold increase of lipid production compared to the initial strain (Xu et al. 2016). In another approach, phosphoketolase pathways were expressed in *Y. lipolytica*, enabling direct production of glyceraldehyde phosphate and acetyl phosphate from xylulose-5-P. Acetyl-P is a new resource to produce acetyl-CoA, the precursor of fatty acid production (see above). Together with other manipulations an "obese" strain was obtained that produced 16.5 g/l lipids, more than eight times higher than the wild type, from agave bagasse hydrolysate (Niehus et al. 2018).

To improve NADPH supply, apart from overexpressing *DGA1* and *ACC1*, metabolism was engineered to generate NADPH instead of NADH during glycolysis by introducing a $NADP^+$-dependent glyceraldehyde-phosphate dehydrogenase (Qiao et al. 2017). NADH kinase, generating NADPH out of NADH was also overexpressed. The resulting strain had a similar yield to the strain constructed by Tai and Stephanopoulos (2013). However, its production rate (1.2 g/l h) was substantially higher. The final lipid concentration in a bioreactor was 99 g/l, which comes close to the values that are postulated to be required for commercial production (Qiao et al. 2017).

Lipids are sensitive against reactive oxygen species (ROS), which can attack unsaturated fatty acids, generating lipid peroxides. From these peroxides, reactive aldehydes can be formed, which are membrane permeable and highly electrophilic, attacking active sites of biomolecules, elicit cellular stress and aging, DNA-replication damage, and protein modification (Zhang et al. 2020). Overexpressing the oxidative stress defence enzymes, glutathione disulphide reductase, glutathione peroxidase, thioredoxin reductase, and glucose-6-P dehydrogenase in combination with aldehyde dehydrogenase to detoxify reactive aldehydes, resulted in a strain that reached a lipid content of 81.4%, with a total lipid concentration of 72.7 g/l and a lipid productivity of 0.97 g/l/h from glucose (Xu et al. 2017).

Interestingly, enhanced transcription of genes involved in responses against oxidative stress were also upregulated in *R. toruloides* when cultivated in a mixture of crude glycerol and hemicellulose hydrolysate, where more rapid lipid accumulation was observed compared to cultivation on crude glycerol alone (Martín-Hernández et al. 2023).

Several attempts of evolutionary engineering have been performed. In one experiment, cells were randomly mutagenised. High-lipid producers were enriched by floating, as high-lipid cells have a lower density and float on top of the culture medium. Through this method, a strain producing 39.1 g/l lipids, with a productivity of 0.56 g/l h and a yield of 0.244 g/g, was obtained (Liu et al. 2015b). In another experiment, cells were cultivated on alternatively changing growth medium, with constantly screening cells. After 77 generations, a strain was obtained that had a 30% higher lipid content than the original strain (Daskalaki et al. 2019).

Metabolic engineering based on whole genome models is a relatively new strategy. However, there are already several reports about successful manipulation of *Y. lipolytica* using this approach. Flux distribution was predicted, and in addition to already known potential targets for manipulation, several new potential targets were identified. Knocking out a gene in one-carbon/methionine metabolism resulted in a mutant producing 45% more lipids than the wild type (Kim et al. 2019).

The red yeast *R. toruloides* can grow on a broader range of carbon sources than *Y. lipolytica* and can accumulate much more lipid than the wild type of *Y. lipolytica*. Until recently, very limited molecular tools for genetic engineering of this yeast were available. Protoplast-PEG transformation of this yeast was demonstrated as early as 1985 (Tully and Gilbert 1985). However, more efficient genetic manipulation techniques have since been developed, using an *Agrobacterium tumefaciens* system (Liu et al. 2013a). A variety of tools for genetic manipulation have been developed, including a KU70 strain enabling targeted gene knock out, a library of promoters, RNA interference, and CRIPR/Cas for gene silencing and targeting (summarised in Yu and Shi 2023). There are several examples of metabolic engineering of this yeast, although the toolbox for manipulation is far less developed compared to *Y. lipolytica*. Overexpressing *DGA1* and *ACC1* significantly increased lipid production. Interestingly, overexpression of *DGA1* had a relative reproducible effect of increased lipid formation, whereas *ACC1* expression in some transformants caused a significant increase in lipid production, while others showed no difference from the wild type. Nevertheless, highest lipid production was demonstrated in a strain overexpressing both genes. On glucose and xylose, respectively, the final concentrations were 16.4 g/l and 9.5 g/l, cellular lipid contents 61.1% and 43.4%, lipid production rates were 0.28 g/l and 0.11 g/l, and lipid yields were 0.23 g/g and 0.14 g/g (Zhang et al. 2016c). Further attempts included the overexpression of enzymes, such as malic enzyme, pyruvate carboxylase, glycerol-3-P dehydrogenase, and stearoyl-CoA-desaturase (SCD) as well as the inactivation of genes involved in β-oxidation. However, only overexpression of SCD and malic enzyme had an effect on lipid production. In a background where *DGA1* and *ACC1* were overexpressed, only overexpression of SCD further increased lipid production. The final lipid concentration reached by this strain was 89.4 g/l (cellular lipid content 75.6%), with a productivity of 0.62 g/l h and a yield of 0.22 g/g (Zhang et al. 2016b). Recently, a high-quality genome sequence reconstructing chromosomes has been published, which will likely enable more efficient genetic manipulations based on genetic models (Martín-Hernández et al. 2021). For other highly lipid-producing yeasts such as *R. babjevae*, no transformation system is currently available. However, there are high-quality genome sequences, which may form the basis of future engineering techniques (Martín-Hernández et al. 2022).

Methods for manipulating oleaginous yeasts other than *Y. lipolytica* became available only some years ago. Therefore, much effort has gone into generating a lipid-producing *S. cerevisiae*. *S. cerevisiae* is usually a non-oleaginous yeast (but there are some oleaginous strains, see below); however, it is an established industrial organism that is stress tolerant and can perform well in industrial fermentations. Molecular manipulation is very well developed in this yeast, so it may have potential for production of biodiesel and other lipid-derived chemicals. In *S. cerevisiae*, the precursor of lipid synthesis, acetyl-CoA, cannot be transferred from the mitochondria to the cytoplasm via the citrate-oxaloacetate shuttle (Beopoulos et al. 2011); it is produced in the cytoplasm by the reactions of pyruvate decarboxylase (PDC), aldehyde dehydrogenase (AlDH), and acetyl-CoA synthase (Holzer and Goedde 1957). In Pdc mutants, C2 compounds such as ethanol or acetate must be provided to enable growth on glucose (Pronk et al. 1996). *S. cerevisiae* is a fermentative yeast, which means that its metabolism is directed towards converting the final products of glycolysis to ethanol rather than to acetyl-CoA. Re-directing the metabolic flux from ethanol towards acetyl-CoA production is thus one of the major challenges when manipulating *S. cerevisiae* for high lipid production (Buijs et al. 2013; Lian and Zhao 2015). When *ALD6*, encoding aldehyde dehydrogenase and a modified ACS from *Salmonella enterica* were over expressed, the level of acetyl-CoA was increased (Shiba et al. 2007). Additional overexpression of *ADH2*, encoding the assimilatory alcohol dehydrogenase of *S. cerevisiae*, also increased the availability of acetyl-CoA (Chen et al. 2013). By overexpressing *FAS1* and *FAS2*, *ACC1*, and *DGA1*, encoding FAS, acetyl-CoA carboxylase, and DAG, a strain was obtained that could accumulate up to 17% lipids per dry matter (Runguphan and Keasling 2014). By further manipulations, including expression of ACL, disruption of isocitrate dehydrogenase genes, overexpression of a fatty acyl-CoA synthetase, overexpression of a modified *DGA1*, and disruption of the transcription factor *SNF2* (which is involved in activation of glucose repressed genes), *S. cerevisiae* could be converted to an oleaginous yeast, accumulating up to 45% lipids per cell dry matter (Kamisaka et al. 2006, 2013; Tang et al. 2013; Zhou et al. 2014). Nevertheless, also non-manipulated *S. cerevisiae* strains were found to be oleaginous. D5A, an isolate from distiller's active dry yeast (Bailey et al. 1982) was found to accumulate 20% lipids under nitrogen limitation (Knoshaug et al. 2018). Nitrogen assimilation via the glutamate-glutamine pathway and amino acid metabolism, as well as fluxes of carbon and NADPH from central carbon metabolism, were crucial to obtain oleaginicity in this strain (He et al. 2018). Another oleaginous *S. cerevisiae* strain was isolated from coconut waste residues in Thailand. This strain could accumulate more than 50% of its cell dry weight as lipids under optimised culture conditions (Watsuntorn et al. 2021). By genetically engineering a strain with initially high level of lipids—specifically by exchanging a Ser to Ala in *ACC1* to remove the target of phosphorylation, overexpressing acyltransferases, and downregulating competitive pathways—a construct was obtained that accumulated TAGs to 65% of its biomass (Arhar et al. 2021).

7.6 Production of Specific Chemicals and Fuels Derived from Lipid Metabolism

Apart from producing triglycerides, and thus a VO-replacement, generation of advanced lipid-derived chemicals and biofuels has been investigated. These compounds include free fatty acids, fatty alcohols, fatty acid ethyl esters, and alkanes (Lian and Zhao 2015). Another important class of compounds are terpenes, which have potential as high-value chemicals and biofuels (Ye et al. 2016).

7.6.1 Fatty Alcohols

About two million tons of fatty alcohols are currently produced worldwide, and their production is expected to grow. The major demand is for shorter chain fatty acids (C11–C14), which are key intermediates for surfactants present in detergents, lubricants, cosmetics, personal care products, and pharmaceuticals. There is also a growing demand for longer chain fatty alcohols (C16–C22), which are used as lubricants bases, emollients, and thickeners in cosmetics and foods. Moreover, they are used as polymerisation agents in plastics. Unsaturated fatty alcohols can also be used for the aforementioned purposes and have some advantages compared to saturated fatty alcohols, as they have a lower melting point, a higher solubility in water, and functional groups can be introduced (Fillet and Adrio 2016; Krishnan et al. 2020). Several approaches have been used to enable and increase the fatty alcohol production in yeasts. In all studies, a fatty acid reductase was expressed in the yeasts. When manipulating *S. cerevisiae*, enzymes improving lipid production were also overexpressed. For instance, Runguphan and Keasling overexpressed *ACC1*, *FAS1*, and *FAS2* from *S. cerevisiae*, a thioesterase (TesA) from *E. coli*, which releases the fatty acid from CoA, and a ME from *M. alpina* (Runguphan and Keasling 2014). At the same time, genes involved in β-oxidation were disrupted. The resulting strain produced 98 mg fatty alcohols per l. In another study with *S. cerevisiae*, apart from overexpressing a fatty acid reductase from barn owl (*Tyto alba*), *ACC1* (from *S. cerevisiae*) and ACL (from *Y. lipolytica*) were overexpressed, and *RPD3*, a negative regulator of the *INO1*-gene involved in phospholipid synthesis, was disrupted. The resulting strain was able to produce 1.1 g/l 1-hexadecanol (Feng et al. 2015). In the same background, a xylose pathway (XR from *Scheffersomyces shehatae*, XDH from *Candida tropicalis*, and XKS from *Komagataella pastoris*) was expressed. Promoter engineering and evolving the strain by serial cultivation on xylose as sole carbon source resulted in 1.2 g/l 1-hexadecanol from xylose (Guo et al. 2016). In a strain engineered for the overproduction of free fatty acids (see below), further manipulation also enabled fatty alcohol production. Overexpressing a fusion protein of carboxylic acid reductase (CAR) from *Mycobacterium marinum* and yeast *ADH5*, along with another copy of the CAR-gene, deletion of *HFD1*, encoding long-chain fatty aldehyde dehydrogenase, and deletion of *ADH6*, resulted in increased titres of fatty alcohols. In a fed-batch culture, 1.5 g/l fatty alcohol was produced (Zhou et al. 2014). Guided by quantitative analyses of metabolic flux and protein levels, enzyme levels and localisation were engineered, together with pull–push engineering of metabolic fluxes and cofactor balancing. Highest activity and ER-localisation were found for a fatty acid reductase from *Mus musculus*. Competing reactions were deleted, acetyl-CoA carboxylase and Δ9-desaturase were overexpressed. NADPH and carbon consumption by glutamate dehydrogenase were limited. The final strain produced 1.2 g/l fatty alcohol in shake flask cultivation and 6 g/l in a fed-batch bioreactor. Remarkably, the strain produced 0.7 g/l from lignocellulosic hydrolysate (switchgrass) (d'Espaux et al. 2017).

Oleaginous yeasts have also been manipulated to produce fatty alcohols. For *Y. lipolytica*, patent literature described fatty alcohol production up to 0.8 g/l (Fillet and Adrio 2016). In another attempt, *Y. lipolytica* was manipulated to produce fatty alcohol by expressing a fatty aldehyde reductase from the bacterium *Marinobacter aquaeolei* (encoded by *Maqu_2220*). The resulting strain produced 167 mg fatty alcohols per l. The same gene was also expressed in *L. starkeyi*, resulting in 770 mg/l fatty alcohol. Overlaying the culture with dodecane resulted in extraction of a major part of the formed fatty alcohols into the dodecane overlay (Wang et al. 2016). In another study, *Maqu_2220* was co-expressed with an *E. coli* fatty acyl-CoA synthetase in *Y. lipolytica*, and the resulting strain produced 2.15 g/l fatty alcohols in a bioreactor-fermentation (Xu et al. 2016). Expressing *Maqu_2220*, in a strain deficient in acyl-CoA degradation, a final titre of 5.8 g/l was obtained (Cordova et al. 2020). Probably, this titre could be further increased by overlaying the culture with dodecane. *Maqu_2220* has been expressed before in *R. toruloides*, resulting in the best-known fatty alcohol production by yeasts observed so far. The resulting strain was able to produce fatty alcohols from glucose, fructose, sucrose, and xylose, as well as from a medium with mixed carbon sources and inhibitors (glucose, xylose, acetic acid, furfural, and formic acid). In fed-batch cultivation with sucrose and corn steep liquor, 8 g/l was obtained. Interestingly, the major proportion of the fatty alcohols was excreted into the medium, making it easy to extract the fatty alcohols (Fillet et al. 2015). Adding surfactants to a fatty alcohol-generating culture enhanced the production, which may be a way to facilitate fatty alcohol production by this yeast (Liu et al. 2020). *L. starkeyi* may also be a good candidate for fatty alcohol production. A strain harbouring six copies of *Maqu_2220*, without further manipulation, produced 4.2 g/l fatty alcohols (Wang et al. 2020b). Recently, the methylotrophic yeast *Ogataea polymorpha* has been manipulated to produce fatty alcohols. Coupling peroxisomal fatty alcohol synthesis and methanol assimilation and enhancing peroxisomal supply of acyl-CoA and NADPH resulted in a strain producing 3.6 g/l fatty alcohols from methanol (Zhai et al. 2023).

Fatty acid ethyl esters (FAEE) can be directly used as biodiesel. For producing biodiesel, triglycerides from VO are transesterified to FAME or FAEE (see above). Direct production of these esters by the production organism would therefore be advantageous. Moreover, FAEE are naturally excreted from yeast cells, and thus no lipid extraction would be required. Various attempts have been made to manipulate *S. cerevisiae* to produce FAEE, most of them based on episomal plasmids, resulting in unstable production patterns (summarised in Lian and Zhao 2015). Based on multi-copy integration of a heterologous wax ester synthase, 34 mg/l could be produced. Loss of the FAEE phenotype in non-selective conditions could be prevented by deleting *RAD52*, which is crucial for homologous recombination. Overexpressing acetyl-CoA binding protein, which is important for the transport of acyl-CoA esters and the reduction of feedback inhibition of acyl-CoA on FAS, ACC and long-chain acyl-CoA synthetase, and a $NADP^+$-dependent glyceraldehyde-3-phosphate dehydrogenase, increased FAEE production to 48 mg/l (Shi et al. 2014). The highest concentrations obtained to date with *S. cerevisiae* have been generated on glycerol and oleic acid as carbon sources. The glycerol assimilation pathway and a glycerol transporter were overexpressed. Moreover, genes for dihydroxyacetone phosphate degradation and glycerol export were deleted. This forced ethanol production from glycerol. Overexpressing a bacterial acyltransferase (from *Acinetobacter baylyi*) resulted in the condensation of the formed ethanol and the externally added oleic acid, with a final ethyl oleate concentration of 0.52 g/l (Yu et al. 2012). *Y. lipolytica* was also engineered to produce FAEE, and the results demonstrated that targeting the enzymes to the respective compartments greatly influences the final product formation. When the *Acinetobacter baylyi* wax ester synthase was expressed in the cytoplasm, only 7.1 mg/l FAEEs was synthesised. However, FAEE production was significantly increased when the enzyme was targeted to ER or peroxisome, achieving FAEE concentrations of 136.5 and 110.9 mg/l, respectively (Xu et al. 2016). In a further study, a wax ester synthase gene from *Marinobacter hydrocarbonoclasticus* was overexpressed, β-oxidation was blocked, and the metabolic flux redirected towards acetyl-CoA. The resulting strain reached a titre of 1.18 g/l FAEE in shake flask culture (Gao et al. 2018). However, best production was obtained by engineering a *R. toruloides* strain. Expression of wax ester synthase from *A. baylyi*, which was modified by a point mutation changing Gly355 to Ile, shifted substrate selectivity towarts shorter chain alcohols. In an optimised bioreactor culture, 9.97 g/l FAEE could be obtained, most of it extracellular (Zhang et al. 2021).

7.6.2 Free Fatty Acids

Free fatty acids are another example of advanced chemicals that can be derived from lipid metabolism. Similar to fatty alcohols, they have a broad potential to be converted to chemicals that can replace oleochemicals derived from fossil resources, including high energy fuels such as diesel or jet fuels. They can also be used for the production of cleaning agents, agrochemicals, biocidal agents, agents for textile processing, and polymer additives (Lennen and Pfleger 2013). It has been discovered that after inactivation of acyl-CoA synthetases, *FAA1* and *FAA4*, yeast cells start to secrete free fatty acids into the growth medium (Scharnewski et al. 2008). Acyl-CoA synthetases (ACS) are enzymes that generate acyl-CoA from free fatty acids, which is essential to further metabolise them. In *S. cerevisiae*, there are four long-chain ACS, encoded by *FAA1-4* (Black and DiRusso 2007). Numerous studies have been performed to engineer *S. cerevisiae* for overproduction of free fatty acids. Simultaneous disruption of *FAA1* and *FAA4*, the genes of the major acyl-CoA synthases, together with overexpression of an acyl-ACP-thioesterase from *E. coli* and a general enhancement of fatty acid synthesis by overexpressing FAS and ACC, increased free fatty acid production up to 400 mg/l (Runguphan and Keasling 2014). ^{13}C-flux analysis of lipid production in this constructed strain revealed further limitations. ACL overexpression alone did not result in a large improvement in lipid production due to the action of malate synthase, which diverted acetyl-CoA to malate synthesis. Knocking out the respective gene (*MLS1*) resulted in a very slow-growing mutant. In contrast, when the gene was only downregulated (under control of a mutated version of the *TEF1*-promoter), the cells grew well and lipid production was significantly enhanced. Further improvement of lipid production was achieved by inactivating *GPD1*, encoding glycerol-3-P dehydrogenase, which catalyses the conversion of dihydroxyacetone phosphate to glycerol-3-P. This reaction is required for triglyceride synthesis but competes with the fatty acid synthesis pathway for acetyl-CoA. Knocking out this gene also increased the flux towards fatty acid production. Together, these manipulations, based on flux analysis, increased the secretion of free fatty acids from the original amount to 780 mg/l (Ghosh et al. 2016). In another approach, the ACS genes *FAA1* and *FAA2* were inactivated, and genes involved in β-oxidation, *FAA2*, *PXA2*, and *POX1*, were disrupted. Faa2 is the ACS in the peroxisomes, Pxa1 is the fatty acid transporter into the peroxisome, and Pox1 degrades fatty acids inside the peroxisome. The resulting strain produced 1.3 g/l of fatty acids. In this mutant strain, *DGA1* and *TGL3* were overexpressed. *TGL3* encodes the major triglyceride lipase. The resulting strain produced 2.2 g/l of free fatty acids (Leber et al. 2015). In another approach, a strain with deleted long-chain fatty aldehyde dehydrogenase (*HFD1*) was further engineered. Additional genes involved in fatty acid degradation were disrupted, including *FAA1* and *-4* and *POX1*. A truncated *E. coli* thioesterase (tesA) was expressed to release free fatty acids from CoA. To increase the amount of acetyl-CoA, a chimeric pathway was expressed, consisting of *Mus musculus* ACL, *R. toruloides* ME, endogenous malate dehydrogenase with removed peroxisomal signal, and a citrate transporter. All over-expressed genes were integrated into the genome. The resulting strain produced 7 g/l free fatty acids in a shake flask fed-batch culture and 10.4 g/l in a glucose limited fed-batch culture in a bioreactor (Zhou et al. 2016). In a further attempt, through various steps of metabolic engineering, process design, and adaptive laboratory evolution, a *S. cerevisiae* strain was reprogrammed from ethanol production towards production of free fatty acids. Subcellular trafficking was changed, NADPH and ATP supply were fine-tuned, and the carbon flux to biomass was reduced. Pyruvate decarboxylase genes were knocked out. Usually, these mutants cannot grow on glucose due to a lack of cytosolic acetyl-CoA and excess of NADH to NADPH. In the initial engineered strain, both problems were solved by (1) expressing ACL providing acetyl-CoA and (2) a transhydrogenase cycle to ensure the regeneration of NAD^+. Still, *pdc*-strains were not able to grow on glucose. By adaptive laboratory evolution with gradual replacement of ethanol by glucose as carbon source, a growing strain was isolated, carrying mutations in the pyruvate kinase genes. Downregulation of pyruvate kinase obviously downregulated the glycolytic flux, preventing the accumulation of cytosolic NADH and abolishing glucose repression of respiration. This indicates that NADH accumulation is a key bottleneck in obtaining high-cell fitness and free fatty acid production. The resulting strain produced a final concentration of 33.4 g/l, which is the highest amount of free fatty acids reported so far. The yield was 0.1 g/g glucose, which is still far below the theoretical maximum and below values obtained with *Y. lipolytica* (see below), indicating that there are still opportunities to obtain higher production of free fatty acids (Yu et al. 2018).

The oleaginous yeast *Y. lipolytica* was also manipulated to produce free fatty acids. Genes involved in triglyceride and sterol ester synthesis (*DGA1*, *DGA2*, *ARE1*, and *LRO1*) and in β-oxidation (*MFE1*) were deleted. Overexpression of several thioesterases increased free fatty acid excretion. Best results were obtained with a codon-optimised thioesterase (*TEII*) from *Rattus norvegicus*. When the cells were cultivated in the presence of 10% decane or dodecane, the excreted fatty acids accumulated in the alkane-phase and were thus removed from the cells, preventing toxic effects of the free fatty acids on the cells. The yeast easily tolerated the alkanes, probably because *Y. lipolytica* is a yeast able to assimilate alkanes. In this study, however, alkanes were not assimilated, possibly because the high sugar concentration in the medium repressed alkane assimilation. In optimised bioreactor cultivation, the engineered strain produced 4.3 g/l of free fatty acids. When 15% dodecane was added to the fermenter, the resulting strain produced 10.4 g/l. Most remarkably, the amount of secreted lipids was higher than the dry cell weight in the fermenter (120%). Thus, by enabling excretion of the free fatty acids, the maximal limit of lipid production being 100% of the biomass was exceeded. The fatty acid yield was 0.2 g/g glucose (Ledesma-Amaro et al. 2016a). Methylotrophic yeasts have also been manipulated to produce free fatty acids. In *Ogataea polymorpha* engineered to produce free fatty acids, cell death with decreased cellular levels of phospholipids was observed. Adaptive laboratory evolution could restore growth. Engineering the pentose-phosphate pathway and gluconeogenesis finally resulted in a production of 15.9 g/l from methanol (Gao et al. 2022). In *Komagataella phaffii*, free fatty acid production was achieved by rewiring the central metabolism and metabolic transforming. Free fatty acid accumulation was achieved by blocking free fatty acid consumption. The supply of acetyl-CoA and NADPH was enhanced, as well as methanol assimilation, together with a decrease in accumulation of formaldehyde. This approach resulted in an accumulation of 23.4 g/l of free fatty acids from methanol (Cai et al. 2022).

The majority of the above-mentioned fatty acids have a chain length of C16–C18. From an industry perspective, especially for the production of biofuels, shorter chains of C6–C10 would be more attractive as they can serve as precursors for gasoline and jet-fuel production (Peralta-Yahya et al. 2012). However, yeasts naturally mainly produce fatty acids with a chain length of C14–C18, and there is a certain cytotoxic effect of short-chain fatty acids, making their microbial production quite challenging (Jarboe et al. 2013). Production of short-chain fatty acids may be achieved when a short-chain thioesterase (TE) removes the growing fatty acid from the FAS. Fungal FAS has a scaffold structure, making it kinetically efficient but rather inflexible for interactions with other enzymes. Thus, TE activity, which removes the growing fatty acids already in a premature stage of synthesis, is difficult to introduce into fungal FAS. In contrast, animal FAS are quite flexible and are less scaffolded (Leibundgut et al. 2008). Therefore, to achieve short-chain fatty acid production, a *Homo sapiens* fatty acid synthase (hFAS), from which the native TE subunit was removed, was expressed. hFAS lacks a 4′-phosphopantetheinyl transferase (PPT), which is required to activate the acyl carrier protein (ACP). Therefore, heterologous PPT, either from *E. coli* (AcpS) or *Bacillus subtilis* (Sfp), were expressed in the yeast. When rat *TEII* was expressed together with the mutated hFAS and the bacterial PPTs, short-chain fatty acids were produced. The best production was obtained in a strain expressing Sfp, where a total of 20 mg/l of short-chain fatty acid were produced, 17 mg/l of which were C8 (Leber and Da Silva 2014). In this context, further optimisations were investigated. Inactivating β-oxidation resulted in increased short-chain fatty acid production. However, specific inactivation of short-chain fatty acid β-oxidation increased short-chain free fatty acid production compared to the strain with complete inactivation of β-oxidation. Specific inactivation of short-chain fatty acid β-oxidation was achieved by disrupting *FAA2*, *ANT1*, and *PEX11* (Leber et al. 2016). These genes encode peroxisomal ACS (*FAA2*), adenosine nucleotide transporter (*ANT1*), which exchange peroxisomal AMP with cytoplasmic ATP, and the peroxisomal membrane protein Pex11 (*PEX11*), which is suggested to be involved in the transport of medium chain free fatty acids (i.e., fatty acids up to a chain length of C12) into the peroxisome. All three genes have been shown to be involved in the degradation of medium chain free fatty acids but not long-chain free fatty acids (Hiltunen et al. 2003). Apart from inactivating these genes, the hFAS-gene was partially codon-optimised (the first 834 bp from the N-terminus; surprisingly, codon-optimising the whole protein rather decreased the amount of short-chain free fatty acids). Moreover, proteinase B, encoded by *PRB1*, was inactivated with the aim to improve the synthase stability. The resulting final strain produced 119 mg/l of short-chain free fatty acids (Leber et al. 2016). However, reprogramming of *S. cerevisiae* FAS was also achieved by specific mutations. The resulting strain, without any further mutation, produced a total of 464 mg/ l of short-chain free fatty acid, mainly hexanoic (C6) and octanoic (C8) acid (Gajewski et al. 2017). In another approach, the endogenous FAS and a bacterial type I FAS were engineered for medium chain fatty acid production (C6–C12). Directed evolution of the membrane transporter Tpo1 and adaptive laboratory evolution were performed to improve the tolerance against toxic effects of the fatty acid. Metabolic flux was engineered towards medium chain fatty acid production, and culture conditions were optimised. Finally, a concentration above 1 g/l was obtained (Zhu et al. 2020).

7.6.3 Synthesis of Polyunsaturated Fatty Acids (PUFAs) by Fungi

Microbial oils have potential as an alternative source of PUFA for human nutrition. Higher plants cannot produce fatty acids longer than C18, and thus they can only provide the essential fatty acids linoleic acid (C18:2) and ALA (C18:3). In principle, animals including humans can synthesise the longer PUFAs AA, EPA, and DHA from linoleic and linolenic acid, but at a slow rate. Specific PUFAs have been demonstrated to have positive effects on preventing cardiovascular and inflammatory diseases, cancer, brain disorders, obesity, and diabetes. Various studies have indicated the importance of a balanced ratio between Ω-6 and Ω-3 fatty acids in a healthy diet, which should ideally range between 1:1 and 1:4. Currently, the Ω-6/Ω-3 ratio is about 15:1 in modern Western diets (Oliver et al. 2020).

Traditionally, fish oils have been the primary sources of Ω-3 fatty acids. However, there are several limitations to extensive utilisation of fish as the main source of Ω-3 fatty acids. Wild-caught fish are a limited resource and their populations are declining (Golden et al. 2016). Farmed fish require the addition of Ω-3 fatty acids to the feed, since fish, like most animals, have a limited capacity to generate longer chain fatty acids. In their natural habitats, fish take up PUFAs from aquatic microorganisms, capable of synthesising long-chain PUFAs, whereas in aquaculture settings, these PUFAs must be supplemented in the feed. Currently, these PUFAs are mainly derived from fish oil; aquaculture alone consumes about 70% of the globally produced fish oil, with approximately 90% of fish oil originating from wild-caught fish. Clearly, given the increasing demand for seafood, this practice is not sustainable (Oliver et al. 2020; Tocher 2015). Fish oils also possess undesirable odour and may contain contaminants such as mutagenic compounds and heavy metals (Bellou et al. 2016). Thus, there is a crucial demand for alternative sources of PUFA, particularly for longer chain Ω-3 fatty acids EPA and DHA.

Microbes, particularly fungi and microalgae, have been regarded as potential PUFA producers. As mentioned earlier, certain filamentous oleaginous zygomycetes are capable of EPA production. However, yeasts have also been recognised as potential PUFA resources. Although yeasts typically do not produce longer chain fatty acids, they can accumulate lipids to much higher amounts compared to filamentous fungi, making them promising candidates for genetic engineering towards the production of long-chain PUFAs (Béligon et al. 2016). In cases where PUFAs are produced by filamentous fungi, production systems are frequently based on solid stage fermentations, implying specific parameters to be optimised (Asadi et al. 2015). Many yeasts demonstrate increased desaturation at lower temperatures, although the mechanisms governing this response vary and can sometimes result in decreased desaturation at lower temperatures (e.g., Olstorpe et al. 2014). However, for several fungi belonging to the genus *Mortierella*, low cultivation temperature (12 °C) during lipid accumulation significantly increased EPA production due to the activation of Δ17 desaturase by cold conditions (Certik and Shimizu 1999; Shimizu et al. 1988). The addition of certain precursors can also increase the production of long-chain PUFAs. For instance, the addition of ALA has been shown to promote EPA production even at standard growth temperatures, due to the temperature-independent activity of the Ω-3 desaturation route, however, lower temperatures additionally increased the production to 1.88 g/l of culture broth (Shimizu et al. 1989).

As mentioned earlier, various genera of Zygomycota have a tremendous ability to produce commercially valuable PUFAs. Among them, the genus *Mortierella* stands out for its ability to produce a variety of PUFAs. The strain *M. alpina* 1S-4 was initially isolated due to its high production of AA. Several desaturase mutants have been generated from this strain through random mutagenesis, such as Δ12 desaturase mutant that overproduces MA, Δ6 desaturase mutant that overproduces dihomo-γ-linolenic acid (DGLA, C20:3 *cis cis cis*-8, 11, 14), a fatty acid of the Ω-6-series, and a Δ5-desaturase mutant that overproduces AA. These strains are not only commercially promising but also contribute to a better understanding of lipid metabolism. Methods for metabolic engineering of this fungus have been developed, including transformation systems based on complementation of a *ura3* mutation or on resistance markers. Transformation has been achieved using biolistic, protoplast-PEG, electroporation, or *Agrobacterium tumefaciens*-based methods (Chang et al. 2022). Overexpression of elongases increased the production of AA, supporting the finding that its production is limited in the elongation step from C18:3 to C20:3 (Takeno et al. 2005a; Wynn and Ratledge 2000). Moreover, RNA interference was introduced in *M. alpina*, and inactivating Δ12-desaturase indeed resulted in a change in the fatty acid profile, demonstrating the potential to influence the fatty acid profile by molecular methods (Takeno et al. 2005b). Overexpression of Δ-17 desaturases resulted in an efficient conversion of AA to EPA (summarised in Chang et al. 2022). An EPA concentration of 3.47 g/l after optimising fermentation by an engineered strain is the highest reported concentration obtained so far (Huang et al. 2020). Further optimisations of lipid production and PUFA-production by *M. alpina* include redox metabolism and cofactor supply (Chang et al. 2022 and references therein).

Several fungi have already been commercialised for the production of PUFAs. From 1985 to 1990, γ-linolenic acid (C18:3, Ω-6) was produced with *M. circinelloides* by J&E Sturge (UK), with an annual volume of 5–10 tons. DHA is commercially produced by a dinoflagellate alga and certain species of thraustochytrids. Upon application of DHA to food, it has been observed that portions of DHA can lose a C2-unit, resulting in EPA. While EPA is generally considered beneficial for health, it was initially contraindicated for its potential effects on infants. However, it has been discovered that the presence of AA can prevent the degradation of DHA to EPA. AA is produced using *M. alpina* and the process has been commercialised by Gist-brocades (The Netherlands), now acquired by DSM (The Netherlands). AA-rich oil is also produced by Suntory (Japan) and Cargill Alking Bioengineering Co. Ltd (China) (Ratledge 2013). EPA is produced by some companies using algae as production organisms. However, an alternative has been developed by DuPont (USA), using genetically modified *Y. lipolytica* (Ratledge 2013; Xue et al. 2013). To achieve this, Δ-9 elongase, a Δ-8 desaturase, a Δ-5 desaturase, and a Δ-17 desaturase were expressed in the yeast, and β-oxidation was impaired by inactivating a peroxisomal gene (*PEX10*). The final strain was able to produce EPA to 15% of the cell dry weight, and the total oil content was 56.6% (Xue et al. 2013). To obtain DHA, copies of Δ-4, Δ-5, Δ-6, Δ-12, and Δ-17 desaturase genes, as well as C16/18-, C18/20-, and C20/22 elongase genes had to be expressed in the according strain. Additionally, the acetyl-CoA oxidase 3 gene *POX3* and an endogenous Δ-12 desaturase were inactivated. This resulted in a production of DHA (5.6% of total fatty acids) (Damude et al. 2009). In an alternative approach, multifunctional polyketide-synthase-like PUFA synthases from myxobacteria were designed for *Y. lipolytica* and expressed. This resulted in the production of a variety of PUFAs, including AA, EPA, DPA, and DHA. In a fed-batch process, the engineered strain produced DHA in concentrations of 350 mg/l from glucose and 300 mg/l from glycerol, with DHA accounting for over 10% of total fatty acids (Gemperlein et al. 2019).

7.6.4 Hydrocarbons

Production of hydrocarbons such as alkanes and alkenes is especially attractive as they are the major components of fossil gasoline and diesel, making them suitable as drop-in fuels. Apart from plants, several microorganisms, including fungi can naturally produce hydrocarbons. The ability of some yeast species to produce long-chain (C_{10}-C_{31}) and volatile (C_2-C_5) n-alkanes was already discovered in the 1960s. Among these are several *Saccharomyces* species, including *S. oviformis* and *S. ludwigii*, which accumulated long-chain n-alkanes up to 10.2% of their cell dry weight, *Candida tropicalis* (up to 0.031%), *R. glutinis*, *Brettanomyces bruxellensis*,

Saccharomyces octosporus, and *Schizosaccharomyces pombe*, which produced ethane in the range from 21.5 to 11.4 ml/l and h, and *Rhodotorula minuta* var. *texenis*, which produced 16.4 ml isobutene/l and h (Fu et al. 2015; Ladygina et al. 2006). These findings have not received much attention, probably because of the rather low amounts of hydrocarbons produced. However, an isolate from a mangrove ecosystem, identified as *Aureobasidium pullulans* var. *melanogenum*, produced 32.5 g/l heavy oil when grown on glucose/corn steep liquor. About 60% of this oil consisted of alkanes—mainly tetradecane (C14), tetracosane (C24), hexacosane (C26), heptacosane (C27), octacosane (C28), and tetratetracontane (C44) (Liu et al. 2014). Not much is known about the metabolic pathways for natural n-alkane production in yeasts (Fu et al. 2015). Head-to-head condensation of fatty acids has been proposed, where two fatty acids are condensed at their carboxylic groups. Another possibility pathway is similar to that in *Arabidopsis thaliana*, where alkanes are produced from very long-chain acyl-CoA (VLC acyl-CoA), which are reduced to aldehydes by VLC-acyl-CoA reductase and then converted to alkanes by VLC-aldehyde decarbonylase (summarised in Passoth 2017a). Interestingly, genes similar to the plant genes encoding VLC-aldehyde decarbonylase have been found in the genome of *A. pullulans* var. *melanogenum* (Gostinčar et al. 2014). The first instance of alkane production in yeast involved the expression of the *A. thalliana* alkane synthesis pathway in *S. cerevisiae*. Genes encoding VLC-acyl-CoA reductase and VLC-aldehyde decarbonylase were expressed, in a strain with a modified fatty acid elongase, producing elongated (C28 and C30) fatty acids. Additionally, long-chain acyl-CoA synthetase I and CYTB5-B, a heme-protein that serves as electron transporting factor, were expressed. The resulting strain produced 86 μg/g yeast biomass (dry weight), mainly nonacosane, C29 (Bernard et al. 2012). However long-chain n-alkanes produced by *A. pullulans* var. *melanogenum* or the engineered *S. cerevisiae* strain are less attractive as biofuels, due to their chain length. Consequently, engineering strategies have been developed that aim for shorter hydrocarbons.

Hydrocarbons can be obtained from fatty acids through various enzyme reactions (Lu et al. 2022). The fatty aldehyde decarbonylation pathway consists of a fatty acyl-ACP/CoA reductase (FAR) and an aldehyde deformylating decarbonylase (FADO). FAR reduces fatty acyl ACPs/CoAs to fatty aldehydes, which are then converted into C_{n-1} alkanes or alkenes by FADO. The FAR gene and a FADO gene from the cyanobacterium *Synechoccocus elongatus* were expressed in *S. cerevisiae*. FAR reduces fatty acyl-CoA to fatty aldehyde, and then FADO converts the fatty aldehydes further to alkanes or alkenes. *E. coli* ferredoxine and ferredoxine reductases were also expressed to provide the required redox cofactors for FADO in the cytoplasm. *HFD1*, encoding a hexadecanal dehydrogenase was disrupted, because it competed with FADO for the fatty aldehydes. The resulting strain produced 22 μg heptadecane per g dry weight (Buijs et al. 2013). To further improve alkane production by *S. cerevisiae*, a carboxylic acid reductase from *M. marinum* was expressed instead of the FAR in a strain overproducing free fatty acids (see above). To activate this enzyme, a 4′-phosphopantetheinyl transferase from *Aspergillus nidulans* was expressed. Deletion of *POX1* further increased alkane production, and deletion of *ADH5* resulted in a lower production of fatty alcohols and increased alkane production. Overexpressing aldehyde deformylating oxygenases from *S. elongatus* and the cyanobacterium *Nostoc punctiforme* finally increased the alkane production to 0.82 mg/l (Zhou et al. 2016).

The 1-alkene biosynthesis pathway is another way to produce hydrocarbons. The key enzyme is a cytochrome P450 peroxygenase, which directly decarboxylates free fatty acids. The enzyme requires heme as cofactor and H_2O_2 as electron donor. The gene *OleTJE* from the bacterium *Jeotgalicoccus* sp., encoding this peroxygenase, was expressed together with a lipase gene from *Thermomyces lanuginosus* (*Tl*Tgl) in *Y. lipolytica* strain engineered for high TAG-production. Both enzymes were localised to ER and lipid bodies, with *OleTJE* additionally localised to the peroxisomes. The engineered strain produced 554.4 mg/ l alkenes (Yang et al. 2019).

A fatty acid photodecarboxylase (*Cv*Fap) from the alga *Chlorella variabilis* has been engineered to convert free fatty acids or acyl-CoA to C_{n-1} alkanes or alkenes. The enzyme belongs to the glucose-methanol-choline oxidoreductase family and harbours a light-capturing FAD-cofactor. Its activity depends on blue light in the wavelength range 400–520 nm (Lu et al. 2022). Overexpression of *Cv*Fap in different engineered *Y. lipolytica* strains resulted in production of hydrocarbons. A strain overproducing free fatty acids was most suitable and produced 10.87 mg/l. After mutational screening, 58.69 mg/l was obtained (Bruder et al. 2019). In another study, it was proven that *Cv*Fap prefers acyl-CoA over free fatty acids. A strain overexpressing multiple copies of *Cv*Fap and inhibited hydrocarbon degradation pathway produced 242 mg/l hydrocarbons from wheat straw hydrolysate and 319 mg/l from acetate (Li et al. 2020).

The soybean lipoxygenase I *Gm*Lox1 cleaves linoleic acid to pentane and a tridecadienoic acid co-product. By overexpressing this gene in a *Y. lipolytica* strain with blocked β-oxidation, a titre of 4.98 mg/l pentane was obtained, which was comparatively high at this time (Blazeck et al. 2013). Producing short-chain hydrocarbons such as pentane would be attractive for obtaining biofuels. However, the narrow substrate range of *Gm*Lox1 and the side product likely make hydrocarbon production by this enzyme rather unattractive, so no further development was pursued in this direction (Lu et al. 2022).

7.6.5 Isoprenoids

Terpenes, also called isoprenoids or isoprenes, are built up from derivatives of the five carbon compound isopren ($CH_2{=}C(CH_3)CH{=}CH_2$). The precursors of all isoprenoids in fungi are synthesised via the methylvalonate (MVA)-pathway. In the first step of this pathway, two cytosolic acetyl-CoA are condensed to aceto-acetyl-CoA by aceto-acetyl-CoA thiolase (in *S. cerevisiae* encoded by *ERG10*). Aceto-acetyl-CoA is then condensed with another acetyl-CoA to 3-hydroxy-3-methylglutaryl-coenzyme A (HMG-CoA) by HMG-CoA synthase (encoded by *HMGS*). HMG-CoA is reduced to mevalonate by HMG-CoA reductase (*HMG1*), consuming two NADPH. Mevalonate is phosphorylated by mevalonate-5-kinase

(*ERG12*) to mevalonate-5-P, which is further phosphorylated to mevalonate pyrophosphate by phosphomevalonate kinase (*ERG8*). Mevalonate pyrophosphate is then decarboxylated to isopentenyl pyrophosphate (IPP) by mevalonate pyrophosphate decarboxylase (*MVD1*), and IPP is converted to dimethylallyl pyrophosphate (DMAPP) by IPP isomerase (*IDI1*) (Lv et al. 2014; Paddon and Keasling 2014; Ye et al. 2016). All terpenoids in fungi are derived from IPP and DMAPP.

Isoprene (C_5H_8; 2-methyl-1,3-butadiene) is the simplest compound of the isoprene family. It has a broad range of industrial applications, including the production of synthetic rubber, adhesives, and speciality elastomers. It also has a potential as fuel additive for gasoline, diesel, or jet fuel. Currently, it is mainly produced from mineral oil. Isoprene is naturally produced by a variety of plants and microorganisms. Fungi also produce terpene compounds (see below); however, current knowledge suggests that they are not able to produce isoprene because they lack the isoprene-synthase, which converts DMAPP to isoprene (Ye et al. 2016). Hong and colleagues introduced a codon-optimised *IspS*-gene encoding an isoprene-synthase from *Pueraria montana* (kudzu vine) into the *S. cerevisiae* genome. Multi-copy integration resulted in the production of about 0.5 mg isoprene and isoprene derivatives in the headspace gas of a one litre culture (Hong et al. 2012). Various attempts have been made to improve isoprene production by *S. cerevisiae*, including overexpressing of the rate-limiting HMGR, downregulating or deleting of metabolic pathways competing for DMAPP, or overexpressing a global transcription factor (summarised in Ye et al. 2016). In another study, several of the above-mentioned approaches were combined with overproduction of acetyl-CoA. An isoprene-synthase gene (*ispS* from *Populus alba*) was overexpressed in *S. cerevisiae*. Then, truncated *HMG1* (*tHMG1*) was overexpressed (Lv et al. 2014). It has been demonstrated that feedback inhibition can be avoided when only the catalytic unit of *HMG1* is expressed (Donald et al. 1997). The activity of farnesyl pyrophosphate synthetase, which removes IPP from the pathway, was decreased by expressing the corresponding gene (*ERG20*) under control of a weak promoter. Attempts to enhance the supply of acetyl-CoA by increasing the activities of assimilatory alcohol dehydrogenase (*ADH2*) and cytosolic aldehyde dehydrogenase (*ALD6*) indicated that the limiting step in acetaldehyde production is the conversion of acetate to acetyl-CoA. Therefore, acetyl-CoA synthetase (*ACS2*) and acetoacetyl-CoA thiolase (*ERG10*) were overexpressed. Further fine-tuning of expression by using appropriate promoters and modifying regulation of some metabolic pathways (deletion of the repressor *GAL80*, since the *GAL1* promoter was used for the expression of some genes), optimising carbon sources (glycerol and sucrose were more conducive than glucose), and oxygen supply yielded a final production of 37 mg/l isoprene (Lv et al. 2014). In a subsequent study, the isoprene-synthase gene was placed under control of the *GAL1*-promoter; simultaneously, the activator *GAL4* was overexpressed, and native *GAL1*, *GAL7*, and *GAL10* promoters were deleted to diminish competition for the transcription factor. Further improvements were obtained by directly selecting enzymes with higher activity. DMAPP has a cytotoxic effect, and cells carrying a very active isoprene-synthase would develop a certain resistance to DMAPP. A mutant library of isoprene-synthase was generated by error-prone PCR, and enzymes with enhanced activity were isolated. These enzymes were expressed in the background described above. Cultivation of the resulting strain in a high-cell density bioreactor resulted in an isoprene production of 3.7 g/l (Wang et al. 2017). Since acetyl-CoA supply is crucial for high production of isoprene, and since cytoplasmic acetyl-CoA production is limited in *S. cerevisiae*, the MVA pathway and the isoprene-synthase were expressed in mitochondria, by fusing the genes with a corresponding mitochondrial import signal sequence. The resulting strain produced more isoprene than strains with manipulated cytoplasmic isoprene-synthesis; however, this strain did not achieve high-cell density in fed-batch cultivation. Best performance was obtained when strains with mitochondrial and cytoplasmic isoprene-synthesis were crossed, forming a diploid strain containing the pathway both in the cytoplasm and mitochondria. The resulting strain produced 2.5 g/l isoprene in a high-cell density fed-batch cultivation, demonstrating the potential of manipulating the metabolism in different cell compartments (Lv et al. 2016).

Apart from isoprene, other, long-chain isoprenoids also have immense biotechnological potential. Synthesis of these isoprenoids is achieved by condensation of IPP and DMAPP to the prenyl-pyrophosphates, geranyl pyrophosphate (GPP, C10), farnesyl pyrophosphate (FPP, C15), and geranyl-geranyl pyrophosphate (GGPP, C20). From these compounds, monoterpenes (C10), sesquiterpenes (C15), or diterpenes (C20) are formed (Buijs et al. 2013; Zhang et al. 2011). A variety of terpene-derived compounds are produced by yeasts, including the membrane sterol ergosterol (Parks and Casey 1995) and carotenoids (Mata-Gómez et al. 2014; Sandmann 2022). A very prominent example of an isoprenoid-derived compound is the anti-malaria drug artemisinin. Artemisinin is a sesquiterpene with a lactone endoperoxide. This compound is naturally produced by a plant, *Artemisia annua*, commonly known as sweet wormwood. Artemisinin production starts with the conversion of FPP to amorphadiene by amorphadiene synthase (ADS). *A. annua* accumulates, likely through the action of cytochrome P450 monooxygenase (also called amorphadiene oxidase), two oxidation products of amorphadiene, artemisinic acid and dihydroartemisinic acid. The latter is the precursor of artemisinin, and it is thought to convert spontaneously in the presence of sunlight (Paddon and Keasling 2014). By engineering the MVA pathway, expressing ADS, and a cytochrome P450 monooxygenase from *A. annua* up to 100 mg/l artemisinic acid could be produced in *S. cerevisiae* S288C (Ro et al. 2006). Artemisinic acid can easily be converted to artemisinin through chemical methods. Further improvements were achieved by additional metabolic engineering and the use of a more suitable yeast strain. The whole MVA pathway, including *tHMG1* and *ERG20,* was overexpressed in CEN. PK2. Three copies of *tHMG1* were integrated into the genome, as this step had previously been identified as limiting for production of artemisinic acid. As the genes were expressed under control of the *GAL1* promoter, galactose metabolism was also manipulated by deleting the *GAL* gene cluster and the repressor *GAL80*. Strains were then constructed containing either the whole pathway for artemisinic acid (ADS, cytochrome P450 monooxygenase, and its cognate reductase, *A. annua* cytochrome P450 reductase) or only ADS. Production of amorphadiene was more strongly enhanced in the latter strain than the production of artemisinic acid.

Therefore, the next study focused on the production of amorphadiene and its subsequent conversion to artemisine through chemical methods. After optimising fermentation conditions, 41 g/l amorphadiene was produced, and the overall yield of converting amorphadiene to dihydroartemisinic acid was 48.4% (Westfall et al. 2012). Further investigation revealed that the low ability of engineered cells to produce artemisinic acid was due to oxidative stress, which was generated due to poor coupling between the cytochrome P450 enzyme and its reductase. This was overcome by fine-tuning their expression and by expression of an *A. annua* cytochrome b5, which has been shown to improve the activity of cytochrome P450 enzymes. The resulting strain, although improved, still did not produce sufficient amounts of artemisinic acid. Moreover, artemisinic aldehyde, an oxidation intermediate, was accumulated, which is likely toxic to the cell. To address this issue, *A. annua* artemisinic aldehyde dehydrogenase was expressed, resulting in substantial increase of artemisinic acid. Expression of an NAD^+-dependent *A. annua* artemisinic alcohol dehydrogenase further increased the artemisinic acid concentration, yielding a final concentration of 25 g/l artemisinic acid (Paddon et al. 2013).

Farnesene has a great economic value in medicine, cosmetics, or as food additive. A strain overexpressing the mevalonate pathway and α-farnesene synthase genes was constructed. After optimising culture conditions, 25.55 g/l farnesene were produced (Liu et al. 2019). Similarly to farnesene, squalene is widely used in food, cosmetics, and pharmacology. Overproduction of *HMG1* could improve squalene production; however, acetyl-CoA was found to be limiting for squalene production. Overexpression of an ACS from *Salmonella enterica* or the endogenous ACL-gene further increased squalene production. Supplementation with citric acid further enhanced production, resulting in a squalene concentration of 10 mg/g cell dry weight, corresponding to a concentration of about 120 mg/l (Huang et al. 2018).

Isoprenoids also hold great potential as biofuels. Hydrolysis of IPP, GPP, and FPP results in alcohols, which can be used as fuel additives (Zhang et al. 2011). These alcohols, especially isopentanol, have been suggested as gasoline additives; they have all advantageous features of ethanol, i.e., increasing the octane number and reducing the concentration of harmful pollutants in the exhaust. Unlike ethanol, they do not increase the vapour pressure, have better tolerance to water, and do not increase fuel consumption (Hull et al. 2006). Farnesol and farnesene have been suggested as diesel fuels (Zhang et al. 2011). *S. cerevisiae* was engineered to produce bisabolene. This was achieved in a strain with overexpressed FPP production pathway and reduced activity of a competing synthesis pathway (squalene synthesis), as well as overexpression of bisabolene synthase. The resulting strain produced more than 900 mg bisabolene per litre. Bisabolene, a sesquiterpene, can be chemically hydrogenated to bisabolane, a biosynthetic alternative to D2 diesel (D2 diesel is usually used as fuel in cars). Bisabolane exhibits similar or even better characteristics compared to D2 diesel (Peralta-Yahya et al. 2012). It has been discovered that bisabolane and α-farnesene provide resistance against the detergent Tween 20, which has been used to isolate mutants with higher production of these compounds (Kirby et al. 2014). Further improvements of isoprenoid production are possible, for instance, by isolating appropriate enzymes such as terpene synthases (Tippmann et al. 2016).

Carotenoids are another group of isoprene-derived compounds. Astaxanthin is one example of a biotechnologically interesting carotenoid that is produced by a yeast. Astaxanthin is used in a variety of cosmetics and other products, and it provides the characteristic colour of salmon. It is naturally produced by marine microbes. Since fish cannot synthesise astaxanthin, it must be added to the feed of farmed salmon, and is actually one of the most expensive ingredients. Currently, most astaxanthin is produced from fossil resources; a little is produced by microbial fermentations using, for instance, the green alga *Haematococcus pluvialis* or the yeast *Phaffia rhodozyma*. This yeast was for several years designated with its teleomorph name *Xanthophyllomyces dendrorhous*, but recently, the priority of the name *P. rhodozyma* has been emphasised (Liu et al. 2015c). Only few yeast species are able to produce astaxanthin including the heterobasidous yeast *P. rhodozyma* and close relatives (Colabella and Libkind 2016). In fermentation experiments, up to 420 mg/l astaxanthin were produced. However, astaxanthin produced by fermentation costs more than three times as much as synthetic astaxanthin. Economically feasible production would require achieving production of 4 mg/g cell dry weight and a final concentration of at least 60 g/l (Rodríguez-Sáiz et al. 2010; Schmidt et al. 2011). There are three specific genes responsible for astaxanthin synthesis in *P. rhodozyma* (Schmidt et al. 2011). Two GGPP are fused to form phytoene by the protein encoded by the fusion gene *crtYB*. This gene encodes a bifunctional enzyme, which acts both in the condensation of two GGPP and later in the cyclisation of lycopene. Then, two double bonds are stepwise introduced into the phytoene molecule by the enzyme encoded by *crtI*, generating neurosporene and finally lycopene. Lycopene is, as mentioned above, converted to β-carotene by a cyclisation catalysed by crtYB (Verdoes et al. 1999a, b). In a side reaction of the pathway, lycopene can be converted by crtI to 3,4-dehydrolycopene, which is then converted by crtYB to torulene. β-Carotene is then converted by insertion of 3-hydroxyl- and 4-keto groups into the ionone rings of β-carotene. This multistep-conversion is achieved by astaxanthin synthase, which is a cytochrome P450 monooxygenase. Cytochrome P450 enzymes typically require an electron-donating system, which is provided by a specific cytochrome P450 reductase (Schmidt et al. 2011 and references therein). Astaxanthin production by *P. rhodozyma* requires sophisticated regulation of fermentation, as the yeast, although showing a high activity of the pentose-phosphate pathway and thus more similarity to respiratory yeasts and filamentous fungi, exhibits a Crabtree-effect—producing ethanol under aerobic conditions, when there is a surplus of glucose (Cannizzaro et al. 2004; Liu and Wu 2008). Methods for metabolic engineering of this yeast have been developed such as system for multiple overexpression of genes. Simultaneous overexpression of the genes of the MVA pathway resulted in a 2.1-fold increase in astaxanthin production compared to the parental strain (Hara et al. 2014). Further improvement could be achieved by deleting both copies of *CYP61* in the diploid genome, encoding C-22 sterol desaturase, which is involved in sterol biosynthesis. Ergosterol has been shown to exert feedback inhibition on the MVA pathway (Yamamoto et al. 2016). Astaxanthin production has been shown to be under glucose repression, and it has recently been demonstrated that a homologue of the *S. cerevisiae MIG1* is involved in this repression. A strain mutated in *P. rhodozyma MIG1* showed enhanced astaxanthin production (Alcaíno et al. 2016).

7.7 Lipid Extraction

Apart from the above-mentioned examples of glycolipids and free fatty acids, neutral lipids are usually produced and stored intracellularly. Therefore, they have to be extracted from the cells. Extraction is one of the major downstream processes, which may substantially contribute to the cost of implementing lipid production as bioprocess. While a variety of studies has been performed on algae, not much is known about efficient lipid extraction from fungi (Ochsenreither et al. 2016). Analytical lipid extraction is often performed with the help of solvents, which both disintegrate the cell and extract the lipids. Chloroform, often mixed with methanol, has been shown to be very efficient for lipid extraction (Folch et al. 1957). Soxhlet extraction is another classical extraction method that was originally used to extract lipids from milk powder. It is currently mainly used as a standard method against which other extraction methods are compared. The method involves drying and grinding the sample, which is then placed on a porous thimble in an extraction chamber. Extraction occurs in several washing steps with an organic solvent under reflux (Soxhlet 1879). A variety of automated Soxhlet systems are commercially available. However, Soxhlet extraction requires large amounts of solvents and is time consuming. Furthermore, extensive use of chloroform and organic solvents is environmentally problematic and should be avoided for food and feed production, due to potential toxic effects. To reduce the utilisation of organic solvents, cell disruption is usually required for lipid extraction, due to the robust cell walls of fungi. One can largely distinguish between mechanical and non-mechanical cell disruption methods. Mechanical methods include bead milling, homogenisation, and ultrasound treatment; for lipid extraction, they are often combined with solvent extraction. Non-mechanical methods can be divided into physical methods, such as decompression or microwave treatment, or drying, or chemical methods such as the application of solvents or acid treatment. Enzyme treatment is also possible (Ochsenreither et al. 2016). In one study, several of these different methods were applied to disrupt cells of the yeast *C. curvatum* and the filamentous fungus *M. isabellina*. Lipid was then extracted using a methanol/chloroform solution. The tested cell disruption methods were autoclaving, microwave treatment, ultrasonication, bead-beating, HCl digestion, and Soxhlet, which included disruption of freeze-dried cells with mortar and pestle and a subsequent methanol/chloroform extraction. All these methods had been tested in other studies for lipid extraction from oleaginous algae. Results showed that for *M. isabellina*, Soxhlet gave the best results, and HCl digestion and bead-beating were still acceptable, whereas ultrasonication, microwaving, and autoclaving gave poor extraction results. For *C. curvatus*, HCl digestion, bead-beating, and ultrasonication performed best, Soxhlet enabled similar extraction efficiencies to the three other, whereas microwaving and autoclaving gave poor yields. Differences are due to different cell structures of yeast and filamentous fungi. Nevertheless, HCl extraction performed well for both organisms; because it is relatively simple and the fatty acid composition was not impaired by the method (i.e., there was no significant decrease in the nutritionally valuable non-saturated fatty acids), HCl-cell disruption may be the method of choice in the future for extracting lipids from fungi (Yu et al. 2015). High-pressure homogenisation (800 bar at 4 °C for several passages) and subsequent hexane extraction have been applied for cell disruption of *C. curvatus* and an oleaginous strain of *Pichia kudriavzevii*. The extraction yield was not directly compared to other methods; however, lipid recoveries of 20% of dry matter (*P. kudriavzevii*) and about 50% (*C. curvatum*) indicate an acceptable efficiency of the method (Sankh et al. 2013; Thiru et al. 2011). Ultrasonic treatment may be another option for cell disruption; however, large-scale cell disruption does not seem to be possible due to difficulties of energy transmission in larger volumes. It may also generate heat, which can influence the quality of the resulting lipids, and it was shown not to be the most efficient method when compared to other methods in fungi (Ochsenreither et al. 2016; Yu et al. 2015). Hydrothermal treatment was used to extract lipids from *C. curvatus*, grown in a 5 l pilot-scale reactor (the lipid content of the yeast biomass was 53%). The biomass was treated at 280 °C and 500 psi for 1 h, and after hexane washing, 43% of the solids, corresponding to 70% of the initial total lipids, was recovered in the hexane fraction. The resulting lipids were free of sulphur and low in salts and nitrogen, making them appropriate for further processing to biofuels. The energy demand of this technique is quite high, thus it would be essential to integrate it into a process, where the energy is recovered via heat-circulation systems (Espinosa-Gonzalez et al. 2014). Treatment with supercritical fluids, especially CO_2, is another option for lipid extraction. Supercritical CO_2 enables good recovery, has a high selectivity to non-polar substances (i.e., salt, water, proteins, polysaccharides, and sugars are not extracted), and can be easily removed after extraction. It is non-toxic, non-flammable, non-polluting, inexpensive, and inert. Extraction with supercritical CO_2 can be performed at relatively low temperatures due to the low critical temperature of CO_2 (31.1 °C). It also protects against oxidation, making it a gentle extraction method, e.g., for polyunsaturated fatty acids (Ochsenreither et al. 2016). When using supercritical CO_2 extraction of oil from the zygomycete *Cunninghamella echinulata*, the best performance was achieved at 30 MPa, 50 °C, an incubation time of 180 min, and a mass flow of 50 kg CO_2/kg dry biomass, and when fungal particles smaller than 0.5 mm were used. Interestingly, CO_2 extraction resulted in a higher recovery of γ-linolenic acid and total lipids than all other tested methods, including Folch and Soxhlet with different solvents (Certik and Horenitzky 1999). In *S. cerevisiae*, good recovery and separation of triglycerides from phospholipids were obtained by CO_2 extraction at 2 MPa and 40 °C with 9% (w/w) ethanol as co-solvent. However, pretreatment was required, and acid hydrolysis followed by methanol maceration was found to be the best method to ensure high-lipid extraction efficiency (Hegel et al. 2011). Enzyme-assisted cell disruption may also be an option for lipid extraction. *R. toruloides* was treated with β-1,3-glucomannanase, after a heat pretreatment with microwaves. Lipids were extracted with ethyl acetate, at room temperature and without dewatering. Enzyme treatments may save energy costs, as they can often be performed at room temperature and at atmospheric pressure (Jin et al. 2012). On the other hand, enzymes may be expensive, so sophisticated methods to recycle enzyme will be necessary. Tailor-made cell wall lytic enzyme against the cell wall of *Schwanniomyces occidentalis* was produced by *Trichoderma harzianum*. A fermentation of *T. harzianum* was continuously fed with *S. occidentalis* as sole carbon source. The supernatant of the *T. harzianum*-culture was then used to degrade the cell wall of *S. occidentalis* for oil extraction. This enzyme-assisted process could avoid the heating step in oil extraction (Heshof et al. 2020).

Lipid extraction and downstream processing are crucial factors for obtaining sustainable biolipid production from fungi (Karlsson et al. 2016; Khot et al. 2020). A variety of lipid extraction methods have been tested in recent years. Investigating the energy efficiency of different cell disruption and lipid extraction methods would help identify the most appropriate methodology for isolating lipids from fungi.

7.8 Conclusions

Lipid metabolism in fungi has been studied for many years. However, significant progress has been achieved over the last decade, driven by a growing interest in the biotechnological utilisation of products from lipid metabolism. Techniques of metabolic engineering and –omics analyses in oleaginous and model organisms have provided novel insights into lipid metabolism, revealing its complexity in terms of biosynthesis pathways and their regulation. Given the necessity to generate fuels, chemicals, and food from non-fossil, sustainable resources, and the growing availability of molecular manipulation and analysis techniques even in non-conventional oleaginous organisms, it is easy to predict that the already rapid development in this field will continue to accelerate in the coming years. Moreover, as lipid metabolism is an essential component of cellular metabolism, research in lipid metabolism in fungi will greatly contribute to understanding general eukaryotic metabolism and thus have a strong impact on fundamental biological research, with implications for human and animal medicine as well as plant production.

Acknowledgement Yeast lipid-related research in our group was/is funded by the Swedish Energy Authority (STEM) (grant number 36649-1), the MicroDrive-program at the Swedish University of Agricultural Sciences, the Swedish Research Council for Environment, Agricultural Sciences and Spatial Planning (Formas) (213-2013-80, 2018-01877 and 2022-02404), NordForsk/Formas (2020-02637 and 2022-02760) and the Visby-program of the Swedish Institute.

References

Abeln F, Chuck CJ (2021) The history, state of the art and future prospects for oleaginous yeast research. Microb Cell Fact 20(1):221

Alcaíno J, Bravo N, Córdova P, Marcoleta AE, Contreras G, Barahona S et al (2016) The involvement of Mig1 from *Xanthophyllomyces dendrorhous* in catabolic repression: an active mechanism contributing to the regulation of carotenoid production. PLoS One 11(9):e0162838

André A, Diamantopoulou P, Philippoussis A, Sarris D, Komaitis M, Papanikolaou S (2010) Biotechnological conversions of bio-diesel derived waste glycerol into added-value compounds by higher fungi: production of biomass, single cell oil and oxalic acid. Ind Crop Prod 31(2):407–416

Arhar S, Gogg-Fassolter G, Ogrizović M, Pačnik K, Schwaiger K, Žganjar M et al (2021) Engineering of *Saccharomyces cerevisiae* for the accumulation of high amounts of triacylglycerol. Microb Cell Fact 20(1):147

Asadi SZ, Khosravi-Darani K, Nikoopour H, Bakhoda H (2015) Evaluation of the effect of process variables on the fatty acid profile of single cell oil produced by Mortierella using solid-state fermentation. Crit Rev Biotechnol 35(1):94–102

Asmer H-J, Lang S, Wagner F, Wray V (1988) Microbial production, structure elucidation and bioconversion of sophorose lipids. J Am Oil Chem Soc 65(9):1460–1466

Bailey RB, Benitez T, Woodward A (1982) *Saccharomyces cerevisiae* mutants resistant to catabolite repression: use in cheese whey hydrolysate fermentation. Appl Environ Microbiol 44(3):631–639

Barnett JA, Kornberg HL (1960) The utilization by yeasts of acids of the tricarboxylic acid cycle. J Gen Microbiol 23:65–82

Beck A, Haitz F, Grunwald S, Preuss L, Rupp S, Zibek S (2019) Influence of microorganism and plant oils on the structure of mannosylerythritol lipid (MEL) biosurfactants revealed by a novel thin layer chromatography mass spectrometry method. J Ind Microbiol Biotechnol 46(8):1191–1204

Béligon V, Christophe G, Fontanille P, Larroche C (2016) Microbial lipids as potential source to food supplements. Curr Opin Food Sci 7:35–42

Bellou S, Triantaphyllidou IE, Aggeli D, Elazzazy AM, Baeshen MN, Aggelis G (2016) Microbial oils as food additives: recent approaches for improving microbial oil production and its polyunsaturated fatty acid content. Curr Opin Biotechnol 37:24–35

Beopoulos A, Cescut J, Haddouche R, Uribelarrea JL, Molina-Jouve C, Nicaud JM (2009) *Yarrowia lipolytica* as a model for bio-oil production. Prog Lipid Res 48(6):375–387

Beopoulos A, Nicaud JM, Gaillardin C (2011) An overview of lipid metabolism in yeasts and its impact on biotechnological processes. Appl Microbiol Biotechnol 90(4):1193–1206

Bernard A, Domergue F, Pascal S, Jetter R, Renne C, Faure JD et al (2012) Reconstitution of plant alkane biosynthesis in yeast demonstrates that *Arabidopsis* ECERIFERUM1 and ECERIFERUM3 are core components of a very-long-chain alkane synthesis complex. Plant Cell 24(7):3106–3118

Beyer R, Rademacher T (2021) Species richness and carbon footprints of vegetable oils: can high yields outweigh palm oil's environmental impact? Sustainability 13(4):1813

Bhutada G, Kavšček M, Ledesma-Amaro R, Thomas S, Rechberger GN, Nicaud J-M et al (2017) Sugar versus fat: elimination of glycogen storage improves lipid accumulation in Yarrowia lipolytica. FEMS Yeast Res 17(3):fox020

Biddy MJ, Davis R, Humbird D, Tao L, Dowe N, Guarnieri MT et al (2016) The techno-economic basis for coproduct manufacturing to enable hydrocarbon fuel production from lignocellulosic biomass. ACS Sustain Chem Eng 4(6):3196–3211

Black PN, DiRusso CC (2007) Yeast acyl-CoA synthetases at the crossroads of fatty acid metabolism and regulation. Biochim Biophys Acta 1771(3):286–298

Blazeck J, Liu L, Knight R, Alper HS (2013) Heterologous production of pentane in the oleaginous yeast *Yarrowia lipolytica*. J Biotechnol 165(3–4):184–194

Blomqvist J, Pickova J, Tilami SK, Sampels S, Mikkelsen N, Brandenburg J et al (2018) Oleaginous yeast as a component in fish feed. Sci Rep 8(1):15945

Brandenburg J, Blomqvist J, Pickova J, Bonturi N, Sandgren M, Passoth V (2016) Lipid production from hemicellulose with *Lipomyces starkeyi* in a pH regulated fed-batch cultivation. Yeast 33(8):451–462

Brandenburg J, Poppele I, Blomqvist J, Puke M, Pickova J, Sandgren M et al (2018) Bioethanol and lipid production from the enzymatic hydrolysate of wheat straw after furfural extraction. Appl Microbiol Biotechnol 102(14):6269–6277

Brandenburg J, Blomqvist J, Shapaval V, Kohler A, Sampels S, Sandgren M et al (2021) Oleaginous yeasts respond differently to carbon sources present in lignocellulose hydrolysate. Biotechnol Biofuels 14(1):124

Bruder S, Moldenhauer EJ, Lemke RD, Ledesma-Amaro R, Kabisch J (2019) Drop-in biofuel production using fatty acid photodecarboxylase from *Chlorella variabilis* in the oleaginous yeast *Yarrowia lipolytica*. Biotechnol Biofuels 12(1):202

Brunel M, Burkina V, Pickova J, Sampels S, Moazzami AA (2022) Oleaginous yeast *Rhodotorula toruloides* biomass effect on the metabolism of Arctic char (*Salvelinus alpinus*). Front Mol Biosci 9:931946

Brunel M, Burkina V, Sampels S, Dahlberg AK, Passoth V, Pickova J (2024) Biomass of the oleaginous yeast *Rhodotorula toruloides* as feed ingredient in the diet of Arctic char (*Salvelinus alpinus*): impact on quality; safety and sensory outcome. Acta Agric Scand A Anim Sci 1–13

Buijs NA, Siewers V, Nielsen J (2013) Advanced biofuel production by the yeast *Saccharomyces cerevisiae*. Curr Opin Chem Biol 17(3):480–488

Cai P, Wu X, Deng J, Gao L, Shen Y, Yao L et al (2022) Methanol biotransformation toward high-level production of fatty acid derivatives by engineering the industrial yeast *Pichia pastoris*. Proc Natl Acad Sci USA 119(29):e2201711119

Cannizzaro C, Christensen B, Nielsen J, von Stockar U (2004) Metabolic network analysis on *Phaffia rhodozyma* yeast using 13C-labeled glucose and gas chromatography-mass spectrometry. Metab Eng 6(4):340–351

Carsanba E, Papanikolaou S, Fickers P, Erten H (2020) Lipids by *Yarrowia lipolytica* strains cultivated on glucose in batch cultures. Microorganisms 8(7)

Cavalier-Smith T, Allsopp MTEP, Chao EE (1994) Thraustochytrids are chromists, not Fungi: 18s rRNA signatures of Heterokonta. Philos Trans R Soc Lond B Biol Sci 346((1318)):387–397

Certik M, Horenitzky R (1999) Supercritical CO2 extraction of fungal oil containing γ-linolenic acid. Biotechnol Tech 13(1):11–15

Certik M, Shimizu S (1999) Biosynthesis and regulation of microbial polyunsaturated fatty acid production. J Biosci Bioeng 87(1):1–14

Chang L, Lu H, Chen H, Tang X, Zhao J, Zhang H et al (2022) Lipid metabolism research in oleaginous fungus *Mortierella alpina*: Current progress and future prospects. Biotechnol Adv 54:107794

Chatzifragkou A, Makri A, Belka A, Bellou S, Mavrou M, Mastoridou M et al (2011) Biotechnological conversions of biodiesel derived waste glycerol by yeast and fungal species. Energy 36(2):1097–1108

Chen X, Li Z, Zhang X, Hu F, Ryu DD, Bao J (2009) Screening of oleaginous yeast strains tolerant to lignocellulose degradation compounds. Appl Biochem Biotechnol 159(3):591–604

Chen Y, Daviet L, Schalk M, Siewers V, Nielsen J (2013) Establishing a platform cell factory through engineering of yeast acetyl-CoA metabolism. Metab Eng 15:48–54

Cheng Y, McNally DJ, Labbé C, Voyer N, Belzile F, Bélanger RR (2003) Insertional mutagenesis of a fungal biocontrol agent led to discovery of a rare cellobiose lipid with antifungal activity. Appl Environ Microbiol 69(5):2595–2602

Chmielarz M, Blomqvist J, Sampels S, Sandgren M, Passoth V (2021) Microbial lipid production from crude glycerol and hemicellulosic hydrolysate with oleaginous yeasts. Biotechnol Biofuels 14(1):65

Ciesielska K, Van Bogaert IN, Chevineau S, Li B, Groeneboer S, Soetaert W et al (2014) Exoproteome analysis of *Starmerella bombicola* results in the discovery of an esterase required for lactonization of sophorolipids. J Proteome 98:159–174

Colabella F, Libkind D (2016) PCR-based method for the rapid identification of astaxanthin-accumulating yeasts (*Phaffia* spp.). Rev Argent Microbiol 48(1):15–20

Cordova LT, Butler J, Alper HS (2020) Direct production of fatty alcohols from glucose using engineered strains of *Yarrowia lipolytica*. Metab Eng Commun 10:e00105

Cui Y, Blackburn JW, Liang Y (2012) Fermentation optimization for the production of lipid by *Cryptococcus curvatus*: Use of response surface methodology. Biomass Bioenergy 47:410–417

Damude HG, Macool DJ, Picataggio SK et al (2009) Docosahexaenoic acid producing strains of *Yarrowia lipolytica*. Patent US7550286B2

Daniel HJ, Otto RT, Binder M, Reuss M, Syldatk C (1999) Production of sophorolipids from whey: development of a two-stage process with *Cryptococcus curvatus* ATCC 20509 and *Candida bombicola* ATCC 22214 using deproteinized whey concentrates as substrates. Appl Microbiol Biotechnol 51(1):40–45

Daskalaki A, Perdikouli N, Aggeli D, Aggelis G (2019) Laboratory evolution strategies for improving lipid accumulation in *Yarrowia lipolytica*. Appl Microbiol Biotechnol 103(20):8585–8596

Davila A-M, Marchal R, Vandecasteele J-P (1992) Kinetics and balance of a fermentation free from product inhibition: sophorose lipid production by *Candida bombicola*. Appl Microbiol Biotechnol 38(1):6–11

Della-Bianca BE, Basso TO, Stambuk BU, Basso LC, Gombert AK (2013) What do we know about the yeast strains from the Brazilian fuel ethanol industry? Appl Microbiol Biotechnol 97(3):979–991

d'Espaux L, Ghosh A, Runguphan W, Wehrs M, Xu F, Konzock O et al (2017) Engineering high-level production of fatty alcohols by *Saccharomyces cerevisiae* from lignocellulosic feedstocks. Metab Eng 42:115–125

Donald KA, Hampton RY, Fritz IB (1997) Effects of overproduction of the catalytic domain of 3-hydroxy-3-methylglutaryl coenzyme A reductase on squalene synthesis in *Saccharomyces cerevisiae*. Appl Environ Microbiol 63(9):3341–3344

Du F, Wang Y-Z, Xu Y-S, Shi T-Q, Liu W-Z, Sun X-M et al (2021) Biotechnological production of lipid and terpenoid from thraustochytrids. Biotechnol Adv 48:107725

Dzurendova S, Losada CB, Dupuy-Galet BX, Fjær K, Shapaval V (2022) Mucoromycota fungi as powerful cell factories for modern biorefinery. Appl Microbiol Biotechnol 106(1):101–115

Espinosa-Gonzalez I, Parashar A, Bressler DC (2014) Hydrothermal treatment of oleaginous yeast for the recovery of free fatty acids for use in advanced biofuel production. J Biotechnol 187:10–15

Evans CT, Ratledge C (1984) Induction of xylulose-5-phosphate phosphoketolase in a variety of yeasts grown ond-xylose: the key to efficient xylose metabolism. Arch Microbiol 139(1):48–52

Fahy E, Cotter D, Sud M, Subramaniam S (2011) Lipid classification, structures and tools. Biochim Biophys Acta 1811(11):637–647

Fakas S, Papanikolaou S, Batsos A, Galiotou-Panayotou M, Mallouchos A, Aggelis G (2009) Evaluating renewable carbon sources as substrates for single cell oil production by *Cunninghamella echinulata* and *Mortierella isabellina*. Biomass Bioenergy 33:573–580

Feng X, Lian J, Zhao H (2015) Metabolic engineering of *Saccharomyces cerevisiae* to improve 1-hexadecanol production. Metab Eng 27:10–19

Fernandes NDAT, Simões LA, Dias DR (2023) Biosurfactants produced by yeasts: fermentation, screening, recovery, purification, characterization, and applications. Fermentation 9(3):207

Filippucci S, Tasselli G, Scardua A, Di Mauro S, Cramarossa MR, Perini D et al (2016) Study of *Holtermanniella wattica*, *Leucosporidium creatinivorum*, *Naganishia adeliensis*, *Solicoccozyma aeria*, and *Solicoccozyma terricola* for their lipogenic aptitude from different carbon sources. Biotechnol Biofuels 9:259

Fillet S, Adrio JL (2016) Microbial production of fatty alcohols. World J Microbiol Biotechnol 32(9):152

Fillet S, Gibert J, Suárez B, Lara A, Ronchel C, Adrio JL (2015) Fatty alcohols production by oleaginous yeast. J Ind Microbiol Biotechnol 42(11):1463–1472

Flores-Cotera LB, Martín R, Sánchez S (2001) Citrate, a possible precursor of astaxanthin in *Phaffia rhodozyma*: influence of varying levels of ammonium, phosphate and citrate in a chemically defined medium. Appl Microbiol Biotechnol 55(3):341–347

Folch J, Lees M, Sloane Stanley GH (1957) A simple method for the isolation and purification of total lipides from animal tissues. J Biol Chem 226(1):497–509

Fridrihsone A, Romagnoli F, Cabulis U (2020) Environmental life cycle assessment of rapeseed and rapeseed oil produced in Northern Europe: a Latvian case study. Sustainability 12(14):5699

Frieman MB, Cormack BP (2004) Multiple sequence signals determine the distribution of glycosylphosphatidylinositol proteins between the plasma membrane and cell wall in *Saccharomyces cerevisiae*. Microbiology (Reading) 150(Pt 10):3105–3114

Fu WJ, Chi Z, Ma ZC, Zhou HX, Liu GL, Lee CF et al (2015) Hydrocarbons, the advanced biofuels produced by different organisms, the evidence that alkanes in petroleum can be renewable. Appl Microbiol Biotechnol 99(18):7481–7494

Gajewski J, Pavlovic R, Fischer M, Boles E, Grininger M (2017) Engineering fungal de novo fatty acid synthesis for short chain fatty acid production. Nat Commun 8(1):14650

Gao Q, Cui Z, Zhang J, Bao J (2014) Lipid fermentation of corncob residues hydrolysate by oleaginous yeast *Trichosporon cutaneum*. Bioresour Technol 152:552–556

Gao Q, Cao X, Huang Y-Y, Yang J-L, Chen J, Wei L-J et al (2018) Overproduction of fatty acid ethyl esters by the oleaginous yeast *Yarrowia lipolytica* through metabolic engineering and process optimization. ACS Synth Biol 7(5):1371–1380

Gao J, Li Y, Yu W, Zhou YJ (2022) Rescuing yeast from cell death enables overproduction of fatty acids from sole methanol. Nat Metab 4(7):932–943

Gemperlein K, Dietrich D, Kohlstedt M, Zipf G, Bernauer HS, Wittmann C et al (2019) Polyunsaturated fatty acid production by *Yarrowia lipolytica* employing designed myxobacterial PUFA synthases. Nat Commun 10(1):4055

Ghosh A, Ando D, Gin J, Runguphan W, Denby C, Wang G et al (2016) (13)C metabolic flux analysis for systematic metabolic engineering of *S. cerevisiae* for overproduction of fatty acids. Front Bioeng Biotechnol 4:76

Golden CD, Allison EH, Cheung WW, Dey MM, Halpern BS, McCauley DJ et al (2016) Nutrition: Fall in fish catch threatens human health. Nature 534(7607):317–320

Gostinčar C, Ohm RA, Kogej T, Sonjak S, Turk M, Zajc J et al (2014) Genome sequencing of four *Aureobasidium pullulans* varieties: biotechnological potential, stress tolerance, and description of new species. BMC Genomics 15:549

Guo W, Sheng J, Zhao H, Feng X (2016) Metabolic engineering of *Saccharomyces cerevisiae* to produce 1-hexadecanol from xylose. Microb Cell Fact 15:24

Haas MJ, McAloon AJ, Yee WC, Foglia TA (2006) A process model to estimate biodiesel production costs. Bioresour Technol 97(4):671–678

Hara KY, Morita T, Mochizuki M, Yamamoto K, Ogino C, Araki M et al (2014) Development of a multi-gene expression system in *Xanthophyllomyces dendrorhous*. Microb Cell Fact 13:175

Hast MA, Nichols CB, Armstrong SM, Kelly SM, Hellinga HW, Alspaugh JA et al (2011) Structures of *Cryptococcus neoformans* protein farnesyltransferase reveal strategies for developing inhibitors that target fungal pathogens. J Biol Chem 286(40):35149–35162

Hasunuma T, Kondo A (2012) Development of yeast cell factories for consolidated bioprocessing of lignocellulose to bioethanol through cell surface engineering. Biotechnol Adv 30(6):1207–1218

He Q, Yang Y, Yang S, Donohoe BS, Van Wychen S, Zhang M et al (2018) Oleaginicity of the yeast strain *Saccharomyces cerevisiae* D5A. Biotechnol Biofuels 11(1):258

Hegel PE, Camy S, Destrac P, Condoret JS (2011) Influence of pretreatments for extraction of lipids from yeast by using supercritical carbon dioxide and ethanol as cosolvent. J Supercrit Fluids 58(1):68–78

Heshof R, Visscher B, van de Zilver E, van de Vondervoort R, van Keulen F, Delahaije RJBM et al (2020) Production of tailor-made enzymes to facilitate lipid extraction from the oleaginous yeast *Schwanniomyces occidentalis*. AMB Express 10(1):41

Hewald S, Linne U, Scherer M, Marahiel MA, Kämper J, Bölker M (2006) Identification of a gene cluster for biosynthesis of mannosylerythritol lipids in the basidiomycetous fungus *Ustilago maydis*. Appl Environ Microbiol 72(8):5469–5477

Hiltunen JK, Mursula AM, Rottensteiner H, Wierenga RK, Kastaniotis AJ, Gurvitz A (2003) The biochemistry of peroxisomal beta-oxidation in the yeast *Saccharomyces cerevisiae*. FEMS Microbiol Rev 27(1):35–64

Hirano H, Kimura Y, Kimura A (2016) Biological significance of co- and post-translational modifications of the yeast 26S proteasome. J Proteome 134:37–46

Holzer H, Goedde HW (1957) Two ways from pyruvate to acetyl-coenzyme A in yeast. Biochem Z 329(3):175–191

Hong SY, Zurbriggen AS, Melis A (2012) Isoprene hydrocarbons production upon heterologous transformation of *Saccharomyces cerevisiae*. J Appl Microbiol 113(1):52–65

Hu C, Zhao X, Zhao J, Wu S, Zhao ZK (2009) Effects of biomass hydrolysis by-products on oleaginous yeast *Rhodosporidium toruloides*. Bioresour Technol 100(20):4843–4847

Huang C, Zong MH, Wu H, Liu QP (2009) Microbial oil production from rice straw hydrolysate by *Trichosporon fermentans*. Bioresour Technol 100(19):4535–4538

Huang C, Wu H, Li RF, Zong MH (2012) Improving lipid production from bagasse hydrolysate with *Trichosporon fermentans* by response surface methodology. New Biotechnol 29(3):372–378

Huang Y-Y, Jian X-X, Lv Y-B, Nian K-Q, Gao Q, Chen J et al (2018) Enhanced squalene biosynthesis in *Yarrowia lipolytica* based on metabolically engineered acetyl-CoA metabolism. J Biotechnol 281:106–114

Huang M, Chen H, Tang X, Lu H, Zhao J, Zhang H et al (2020) Two-stage pH control combined with oxygen-enriched air strategies for the highly efficient production of EPA by *Mortierella alpina* CCFM698 with fed-batch fermentation. Bioprocess Biosyst Eng 43(9):1725–1733

Hull A, Golubkov I, Kronberg B, Marandzheva T, van Stam J (2006) An alternative fuel for spark ignition engines. Int J Engine Res 7(3):203–214

Hyvönen R, Olsson BA, Lundkvist H, Staaf H (2000) Decomposition and nutrient release from *Picea abies* (L.) Karst. and *Pinus sylvestris* L. logging residues. Forest Ecol Manag 126(2):97–112

Im JH, Yanagishita H, Ikegami T, Takeyama Y, Idemoto Y, Koura N et al (2003) Mannosylerythritol lipids, yeast glycolipid biosurfactants, are potential affinity ligand materials for human immunoglobulin G. J Biomed Mater Res A 65(3):379–385

Inoue S, Ito S (1982) Sophorolipids from *Torulopsis bombicola* as microbial surfactants in alkane fermentations. Biotechnol Lett 4(1):3–8

Jackson FM, Fraser TC, Smith MA, Lazarus C, Stobart AK, Griffiths G (1998) Biosynthesis of C18 polyunsaturated fatty acids in microsomal membrane preparations from the filamentous fungus *Mucor circinelloides*. Eur J Biochem 252(3):513–519

Jagtap SS, Rao CV (2018) Production of d-arabitol from d-xylose by the oleaginous yeast *Rhodosporidium toruloides* IFO0880. Appl Microbiol Biotechnol 102(1):143–151

Jagtap SS, Deewan A, Liu J-J, Walukiewicz HE, Yun EJ, Jin Y-S et al (2021) Integrating transcriptomic and metabolomic analysis of the oleaginous yeast *Rhodosporidium toruloides* IFO0880 during growth under different carbon sources. Appl Microbiol Biotechnol 105(19):7411–7425

Jarboe LR, Royce LA, Liu P (2013) Understanding biocatalyst inhibition by carboxylic acids. Front Microbiol 4:272

Jezierska S, Claus S, Van Bogaert I (2018) Yeast glycolipid biosurfactants. FEBS Lett 592(8):1312–1329

Jiang H, Zhang X, Chen X, Aramsangtienchai P, Tong Z, Lin H (2018) Protein lipidation: occurrence, mechanisms, biological functions, and enabling technologies. Chem Rev 118(3):919–988

Jimoh AA, Lin J (2019) Biosurfactant: A new frontier for greener technology and environmental sustainability. Ecotoxicol Environ Saf 184:109607

Jin G, Yang F, Hu C, Shen H, Zhao ZK (2012) Enzyme-assisted extraction of lipids directly from the culture of the oleaginous yeast *Rhodosporidium toruloides*. Bioresour Technol 111:378–382

Jin M, Slininger PJ, Dien BS, Waghmode S, Moser BR, Orjuela A et al (2015) Microbial lipid-based lignocellulosic biorefinery: feasibility and challenges. Trends Biotechnol 33(1):43–54

Johnson DT, Taconi KA (2007) The glycerin glut: Options for the value-added conversion of crude glycerol resulting from biodiesel production. Environ Prog 26(4):338–348

Jönsson LJ, Alriksson B, Nilvebrant NO (2013) Bioconversion of lignocellulose: inhibitors and detoxification. Biotechnol Biofuels 6(1):16

Kamisaka Y, Noda N, Tomita N, Kimura K, Kodaki T, Hosaka K (2006) Identification of genes affecting lipid content using transposon mutagenesis in *Saccharomyces cerevisiae*. Biosci Biotechnol Biochem 70(3):646–653

Kamisaka Y, Tomita N, Kimura K, Kainou K, Uemura H (2007) DGA1 (diacylglycerol acyltransferase gene) overexpression and leucine biosynthesis significantly increase lipid accumulation in the Deltasnf2 disruptant of *Saccharomyces cerevisiae*. Biochem J 408(1):61–68

Kamisaka Y, Kimura K, Uemura H, Yamaoka M (2013) Overexpression of the active diacylglycerol acyltransferase variant transforms *Saccharomyces cerevisiae* into an oleaginous yeast. Appl Microbiol Biotechnol 97(16):7345–7355

Kamiya Y, Sakurai A, Tamura S, Takahashi N (1978) Structure of rhodotorucine A, a novel lipopeptide, inducing mating tube formation in *Rhodosporidium toruloides*. Biochem Biophys Res Commun 83(3):1077–1083

Karlsson H, Ahlgren S, Sandgren M, Passoth V, Wallberg O, Hansson PA (2016) A systems analysis of biodiesel production from wheat straw using oleaginous yeast: process design, mass and energy balances. Biotechnol Biofuels 9:229

Kendrick A, Ratledge C (1992) Desaturation of polyunsaturated fatty acids in *Mucor circinelloides* and the involvement of a novel membrane-bound malic enzyme. Eur J Biochem 209(2):667–673

Khot M, Raut G, Ghosh D, Alarcón-Vivero M, Contreras D, Ravikumar A (2020) Lipid recovery from oleaginous yeasts: Perspectives and challenges for industrial applications. Fuel 259:116292

Kim M, Park BG, Kim E-J, Kim J, Kim B-G (2019) In silico identification of metabolic engineering strategies for improved lipid production in *Yarrowia lipolytica* by genome-scale metabolic modeling. Biotechnol Biofuels 12(1):187

Kirby J, Nishimoto M, Chow RW, Pasumarthi VN, Chan R, Chan LJ et al (2014) Use of nonionic surfactants for improvement of terpene production in *Saccharomyces cerevisiae*. Appl Environ Microbiol 80(21):6685–6693

Kitamoto D, Yanagishita H, Shinbo T, Nakane T, Kamisawa C, Nakahara T (1993) Surface active properties and antimicrobial activities of mannosylerythritol lipids as biosurfactants produced by *Candida antarctica*. J Biotechnol 29(1):91–96

Kitamoto D, Ghosh S, Ourisson G, Nakatani Y (2000) Formation of giant vesicles from diacylmannosylerythritols, and their binding to concanavalin A. Chem Commun 10:861–862. https://doi.org/10.1039/B000968G

Kitamoto D, Ikegami T, Suzuki GT, Sasaki A, Takeyama Y-i, Idemoto Y et al (2001) Microbial conversion of n-alkanes into glycolipid biosurfactants, mannosylerythritol lipids, by *Pseudozyma* (*Candida antarctica*). Biotechnol Lett 23(20):1709–1714

Klug YA, Deme JC, Corey RA, Renne MF, Stansfeld PJ, Lea SM et al (2021) Mechanism of lipid droplet formation by the yeast Sei1/Ldb16 Seipin complex. Nat Commun 12(1):5892

Knoshaug EP, Van Wychen S, Singh A, Zhang M (2018) Lipid accumulation from glucose and xylose in an engineered, naturally oleaginous strain of *Saccharomyces cerevisiae*. Biofuel Res J 5(2):800–805

Kolouchová I, Schreiberová O, Sigler K, Masák J, Řezanka T (2015) Biotransformation of volatile fatty acids by oleaginous and non-oleaginous yeast species. FEMS Yeast Res 15(7)

Krishnan A, McNeil BA, Stuart DT (2020) Biosynthesis of fatty alcohols in engineered microbial cell factories: advances and limitations. Front Bioeng Biotechnol 8:610936
Kulakovskaya TV, Golubev WI, Tomashevskaya MA, Kulakovskaya EV, Shashkov AS, Grachev AA et al (2010) Production of antifungal cellobiose lipids by *Trichosporon porosum*. Mycopathologia 169(2):117–123
Kurtzman CP, Price NP, Ray KJ, Kuo TM (2010) Production of sophorolipid biosurfactants by multiple species of the *Starmerella (Candida) bombicola* yeast clade. FEMS Microbiol Lett 311(2):140–146
Kurz M, Eder C, Isert D, Li Z, Paulus EF, Schiell M et al (2003) Ustilipids, acylated beta-D-mannopyranosyl D-erythritols from *Ustilago maydis* and *Geotrichum candidum*. J Antibiot (Tokyo) 56(2):91–101
Lachance M-A, Dobson J, Wijayanayaka DN, Smith AME (2010) The use of parsimony network analysis for the formal delineation of phylogenetic species of yeasts: *Candida apicola*, *Candida azyma*, and *Candida parazyma* sp. nov., cosmopolitan yeasts associated with floricolous insects. Antonie Van Leeuwenhoek 97(2):155–170
Ladygina N, Dedyukhina EG, Vainshtein MB (2006) A review on microbial synthesis of hydrocarbons. Process Biochem 41(5):1001–1014
Leber C, Da Silva NA (2014) Engineering of *Saccharomyces cerevisiae* for the synthesis of short chain fatty acids. Biotechnol Bioeng 111(2):347–358
Leber C, Polson B, Fernandez-Moya R, Da Silva NA (2015) Overproduction and secretion of free fatty acids through disrupted neutral lipid recycle in *Saccharomyces cerevisiae*. Metab Eng 28:54–62
Leber C, Choi JW, Polson B, Da Silva NA (2016) Disrupted short chain specific β-oxidation and improved synthase expression increase synthesis of short chain fatty acids in *Saccharomyces cerevisiae*. Biotechnol Bioeng 113(4):895–900
Ledesma-Amaro R, Nicaud JM (2016) Metabolic engineering for expanding the substrate range of *Yarrowia lipolytica*. Trends Biotechnol 34(10):798–809
Ledesma-Amaro R, Dulermo R, Niehus X, Nicaud JM (2016a) Combining metabolic engineering and process optimization to improve production and secretion of fatty acids. Metab Eng 38:38–46
Ledesma-Amaro R, Lazar Z, Rakicka M, Guo Z, Fouchard F, Coq AC et al (2016b) Metabolic engineering of *Yarrowia lipolytica* to produce chemicals and fuels from xylose. Metab Eng 38:115–124
Leibundgut M, Maier T, Jenni S, Ban N (2008) The multienzyme architecture of eukaryotic fatty acid synthases. Curr Opin Struct Biol 18(6):714–725
Lennen RM, Pfleger BF (2013) Microbial production of fatty acid-derived fuels and chemicals. Curr Opin Biotechnol 24(6):1044–1053
Leyland B, Leu S, Boussiba S (2017) Are Thraustochytrids algae? Fungal Biol 121(10):835–840
Li Y, Zhao Z, Bai F (2007) High-density cultivation of oleaginous yeast *Rhodosporidium toruloides* Y4 in fed-batch culture. Enzym Microb Technol 41(3):312–317
Li J, Ma Y, Liu N, Eser BE, Guo Z, Jensen PR et al (2020) Synthesis of high-titer alka(e)nes in *Yarrowia lipolytica* is enabled by a discovered mechanism. Nat Commun 11(1):6198
Lian J, Zhao H (2015) Recent advances in biosynthesis of fatty acids derived products in *Saccharomyces cerevisiae* via enhanced supply of precursor metabolites. J Ind Microbiol Biotechnol 42(3):437–451
Liang Y, Cui Y, Trushenski J, Blackburn JW (2010) Converting crude glycerol derived from yellow grease to lipids through yeast fermentation. Bioresour Technol 101(19):7581–7586
Liu YS, Wu JY (2008) Modeling of *Xanthophyllomyces dendrorhous* growth on glucose and overflow metabolism in batch and fed-batch cultures for astaxanthin production. Biotechnol Bioeng 101(5):996–1004
Liu Y, Koh CMJ, Sun L, Hlaing MM, Du M, Peng N et al (2013a) Characterization of glyceraldehyde-3-phosphate dehydrogenase gene RtGPD1 and development of genetic trans-

formation method by dominant selection in oleaginous yeast *Rhodosporidium toruloides*. Appl Microbiol Biotechnol 97(2):719–729

Liu Z, Gao Y, Chen J, Imanaka T, Bao J, Hua Q (2013b) Analysis of metabolic fluxes for better understanding of mechanisms related to lipid accumulation in oleaginous yeast *Trichosporon cutaneum*. Bioresour Technol 130:144–151

Liu YY, Chi Z, Wang ZP, Liu GL, Chi ZM (2014) Heavy oils, principally long-chain n-alkanes secreted by *Aureobasidium pullulans* var. *melanogenum* strain P5 isolated from mangrove system. J Ind Microbiol Biotechnol 41(9):1329–1337

Liu L, Markham K, Blazeck J, Zhou N, Leon D, Otoupal P et al (2015a) Surveying the lipogenesis landscape in *Yarrowia lipolytica* through understanding the function of a Mga2p regulatory protein mutant. Metab Eng 31:102–111

Liu L, Pan A, Spofford C, Zhou N, Alper HS (2015b) An evolutionary metabolic engineering approach for enhancing lipogenesis in *Yarrowia lipolytica*. Metab Eng 29:36–45

Liu XZ, Wang QM, Göker M, Groenewald M, Kachalkin AV, Lumbsch HT et al (2015c) Towards an integrated phylogenetic classification of the *Tremellomycetes*. Stud Mycol 81:85–147

Liu Y, Jiang X, Cui Z, Wang Z, Qi Q, Hou J (2019) Engineering the oleaginous yeast *Yarrowia lipolytica* for production of α-farnesene. Biotechnol Biofuels 12(1):296

Liu D, Geiselman GM, Coradetti S, Cheng Y-F, Kirby J, Prahl J-P et al (2020) Exploiting nonionic surfactants to enhance fatty alcohol production in *Rhodosporidium toruloides*. Biotechnol Bioeng 117(5):1418–1425

Lodge JK, Jackson-Machelski E, Higgins M, McWherter CA, Sikorski JA, Devadas B et al (1998) Genetic and biochemical studies establish that the fungicidal effect of a fully depeptidized inhibitor of *Cryptococcus neoformans* myristoyl-CoA:protein N-myristoyltransferase (Nmt) is Nmt-dependent. J Biol Chem 273(20):12482–12491

Lu R, Shi T-Q, Lin L, Ledesma-Amaro R, Ji X-J, Huang H (2022) Advances in metabolic engineering of yeasts for the production of fatty acid-derived hydrocarbon fuels. Green Chem Eng 3(4):289–303

Lucantonio S, Di Giuliano A, Rossi L, Gallucci K (2023) Green diesel production via deoxygenation process: a review. Energies 16(2):844

Lundin H (1950) Fat synthesis by micro-organisms and its possible applications in industry. J Inst Brew 56(1):17–28

Lv X, Xie W, Lu W, Guo F, Gu J, Yu H et al (2014) Enhanced isoprene biosynthesis in *Saccharomyces cerevisiae* by engineering of the native acetyl-CoA and mevalonic acid pathways with a push-pull-restrain strategy. J Biotechnol 186:128–136

Lv X, Wang F, Zhou P, Ye L, Xie W, Xu H et al (2016) Dual regulation of cytoplasmic and mitochondrial acetyl-CoA utilization for improved isoprene production in *Saccharomyces cerevisiae*. Nat Commun 7:12851

Ma P, Takashima S, Fujita C, Yamada S, Oshima Y, Cai HL et al (2021) Fatty acid composition of the methylotrophic yeast *Komagataella phaffii* grown under low- and high-methanol conditions. Yeast 38(10):541–548

Mann PA, McLellan CA, Koseoglu S, Si Q, Kuzmin E, Flattery A et al (2015) Chemical genomics-based antifungal drug discovery: targeting glycosylphosphatidylinositol (GPI) precursor biosynthesis. ACS Infect Dis 1(1):59–72

Martín-Hernández GC, Müller B, Chmielarz M, Brandt C, Hölzer M, Viehweger A et al (2021) Chromosome-level genome assembly and transcriptome-based annotation of the oleaginous yeast *Rhodotorula toruloides* CBS 14. Genomics 113(6):4022–4027

Martín-Hernández GC, Müller B, Brandt C, Hölzer M, Viehweger A, Passoth V (2022) Near chromosome-level genome assembly and annotation of *Rhodotorula babjevae* strains reveals high intraspecific divergence. J Fungi 8(4):323

Martín-Hernández GC, Chmielarz M, Müller B, Brandt C, Viehweger A, Hölzer M et al (2023) Enhanced glycerol assimilation and lipid production in *Rhodotorula toruloides* CBS14 upon addition of hemicellulose primarily correlates with early transcription of energy-metabolism-related genes. Biotechnol Biofuels Bioprod 16(1):42

Mata-Gómez LC, Montañez JC, Méndez-Zavala A, Aguilar CN (2014) Biotechnological production of carotenoids by yeasts: an overview. Microb Cell Fact 13:12

McLellan CA, Whitesell L, King OD, Lancaster AK, Mazitschek R, Lindquist S (2012) Inhibiting GPI anchor biosynthesis in fungi stresses the endoplasmic reticulum and enhances immunogenicity. ACS Chem Biol 7(9):1520–1528

Michaelis S, Barrowman J (2012) Biogenesis of the *Saccharomyces cerevisiae* pheromone a-factor, from yeast mating to human disease. Microbiol Mol Biol Rev 76(3):626–651

Mnif I, Ghribi D (2016) Glycolipid biosurfactants: main properties and potential applications in agriculture and food industry. J Sci Food Agric 96(13):4310–4320

Morita T, Fukuoka T, Konishi M, Imura T, Yamamoto S, Kitagawa M et al (2009) Production of a novel glycolipid biosurfactant, mannosylmannitol lipid, by *Pseudozyma parantarctica* and its interfacial properties. Appl Microbiol Biotechnol 83(6):1017–1025

Morita T, Fukuoka T, Imura T, Kitamoto D (2012) Formation of the two novel glycolipid biosurfactants, mannosylribitol lipid and mannosylarabitol lipid, by *Pseudozyma parantarctica* JCM 11752T. Appl Microbiol Biotechnol 96(4):931–938

Mulligan CN (2005) Environmental applications for biosurfactants. Environ Pollut 133(2):183–198

Muñiz M, Zurzolo C (2014) Sorting of GPI-anchored proteins from yeast to mammals-common pathways at different sites? J Cell Sci 127(Pt 13):2793–2801

Münßinger S, Beck A, Oraby A, Zibek S (2024) Past, present and future of glycolipids from Ustilaginaceae—A review on cellobiose lipids and mannosylerythritol lipids. J Surfactant Deterg 27:647–689

Ng YJ, Chan SS, Khoo KS, Munawaroh HSH, Lim HR, Chew KW et al (2023) Recent advances and discoveries of microbial-based glycolipids: Prospective alternative for remediation activities. Biotechnol Adv 68:108198

Nguyen UT, Goody RS, Alexandrov K (2010) Understanding and exploiting protein prenyltransferases. Chembiochem 11(9):1194–1201

Nichols CB, Ost KS, Grogan DP, Pianalto K, Hasan S, Alspaugh JA (2015) Impact of protein palmitoylation on the virulence potential of *Cryptococcus neoformans*. Eukaryot Cell 14(7):626–635

Niehus X, Crutz-Le Coq A-M, Sandoval G, Nicaud J-M, Ledesma-Amaro R (2018) Engineering *Yarrowia lipolytica* to enhance lipid production from lignocellulosic materials. Biotechnol Biofuels 11(1):11

Nitschke M, Costa SGVAO (2007) Biosurfactants in food industry. Trends Food Sci Technol 18(5):252–259

Ochsenreither K, Glück C, Stressler T, Fischer L, Syldatk C (2016) Production strategies and applications of microbial single cell oils. Front Microbiol 7:1539

OECD and Food and Agriculture Organization of the United Nations (2016) OECD-FAO agricultural outlook 2016-2025

Oliver L, Dietrich T, Marañón I, Villarán MC, Barrio RJ (2020) Producing omega-3 polyunsaturated fatty acids: a review of sustainable sources and future trends for the EPA and DHA market. Resources 9(12):148

Olstorpe M, Pickova J, Kiessling A, Passoth V (2014) Strain- and temperature-dependent changes of fatty acid composition in *Wickerhamomyces anomalus* and *Blastobotrys adeninivorans*. Biotechnol Appl Biochem 61(1):45–50

Omer CA, Gibbs JB (1994) Protein prenylation in eukaryotic microorganisms: genetics, biology and biochemistry. Mol Microbiol 11(2):219–225

Oraby A, Werner N, Sungur Z, Zibek S (2020) Factors affecting the synthesis of cellobiose lipids by *Sporisorium scitamineum*. Front Bioeng Biotechnol 8:555647

Orlean P, Menon AK (2007) Thematic review series: lipid posttranslational modifications. GPI anchoring of protein in yeast and mammalian cells, or: how we learned to stop worrying and love glycophospholipids. J Lipid Res 48(5):993–1011

Paddon CJ, Keasling JD (2014) Semi-synthetic artemisinin: a model for the use of synthetic biology in pharmaceutical development. Nat Rev Microbiol 12(5):355–367

Paddon CJ, Westfall PJ, Pitera DJ, Benjamin K, Fisher K, McPhee D et al (2013) High-level semi-synthetic production of the potent antimalarial artemisinin. Nature 496(7446):528–532
Papanikolaou S, Aggelis G (2011) Lipids of oleaginous yeasts. Part I: Biochemistry of single cell oil production. Eur J Lipid Sci Technol 113(8):1031–1051
Park Y-K, Bordes F, Letisse F, Nicaud J-M (2021) Engineering precursor pools for increasing production of odd-chain fatty acids in *Yarrowia lipolytica*. Metab Eng Commun 12:e00158
Parks LW, Casey WM (1995) Physiological implications of sterol biosynthesis in yeast. Ann Rev Microbiol 49:95–116
Passoth V (2014) Molecular mechanisms in yeast carbon metabolism: bioethanol and other biofuels. In: Piškur J, Compagno C (eds) Molecular mechanisms in yeast carbon metabolism. Springer, Berlin, pp 217–259
Passoth V (2017a) Conventional and non-conventional yeasts for the production of biofuels. In: Satyanarayana T, Kunze G (eds) Yeast diversity in human welfare. Springer, Singapore
Passoth V (2017b) Lipids of yeasts and filamentous fungi and their importance for biotechnology. In: Sibirny A (ed) Biotechnology of yeasts and filamentous fungi. Springer, Cham, pp 149–204
Passoth V, Sandgren M (2019) Biofuel production from straw hydrolysates: current achievements and perspectives. Appl Microbiol Biotechnol 103(13):5105–5116
Passoth V, Schnürer J (2003) Non-conventional yeasts in antifungal application. In: de Winde JH (ed) Functional genetics of industrial yeasts. Springer, Berlin, pp 297–329
Passoth V, Brandenburg J, Chmielarz M, Martín-Hernández GC, Nagaraj Y, Müller B et al (2023) Oleaginous yeasts for biochemicals, biofuels and food from lignocellulose-hydrolysate and crude glycerol. Yeast 40(8):290–302
Paulino BN, Pessôa MG, Mano MC, Molina G, Neri-Numa IA, Pastore GM (2016) Current status in biotechnological production and applications of glycolipid biosurfactants. Appl Microbiol Biotechnol 100(24):10265–10293
Peralta-Yahya PP, Zhang F, del Cardayre SB, Keasling JD (2012) Microbial engineering for the production of advanced biofuels. Nature 488(7411):320–328
Pronk JT, Yde Steensma H, Van Dijken JP (1996) Pyruvate metabolism in *Saccharomyces cerevisiae*. Yeast 12(16):1607–1633
Pugh EL, Kates M (1979) Membrane-bound phospholipid desaturases. Lipids 14(2):159–165
Qiao K, Imam Abidi SH, Liu H, Zhang H, Chakraborty S, Watson N et al (2015) Engineering lipid overproduction in the oleaginous yeast *Yarrowia lipolytica*. Metab Eng 29:56–65
Qiao K, Wasylenko TM, Zhou K, Xu P, Stephanopoulos G (2017) Lipid production in *Yarrowia lipolytica* is maximized by engineering cytosolic redox metabolism. Nat Biotechnol 35(2):173–177
Ratledge C (2004) Fatty acid biosynthesis in microorganisms being used for single cell oil production. Biochimie 86(11):807–815
Ratledge C (2013) Microbial oils: an introductory overview of current status and future prospects. OCL 20(6):D602
Ratledge C (2014) The role of malic enzyme as the provider of NADPH in oleaginous microorganisms: a reappraisal and unsolved problems. Biotechnol Lett 36(8):1557–1568
Ratledge C, Wynn JP (2002) The biochemistry and molecular biology of lipid accumulation in oleaginous microorganisms. Adv Appl Microbiol 51:1–51
Rau U, Nguyen LA, Roeper H, Koch H, Lang S (2005) Fed-batch bioreactor production of mannosylerythritol lipids secreted by *Pseudozyma aphidis*. Appl Microbiol Biotechnol 68(5):607–613
Reinertsen SA, Elliott LF, Cochran VL, Campbell GS (1984) Role of available carbon and nitrogen in determining the rate of wheat straw decomposition. Soil Biol Biochem 16:459–464
Renne MF, Klug YA, Carvalho P (2020) Lipid droplet biogenesis: A mystery "unmixing"? Semin Cell Dev Biol 108:14–23
Ro DK, Paradise EM, Ouellet M, Fisher KJ, Newman KL, Ndungu JM et al (2006) Production of the antimalarial drug precursor artemisinic acid in engineered yeast. Nature 440(7086):940–943

Robles-Iglesias R, Naveira-Pazos C, Fernández-Blanco C, Veiga MC, Kennes C (2023) Factors affecting the optimisation and scale-up of lipid accumulation in oleaginous yeasts for sustainable biofuels production. Renew Sust Energ Rev 171:113043

Rodriguez GM, Hussain MS, Gambill L, Gao D, Yaguchi A, Blenner M (2016) Engineering xylose utilization in *Yarrowia lipolytica* by understanding its cryptic xylose pathway. Biotechnol Biofuels 9:149

Rodríguez-Frómeta RA, Gutiérrez A, Torres-Martínez S, Garre V (2013) Malic enzyme activity is not the only bottleneck for lipid accumulation in the oleaginous fungus *Mucor circinelloides*. Appl Microbiol Biotechnol 97(7):3063–3072

Rodríguez-Sáiz M, de la Fuente JL, Barredo JL (2010) *Xanthophyllomyces dendrorhous* for the industrial production of astaxanthin. Appl Microbiol Biotechnol 88(3):645–658

Roelants SL, De Maeseneire SL, Ciesielska K, Van Bogaert IN, Soetaert W (2014) Biosurfactant gene clusters in eukaryotes: regulation and biotechnological potential. Appl Microbiol Biotechnol 98(8):3449–3461

Runguphan W, Keasling JD (2014) Metabolic engineering of *Saccharomyces cerevisiae* for production of fatty acid-derived biofuels and chemicals. Metab Eng 21:103–113

Saenge C, Cheirsilp B, Suksaroge TT et al (2011) Potential use of oleaginous red yeast *Rhodotorula glutinis* for the bioconversion of crude glycerol from biodiesel plant to lipids and carotenoids. Process Biochem 46(1):210–218

Saerens KM, Roelants SL, Van Bogaert IN, Soetaert W (2011a) Identification of the UDP-glucosyltransferase gene UGTA1, responsible for the first glucosylation step in the sophorolipid biosynthetic pathway of *Candida bombicola* ATCC 22214. FEMS Yeast Res 11(1):123–132

Saerens KM, Saey L, Soetaert W (2011b) One-step production of unacetylated sophorolipids by an acetyltransferase negative *Candida bombicola*. Biotechnol Bioeng 108(12):2923–2931

Saerens KM, Zhang J, Saey L, Van Bogaert IN, Soetaert W (2011c) Cloning and functional characterization of the UDP-glucosyltransferase UgtB1 involved in sophorolipid production by *Candida bombicola* and creation of a glucolipid-producing yeast strain. Yeast 28(4):279–292

Sagnak R, Cochot S, Molina-Jouve C, Nicaud J-M, Guillouet SE (2018) Modulation of the Glycerol Phosphate availability led to concomitant reduction in the citric acid excretion and increase in lipid content and yield in *Yarrowia lipolytica*. J Biotechnol 265:40–45

Salvador López JM, Vandeputte M, Van Bogaert INA (2022) Oleaginous yeasts: Time to rethink the definition? Yeast 39(11-12):553–606

Sandmann G (2022) Carotenoids and their biosynthesis in fungi. Molecules 27(4):1431

Sankh S, Thiru M, Saran S, Rangaswamy V (2013) Biodiesel production from a newly isolated *Pichia kudriavzevii* strain. Fuel 106:690–696

Santiago-Tirado FH, Doering TL (2016) All about that fat: Lipid modification of proteins in *Cryptococcus neoformans*. J Microbiol 54(3):212–222

Santos ARO, Leon MP, Barros KO, Freitas LFD, Hughes AFS, Morais PB et al (2018) *Starmerella camargoi* f.a., sp. nov., *Starmerella ilheusensis* f.a., sp. nov., *Starmerella litoralis* f.a., sp. nov., *Starmerella opuntiae* f.a., sp. nov., *Starmerella roubikii* f.a., sp. nov. and Starmerella vitae f.a., sp. nov., isolated from flowers and bees, and transfer of related Candida species to the genus Starmerella as new combinations. Int J Syst Evol Microbiol 68(4):1333–1343

Scharnewski M, Pongdontri P, Mora G, Hoppert M, Fulda M (2008) Mutants of *Saccharomyces cerevisiae* deficient in acyl-CoA synthetases secrete fatty acids due to interrupted fatty acid recycling. FEBS J 275(11):2765–2778

Schmidt JH (2015) Life cycle assessment of five vegetable oils. J Clean Prod 87:130–138

Schmidt I, Schewe H, Gassel S, Jin C, Buckingham J, Hümbelin M et al (2011) Biotechnological production of astaxanthin with *Phaffia rhodozyma/Xanthophyllomyces dendrorhous*. Appl Microbiol Biotechnol 89(3):555–571

Shah V, Doncel GF, Seyoum T, Eaton KM, Zalenskaya I, Hagver R et al (2005) Sophorolipids, microbial glycolipids with anti-human immunodeficiency virus and sperm-immobilizing activities. Antimicrob Agents Chemother 49(10):4093–4100

Shi S, Valle-Rodríguez JO, Siewers V, Nielsen J (2014) Engineering of chromosomal wax ester synthase integrated *Saccharomyces cerevisiae* mutants for improved biosynthesis of fatty acid ethyl esters. Biotechnol Bioeng 111(9):1740–1747

Shiba Y, Paradise EM, Kirby J, Ro DK, Keasling JD (2007) Engineering of the pyruvate dehydrogenase bypass in *Saccharomyces cerevisiae* for high-level production of isoprenoids. Metab Eng 9(2):160–168

Shibasaki S, Aoki W, Nomura T, Miyoshi A, Tafuku S, Sewaki T et al (2013) An oral vaccine against candidiasis generated by a yeast molecular display system. Pathog Dis 69(3):262–268

Shimizu S, Shinmen Y, Kawashima H, Akimoto K, Yamada H (1988) Fungal mycelia as a novel source of eicosapentaenoic acid. Activation of enzyme(s) involved in eicosapentaenoic acid production at low temperature. Biochem Biophys Res Commun 150(1):335–341

Shimizu S, Kawashima H, Akimoto K, Shinmen Y, Yamada H (1989) Conversion of linseed oil to an eicosapentaenoic acid-containing oil by *Mortierella alpina* 1S-4 at low temperature. Appl Microbiol Biotechnol 32(1):1–4

Sigtryggsson C, Karlsson Potter H, Passoth V, Hansson P-A (2023) From straw to salmon: a technical design and energy balance for production of yeast oil for fish feed from wheat straw. Biotechnol Biofuels Bioprod 16(1):140

Sitepu IR, Garay LA, Sestric R, Levin D, Block DE, German JB et al (2014) Oleaginous yeasts for biodiesel: current and future trends in biology and production. Biotechnol Adv 32(7):1336–1360

Slininger PJ, Dien BS, Kurtzman CP, Moser BR, Bakota EL, Thompson SR et al (2016) Comparative lipid production by oleaginous yeasts in hydrolyzates of lignocellulosic biomass and process strategy for high titers. Biotechnol Bioeng 113(8):1676–1690

Soccol CR, Dalmas Neto CJ, Soccol VT, Sydney EB, da Costa ESF, Medeiros ABP et al (2017) Pilot scale biodiesel production from microbial oil of *Rhodosporidium toruloides* DEBB 5533 using sugarcane juice: Performance in diesel engine and preliminary economic study. Bioresour Technol 223:259–268

Sorger D, Daum G (2003) Triacylglycerol biosynthesis in yeast. Appl Microbiol Biotechnol 61(4):289–299

Soxhlet F (1879) Die gewichtsanalytische Bestimmung des Milchfettes. Dinglers Polytechnisches Journal 232:461–465

Tai M, Stephanopoulos G (2013) Engineering the push and pull of lipid biosynthesis in oleaginous yeast *Yarrowia lipolytica* for biofuel production. Metab Eng 15:1–9

Takeno S, Sakuradani E, Murata S, Inohara-Ochiai M, Kawashima H, Ashikari T et al (2005a) Molecular evidence that the rate-limiting step for the biosynthesis of arachidonic acid in *Mortierella alpina* is at the level of an elongase. Lipids 40(1):25–30

Takeno S, Sakuradani E, Tomi A, Inohara-Ochiai M, Kawashima H, Ashikari T et al (2005b) Improvement of the fatty acid composition of an oil-producing filamentous fungus, *Mortierella alpina* 1S-4, through RNA interference with delta12-desaturase gene expression. Appl Environ Microbiol 71(9):5124–5128

Tang W, Zhang S, Tan H, Zhao ZK (2010) Molecular cloning and characterization of a malic enzyme gene from the oleaginous yeast *Lipomyces starkeyi*. Mol Biotechnol 45(2):121–128

Tang X, Feng H, Chen WN (2013) Metabolic engineering for enhanced fatty acids synthesis in *Saccharomyces cerevisiae*. Metab Eng 16:95–102

Tang X, Chen H, Chen YQ, Chen W, Garre V, Song Y et al (2015) Comparison of biochemical activities between high and low lipid-producing strains of *Mucor circinelloides*: an explanation for the high oleaginicity of strain WJ11. PLoS One 10(6):e0128396

Tasselli G, Filippucci S, Borsella E, D'Antonio S, Gelosia M, Cavalaglio G et al (2018) Yeast lipids from cardoon stalks, stranded driftwood and olive tree pruning residues as possible extra sources of oils for producing biofuels and biochemicals. Biotechnol Biofuels 11(1):147

Tchakouteu SS, Kalantzi O, Gardeli C, Koutinas AA, Aggelis G, Papanikolaou S (2015) Lipid production by yeasts growing on biodiesel-derived crude glycerol: strain selection and impact of substrate concentration on the fermentation efficiency. J Appl Microbiol 118(4):911–927

Tehlivets O, Scheuringer K, Kohlwein SD (2007) Fatty acid synthesis and elongation in yeast. Biochim Biophys Acta 1771(3):255–270

Teichmann B, Linne U, Hewald S, Marahiel MA, Bölker M (2007) A biosynthetic gene cluster for a secreted cellobiose lipid with antifungal activity from *Ustilago maydis*. Mol Microbiol 66(2):525–533

Teichmann B, Labbé C, Lefebvre F, Bölker M, Linne U, Bélanger RR (2011) Identification of a biosynthesis gene cluster for flocculosin a cellobiose lipid produced by the biocontrol agent *Pseudozyma flocculosa*. Mol Microbiol 79(6):1483–1495

Teymennet-Ramírez KV, Martínez-Morales F, Trejo-Hernández MR (2021) Yeast surface display system: strategies for improvement and biotechnological applications. Front Bioeng Biotechnol 9:794742

Thevenieau F, Beopoulos A, Desfougeres T, Sabirova J, Albertin K, Zinjarde S et al (2018) Uptake and assimilation of hydrophobic substrates by the oleaginous yeast *Yarrowia lipolytica*. In: Krell T (ed) Cellular ecophysiology of microbe: hydrocarbon and lipid interactions. Springer, Cham, pp 59–74

Thiru M, Sankh S, Rangaswamy V (2011) Process for biodiesel production from *Cryptococcus curvatus*. Bioresour Technol 102(22):10436–10440

Tippmann S, Scalcinati G, Siewers V, Nielsen J (2016) Production of farnesene and santalene by *Saccharomyces cerevisiae* using fed-batch cultivations with RQ-controlled feed. Biotechnol Bioeng 113(1):72–81

Tocher DR (2015) Omega-3 long-chain polyunsaturated fatty acids and aquaculture in perspective. Aquaculture 449:94–107

Tomás-Pejó E, González-Fernández C, Greses S, Kennes C, Otero-Logilde N, Veiga MC et al (2023) Production of short-chain fatty acids (SCFAs) as chemicals or substrates for microbes to obtain biochemicals. Biotechnol Biofuels Bioprod 16(1):96

Tully M, Gilbert HJ (1985) Transformation of *Rhodosporidium toruloides*. Gene 36(3):235–240

Uçkun Kiran E, Trzcinski A, Webb C (2013) Microbial oil produced from biodiesel by-products could enhance overall production. Bioresour Technol 129:650–654

Uprety BK, Dalli SS, Rakshit SK (2017) Bioconversion of crude glycerol to microbial lipid using a robust oleaginous yeast *Rhodosporidium toruloides* ATCC 10788 capable of growing in the presence of impurities. Energy Convers Manag 135:117–128

Van Bogaert IN, De Maeseneire SL, Develter D, Soetaert W, Vandamme EJ (2008a) Cloning and characterisation of the glyceraldehyde 3-phosphate dehydrogenase gene of *Candida bombicola* and use of its promoter. J Ind Microbiol Biotechnol 35(10):1085–1092

Van Bogaert IN, De Maeseneire SL, Develter D, Soetaert W, Vandamme EJ (2008b) Development of a transformation and selection system for the glycolipid-producing yeast *Candida bombicola*. Yeast 25(4):273–278

Van Bogaert INA, Zhang J, Soetaert W (2011) Microbial synthesis of sophorolipids. Process Biochem 46(4):821–833

Van Bogaert IN, Holvoet K, Roelants SL, Li B, Lin YC, Van de Peer Y et al (2013) The biosynthetic gene cluster for sophorolipids: a biotechnological interesting biosurfactant produced by *Starmerella bombicola*. Mol Microbiol 88(3):501–509

Venkataraman S, Rajendran DS, Kumar PS, Vo D-VN, Vaidyanathan VK (2022) Extraction, purification and applications of biosurfactants based on microbial-derived glycolipids and lipopeptides: a review. Environ Chem Lett 20(1):949–970

Verdoes JC, Krubasik KP, Sandmann G, van Ooyen AJ (1999a) Isolation and functional characterisation of a novel type of carotenoid biosynthetic gene from *Xanthophyllomyces dendrorhous*. Mol Gen Genet 262(3):453–461

Verdoes JC, Misawa N, van Ooyen AJ (1999b) Cloning and characterization of the astaxanthin biosynthetic gene encoding phytoene desaturase of *Xanthophyllomyces dendrorhous*. Biotechnol Bioeng 63(6):750–755

Vongsangnak W, Ruenwai R, Tang X, Hu X, Zhang H, Shen B et al (2013) Genome-scale analysis of the metabolic networks of oleaginous Zygomycete fungi. Gene 521(1):180–190

Vorapreeda T, Thammarongtham C, Cheevadhanarak S, Laoteng K (2013) Repertoire of malic enzymes in yeast and fungi: insight into their evolutionary functional and structural significance. Microbiology (Reading) 159(Pt 12):2548–2557

Wang CW (2015) Lipid droplet dynamics in budding yeast. Cell Mol Life Sci 72(14):2677–2695

Wang QM, Yurkov AM, Göker M, Lumbsch HT, Leavitt SD, Groenewald M et al (2015) Phylogenetic classification of yeasts and related taxa within *Pucciniomycotina.* Stud Mycol 81:149–189

Wang W, Wei H, Knoshaug E, Van Wychen S, Xu Q, Himmel ME et al (2016) Fatty alcohol production in *Lipomyces starkeyi* and *Yarrowia lipolytica.* Biotechnol Biofuels 9:227

Wang F, Lv X, Xie W, Zhou P, Zhu Y, Yao Z et al (2017) Combining Gal4p-mediated expression enhancement and directed evolution of isoprene synthase to improve isoprene production in *Saccharomyces cerevisiae.* Metab Eng 39:257–266

Wang L-j, Xiao T, Xu C, Li J, Liu G-z, Yin K et al (2018) Protective immune response against *Toxoplasma gondii* elicited by a novel yeast-based vaccine with microneme protein 16. Vaccine 36(27):3943–3948

Wang J, Ledesma-Amaro R, Wei Y, Ji B, Ji X-J (2020a) Metabolic engineering for increased lipid accumulation in *Yarrowia lipolytica*—a review. Bioresour Technol 313:123707

Wang W, Knoshaug EP, Wei H, Van Wychen S, Lin C-Y, Wall TV et al (2020b) High titer fatty alcohol production in *Lipomyces starkeyi* by fed-batch fermentation. Curr Res Biotechnol 2:83–87

Wang Q, Zhang Y, Hui R, Zhu Y (2024) Marine thraustochytrid: exploration from taxonomic challenges to biotechnological applications [review]. Front Mar Sci 11

Watanabe NA, Miyazaki M, Horii T, Sagane K, Tsukahara K, Hata K (2012) E1210, a new broad-spectrum antifungal, suppresses *Candida albicans* hyphal growth through inhibition of glycosylphosphatidylinositol biosynthesis. Antimicrob Agents Chemother 56(2):960–971

Watsuntorn W, Chuengcharoenphanich N, Niltaya P, Butkumchote C, Theerachat M, Glinwong C et al (2021) A novel oleaginous yeast *Saccharomyces cerevisiae* CU-TPD4 for lipid and biodiesel production. Chemosphere 280:130782

Westfall PJ, Pitera DJ, Lenihan JR, Eng D, Woolard FX, Regentin R et al (2012) Production of amorphadiene in yeast, and its conversion to dihydroartemisinic acid, precursor to the antimalarial agent artemisinin. Proc Natl Acad Sci USA 109(3):E111–E118

Wu S, Hu C, Jin G, Zhao X, Zhao ZK (2010) Phosphate-limitation mediated lipid production by *Rhodosporidium toruloides.* Bioresour Technol 101(15):6124–6129

Wu S, Zhao X, Shen H, Wang Q, Zhao ZK (2011) Microbial lipid production by *Rhodosporidium toruloides* under sulfate-limited conditions. Bioresour Technol 102(2):1803–1807

Wynn JP, Ratledge C (2000) Evidence that the rate-limiting step for the biosynthesis of arachidonic acid in Mortierella alpina is at the level of the 18:3 to 20:3 elongase. Microbiology (Reading) 146(Pt 9):2325–2331

Wynn JP, Hamid AA, Li Y, Ratledge C (2001) Biochemical events leading to the diversion of carbon into storage lipids in the oleaginous fungi *Mucor circinelloides* and *Mortierella alpina.* Microbiology (Reading) 147(Pt 10):2857–2864

Xu P, Qiao K, Ahn WS, Stephanopoulos G (2016) Engineering *Yarrowia lipolytica* as a platform for synthesis of drop-in transportation fuels and oleochemicals. Proc Natl Acad Sci USA 113(39):10848–10853

Xu P, Qiao K, Stephanopoulos G (2017) Engineering oxidative stress defense pathways to build a robust lipid production platform in *Yarrowia lipolytica.* Biotechnol Bioeng 114(7):1521–1530

Xue Z, Sharpe PL, Hong SP, Yadav NS, Xie D, Short DR et al (2013) Production of omega-3 eicosapentaenoic acid by metabolic engineering of *Yarrowia lipolytica.* Nat Biotechnol 31(8):734–740

Yamamoto K, Hara KY, Morita T, Nishimura A, Sasaki D, Ishii J et al (2016) Enhancement of astaxanthin production in *Xanthophyllomyces dendrorhous* by efficient method for the complete deletion of genes. Microb Cell Fact 15(1):155

Yang K, Qiao Y, Li F, Xu Y, Yan Y, Madzak C et al (2019) Subcellular engineering of lipase dependent pathways directed towards lipid related organelles for highly effectively compartmentalized biosynthesis of triacylglycerol derived products in *Yarrowia lipolytica*. Metab Eng 55:231–238

Yang Q, Shen L, Yu F, Zhao M, Jin M, Deng S et al (2023) Enhanced fermentation of biosurfactant mannosylerythritol lipids on the pilot scale under efficient foam control with addition of soybean oil. Food Bioprod Process 138:60–69

Ye L, Lv X, Yu H (2016) Engineering microbes for isoprene production. Metab Eng 38:125–138

Yen HW, Yang YC, Yu YH (2012) Using crude glycerol and thin stillage for the production of microbial lipids through the cultivation of *Rhodotorula glutinis*. J Biosci Bioeng 114(4):453–456

Yu Y, Shi S (2023) Development and perspective of *Rhodotorula toruloides* as an efficient cell factory. J Agric Food Chem 71(4):1802–1819

Yu KO, Jung J, Kim SW, Park CH, Han SO (2012) Synthesis of FAEEs from glycerol in engineered *Saccharomyces cerevisiae* using endogenously produced ethanol by heterologous expression of an unspecific bacterial acyltransferase. Biotechnol Bioeng 109(1):110–115

Yu X, Dong T, Zheng Y, Miao C, Chen S (2015) Investigations on cell disruption of oleaginous microorganisms: Hydrochloric acid digestion is an effective method for lipid extraction. Eur J Lipid Sci Technol 117(5):730–737

Yu T, Zhou YJ, Huang M, Liu Q, Pereira R, David F et al (2018) Reprogramming yeast metabolism from alcoholic fermentation to lipogenesis. Cell 174(6):1549–1558.e1514

Zaballa ME, van der Goot FG (2018) The molecular era of protein S-acylation: spotlight on structure, mechanisms, and dynamics. Crit Rev Biochem Mol Biol 53(4):420–451

Zhai X, Gao J, Li Y, Grininger M, Zhou YJ (2023) Peroxisomal metabolic coupling improves fatty alcohol production from sole methanol in yeast. Proc Natl Acad Sci USA 120(12):e2220816120

Zhang Y, Adams IP, Ratledge C (2007) Malic enzyme: the controlling activity for lipid production? Overexpression of malic enzyme in Mucor circinelloides leads to a 2.5-fold increase in lipid accumulation. Microbiology (Reading) 153(Pt 7):2013–2025

Zhang F, Rodriguez S, Keasling JD (2011) Metabolic engineering of microbial pathways for advanced biofuels production. Curr Opin Biotechnol 22(6):775–783

Zhang H, Zhang L, Chen H, Chen YQ, Ratledge C, Song Y et al (2013) Regulatory properties of malic enzyme in the oleaginous yeast, *Yarrowia lipolytica*, and its non-involvement in lipid accumulation. Biotechnol Lett 35(12):2091–2098

Zhang H, Wu C, Wu Q, Dai J, Song Y (2016a) Metabolic flux analysis of lipid biosynthesis in the yeast *Yarrowia lipolytica* using 13C-labled glucose and gas chromatography-mass spectrometry. PLoS One 11(7):e0159187

Zhang S, Ito M, Skerker JM, Arkin AP, Rao CV (2016b) Metabolic engineering of the oleaginous yeast *Rhodosporidium toruloides* IFO0880 for lipid overproduction during high-density fermentation. Appl Microbiol Biotechnol 100(21):9393–9405

Zhang S, Skerker JM, Rutter CD, Maurer MJ, Arkin AP, Rao CV (2016c) Engineering *Rhodosporidium toruloides* for increased lipid production. Biotechnol Bioeng 113(5):1056–1066

Zhang S, He Y, Sen B, Wang G (2020) Reactive oxygen species and their applications toward enhanced lipid accumulation in oleaginous microorganisms. Bioresour Technol 307:123234

Zhang Y, Peng J, Zhao H, Shi S (2021) Engineering oleaginous yeast *Rhodotorula toruloides* for overproduction of fatty acid ethyl esters. Biotechnol Biofuels 14(1):115

Zhao L, Zhang H, Wang L, Chen H, Chen YQ, Chen W et al (2015) (13)C-metabolic flux analysis of lipid accumulation in the oleaginous fungus *Mucor circinelloides*. Bioresour Technol 197:23–29

Zhao H, Lv M, Liu Z, Zhang M, Wang Y, Ju X et al (2021) High-yield oleaginous fungi and high-value microbial lipid resources from *Mucoromycota*. BioEnergy Res 14(4):1196–1206

Zhou YJ, Buijs NA, Siewers V, Nielsen J (2014) Fatty acid-derived biofuels and chemicals production in *Saccharomyces cerevisiae*. Front Bioeng Biotechnol 2:32

Zhou YJ, Buijs NA, Zhu Z, Qin J, Siewers V, Nielsen J (2016) Production of fatty acid-derived oleochemicals and biofuels by synthetic yeast cell factories. Nat Commun 7:11709

Zhu Z, Ding Y, Gong Z, Yang L, Zhang S, Zhang C et al (2015) Dynamics of the lipid droplet proteome of the Oleaginous yeast *Rhodosporidium toruloides*. Eukaryot Cell 14(3):252–264

Zhu Z, Hu Y, Teixeira PG, Pereira R, Chen Y, Siewers V et al (2020) Multidimensional engineering of *Saccharomyces cerevisiae* for efficient synthesis of medium-chain fatty acids. Nat Catal 3(1):64–74

Chapter 8
Biotechnological Production and Applications of Fungal Carotenoids

Javier Avalos, Javier Pardo-Medina, Julia Marente, and M. Carmen Limón

Contents

8.1 Introduction ... 230
8.2 Carotenoids in Yeasts and Filamentous Fungi ... 231
8.3 Biotechnological Applications of Fungal Carotenoids ... 233
8.4 Production and Biotechnology of Fungal Carotenes ... 234
8.4.1 β-Carotene ... 234
8.4.2 Lycopene ... 236
8.4.3 Phytoene ... 237
8.4.4 Torulene ... 237
8.5 Production and Biotechnology of Fungal Xanthophylls ... 238
8.5.1 Astaxanthin ... 238
8.5.2 Neurosporaxanthin ... 240
8.5.3 Torularhodin ... 243
8.5.4 Other Xanthophylls ... 245
8.6 Heterologous Carotenoid Production in Fungal Hosts ... 247
8.6.1 Yeasts ... 247
8.6.2 Filamentous Fungi ... 250
8.7 Conclusions ... 250
References ... 251

Abstract Carotenoids are lipophilic terpenoid pigments with biotechnological applications. They are ubiquitous in photosynthetic species and are also produced by many non-photosynthetic bacteria, yeasts, and filamentous fungi, which are amenable to genetic manipulations and can be easily cultured in large fermenters. The biotechnological interest of carotenoids derives from their healthy properties due to their antioxidant activity or to physiological functions. Some carotenoids can be used as a source of vitamin A, chromophore of visual opsins and the origin of the

J. Avalos (✉) · J. Pardo-Medina · J. Marente · M. C. Limón
Department of Genetics, Faculty of Biology, University of Seville, Sevilla, Spain
e-mail: avalos@us.es

A. A. Sibirny (ed.), *Biotechnology of Yeasts and Filamentous Fungi*,
Grand Challenges in Biology and Biotechnology,
https://doi.org/10.1007/978-3-031-74726-7_8

retinoid metabolism associated with gene regulation functions in chordates, and others are related to the maintenance of the macula in the eye. In general, consumption of carotenoids is associated with a decrease in the possibilities of suffering from numerous diseases, from cancer to alterations in the immune system. Fungal carotenoid biosynthetic pathways are relatively simpler than those found in photosynthetic organisms, especially in plants, and not many carotenoids are usually found in these organisms. These include carotenes, such as lycopene and β-carotene, and xanthophylls, such as astaxanthin, neurosporaxanthin, or torularhodin. Some of these carotenoids are produced industrially, but it is to be hoped that their applied use will be extended to others. This chapter reviews the biotechnological and production possibilities of the best-known fungal carotenoids. The scientific literature on this subject is very extensive, especially in the areas of the biological effects of carotenoids on humans and the improvement of production for biotechnological purposes. In this review, we select significant contributions, recent or historical, that serve to make a comprehensive review of fungal carotenoids, focusing on their actual or potential production and their possible applications.

8.1 Introduction

Many fungi, especially filamentous fungi, are remarkable for the enormous complexity of their secondary metabolism (Avalos and Limón 2022). Among the numerous metabolites produced, there are many soluble compounds secreted into their environment, frequently toxic for other organisms, which gives them some advantage for their survival in nature. In other cases, the metabolites accumulate inside the cell and may have other functions. Among them are carotenoids, hydrophobic pigments of tetraterpenoid origin. These molecules have a characteristic polyene chain of conjugated double bonds that acts as a chromophore and provides them with bright colors, usually in the yellow, orange, and red range.

Carotenoids play essential roles in plants, where they act as accessory pigments in photosynthesis. Their importance is evidenced by the herbicidal effect of certain inhibitors of their synthesis (Bramley 1994). As mechanisms of action, carotenoids capture light and transmit energy to chlorophyll, thus aiding in photosynthesis and absorbing excess energy from chlorophyll under high sun exposure. In addition, carotenoids quench reactive oxygen species produced in photosynthesis, thus playing a dual protective role, as well as being an additional antenna pigment to chlorophyll (Maoka 2020). An important protective role is played by the xanthophyll cycle, which involves the removal of epoxy groups from carotenoids such as violaxanthin, zeaxanthin, or epoxy lutein to produce zeaxanthin or lutein, which can regain epoxy groups during light stress. Carotenoids are also present in non-photosynthetic organs of plants, providing color to flowers and fruits to make them more attractive to pollinators or to animals that consume the fruit and spread their seeds. In addition, carotenoids are precursors for two kinds of phytohormones, abscisic acid (Nambara and Marion-Poll 2005) and strigolactones (Al-Babili and Bouwmeester 2015).

Carotenoids have enormous biotechnological potential, as their consumption confers many benefits (Eggersdorfer and Wyss 2018). Some carotenoids, especially β-carotene, are a source of vitamin A (Weber and Grune 2012), which is necessary both for the proper functioning of vision and for the maintenance of apocarotenoid metabolism, including the formation of retinoic acid, which is used as a regulatory signal for gene expression in chordates. In addition, carotenoids provide many health benefits to those who consume them in sufficient quantities, which are described for specific carotenoids in next sections. It is worth mentioning the protective effects that different carotenoids have against the development of various types of cancer (Milani et al. 2017), and their efficacy increases if formulations that improve their accessibility to the tumor tissue are used (Zare et al. 2021).

8.2 Carotenoids in Yeasts and Filamentous Fungi

Carotenoids can play different roles in fungi. Protection against oxidative stress is a widely accepted effect, but depending on the species they can also serve as precursors by oxidative cleavage of molecules with other functions, such as retinal, the chromophore of opsins, and trisporic acids, sexual hormones in zygomycete fungi (Avalos and Limón 2015), or plant hormones such as abscisic acid (Takino et al. 2019), with a role in infection. However, fungi are not photosynthetic organisms, and therefore carotenoid synthesis is not an essential process in them. For this reason, not all fungi produce these compounds. This is the case for two models widely used in research, the yeast *Saccharomyces cerevisiae* and the filamentous fungus *Aspergillus nidulans*. However, the ability to synthesize carotenoids is a frequent trait in filamentous fungi and yeasts, and some of them have been used as models in the investigation of their biosynthetic pathways.

The carotenoids most frequently found in fungi are highly hydrophobic molecules such as β-carotene or its precursor lycopene, and xanthophylls with some oxygenated polar groups, such as astaxanthin, neurosporaxanthin, or torularhodin (Avalos et al. 2017a; Rapoport et al. 2021; Sandmann 2022). The knowledge of carotenoid pathways in fungi has revealed some features conserved in all of them, indicating that the ability to synthesize carotenoids was a character acquired by a very early ancestor, common to most present-day fungi. Among the unique features are the use of a single desaturase to carry out all desaturations and the coincidence in the same enzyme of the cyclase and phytoene synthase activities. In contrast, in bacteria and photosynthetic organisms, the desaturations are carried out by two desaturases that act in a sequential manner, and the synthesis of phytoene and the cyclization steps are carried out by independent enzymes (Sandmann 2009; Domonkos et al. 2013). In summary, fungi have fewer genes dedicated to the pathway, and, fungal biosynthetic pathways are simpler than those of algae or plants.

Fungal carotenoid pathways have been reviewed (Avalos et al. 2017a; Sandmann 2022; Naz et al. 2023). In all investigated carotenogenic fungi, synthesis begins by the production of isopentenyl pyrophosphate (IPP) from mevalonate and continues with the usual elongation in terpenoids by successive additions of IPP until the formation of geranylgeranyl diphosphate (GGPP). The first enzyme specific for carotenoid biosynthesis, phytoene synthase, binds two GGPP molecules in an inverted fashion to generate the colorless carotenoid precursor, phytoene (Fig. 8.1). Subsequently, desaturase introduces three, four, and five conjugated double bonds to generate neurosporene, lycopene, and 3–4-didehydrolycopene, respectively. The cyclase creates β-rings at one end of any of the molecules depending on the fungus, to generate β-zeacarotene, γ-carotene, or torulene, respectively. In some fungi, the pathway is directed to the synthesis of β-carotene by a second cyclization on γ-carotene, which may be the major accumulated product, as occurs in various Mucorales, such as *Phycomyces blakesleeanus*, *Blakeslea trispora*, or *Mucor circinelloides*. In other fungi, such as the yeast *Xanthophyllomyces dendrorhous*, the β-rings of β-carotene can undergo ketolation and hydroxylation, giving rise to astaxanthin. In some yeasts, as *Rhodotorula* spp., torulene can be oxidized by an unknown enzyme up to torularhodin, while in others, as the ascomycetes *Verticillium* spp., *Fusarium* spp., or *Neurospora* spp., torulene can be cleaved by an oxygenase to produce neurosporaxanthin in two steps.

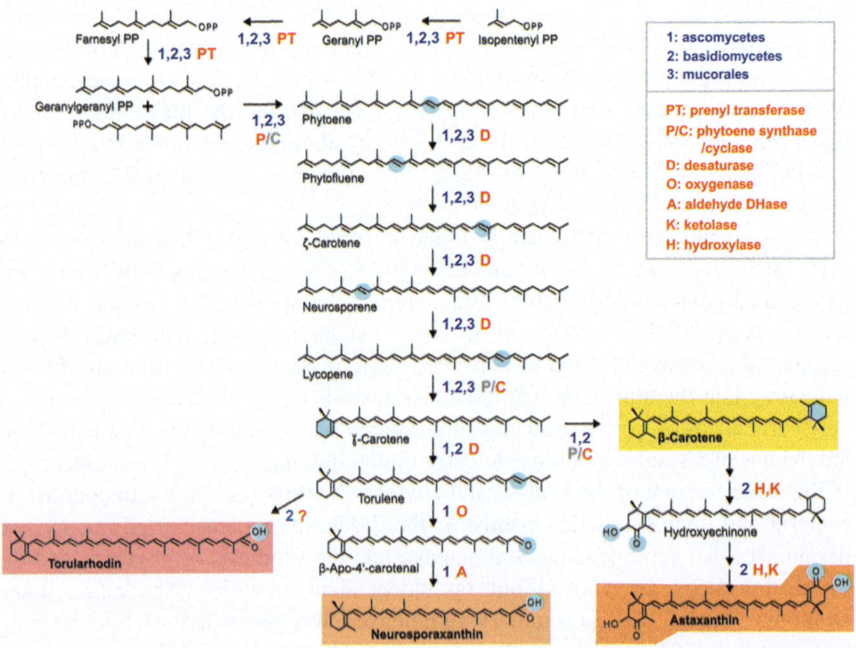

Fig. 8.1 Integrated view of best-known carotenoid biosynthetic pathways in fungi

8.3 Biotechnological Applications of Fungal Carotenoids

Carotenoids are produced by two major means, chemical (Pfander et al. 1997; Ito et al. 2009) and biological syntheses (Sandmann 2015; Saini et al. 2022). Chemical synthesis has been predominant for many years, but growing concerns about its polluting effects have boosted the development of biological production methods, accepted as less aggressive with the environment. Biological sources are of two types: extraction from plant tissues rich in carotenoids or production by microorganisms, including algae and fungi (Martínez-Cámara et al. 2021). Production by algae is usually based on large-scale outdoor cultivation, such as those used to produce β-carotene with *Dunaliella* sp. (Lamers et al. 2008) or astaxanthin with *Haematococcus* sp. (Shah et al. 2016). Although these organisms accumulate very high amounts in cellular forms of resistance, the cultures require high solar exposure and are therefore dependent on appropriate climatic conditions. In contrast, fungi are grown in large fermenters under conditions that are controllable and easily scalable and can compete in yield with other production systems (Adrio and Demain 2003). Much effort has been devoted to improving the production levels of different fungi, and some of them are currently used for industrial production of β-carotene, lycopene, and astaxanthin, while other carotenoids are still awaiting development. For these, the study of their biological properties and possible applications is of special interest.

Fungal carotenoids have applications in several markets, as feed or food additives, as nutritional supplements, and in the pharmaceutical and cosmetic industries (Rapoport et al. 2021). Their use in food has four benefits or utilities, as colorants (Luzardo-Ocampo et al. 2021), as sources of vitamin A (Weber and Grune 2012), as components of the macula (Khoo et al. 2019), and as healthy compounds due to their antioxidant properties (Stahl and Sies 2003; Skibsted 2012). Not all of them have the potential to be sources of vitamin A, for which they must have at least an unmodified β-ring. On the other hand, only three xanthophylls, lutein, zeaxanthin, and meso-zeaxanthin, are involved in the maintenance of the macula (Li et al. 2023b), but none of them are sources of vitamin A. The other two applications are general to all carotenoids, although there are differences between them in their antioxidant potential and in the type of pigmentation they produce. The applications of fungal carotenoids, confirmed or potential, as well as their production possibilities from suitable species, are described below.

8.4 Production and Biotechnology of Fungal Carotenes

8.4.1 *β*-Carotene

The fungus mainly used for the industrial production of β-carotene is the mucoral *B. trispora* (Avalos and Cerdá-Olmedo 2004). Research with this fungus was historically preceded by genetic studies with its close relative *P. blakesleeanus*, which allowed advancing in the knowledge of the genetics and biochemistry of the β-carotene biosynthetic pathway and its regulation (Cerdá-Olmedo 1987). This biosynthetic pathway, the first one described for carotenoids (Porter and Lincoln 1950), is the simplest among carotenogenic organisms, and it consists of the synthesis of phytoene from GGPP, and the sequential introduction of four desaturations and two terminal β-cyclizations (Fig. 8.1). The facilities offered by *P. blakesleeanus* as a genetic tool made it possible to identify the genes *carB* and *carRA* responsible for this pathway using mutants (Cerdá-Olmedo 1987) and revealed the functioning as a multimeric complex of two of their enzymes, desaturase and cyclase, through the skillful use of heterokaryons, i.e., strains with two genetically different nuclei in their cells (De la Guardia et al. 1971; Aragón et al. 1976). Genetic analysis also allowed the identification of a dual function for *carRA* gene, which was initially associated with a cyclase enzyme, encoded in segment *carR*, and an auxiliary protein tentatively involved in substrate transfer, encoded by segment *carA* (Torres-Martínez et al. 1980). Subsequently, it was found that the CarA domain corresponds to phytoene synthase and that it is presumably cleaved by a protease from the CarR cyclase polypeptide (Arrach et al. 2001). The identification of genes at the molecular level had to wait for their discovery in a fungus more amenable to genetic manipulation, *Neurospora crassa* (Avalos and Corrochano 2013), which allowed cloning the orthologous genes in *P. blakesleeanus* (Ruiz-Hidalgo et al. 1997; Sanz et al. 2011) and *B. trispora* (Rodríguez-Sáiz et al. 2004).

From the biotechnological point of view, the most remarkable characteristic possessed by *P. blakesleeanus* and *B. trispora* is the stimulation of β-carotene synthesis by sexual interaction between strains of opposite sexes (Mehta et al. 2003; Kuzina and Cerdá-Olmedo 2006; Breitenbach et al. 2012), a response mediated by the production of carotene-derived sexual hormones, the trisporic acids (Sutter et al. 1974; Govind and Cerdá-Olmedo 1986). Such stimulation, which takes place at the level of transcriptional regulation of their two structural genes *carRA* and *carB* (Schmidt et al. 2005; Almeida and Cerdá-Olmedo 2008), is exploited for increased production on an industrial scale by using mixed cultures, i.e., those in which both sexual types are grown together. Moreover, in the two fungi, β-carotene super-producing mutants are obtained by conventional mutagenesis in both sexual backgrounds, which allow strong increases in production in mixed cultures (Mehta et al. 1997, 2003). The coenocytic nature of these fungi makes it possible to use heterokaryons with nuclei of both sexes to improve stimulation by sexual interaction (Murillo et al. 1978; Mehta et al. 2003). This strategy has the disadvantage that the nuclear ratios may oscillate randomly and alter the stimulation efficiency, although in the case of

P. blakesleeanus, stability was enhanced using balanced lethal mutations. It turned out that *P. blakesleeanus* mutants (Mehta et al. 1997) were better producers than *B. trispora* mutants, but induction in mixed cultures was more efficient in the latter (Mehta et al. 2003). This fact, together with a greater amenability of *B. trispora* for carotenoid production in submerged cultures, explains why it was the organism of choice for industrial use. Subsequently, the progressive development of improved strains and culture conditions, exploited under sexual stimulation, allowed the use of *B. trispora* for industrial production to be competitive with other sources of production, such as chemical synthesis or obtaining from algae.

A drawback of *B. trispora* and *P. blakesleeanus* is that they are not easy fungi to manipulate genetically in a targeted manner. Although exogenous DNA sequences have been detected in *P. blakesleeanus* transformants (Suárez and Eslava 1988; Obraztsova et al. 2004), and there are patents for transformation of *B. trispora*, no works have been published in which any transformants of these fungi have been used as research tools. However, a closely related species, *M. circinelloides*, is amenable to genetic modification by transformation (Van Heeswijck and Roncero 1984), and the availability of techniques to obtain targeted gene alterations stimulated their use in molecular studies of carotenogenesis, involving both structural and regulatory genes (Velayos et al. 2000a, b; Navarro et al. 2001). Although its characteristics and production capacity do not indicate that it can surpass *B. trispora* for the industrial production of β-carotene, it has been an important source of molecular information, applicable to its relatives *P. blakesleeanus* and *B. trispora*. As an example, the functions of the *carRA* and *carB* genes of *B. trispora* were confirmed by complementation of mutants of both genes in *M. circinelloides* (Rodríguez-Sáiz et al. 2004).

β-Carotene is, along with astaxanthin, described in Sect. 8.5.1, the carotenoid with the most established biotech market, and its consumption as a food additive is very widespread as a healthy colorant and a source of vitamin A (Anand et al. 2022). β-Carotene is a normal component of human milk, where it contributes to antioxidant defenses in the newborn (Ostrea et al. 1986). The scarcity of vitamin A in the diet is a major nutritional deficiency in many regions of the world. β-Carotene is the main source of vitamin A and therefore of retinoid metabolism in animals and humans (Grune et al. 2010). It has numerous applications as an ingredient in vitamin complexes, in cosmetic products, or as a colorant in food products, to which it not only provides color but also antioxidant effects. It is also used in sunscreen creams, helping to protect against the harmful effects of excessive exposure to solar radiation, such as the formation of carcinomas (Eggersdorfer and Wyss 2018). However, although there were studies indicating that its consumption provides protective effects against other types of cancer (van Poppel and Goldbohm 1995), there has been much controversy about this protective effect (Patrick 2000), often not supported by data or even with counterproductive effects found in smokers (Jeon et al. 2011; Zhang et al. 2023b).

8.4.2 Lycopene

Most of the lycopene provided in the diet comes from fresh tomato or tomato-derived products (Khan et al. 2021), but this carotene is also obtained from biotechnological sources. As intermediates in the synthesis of β-carotene, the biotechnological production of lycopene takes advantage of the conditions developed for β-carotene-producing organisms by preventing lycopene cyclization through chemical inhibition or genetic blocking. Cyclase mutants of the fungal bifunctional phytoene synthase/cyclase gene affected in the cyclase domain, without altering the activity of the neighboring phytoene synthase domain, are easy to identify due to their characteristic reddish pigmentation. Mutants affected in the different domains of this gene were originally described in *P. blakesleeanus* (Torres-Martínez et al. 1980). As indicated above, the carboxyl domain of the gene was initially associated with a substrate transfer function, and its phytoene synthase activity was identified later (Arrach et al. 2001). Lycopene-accumulating mutants have also been described in *M. circinelloides* (Navarro et al. 1995; Velayos et al. 1997) and *B. trispora* (Mehta and Cerdá-Olmedo 1995), and in the latter, both mixed and heterokaryon cultures have been used to improve their production with cyclase mutants under laboratory conditions (Mehta et al. 2003). Chemical inhibition is based on the use of different nitrogenated inhibitors, such as 2-(4-chlorophenylthio) ethylamine (CPTA), nicotine, α-picoline (Mehta and Cerdá-Olmedo 1999), or tertiary amines, such as tripropylamine (Pegklidou et al. 2008). Chemical inhibition has been used to develop lycopene production on a semi-industrial scale using raw materials (López-Nieto et al. 2004). The use of inhibitors has the advantage of not requiring strains with the appropriate mutations, but it is less advantageous than the use of mutants, since complete inhibition of cyclase activity is not easy to achieve by chemical means, and chemical inhibitors always have other side effects.

Lycopene has biological properties that distinguish it from other carotenoids. It stands out for its high efficacy as a singlet oxygen quencher (Di Mascio et al. 1989). Numerous studies have described health benefits of lycopene consumption (Przybylska 2020), including effects such as prevention of obesity or different types of cancer, beneficial effects on liver and kidney functions, digestive system or skin health, protection against osteoporosis or cardiovascular or neurodegenerative diseases, anti-diabetic or anti-inflammatory effects, improvement of fertility and spermatogenesis, and protection of eyesight (reviewed by Shafe et al. 2024). Its beneficial effects are not only a direct consequence of its antioxidant activity but are also due to indirect regulatory effects. Thus, lycopene acts on the signal transduction chain that regulates the cellular defense system against oxidative damage, as well as on regulatory cascades that control lipid metabolism, explaining its effects against obesity. The causes of their regulatory effects on the cell cycle, apoptosis, inflammation, and angiogenesis, associated with their protective effects against cancer, are also known, in addition to those that explain their anti-inflammatory activity or effects on the immune system. The signal transduction pathways on which lycopene acts that explain these effects have recently been reviewed (Long et al. 2024). The beneficial effects of lycopene also apply to the breeding of food animals, both ruminants and poultry, as its supply in the diet can lead to increases in meat production and meat quality, as well as positively influencing their reproduction rates (Long et al. 2024).

8.4.3 *Phytoene*

Phytoene is present as a secondary carotene in different foods rich in carotenoids, such as tomatoes, carrots, or apricots, but its chemical structure gives it a higher bioaccessibility after ingestion than other carotenoids present in these foods or in derivatives, such as their juices (Mapelli-Brahm et al. 2017). Like lycopene, phytoene accumulation can be obtained from carotenogenic fungi by chemical inhibition of the desaturase enzyme or by mutation of the encoding gene. A well-known desaturase inhibitor is diphenylamine (DPA), effective in industrial fungi, as *Blakeslea trispora* (Thomas and Goodwin 1967) and *Phaffia rhodozyma* (Chumpolkulwong et al. 1997). Albino mutants of these fungi are candidates for possessing a defect in desaturase, which can be corroborated by the expected accumulation of phytoene, as is the case with *al-1* gene mutants of *N. crassa* (Goldie and Subden 1973). Such mutants are also known in *X. dendrorhous* (Barbachano-Torres et al. 2014) and *B. trispora* (Mehta and Cerdá-Olmedo 1995; Rodríguez-Sáiz et al. 2004), and therefore, systems developed for their respective β-carotene and astaxanthin productions should be applicable to produce phytoene. However, unlike lycopene, phytoene is not a pigment and has received much less attention for its industrial exploitation. This may change in the future, as there is currently an increasing focus on colorless carotenoids, such as phytoene, due to an increased ability to protect against UV radiation. This, together with their high bioavailability after ingestion, gives them a potential to be considered in nutrition because of their relationship to health (Meléndez-Martínez et al. 2018; Olmedilla-Alonso et al. 2021). In support, phytoene accumulates in the skin in small concentrations, together with other carotenoids, with presumed effects in preventing skin cancer or improving its appearance and properties (Meléndez-Martínez 2019; So and Kwon 2023).

8.4.4 *Torulene*

Torulene is found in minor amounts as an intermediate carotene in fungi that produce torularhodin or neurosporaxanthin (Fig. 8.1, Sects. 8.5.2 and 8.5.3). The end of the torulene polyene chain with the fifth desaturation is not recognized by the cyclase to introduce the β-ring and will be subject to other chemical reactions depending on the organism. Mutants of the gene encoding the carotenoid oxygenase that introduces an oxidative cleavage of torulene have been described in *Fusarium fujikuroi* (Prado-Cabrero et al. 2007a) and *N. crassa* (Saelices et al. 2007), whereby these mutants accumulate torulene as a major carotenoid. However, torulene can be naturally accumulated as a predominant carotenoid under specific culture conditions by some torularhodin-producing yeasts, such as *Sporidiobolus pararoseus* (Li et al. 2017; Wei et al. 2020). As usual for this type of carotenoids, torulene has antioxidant (Dimitrova et al. 2013) and anticancer (Du et al. 2016) properties, making it a carotenoid with a considerable applied potential, although not yet commercially exploited. The conditions developed to produce torularhodin in this yeast should be potentially useful for producing torulene in mutants that are not able to modify this carotene, such as those described in *F. fujikuroi* and *N. crassa*, mentioned above.

8.5 Production and Biotechnology of Fungal Xanthophylls

8.5.1 Astaxanthin

Astaxanthin is a xanthophyll found in many animals because they consume it through their diet, which includes algae that produce it (Schmidt et al. 2011). This is the case of some river fish, such as trout and salmon, to which it provides the orange pigmentation to their flesh, or of crustaceans for human consumption such as shrimps, crayfish, and lobsters, which owe to this xanthophyll the typical orange pigmentation they acquire when boiled. In the latter, the color is not visible under natural conditions due to the reactivity of the oxygenated groups of astaxanthin, which cause them to be conjugated with proteins or esterified to fatty acids (Hussein et al. 2006).

As with β-carotene, most of the astaxanthin marketed comes from chemical synthesis, but there is a trend toward greater use of astaxanthin of biological origin. The highest levels of production are obtained from outdoor cultivation of the alga *Haematococcus pluvialis*, in which astaxanthin reaches 3.8% of its dry weight under strong solar radiation (Ambati et al. 2014). In fungi, the best-known producer is *X. dendrorhous*, originally known as *Phaffia rhodozyma* (Schmidt et al. 2011). Although the quantities accumulated by this yeast are smaller, its facilities to be cultivated in fermenters in a more controlled manner and occupying less land than algae cultures make this yeast a biotechnologically attractive alternative. Depending on the orientation, hydroxylic (OH) groups may be in two different isomeric forms in the β-rings, R or S (Andrewes and Starr 1976). Biological synthesis has the advantage of the production of more natural isomeric forms, since chemically synthesized astaxanthin contains a 1:2:1 mixture of [3S, 3′S], [3R, 3′S], [3R, 3′R] isomers, whereas those of biological synthesis are usually of only one type, [3S, 3′S] when obtained from *H. pluvialis* and [3R, 3′R] when obtained from *X. dendrorhous* (Moretti et al. 2006).

Because of its amenability for genetic manipulation, *X. dendrorhous* has been the subject of numerous research studies on its carotenoid pathway. A major goal was the identification of the genes responsible for the enzymes, such as *crtI*, which determines phytoene synthase (Verdoes et al. 1999b), and *crtYB*, which encodes phytoene synthase/cyclase (Verdoes et al. 1999a). It was in this yeast that the bifunctional nature of this enzyme was first identified in fungi by means of *crtYB* heterologous expression in *E. coli* engineered strains that produce different carotenoids through the expression of other bacterial genes. Finally, the introduction of keto and hydroxyl groups on the β-carotene rings is carried out by an enzyme of the P450 monooxygenase family, whose gene was discovered in parallel by two research groups, who called it *asy* (Ojima et al. 2006) and *crtS* (Martín et al. 2008). This enzyme works in cooperation with a cytochrome P450 reductase, product of the *crtR* gene (Alcaíno et al. 2008, 2012), and introduces first a keto group and then a hydroxyl group on each ring. In *X. dendrorhous*, torulene and the hydroxylated and ketolated version of this carotenoid are also detected, indicating the existence of a side branch by desaturation of γ-carotene to torulene and the recognition of its β-ring by the CrtS/CrtR system (Fig. 8.2). Moreover, 3-hydroxy-torulene-4-one may be subject of further reactions to yield 3,3′-dihydroxy-β,ψ carotene-4,4′dione (DCD) (An et al. 1999).

Consumption of astaxanthin has been associated with multiple health benefits and has been the subject of numerous patents (Ambati et al. 2014). The market for its production is mainly for use as a feed additive for animals destined for human food, and its health effects have therefore been the subject of detailed attention (Stachowiak and Szulc 2021). Among the numerous benefits described are anti-inflammatory activity, antidiabetic effects, prevention of cardiovascular diseases, stimulation of the immune system, neuroprotection, improvement of eye health, and skin protection, among others (Yuan et al. 2011). Similarly to lycopene, the wide range of beneficial effects detected may be partly due to its high antioxidant capacity, supported in both in vivo and in vitro studies, among which its high anti-lipid peroxidation activity stands out (Lee et al. 2011). This activity exceeds that of

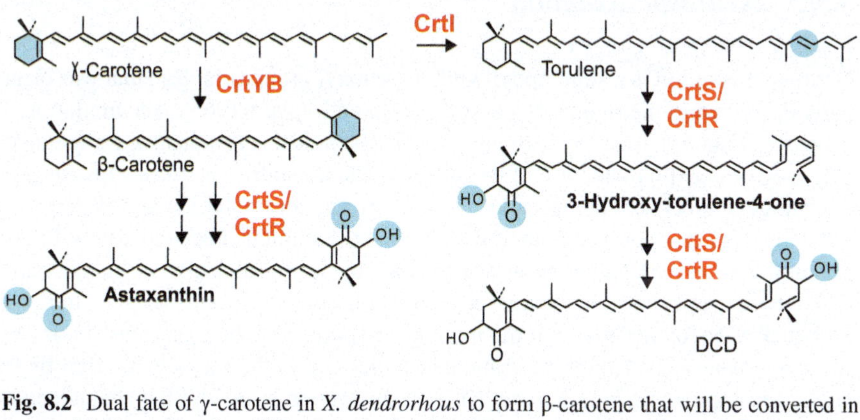

Fig. 8.2 Dual fate of γ-carotene in *X. dendrorhous* to form β-carotene that will be converted in astaxanthin or torulene that will be converted to 3-hydroxy-torulene-4-one and its DCD derivative

β-carotene, which is attributed to its perpendicular location in the membranes forced by its extreme polar groups, which facilitates lipid protection. For this reason, there is also a market for astaxanthin as a dietary supplement and as an additive in cosmetics.

The applied interest in astaxanthin has led to numerous studies exploring ways to improve production with *X. dendrorhous*. For years, different publications have described the generation of mutants with increased production of carotenoids and astaxanthin (An et al. 1989; Sun et al. 2004; Liu et al. 2008), including the development of potential screening methodologies in yeast, such as cell sorting (An et al. 1991; Ukibe et al. 2008). Mutant searches not only generate astaxanthin overproducers but also strains that accumulate instead β-carotene or phytoene or other changes in carotenoid ratios (Barbachano-Torres et al. 2014). As a particularly successful example of improved astaxanthin production, screening by mutagenesis yielded strains with at least 10-fold increases in astaxanthin content, which reached 50-fold by introducing additional copies of the *crtYB* and *asy* genes into one of these mutants (Gassel et al. 2013). Other studies focused on improving astaxanthin synthesis by metabolic alterations, such as the use of more potent promoters to increase the expression of genes for limiting steps of the synthesis, the *crtE* gene, responsible for the synthesis of GGPP (Hara et al. 2014a), or the *acaT*, *hmgS*, and *hmgR* genes, necessary for the synthesis of the precursor mevalonate (Hara et al. 2014b). Substrate flux to carotenoids has also been increased by deleting genes involved in negative regulation of mevalonate synthesis by sterol (Yamamoto et al. 2016). These and other approaches, which have been recently revised, allow variable increases in astaxanthin production which can be combined with other regulatory mutations to achieve more significant improvements (Torres-Haro et al. 2021). Many efforts have also been made to improve culture conditions by checking different media compositions, especially useful when using low-cost raw materials derived from other industries (Rodríguez-Sáiz et al. 2010).

8.5.2 Neurosporaxanthin

Neurosporaxanthin is a xanthophyll with a carboxyl group (Fig. 8.1) that was originally described in the fungus *N. crassa* (Aasen and Jensen 1965). Accumulation of neurosporaxanthin, together with precursor carotenoids, is responsible for the characteristic orange pigmentation of its cultures. The aerial hyphae of this fungus produce large numbers of spores, called conidia, in which carotenoids accumulate by a regulatory process coupled to their development (Schmidhauser et al. 1990; Li and Schmidhauser 1995). Conidiation is stimulated by light, which also stimulates carotenogenesis, and thus the orange pigmentation of illuminated aerial cultures is accentuated by increasing both the number of spores and their carotenoid content. *N. crassa* is an outstanding experimental model due to the facilities offered by its genetic manipulation, which, together with its easy crossing and the abundance of genetic techniques, has made it one of the most widely used fungi in basic research (Roche et al. 2014). In the classic experiments to identify mutants of this fungus, the albino mutants stood out as excellent markers for their lack of pigmentation and were assigned by genetic analysis to three genes, *al-1*, *al-2*, and *al-3*. Successive cloning of these genes made it possible to identify for the first time the enzymes responsible for desaturase, AL-1 (Schmidhauser et al. 1990); phytoene synthase, AL-2 (Schmidhauser et al. 1994); and GGPP synthase, AL-3 (Nelson et al. 1989). Subsequently, the bifunctional cyclase/phytoene synthase nature of AL-2 was elucidated (Arrach et al. 2002), based on the previous analysis of its ortholog in the yeast *X. dendrorhous*. Finally, the genes responsible for the last steps of the pathway were identified, encoding an oxygenase, CAO-2 (Saelices et al. 2007), and an aldehyde dehydrogenase, YLO-1 (Estrada et al. 2008b).

In parallel to the studies in *N. crassa*, and prior to the identification of its genes, the presence of neurosporaxanthin was determined in *Fusarium aquaeductuum*, in this case used as a model of regulation by light as an agent stimulating its synthesis (Bindl ct al. 1970). The work in *F. aquaeductuum* was followed by more detailed studies of neurosporaxanthin synthesis in *F. fujikuroi* (Avalos et al. 2017b). In contrast to *F. aquaeductuum*, *F. fujikuroi* was studied at the genetic level by obtaining and characterizing mutants with altered pigmentation, including those expected due to mutations in structural genes of the carotenoid pathway (Avalos and Cerdá-Olmedo 1987). Later, based on previous knowledge of *al-1*, *al-2*, *al-3*, and *ylo-1* genes in *N. crassa*, the *F. fujikuroi* orthologs *carB* (Fernández-Martín et al. 2000), *carRA* (Thewes et al. 2005), *ggs1* (Mende et al. 1997), and *carD* (Díaz-Sánchez et al. 2011) were identified. In contrast, the oxygenase gene responsible for oxidative cleavage of torulene, *carT*, was originally identified in *F. fujikuroi* (Prado-Cabrero et al. 2007a), and its discovery allowed the identification of the ortholog in *N. crassa*, *cao-2* (Saelices et al. 2007).

Although these two fungi produce neurosporaxanthin with a similar set of enzymes, there are differences in the order in which the reactions can be performed and in the substrate preferences of the enzymes, which may result in differences in the intermediates accumulated (Fig. 8.3). This is especially striking in the case of *N. crassa*, where a different order of reactions has been observed when its cultures are stimulated by light at low temperature (Estrada et al. 2008a), conditions under which neurosporaxanthin production is more efficient. Moreover, in *N. crassa*, AL-1 desaturase does not recognize γ-carotene as a substrate (Hausmann and Sandmann 2000), so the preferential pathway is through lycopene and 3–4-didehydrolycopene. On the other hand, a β-carotene oxygenase, CarX, which generates retinal, used as a chromophore for fungal rhodopsins, has been described in *F. fujikuroi* (Prado-Cabrero et al. 2007b), whereas no such activity has been identified in *N. crassa* (Díaz-Sánchez et al. 2013), despite also possessing a retinal-binding rhodopsin (Bieszke et al. 1999).

Neurosporaxanthin production appears ubiquitous in *Neurospora* (Luque et al. 2012) and *Fusarium* (Avalos et al. 2017b) genera, the latter a particularly complex taxonomic group covering hundreds of species (Aoki et al. 2014). However, this carotenoid is not frequent in other fungi and has only been described to date in few

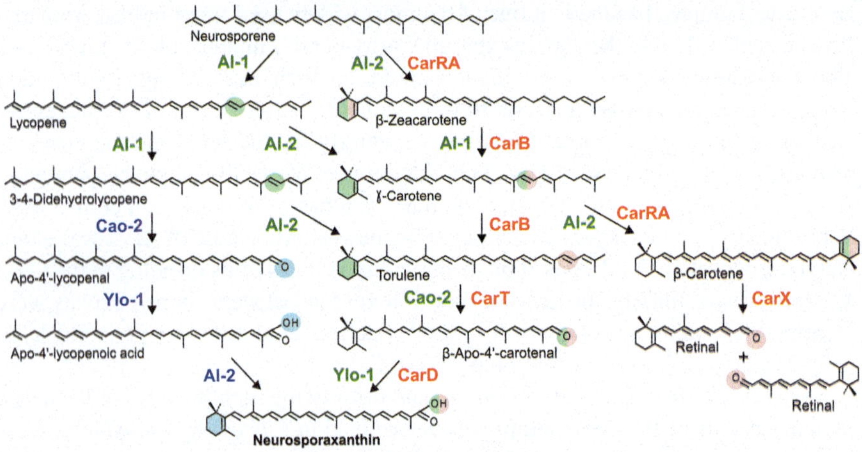

Fig. 8.3 Diversity of chemical reactions from neurosporene in *N. crassa* (enzymatic steps at normal temperature in green and predominant enzymatic steps at low temperature in blue) and *F. fujikuroi* (enzymatic steps in red). Shaded areas indicate chemical modifications carried out by the preceding enzyme of the same color. β-Carotene is a minority product in *N. crassa*, and retinal production has only been detected in *F. fujikuroi*. Reactions deduced from carotenoid contents in *N. crassa* conidia and mycelia and *F. fujikuroi* mycelia under different culture conditions (Estrada et al. 2008a; Hornero-Méndez et al. 2018)

other ascomycetes, such as *Arthrobotrys oligospora*, (Valadon and Cooke 1963), *Verticillium sp.* (Valadon and Heale 1965; Osman and Valadon 1978), and *Podospora anserina* (Strobel et al. 2009). However, from a biotechnological point of view, *N. crassa* and *F. fujikuroi* are the only species for which there is a sufficiently deep genetic knowledge of the biosynthetic pathway to consider its suitability for overproduction of this carotenoid. *N. crassa* does not appear to be an appropriate fungus for this purpose. In submerged cultures, neurosporaxanthin production by *N. crassa* and other *Neurospora* species is low (Luque et al. 2012), and no mutans have been described increasing the wild-type carotenoid levels in the dark. A significant increase is attained through stimulation by light, but it does not reach very high values due to the existence of a photoadaptation mechanism. In contrast, surface cultures can accumulate higher amounts of carotenoids, but both the use of surface culture conditions and the need to extract carotenoids from the spore masses do not appear ideal conditions for biotechnological production. In addition, the carotenoid contents in its spores are complex (Hornero-Méndez et al. 2018), and their concentrations, although higher, are not high enough to be considered a suitable source for industrial production. At least 3 mutants have been described that increase 2–3 times the amount of carotenoid in this fungus in the light, the *cbb-1* mutant, possibly affected in a negative regulator of the structural genes (Linden et al. 1997), the *ovc* mutant (Youssar et al. 2005), which has a deletion affecting 21 genes and whose causes of overproduction are unknown (Youssar and Avalos 2007), and the *vvd* mutant (from vivid), whose increased production is due to the loss of the photoadaptation mechanism to the light signal (Schwerdtfeger and Linden 2003). Moreover, under submerged conditions, the *ovc* and *vvd* mutants only overproduce carotenoids under illumination (Youssar et al. 2005), which limits the development of production conditions in fermenters. Even so, production levels are not very high.

The scenario is different in the fungus *F. fujikuroi*, in which despite the similarities of the pathway, it has different regulatory characteristics. Thus, while light induction of *N. crassa* carotenogenesis is mediated by the White-collar complex, a heterodimer formed by the WC-1 photoreceptor and its partner WC-2 (Corrochano and Avalos 2010), light regulation in *F. fujikuroi* depends on more than one photoreceptor, including the WC-1 ortholog WcoA (Castrillo and Avalos 2015). However, the most remarkable feature of *F. fujikuroi* carotenoid regulation is the existence of a repressor gene, the RING-finger protein CarS (Rodríguez-Ortiz et al. 2013), whose loss of function results in a carotenoid overproducing phenotype due to a strong increase in the transcription of structural genes (Avalos et al. 2017b; Ruger-Herreros et al. 2019). Notably, such derepression also occurs in the dark, which makes it possible to obtain mycelia enriched in carotenoids in submerged cultures. Synthesis in this fungus is also stimulated by nitrogen starvation (Rodríguez-Ortiz et al. 2009; Ruger-Herreros et al. 2022), which has allowed the development of cultivation methods that have made it possible to reach 8 mg of carotenoids per g of dry mass in a *carS* mutant, of which at least 95% is neurosporaxanthin (Parra-Rivero et al. 2020). It is to be hoped that these levels can be increased with future manipulations of the culture conditions, e.g., by combining nitrogen starvation with oxidative stress, which also stimulates carotenoid synthesis in this fungus (Ruger-Herreros et al. 2022).

The existence of conditions of overproduction of neurosporaxanthin has made it possible to obtain this compound in high quantities to feed mice and analyze its properties (Miller et al. 2023). The results showed that neurosporaxanthin has a high incorporation into the body, possibly favored by the amphipathic character of the molecule due to its terminal carboxyl group, and is recognized as a substrate by the enzyme BCO-1, which allows its conversion into retinal and its incorporation into the body's retinoid metabolism. The ability of BCO-1 to cleave neurosporaxanthin was also demonstrated in this work by in vitro experiments, incubating this xanthophyll with the enzyme. The availability of samples enriched in neurosporaxanthin also made it possible to prove that it has a high antioxidant capacity and that it is much more effective than β-carotene in protecting membranes from oxidation (Parra-Rivero et al. 2020). All these characteristics make neurosporaxanthin a particularly promising xanthophyll for future biotechnological exploitation.

8.5.3 Torularhodin

Torularhodin is a carotenoid with a carboxyl-terminal end, like neurosporaxanthin but with 40 carbon atoms, in whose synthesis there is no oxidative cleavage (Fig. 8.1). This xanthophyll is produced mainly by certain yeasts, best known in those of the genus *Rhodotorula* (Tang et al. 2019), although it is also synthesized by other yeasts, such as those of the genera *Rhodosporidium*, *Sporobolomyces*, and *Sporidiobolus*, which produce them in different proportions with torulene and β-carotene (Buzzini et al. 2007). The synthesis of torularhodin derives from torulene, which is formed by the same enzymatic steps that give rise to this carotene in neurosporaxanthin-producing fungi (Fig. 8.1).

Unlike *Neurospora* or *Fusarium*, the genes responsible for the synthesis of torularhodin and its precursors have received little attention in these yeasts, it being assumed that the enzymes are similar to those of other fungi. Desaturase and phytoene synthase/cyclase genes have been identified because of their similarity to other fungal counterparts. The desaturase gene, *crtI*, was identified in *S. pararoseus*, and its enzymatic activity was confirmed by heterologous complementation in an engineered *E. coli* strain (Li et al. 2016). Sequencing of genomes from this group of torularhodin-producing yeasts revealed a similar genomic organization of the *crtI* and *crtYB* genes as a cluster together with a gene for an oxygenase, *crtX*, which is presumably responsible for the cleavage of β-carotene to produce retinal. The functions of the desaturase and phytoene synthase/cyclase genes *crtI* and *crtYB* from *Rhodotorula glutinis* were confirmed by heterologous expression in an engineered *S. cerevisiae* strain (Bo et al. 2022). However, the genes responsible for the conversion of torulene to torularhodin, to which hydroxylation or oxidation reactions are attributed (Tang et al. 2019), have not been identified in any of the yeasts producing this xanthophyll. In contrast to the synthesis of neurosporaxanthin, in which oxidative cleavage of torulene generates the aldehyde group at the cleaved end, the enzymes involved in torularhodin synthesis must act directly on the uncyclized end of torulene, and the type of enzymes that perform this reaction is unknown.

Similar to what was observed for astaxanthin synthesis in *X. dendrorhous*, oxidative stress is a stimulatory agent for torularhodin synthesis in these yeasts. Thus, chemical agents generating singlet oxygen, superoxide anions, and peroxy radicals, respectively, stimulate the accumulation of torularhodin and its precursor torulene in *R. glutinis*, while they hardly affect the accumulation of β-carotene (Sakaki et al. 2002). Illumination also produces an increase in torularhodin synthesis in this yeast, which was interpreted as a response to oxidative stress produced by light (Sakaki et al. 2001). The stimulatory effect of light on torularhodin synthesis had been previously observed in another yeast, *Epicoccum nigrum* (Gribanovski-Sassu and Foppen 1968).

Torularhodin is not yet in commercial use, as neurosporaxanthin, but it has very promising properties that give it considerable applied potential, such as antioxidant power and anticancer and antimicrobial activities (Kot et al. 2018). In addition, the dietary supply of the mixture of carotenoids produced by *R. glutinis* for long periods of time to mice did not show any toxicity, so it is presumed that it could be used as an additive for human use (Latha and Jeevaratanm 2012). Although it has not been shown to be recognized as a substrate by the Bco1 enzyme to produce vitamin A, the high structural similarity to neurosporaxanthin, which is cleaved by Bco1 (Miller et al. 2023), makes it very likely that torularhodin also serves this purpose by ingestion in mammals.

The favorable properties of torularhodin, although it is usually mixed with torulene and β-carotene, have led to numerous works that have explored different ways to improve its production, mainly with *R. glutinis*, but also with other species of *Rhodotorula*, or with species of the genera *Sporobolomyces* and *Sporidiobolus* (Kot et al. 2018; Tang et al. 2019; Mussagy et al. 2023). These studies have employed different strains, including mutants with higher production levels, and cultivation conditions, ranging from the use of synthetic media to low-cost substrates or by-products from other industries. Although most of the work has focused on *Rhodotorula*, particularly high yields of torularhodin production have been achieved with other species. A good example is provided by the 350 mg/L of torularhodin obtained after only 2.5 days of culture of *Sporobolomyces ruberrimus* in medium with raw glycerol as the carbon source (Cardoso et al. 2016), at least tenfold higher than the best values obtained with *Rhodotorula* spp.

8.5.4 Other Xanthophylls

Most of the knowledge about xanthophylls in fungi has been centered on the production of astaxanthin, neurosporaxanthin, and torularhodin. Precursors of these xanthophylls, such as hydroxyechinone in the synthesis of astaxanthin, and probably other intermediates (Barredo et al. 2017), and β-apo-4′-carotenal in the synthesis of neurosporaxanthin (Díaz-Sánchez et al. 2011), are not considered here. Some of these compounds may be unstable, as is the case with β-apo-4′-carotenal in *F. fujikuroi*, whose accumulation in the *carD* gene deletion mutant results in its rapid degradation (Díaz-Sánchez et al. 2011). In the case of torularhodin, a side torulene product, β-apo-2′-carotenal, and two intermediate xanthophylls presumably involved in torularhodin synthesis, 16′-hydroxytorulene and torularhodinaldehyde (Fig. 8.4), were found in the red yeast *Cystofilobasidium infirmominiatum*

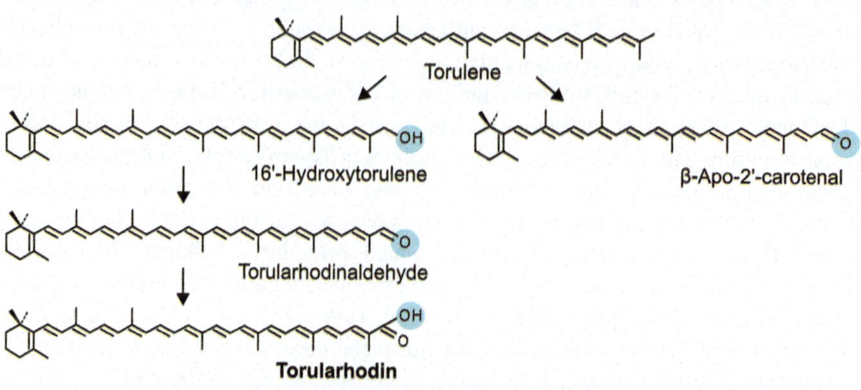

Fig. 8.4 Proposed xanthophylls derived from torulene in the red yeast *C. infirmominiatum* (Herz et al. 2007)

(Herz et al. 2007), but their occurrence has not been confirmed in other torularhodin-producing yeasts.

Mention has already been made of the existence of a lateral branch in the *X. dendrorhous* carotenoid pathway leading to 3-hydroxy-torulene-4-one and CDC (Fig. 8.2), but the possible applied interest of these unusual xanthophylls has not been explored, presumably because of its low concentrations. However, its production could be increased by improving the ability of the pathway to reach the fifth desaturation step. A key point of the astaxanthin pathway is the recognition of γ-carotene as a substrate by two competing enzymes, CrtYB cyclase to produce β-carotene and thus drive the pathway to astaxanthin and CrtI desaturase to produce torulene and drive the pathway to 3-hydroxy-torulene-4-one. The carotenoids found in this yeast indicate that most of the γ-carotene is converted to β-carotene. It would be possible to redirect the pathway to torulene by increasing the efficiency of the desaturase. The fifth desaturation can be carried out with varying efficiencies by desaturases from different fungi. That of *F. fujikuroi* seems to be especially efficient in carrying out this reaction, and it could be a promising strategy to replace the endogenous *crtI* gene with the *Fusarium carB* gene and increase its expression with a potent promoter. At the same time, a decrease in cyclase activity by reducing expression of the *crtYB* gene and/or additional expression of a phytoene synthase enzyme of plant or bacteria origin could further promote this pathway bypass.

Canthaxanthin, a derivative of β-carotene with keto groups on its two rings (Fig. 8.4), is an industrial carotenoid used as an additive by the feed industry (Rebelo et al. 2020). Among other applications, this reddish xanthophyll is widely used in poultry feeding to color egg yolks. In addition, its consumption has beneficial effects on health. Besides its direct antioxidant properties, usual in xanthophylls, positive effects have been described on the immune system, increasing the proliferation and function of immunocompetent cells, or on the response to oxidative stress, inducing catalase and superoxide dismutase activities (Esatbeyoglu and Rimbach 2017). Canthaxanthin is very rare in fungi, and so far, it has been described only in the basidiomycete *Cantharellus cinnabarinus* (Haxo 1950), where it seems to be an exception, since in other *Cantharellus* species β-carotene is the main carotenoid (Mui et al. 1998). However, canthaxanthin is found in many algae and bacteria, and it is also produced chemically (Rebelo et al. 2020). Despite its consolidated applied interest, the industrial production of canthaxanthin in fungi has not been developed except by heterologous expression (see the next section). In addition to canthaxanthin, and disregarding astaxanthin, neurosporaxanthin, and torularhodin, mentioned above, few other xanthophylls have been found in fungi, and none of them have been investigated enough to anticipate possible biotechnological applications. These include plectaniaxanthin, which provides a reddish color to the *Dioszegia* yeast (Madhour et al. 2005), and phyllipsiaxanthin, a lycopene derivative produced by *Phillipsia carminea* (Arpin et al. 1966) (Fig. 8.4). Other examples of unusual fungal xanthophylls can be found in the carotenoid database through its taxonomic section (http://carotenoiddb.jp/ORGANISMS/all_org.html).

8.6 Heterologous Carotenoid Production in Fungal Hosts

Fungi produce only a small minority of the more than 1000 carotenoids identified in nature (Yabuzaki 2017). However, due to the facilities offered by their genetic manipulation and cultivation under industrial conditions, the most manageable ones have been used for the introduction or modification of carotenoid pathways by heterologous expression of genes from other biological origins, frequently from bacteria. Most of these works have been achieved with yeasts but, in some cases, also with filamentous fungi (Naz et al. 2023).

8.6.1 Yeasts

The huge biotechnological development available for *S. cerevisiae*, whose wild strains lack carotenoid biosynthesis, has made this yeast an idoneous system for engineered carotenoid production. Many efforts have also focused on another carotenoid-lacking yeast, *Yarrowia lipolytica*, an oleaginous species particularly suitable for industrial purposes (Park and Ledesma-Amaro 2023). Numerous works describing the engineered production of β-carotene by expression of appropriate heterologous genes in both yeasts have recently been reviewed (Guo et al. 2024). This also led to the use of these yeasts to produce lycopene. From the first report describing β-carotene and lycopene production in *S. cerevisiae* through the heterologous expression of a complete set of bacterial genes (Yamano et al. 1994), different works have devoted increasingly sophisticated efforts to improve lycopene production in this yeast. The most recent combines a multi-modular expression system, downregulation of competitive pathways, and efflux optimization strategy (Huang et al. 2024). Efforts have also been spent to develop lycopene production with *Y. lipolytica*, which started with the heterologous expression of *Pantoea ananatis crtB* and *crtI* genes together with the enhanced expression of endogenous genes for terpenoid synthesis (Matthäus et al. 2014), and later works refined the engineering approaches to improve lycopene yields (Schwartz et al. 2017; Zhang et al. 2019; Luo et al. 2023). Another industrial organism well-suited for heterologous expression is the yeast *Pichia pastoris* (Karbalaei et al. 2020), which has also been engineered to synthesize carotenoids (Araya-Garay et al. 2012a, b; Veiga-Crespo et al. 2018) and has been recently improved for lycopene production (Zhang et al. 2023a).

The engineered β-carotene-producing yeasts are useful platforms for further modifications to produce other carotenoids. To mention some relevant examples, in an attempt to compete with chemical synthesis, canthaxanthin producing *S. cerevisiae* strains were developed by expressing an improved *H. pluvialis* ketolase gene (Chen et al. 2022) or a *Bradyrhizobium* sp. ketolase gene (Promdonkoy et al. 2024), in both cases combined with other genetic manipulations, resulting in high yield productions. Similarly, a β-carotene-producing *S. cerevisiae* strain was modified to produce violaxanthin (Fig. 8.5), a plant epoxidated carotenoid with strong

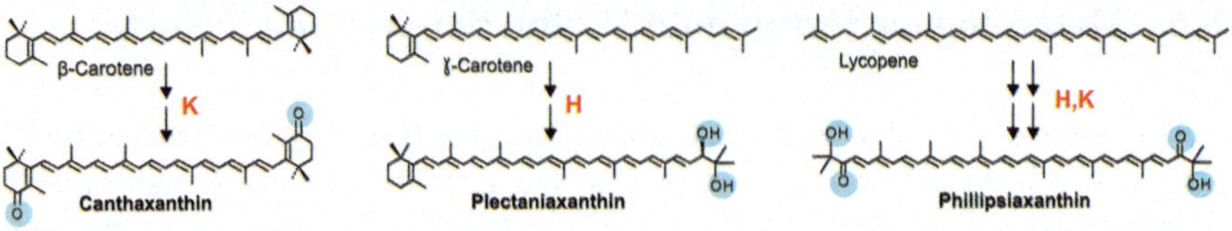

Fig. 8.5 Synthesis of some unusual fungal xanthophylls from their predictable precursors. The reactions leading to these precursors are as indicated in Fig. 8.1. Putative enzyme activities (K, ketolase; H, hydroxylase) are indicated

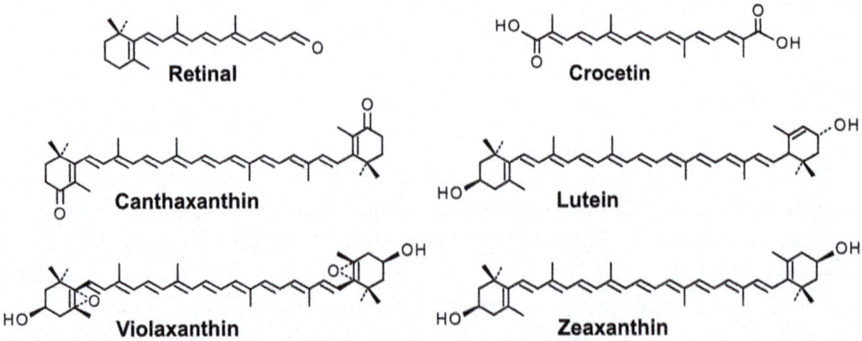

Fig. 8.6 Some carotenoids produced by heterologous gene expression in biotechnological yeasts

antioxidant activity and a key precursor of valuable compounds (Takemura et al. 2021). This was achieved by expressing two bacterial genes, encoding a hydroxylase from *Pantoea ananatis* and a truncated epoxidase from *Haematococcus lacustris*, and further optimizing redox partner systems, which allowed the highest violaxanthin production yield achieved so far with a microorganism (Cataldo et al. 2020). However, a further increase in violaxanthin yields in *S. cerevisiae* was recently obtained by screening of epoxidase alternatives, combined with cytosol redox state engineering, and improving regeneration of nicotinamide adenine dinucleotide phosphate (NADPH) (Wang et al. 2024). As one more example, crocetin (Fig. 8.5), a plant carotenoid from *Crocus* species with applications in medical and food fields (Guo et al. 2021), was produced in *S. cerevisiae* by combinatorial manipulation of CrtZ hydroxylases from nine different species, *Crocus* CCD carotenoid cleaving dioxygenases, and ALD from *Bixa orellana* and *Synechocystis* sp. (Chai et al. 2017).

S. cerevisiae and *Y. lipolytica* have been engineered to synthesize other commercial xanthophylls, such as astaxanthin, lutein, and zeaxanthin (Fig. 8.6). The synthesis of astaxanthin in both yeasts by heterologous gene expression has been recently reviewed (Zhu et al. 2023; Zhou et al. 2024). In the case of *S. cerevisiae*, its genetic versatility has fostered the application of sophisticated molecular approaches to improve the production, such as promoter-terminator shuffling (Cautereels et al. 2024), recombination of heterologous gene modules (Qi et al. 2020), or directed evolution of a fusion enzyme (Ding et al. 2023). Lutein production in *S. cerevisiae* was optimized by coupling the expression of a temperature-sensitive β-cyclase with constitutive expression of a ε-cyclase to direct flux toward α-carotene and by engineering ε-cyclase to localize to the plasma membrane, thus overcoming the limiting step of lycopene cyclization (Bian et al. 2021). Lutein has also been produced in *Y. lipolytica* by using three modules, a lycopene producing module that included the genes *carB* and a mutated *carRP* version from *M. circinelloides*, an α-carotene synthase module consisting of LCY from *Ostreococcus lucimarinus* targeted to subcellular organelles to enhance α-carotene synthesis, and a lutein module consisting of β-hydroxylase LUT5 and ε-hydroxylase LUT1 from *Arabidopsis thaliana*. The combined expression of these modules led to the highest lutein production in a microorganism obtained so far (Qin et al. 2024). Finally, zeaxanthin production was achieved in *S. cerevisiae* by expression of a CrtZ hydroxylase variant from *P. ananatis* (Promdonkoy et al. 2024), with other targeted genetic alterations, and in *Y. lipolytica* by heterologous expression of the *carB*/*carRP* genes from *M. circinelloides*, and the same *P. ananatis* hydroxylase gene, together with other genes engineered to improve MVA metabolism. As astaxanthin, beneficial effects on health of lutein and zeaxanthin consumption are well documented (Mitra et al. 2021; Bouyahya et al. 2021).

β-Carotene production in *S. cerevisiae* has also been exploited for its cleavage to produce retinoids by expressing additional heterologous genes and introducing other modifications. Retinol (Fig. 8.5) (vitamin A) plays a key role in vision (Martin Ask et al. 2021) and is widely used in the food and pharmaceutical industries. Retinal production was achieved in *S. cerevisiae* through the expression of the human β-carotene 15,15′-dioxygenase gene (BCMO), which symmetrically cleaves β-carotene, in a xylose-fermenting strain (Sun et al. 2019). Subsequently, selective production of retinol was obtained by adding genes for human retinal dehydrogenase RDH12 and *Lactococcus lactis* NADH oxidase NoxE, which significantly reduced glycerol accumulation and increased retinol yield (Lee et al. 2022). This engineered strain produced retinol exclusively, and optimal production conditions were identified to address retinol instability, providing an efficient microbial alternative to chemical synthesis. Lately, a new de novo retinol synthesis pathway has been developed that has resulted in a highly efficient retinol-producing strain by systematic metabolic engineering, transporter engineering, semi-rational enzyme transformation, and optimization of fermentation conditions (Wang et al. 2023).

The astaxanthin-producing yeast *X. dendrorhous* has also been used as a platform for heterologous carotenoid production (Sandmann et al. 2021). A β-carotene-producing strain of this yeast, blocked at the *asy* gene, was engineered to produce zeaxanthin by expressing the bacterial *crtZ* hydroxylase genes from *P. ananatis* or *Brevundimonas* sp. and overexpressing three genes for limiting steps, *hmgR*, *crtE*, and *crtYB*, to increase the final yield (Pollmann et al. 2016). Zeaxanthin yield was considerably increased combining four strategies: obtaining a β-carotene overproducer by standard mutagenesis, introducing a high copy number of the *crtZ* hydroxylase gene, enhancing phytoene synthase activity, and substituting in the culture medium glucose with a xylose-rich hydrolysate of the waste product hemicellulose (Breitenbach et al. 2019).

8.6.2 Filamentous Fungi

Some filamentous fungi have also been used as biological platforms for engineered carotenoid production. Expression of *crtW* ketolase and *crtZ* hydroxylase genes from *Agrobacterium aurantiacum* in *M. circinelloides* led to the production of β-cryptoxanthin, echinenone, canthaxanthin, and a small amount of astaxanthin (Papp et al. 2006), and the yields of these carotenoids were improved by expressing the *crtR* and *crtS* genes from *X. dendrorhous* instead (Csernetics et al. 2015). In addition, expression in *M. circinelloides* of only the *crtS X. dendrorhous* gene allowed obtaining β-cryptoxanthin- and zeaxanthin-producing transformants (Rodríguez-Sáiz et al. 2012). Canthaxanthin production in this fungus was enhanced by expression of the *crtW* ketolase gene from the bacterium *Paracoccus* sp. (Papp et al. 2013), and canthaxanthin yield was further improved by overexpression of β-carotene ketolase gene *bkt* from *Haematococcus pluvialis* combined with disruption of the *crgA* gene, which encodes a negative regulator of carotenoid synthesis (Naz et al. 2021).

Trichoderma reesei, a fungus lacking carotenoids and traditionally used as a cell factory for cellulase production, has been also engineered to produce β-carotene by overexpressing *M. circinelloides carRP* and *carB* genes and *X. dendrorhous crtE* gene. The production was optimized by multicopy integrations and the overexpression of endogenous enzyme genes *hmg1* and *ggs1* (Li et al. 2023a). In the dimorphic fungus *Ustilago maydis*, phytoene synthase function was restored in a strain lacking the *car2* gene, which encodes the bifunctional phytoene synthase/cyclase, by heterologous expression of *Agrobacterium aurantiacum* phytoene synthase *crtB* gene. In this case, the purpose was to accumulate lycopene instead of β-carotene as a more effective visual marker of mevalonate pathway activity (Lee et al. 2020).

8.7 Conclusions

Fungi offer alternative and efficient sources to obtain widely used industrial carotenoids, such as β-carotene, lycopene, and astaxanthin. They do not produce as wide a range of carotenoids as bacteria or photosynthetic eukaryotes, but their biosynthetic pathways are simpler and therefore more easily directed toward a certain product of interest, as well as offering important advantages in terms of handling and cultivation. On the other hand, two carboxylic xanthophylls, neurosporaxanthin and torularhodin, although not yet used commercially, show very promising properties for future use in human food and animal feed or in other commercial areas, such as cosmetics. In addition, the advances obtained in the development of strains producing high amounts of different carotenes and xanthophylls with industrial fungi, such as *S. cerevisiae* or *Y. lipolytica*, constitute a powerful biotechnology platform that is expected to be increasingly exploited in the future. While the applied potential of most carotenoids found in nature remains to be investigated, unusual xanthophylls found in some fungi remind us that the arsenal of carotenoids in this heterogeneous taxonomic kingdom may yet yield the discovery of compounds with unique, and presumably beneficial or healthy, properties. The large number of publications on advances in the production of carotenoids by fungi, of which only a selection is cited in this review, indicates that the market for fungal carotenoids is expanding and should bring more wealth to the natural resources exploited for human consumption.

References

Aasen AJ, Jensen SL (1965) Fungal carotenoids II. The structure of the carotenoid acid neurosporaxanthin. Acta Chem Scand 19:1843–1853. https://doi.org/10.3891/acta.chem.scand.19-1843

Adrio JL, Demain AL (2003) Fungal biotechnology. Int Microbiol Off J Span Soc Microbiol 6:191–199. https://doi.org/10.1007/s10123-003-0133-0

Al-Babili S, Bouwmeester HJ (2015) Strigolactones, a novel carotenoid-derived plant hormone. Annu Rev Plant Biol 66:161–186. https://doi.org/10.1146/annurev-arplant-043014-114759

Alcaíno J, Barahona S, Carmona M et al (2008) Cloning of the cytochrome p450 reductase (*crtR*) gene and its involvement in the astaxanthin biosynthesis of *Xanthophyllomyces dendrorhous*. BMC Microbiol 8:169. https://doi.org/10.1186/1471-2180-8-169

Alcaíno J, Fuentealba M, Cabrera R et al (2012) Modeling the interfacial interactions between CrtS and CrtR from *Xanthophyllomyces dendrorhous*, a P450 system involved in astaxanthin production. J Agric Food Chem 60:8640–8647. https://doi.org/10.1021/jf302287f

Almeida ER, Cerdá-Olmedo E (2008) Gene expression in the regulation of carotene biosynthesis in *Phycomyces*. Curr Genet 53:129–137. https://doi.org/10.1007/s00294-007-0170-x

Ambati RR, Phang SM, Ravi S, Aswathanarayana RG (2014) Astaxanthin: sources, extraction, stability, biological activities and its commercial applications–a review. Mar Drugs 12:128–152. https://doi.org/10.3390/md12010128

An GH, Schuman DB, Johnson EA (1989) Isolation of *Phaffia rhodozyma* mutants with increased astaxanthin content. Appl Environ Microbiol 55:116–124. https://doi.org/10.1128/aem.55.1.116-124.1989

An GH, Bielich J, Auerbach R, Johnson EA (1991) Isolation and characterization of carotenoid hyperproducing mutants of yeast by flow cytometry and cell sorting. Biotechnology 9:70–73. https://doi.org/10.1038/nbt0191-70

An GH, Cho MH, Johnson EA (1999) Monocyclic carotenoid biosynthetic pathway in the yeast *Phaffia rhodozyma* (*Xanthophyllomyces dendrorhous*). J Biosci Bioeng 88:189–193. https://doi.org/10.1016/S1389-1723(99)80200-X

Anand R, Mohan L, Bharadvaja N (2022) Disease prevention and treatment using β-carotene: the ultimate provitamin A. Rev Bras Farmacogn 32:491–501. https://doi.org/10.1007/s43450-022-00262-w

Andrewes AG, Starr MP (1976) (3R,3′R)-astaxanthin from the yeast *Phaffia rhodozyma*. Phytochemistry 15:1009–1011. https://doi.org/10.1016/S0031-9422(00)84391-5

Aoki T, O'Donnell K, Geiser DM (2014) Systematics of key phytopathogenic *Fusarium* species: current status and future challenges. J Gen Plant Pathol 80:189–201. https://doi.org/10.1007/s10327-014-0509-3

Aragón CM, Murillo FJ, de la Guardia MD, Cerdá-Olmedo E (1976) An enzyme complex for the dehydrogenation of phytoene in *Phycomyces*. Eur J Biochem FEBS 63:71–75. https://doi.org/10.1111/j.1432-1033.1976.tb10208.x

Araya-Garay JM, Ageitos JM, Vallejo JA et al (2012a) Construction of a novel *Pichia pastoris* strain for production of xanthophylls. AMB Express 2:24. https://doi.org/10.1186/2191-0855-2-24

Araya-Garay JM, Feijoo-Siota L, Rosa-dos-Santos F et al (2012b) Construction of new *Pichia pastoris* X-33 strains for production of lycopene and β-carotene. Appl Microbiol Biotechnol 93:2483–2492. https://doi.org/10.1007/s00253-011-3764-7

Arpin N, Lebreton P, Fiasson JL (1966) Chemotaxonomic research on fungi. II. Carotenoids of *Peniophora aurantiaca* (Bres.) (Basidiomycete). Bull Société Mycol Fr 82:450–459

Arrach N, Fernández-Martín R, Cerdá-Olmedo E, Avalos J (2001) A single gene for lycopene cyclase, phytoene synthase, and regulation of carotene biosynthesis in *Phycomyces*. Proc Natl Acad Sci U S A 98:1687–1692. https://doi.org/10.1073/pnas.98.4.1687

Arrach N, Schmidhauser TJ, Avalos J (2002) Mutants of the carotene cyclase domain of *al-2* from *Neurospora crassa*. Mol Genet Genomics 266:914–921. https://doi.org/10.1007/s00438-001-0626-5

Avalos J, Cerdá-Olmedo E (1987) Carotenoid mutants of *Gibberella fujikuroi*. Curr Genet 11:505–511. https://doi.org/10.1007/BF00384613

Avalos J, Cerdá-Olmedo E (2004) Fungal carotenoid production. In: Arora DK (ed) Handbook of fungal biotechnology, 2nd edn. Marcel Dekker, New York, pp 367–378

Avalos J, Corrochano LM (2013) Carotenoid biosynthesis in *Neurospora*. In: Kasbekar DP, McCluskey K (eds) *Neurospora*: genomics and molecular biology. Caister Academic Press, Norfolk, pp 227–241

Avalos J, Limón MC (2015) Biological roles of fungal carotenoids. Curr Genet 61:309–324. https://doi.org/10.1007/s00294-014-0454-x

Avalos J, Limón MC (2022) Fungal secondary metabolism. Encyclopedia 2:1–13. https://doi.org/10.3390/encyclopedia2010001

Avalos J, Nordzieke S, Parra O et al (2017a) Carotenoid production by filamentous fungi and yeasts. In: Sibirny AA (ed) Biotechnology of yeasts and filamentous fungi. Springer, Cham, pp 225–279. https://doi.org/10.1007/978-3-319-58829-2_8

Avalos J, Pardo-Medina J, Parra-Rivero O et al (2017b) Carotenoid biosynthesis in *Fusarium*. J Fungi 3(39):1–16. https://doi.org/10.3390/jof3030039

Barbachano-Torres A, Castelblanco-Matiz LM, Ramos-Valdivia AC et al (2014) Analysis of proteomic changes in colored mutants of *Xanthophyllomyces dendrorhous* (*Phaffia rhodozyma*). Arch Microbiol 196:411–421. https://doi.org/10.1007/s00203-014-0979-x

Barredo JL, García-Estrada C, Kosalkova K, Barreiro C (2017) Biosynthesis of astaxanthin as a main carotenoid in the heterobasidiomycetous yeast *Xanthophyllomyces dendrorhous*. J Fungi Basel Switz 3:44. https://doi.org/10.3390/jof3030044

Bian Q, Zhou P, Yao Z et al (2021) Heterologous biosynthesis of lutein in *S. cerevisiae* enabled by temporospatial pathway control. Metab Eng 67:19–28. https://doi.org/10.1016/j.ymben.2021.05.008

Bieszke JA, Spudich EN, Scott KL et al (1999) A eukaryotic protein, NOP-1, binds retinal to form an archaeal rhodopsin-like photochemically reactive pigment. Biochemistry 38:14138–14145. https://doi.org/10.1021/bi9916170

Bindl E, Lang W, Rau W (1970) Untersuchungen über die lichtabhängige Carotinoidsynthese. VI. Zeitlicher Verlauf der Synthese der einzelnen Carotinoide bei *Fusarium aquaeductuum* unter verschiedenen Induktionsbedingungen. Planta 94:156–174. https://doi.org/10.1007/BF00384771

Bo S, Ni X, Guo J et al (2022) Carotenoid biosynthesis: genome-wide profiling, pathway identification in *Rhodotorula glutinis* X-20, and high-level production. Front Nutr 9:918240. https://doi.org/10.3389/fnut.2022.918240

Bouyahya A, El Omari N, Hakkur M et al (2021) Sources, health benefits, and biological properties of zeaxanthin. Trends Food Sci Technol 118:519–538. https://doi.org/10.1016/j.tifs.2021.10.017

Bramley PM (1994) Carotenoid biosynthesis: a target site for bleaching herbicides. Biochem Soc Trans 22:625–629. https://doi.org/10.1042/bst0220625

Breitenbach J, Fraser PD, Sandmann G (2012) Carotenoid synthesis and phytoene synthase activity during mating of *Blakeslea trispora*. Phytochemistry 76:40–45. https://doi.org/10.1016/j.phytochem.2011.12.017

Breitenbach J, Pollmann H, Sandmann G (2019) Genetic modification of the carotenoid pathway in the red yeast *Xanthophyllomyces dendrorhous*: Engineering of a high-yield zeaxanthin strain. J Biotechnol 289:112–117. https://doi.org/10.1016/j.jbiotec.2018.11.019

Buzzini P, Innocenti M, Turchetti B et al (2007) Carotenoid profiles of yeasts belonging to the genera *Rhodotorula*, *Rhodosporidium*, *Sporobolomyces*, and *Sporidiobolus*. Can J Microbiol 53:1024–1031. https://doi.org/10.1139/W07-068

Cardoso L, Jäckel S, Karp SG et al (2016) Improvement of *Sporobolomyces ruberrimus* carotenoids production by the use of raw glycerol. Bioresour Technol 200:374–379. https://doi.org/10.1016/j.biortech.2015.09.108

Castrillo M, Avalos J (2015) The flavoproteins CryD and VvdA cooperate with the white collar protein WcoA in the control of photocarotenogenesis in *Fusarium fujikuroi*. PLoS One 10:e0119785. https://doi.org/10.1371/journal.pone.0119785

Cataldo VF, Arenas N, Salgado V et al (2020) Heterologous production of the epoxycarotenoid violaxanthin in *Saccharomyces cerevisiae*. Metab Eng 59:53–63. https://doi.org/10.1016/j.ymben.2020.01.006

Cautereels C, Smets J, Bircham P et al (2024) Combinatorial optimization of gene expression through recombinase-mediated promoter and terminator shuffling in yeast. Nat Commun 15:1112. https://doi.org/10.1038/s41467-024-44997-7

Cerdá-Olmedo E (1987) Carotene. In: Cerdá-Olmedo E, Lipson ED (eds) Phycomyces. Cold Spring Harbor Laboratory Press, Cold Spring Harbor, pp 199–222

Chai F, Wang Y, Mei X et al (2017) Heterologous biosynthesis and manipulation of crocetin in *Saccharomyces cerevisiae*. Microb Cell Fact 16:54. https://doi.org/10.1186/s12934-017-0665-1

Chen M, Li M, Ye L, Yu H (2022) Construction of canthaxanthin-producing yeast by combining spatiotemporal regulation and pleiotropic drug resistance engineering. ACS Synth Biol 11:325–333. https://doi.org/10.1021/acssynbio.1c00437

Chumpolkulwong N, Kakizono T, Nagai S, Nishio N (1997) Increased astaxanthin production by *Phaffia rhodozyma* mutants isolated as resistant to diphenylamine. J Ferment Bioeng 83:429–434. https://doi.org/10.1016/S0922-338X(97)82996-0

Corrochano LM, Avalos J (2010) Light sensing. In: Borkovich K, Ebbole D, Momany M (eds) Cellular and molecular biology of filamentous fungi. ASM, Washington, pp 417–441

Csernetics Á, Tóth E, Farkas A et al (2015) Expression of *Xanthophyllomyces dendrorhous* cytochrome-P450 hydroxylase and reductase in *Mucor circinelloides*. World J Microbiol Biotechnol 31:321–336. https://doi.org/10.1007/s11274-014-1784-z

De la Guardia MD, Aragón CM, Murillo FJ, Cerdá-Olmedo E (1971) A carotenogenic enzyme aggregate in *Phycomyces*: evidence from quantitive complementation. Proc Natl Acad Sci U S A 68:2012–2015. https://doi.org/10.1073/pnas.68.9.2012

Díaz-Sánchez V, Estrada AF, Trautmann D et al (2011) The gene *carD* encodes the aldehyde dehydrogenase responsible for neurosporaxanthin biosynthesis in *Fusarium fujikuroi*. FEBS J 278:3164–3176. https://doi.org/10.1111/j.1742-4658.2011.08242.x

Díaz-Sánchez V, Estrada AF, Limón MC et al (2013) The oxygenase CAO-1 of *Neurospora crassa* is a resveratrol cleavage enzyme. Eukaryot Cell 12:1305–1314. https://doi.org/10.1128/EC.00084-13

Di Mascio P, Kaiser S, Sies H (1989) Lycopene as the most efficient biological carotenoid singlet oxygen quencher. Arch Biochem Biophys 274:532–538. https://doi.org/10.1016/0003-9861(89)90467-0

Dimitrova S, Pavlova K, Lukanov L et al (2013) Production of metabolites with antioxidant and emulsifying properties by Antarctic strain *Sporobolomyces salmonicolor* AL_1. Appl Biochem Biotechnol 169:301–311. https://doi.org/10.1007/s12010-012-9983-2

Ding Y-W, Lu C-Z, Zheng Y et al (2023) Directed evolution of the fusion enzyme for improving astaxanthin biosynthesis in *Saccharomyces cerevisiae*. Synth Syst Biotechnol 8:46–53. https://doi.org/10.1016/j.synbio.2022.10.005

Domonkos I, Kis M, Gombos Z, Ughy B (2013) Carotenoids, versatile components of oxygenic photosynthesis. Prog Lipid Res 52:539–561. https://doi.org/10.1016/j.plipres.2013.07.001

Du C, Li Y, Guo Y et al (2016) The suppression of torulene and torularhodin treatment on the growth of PC-3 xenograft prostate tumors. Biochem Biophys Res Commun 469:1146–1152. https://doi.org/10.1016/j.bbrc.2015.12.112

Eggersdorfer M, Wyss A (2018) Carotenoids in human nutrition and health. Arch Biochem Biophys 652:18–26. https://doi.org/10.1016/j.abb.2018.06.001

Esatbeyoglu T, Rimbach G (2017) Canthaxanthin: from molecule to function. Mol Nutr Food Res 61. https://doi.org/10.1002/mnfr.201600469

Estrada AF, Maier D, Scherzinger D et al (2008a) Novel apocarotenoid intermediates in *Neurospora crassa* mutants imply a new biosynthetic reaction sequence leading to neurosporaxanthin formation. Fungal Genet Biol 45:1497–1505. https://doi.org/10.1016/j.fgb.2008.09.001

Estrada AF, Youssar L, Scherzinger D et al (2008b) The *ylo-1* gene encodes an aldehyde dehydrogenase responsible for the last reaction in the *Neurospora* carotenoid pathway. Mol Microbiol 69:1207–1220. https://doi.org/10.1111/j.1365-2958.2008.06349.x

Fernández-Martín R, Cerdá-Olmedo E, Avalos J (2000) Homologous recombination and allele replacement in transformants of *Fusarium fujikuroi*. Mol Gen Genet 263:838–845. https://doi.org/10.1007/s004380000249

Gassel S, Schewe H, Schmidt I et al (2013) Multiple improvement of astaxanthin biosynthesis in *Xanthophyllomyces dendrorhous* by a combination of conventional mutagenesis and metabolic pathway engineering. Biotechnol Lett 35:565–569. https://doi.org/10.1007/s10529-012-1103-4

Goldie AH, Subden RE (1973) The neutral carotenoids of wild-type and mutant strains of *Neurospora crassa*. Biochem Genet 10:275–284. https://doi.org/10.1007/BF00485705

Govind NS, Cerdá-Olmedo E (1986) Sexual activation of carotenogenesis in *Phycomyces blakesleeanus*. J Gen Microbiol 132:2775–2780. https://doi.org/10.1099/00221287-132-10-2775

Gribanovski-Sassu O, Foppen FH (1968) Light and temperature effect on *Epicoccum nigrum*. Phytochemistry 7:1605–1612. https://doi.org/10.1016/S0031-9422(00)88613-6

Grune T, Lietz G, Palou A et al (2010) β-carotene is an important vitamin A source for humans. J Nutr 140:2268S–2285S. https://doi.org/10.3945/jn.109.119024

Guo Z-L, Li M-X, Li X-L et al (2021) Crocetin: a systematic review. Front Pharmacol 12:745683. https://doi.org/10.3389/fphar.2021.745683

Guo Q, Peng Q-Q, Li Y-W et al (2024) Advances in the metabolic engineering of *Saccharomyces cerevisiae* and *Yarrowia lipolytica* for the production of β-carotene. Crit Rev Biotechnol 44:337–351. https://doi.org/10.1080/07388551.2023.2166809

Hara KY, Morita T, Endo Y et al (2014a) Evaluation and screening of efficient promoters to improve astaxanthin production in *Xanthophyllomyces dendrorhous*. Appl Microbiol Biotechnol 98:6787–6793. https://doi.org/10.1007/s00253-014-5727-2

Hara KY, Morita T, Mochizuki M et al (2014b) Development of a multi-gene expression system in *Xanthophyllomyces dendrorhous*. Microb Cell Factories 13:175. https://doi.org/10.1186/s12934-014-0175-3

Hausmann A, Sandmann G (2000) A single five-step desaturase is involved in the carotenoid biosynthesis pathway to β-carotene and torulene in *Neurospora crassa*. Fungal Genet Biol 30:147–153. https://doi.org/10.1006/fgbi.2000.1212

Haxo F (1950) Carotenoids of the mushroom *Cantharellus cinnabarinus*. Bot Gaz 112:228–232. https://doi.org/10.1086/335653

Herz S, Weber RW, Anke H et al (2007) Intermediates in the oxidative pathway from torulene to torularhodin in the red yeasts *Cystofilobasidium infirmominiatum* and *C. capitatum* (Heterobasidiomycetes, Fungi). Phytochemistry 68:2503–2511. https://doi.org/10.1016/j.phytochem.2007.05.019

Hornero-Méndez D, Limón MC, Avalos J (2018) HPLC analysis of carotenoids in neurosporaxanthin-producing fungi. In: Barreiro C, Barredo J-L (eds) Microbial carotenoids: methods and protocols. Springer, New York, pp 269–281. https://doi.org/10.1007/978-1-4939-8742-9_16

Huang G, Li J, Lin J et al (2024) Multi-modular metabolic engineering and efflux engineering for enhanced lycopene production in recombinant *Saccharomyces cerevisiae*. J Ind Microbiol Biotechnol 51:kuae015. https://doi.org/10.1093/jimb/kuae015

Hussein G, Sankawa U, Goto H et al (2006) Astaxanthin, a carotenoid with potential in human health and nutrition. J Nat Prod 69:443–449. https://doi.org/10.1021/np050354+

Ito M, Yamano Y, Tode C, Wada A (2009) Carotenoid synthesis: retrospect and recent progress. Arch Biochem Biophys 483:224–228. https://doi.org/10.1016/j.abb.2008.11.021

Jeon Y-J, Myung S-K, Lee E-H et al (2011) Effects of beta-carotene supplements on cancer prevention: meta-analysis of randomized controlled trials. Nutr Cancer 63:1196–1207. https://doi.org/10.1080/01635581.2011.607541

Karbalaei M, Rezaee SA, Farsiani H (2020) *Pichia pastoris*: A highly successful expression system for optimal synthesis of heterologous proteins. J Cell Physiol 235:5867–5881. https://doi.org/10.1002/jcp.29583

Khan UM, Sevindik M, Zarrabi A et al (2021) Lycopene: food sources, biological activities, and human health benefits. Oxid Med Cell Longev 2021:2713511. https://doi.org/10.1155/2021/2713511

Khoo HE, Ng HS, Yap W-S et al (2019) Nutrients for prevention of macular degeneration and eye-related diseases. Antioxid Basel Switz 8. https://doi.org/10.3390/antiox8040085

Kot AM, Błażejak S, Gientka I et al (2018) Torulene and torularhodin: "new" fungal carotenoids for industry? Microb Cell Factories 17:49. https://doi.org/10.1186/s12934-018-0893-z

Kuzina V, Cerdá-Olmedo E (2006) Modification of sexual development and carotene production by acetate and other small carboxylic acids in *Blakeslea trispora* and *Phycomyces blakesleeanus*. Appl Environ Microbiol 72:4917–4922. https://doi.org/10.1128/AEM.02845-05

Lamers PP, Janssen M, De Vos RCH et al (2008) Exploring and exploiting carotenoid accumulation in *Dunaliella salina* for cell-factory applications. Trends Biotechnol 26:631–638. https://doi.org/10.1016/j.tibtech.2008.07.002

Latha BV, Jeevaratanm K (2012) Thirteen-week oral toxicity study of carotenoid pigment from *Rhodotorula glutinis* DFR-PDY in rats. Indian J Exp Biol 50:645–651

Lee J-S, Park S-A, Chung D, Lee HG (2011) Encapsulation of astaxanthin-rich *Xanthophyllomyces dendrorhous* for antioxidant delivery. Int J Biol Macromol 49:268–273. https://doi.org/10.1016/j.ijbiomac.2011.04.021

Lee J, Hilgers F, Loeschke A et al (2020) *Ustilago maydis* serves as a novel production host for the synthesis of plant and fungal sesquiterpenoids. Front Microbiol 11:1655. https://doi.org/10.3389/fmicb.2020.01655

Lee Y-G, Kim C, Sun L et al (2022) Selective production of retinol by engineered *Saccharomyces cerevisiae* through the expression of retinol dehydrogenase. Biotechnol Bioeng 119:399–410. https://doi.org/10.1002/bit.28004

Li C, Schmidhauser TJ (1995) Developmental and photoregulation of *al-1* and *al-2*, structural genes for two enzymes essential for carotenoid biosynthesis in *Neurospora*. Dev Biol 169:90–95. https://doi.org/10.1006/dbio.1995.1129

Li C, Zhang N, Song J et al (2016) A single desaturase gene from red yeast *Sporidiobolus pararoseus* is responsible for both four- and five-step dehydrogenation of phytoene. Gene 590:169–176. https://doi.org/10.1016/j.gene.2016.06.042

Li C, Zhang N, Li B et al (2017) Increased torulene accumulation in red yeast *Sporidiobolus pararoseus* NGR as stress response to high salt conditions. Food Chem 237:1041–1047. https://doi.org/10.1016/j.foodchem.2017.06.033

Li J, Chen Y, Gao A et al (2023a) Simultaneous production of cellulase and β-carotene in the filamentous fungus *Trichoderma reesei*. J Agric Food Chem 71:6358–6365. https://doi.org/10.1021/acs.jafc.3c00690

Li X, Holt RR, Keen CL et al (2023b) Potential roles of dietary zeaxanthin and lutein in macular health and function. Nutr Rev 81:670–683. https://doi.org/10.1093/nutrit/nuac076

Linden H, Rodríguez-Franco M, Macino G (1997) Mutants of *Neurospora crassa* defective in regulation of blue light perception. Mol Gen Genet 254:111–118. https://doi.org/10.1007/s004380050398

Liu Z-Q, Zhang J-F, Zheng Y-G, Shen Y-C (2008) Improvement of astaxanthin production by a newly isolated *Phaffia rhodozyma* mutant with low-energy ion beam implantation. J Appl Microbiol 104:861–872. https://doi.org/10.1111/j.1365-2672.2007.03603.x

Long Y, Paengkoum S, Lu S et al (2024) Physicochemical properties, mechanism of action of lycopene and its application in poultry and ruminant production. Front Vet Sci 11:1364589. https://doi.org/10.3389/fvets.2024.1364589

López-Nieto MJ, Costa J, Peiro E et al (2004) Biotechnological lycopene production by mated fermentation of *Blakeslea trispora*. Appl Microbiol Biotechnol 66:153–159. https://doi.org/10.1007/s00253-004-1669-4

Luo Z, Shi J-T, Chen X-L et al (2023) Iterative gene integration mediated by 26S rDNA and non-homologous end joining for the efficient production of lycopene in *Yarrowia lipolytica*. Bioresour Bioprocess 10:83. https://doi.org/10.1186/s40643-023-00697-6

Luque EM, Gutiérrez G, Navarro-Sampedro L et al (2012) A relationship between carotenoid accumulation and the distribution of species of the fungus *Neurospora* in Spain. PLoS One 7:e33658. https://doi.org/10.1371/journal.pone.0033658. PONE-D-11-14514 [pii]

Luzardo-Ocampo I, Ramírez-Jiménez AK, Yañez J et al (2021) Technological applications of natural colorants in food systems: a review. Foods Basel Switz 10:634. https://doi.org/10.3390/foods10030634

Madhour A, Anke H, Mucci A et al (2005) Biosynthesis of the xanthophyll plectaniaxanthin as a stress response in the red yeast *Dioszegia* (Tremellales, Heterobasidiomycetes, Fungi). Phytochemistry 66:2617–2626. https://doi.org/10.1016/j.phytochem.2005.09.010

Maoka T (2020) Carotenoids as natural functional pigments. J Nat Med 74:1–16. https://doi.org/10.1007/s11418-019-01364-x

Mapelli-Brahm P, Corte-Real J, Meléndez-Martínez AJ, Bohn T (2017) Bioaccessibility of phytoene and phytofluene is superior to other carotenoids from selected fruit and vegetable juices. Food Chem 229:304–311. https://doi.org/10.1016/j.foodchem.2017.02.074

Martín JF, Gudiña E, Barredo JL (2008) Conversion of β-carotene into astaxanthin: Two separate enzymes or a bifunctional hydroxylase-ketolase protein? Microb Cell Factor 7:3. https://doi.org/10.1186/1475-2859-7-3

Martin Ask N, Leung M, Radhakrishnan R, Lobo GP (2021) Vitamin A transporters in visual function: a mini review on membrane receptors for dietary vitamin A uptake, storage, and transport to the eye. Nutrients 13:3987. https://doi.org/10.3390/nu13113987

Martínez-Cámara S, Ibañez A, Rubio S et al (2021) Main carotenoids produced by microorganisms. Encyclopedia 1:1223–1245. https://doi.org/10.3390/encyclopedia1040093

Matthäus F, Ketelhot M, Gatter M, Barth G (2014) Production of lycopene in the non-carotenoid-producing yeast *Yarrowia lipolytica*. Appl Environ Microbiol 80:1660–1669. https://doi.org/10.1128/AEM.03167-13

Mehta BJ, Cerdá-Olmedo E (1995) Mutants of carotene production in *Blakeslea trispora*. Appl Microbiol Biotechnol 42:836–838. https://doi.org/10.1007/BF00191178

Mehta BJ, Cerdá-Olmedo E (1999) Lycopene cyclization in *Blakeslea trispora*. Mycoscience 40:307–310. https://doi.org/10.1007/BF02463967

Mehta BJ, Salgado LM, Bejarano ER, Cerdá-Olmedo E (1997) New mutants of *Phycomyces blakesleeanus* for β-carotene production. Appl Environ Microbiol 63:3657–3661. https://doi.org/10.1128/aem.63.9.3657-3661.1997

Mehta BJ, Obraztsova IN, Cerdá-Olmedo E (2003) Mutants and intersexual heterokaryons of *Blakeslea trispora* for production of β-carotene and lycopene. Appl Environ Microbiol 69:4043–4048. https://doi.org/10.1128/AEM.69.7.4043-4048.2003

Meléndez-Martínez AJ (2019) An overview of carotenoids, apocarotenoids, and vitamin A in agro-food, nutrition, health, and disease. Mol Nutr Food Res 63:e1801045. https://doi.org/10.1002/mnfr.201801045

Meléndez-Martínez AJ, Mapelli-Brahm P, Stinco CM (2018) The colourless carotenoids phytoene and phytofluene: From dietary sources to their usefulness for the functional foods and nutricosmetics industries. J Food Compos Anal 67:91–103. https://doi.org/10.1016/j.jfca.2018.01.002

Mende K, Homann V, Tudzynski B (1997) The geranylgeranyl diphosphate synthase gene of *Gibberella fujikuroi*: isolation and expression. Mol Gen Genet 255:96–105. https://doi.org/10.1007/s004380050477

Milani A, Basirnejad M, Shahbazi S, Bolhassani A (2017) Carotenoids: biochemistry, pharmacology and treatment. Br J Pharmacol 174:1290–1324. https://doi.org/10.1111/bph.13625

Miller AP, Hornero-Méndez D, Bandara S et al (2023) Bioavailability and provitamin A activity of neurosporaxanthin in mice. Commun Biol 6:1068. https://doi.org/10.1038/s42003-023-05446-1

Mitra S, Rauf A, Tareq AM et al (2021) Potential health benefits of carotenoid lutein: An updated review. Food Chem Toxicol Int J Publ Br Ind Biol Res Assoc 154:112328. https://doi.org/10.1016/j.fct.2021.112328

Moretti VM, Mentasti T, Bellagamba F et al (2006) Determination of astaxanthin stereoisomers and colour attributes in flesh of rainbow trout (*Oncorhynchus mykiss*) as a tool to distinguish the dietary pigmentation source. Food Addit Contam 23:1056–1063. https://doi.org/10.1080/02652030600838399

Mui D, Feibelman T, Bennett JW (1998) A preliminary study of the carotenoids of some north American species of *Cantharellus*. Int J Plant Sci 159:244–248. https://doi.org/10.1086/297545

Murillo FJ, Calderón IL, López-Díaz I, Cerdá-Olmedo E (1978) Carotene-superproducing strains of *Phycomyces*. Appl Environ Microbiol 36:639–642. https://doi.org/10.1128/aem.36.5.639-642.1978

Mussagy CU, Gonzalez-Miquel M, Santos-Ebinuma VC, Pereira JFB (2023) Microbial torularhodin—a comprehensive review. Crit Rev Biotechnol 43:540–558. https://doi.org/10.1080/07388551.2022.2041540

Nambara E, Marion-Poll A (2005) Abscisic acid biosynthesis and catabolism. Annu Rev Plant Biol 56:165–185. https://doi.org/10.1146/annurev.arplant.56.032604.144046

Navarro E, Sandmann G, Torres-Martínez S (1995) Mutants of the carotenoid biosynthetic pathway of *Mucor circinelloides*. Exp Mycol 19:186–190. https://doi.org/10.1006/emyc.1995.1023

Navarro E, Lorca-Pascual JM, Quiles-Rosillo MD et al (2001) A negative regulator of light-inducible carotenogenesis in *Mucor circinelloides*. Mol Genet Genomics 266:463–470. https://doi.org/10.1007/s004380100558

Naz T, Yang J, Nosheen S et al (2021) Genetic modification of *Mucor circinelloides* for canthaxanthin production by heterologous expression of β-carotene ketolase gene. Front Nutr 8:756218. https://doi.org/10.3389/fnut.2021.756218

Naz T, Ullah S, Nazir Y et al (2023) Industrially important fungal carotenoids: advancements in biotechnological production and extraction. J Fungi Basel Switz 9:578. https://doi.org/10.3390/jof9050578

Nelson MA, Morelli G, Carattoli A et al (1989) Molecular cloning of a *Neurospora crassa* carotenoid biosynthetic gene (albino-3) regulated by blue light and the products of the white collar genes. Mol Cell Biol 9:1271–1276. https://doi.org/10.1128/mcb.9.3.1271-1276.1989

Obraztsova IN, Prados N, Holzmann K et al (2004) Genetic damage following introduction of DNA in *Phycomyces*. Fungal Genet Biol 41:168–180. https://doi.org/10.1016/j.fgb.2003.09.007

Ojima K, Breitenbach J, Visser H et al (2006) Cloning of the astaxanthin synthase gene from *Xanthophyllomyces dendrorhous* (*Phaffia rhodozyma*) and its assignment as a β-carotene 3-hydroxylase/4-ketolase. Mol Genet Genomics 275:148–158. https://doi.org/10.1007/s00438-005-0072-x

Olmedilla-Alonso B, Benítez-González AM, Estévez-Santiago R et al (2021) Assessment of food sources and the intake of the colourless carotenoids phytoene and phytofluene in Spain. Nutrients 13:4436. https://doi.org/10.3390/nu13124436

Osman M, Valadon LR (1978) Studies on the near-UV effect on carotenogenesis in *Verticillium agaricinum*. Microbios 23:53–64

Ostrea EM, Balun JE, Winkler R, Porter T (1986) Influence of breast-feeding on the restoration of the low serum concentration of vitamin E and beta-carotene in the newborn infant. Am J Obstet Gynecol 154:1014–1017. https://doi.org/10.1016/0002-9378(86)90740-4

Papp T, Velayos A, Bartok T et al (2006) Heterologous expression of astaxanthin biosynthesis genes in *Mucor circinelloides*. Appl Microbiol Biotechnol 69:526–531. https://doi.org/10.1007/s00253-005-0026-6

Papp T, Csernetics A, Nagy G et al (2013) Canthaxanthin production with modified *Mucor circinelloides* strains. Appl Microbiol Biotechnol 97:4937–4950. https://doi.org/10.1007/s00253-012-4610-2

Park Y-K, Ledesma-Amaro R (2023) What makes *Yarrowia lipolytica* well suited for industry? Trends Biotechnol 41:242–254. https://doi.org/10.1016/j.tibtech.2022.07.006

Parra-Rivero O, de Barros MP, Prado MDM et al (2020) Neurosporaxanthin overproduction by *Fusarium fujikuroi* and evaluation of its antioxidant properties. Antioxid Basel Switz 9:1–27. https://doi.org/10.3390/antiox9060528

Patrick L (2000) Beta-carotene: the controversy continues. Altern Med Rev J Clin Ther 5:530–545

Pegklidou K, Mantzouridou F, Tsimidou MZ (2008) Lycopene production using *Blakeslea trispora* in the presence of 2-methyl imidazole: yield, selectivity, and safety aspects. J Agric Food Chem 56:4482–4490. https://doi.org/10.1021/jf800272k

Pfander H, Lanz M, Traber B (1997) Synthesis of carotenoids. Stud Nat Prod Chem 20:561–612. Elsevier. https://doi.org/10.1016/S1572-5995(97)80038-5

Pollmann H, Breitenbach J, Sandmann G (2016) Engineering of the carotenoid pathway in *Xanthophyllomyces dendrorhous* leading to the synthesis of zeaxanthin. Appl Microbiol Biotechnol. https://doi.org/10.1007/s00253-016-7769-0

Porter JW, Lincoln RE (1950) Lycopersicon selections containing a high content of carotenes and colorless polyenes; the mechanism of carotene biosynthesis. Arch Biochem 27:390–403

Prado-Cabrero A, Estrada AF, Al-Babili S, Avalos J (2007a) Identification and biochemical characterization of a novel carotenoid oxygenase: elucidation of the cleavage step in the *Fusarium* carotenoid pathway. Mol Microbiol 64:448–460. https://doi.org/10.1111/j.1365-2958.2007.05665.x

Prado-Cabrero A, Scherzinger D, Avalos J, Al-Babili S (2007b) Retinal biosynthesis in fungi: Characterization of the carotenoid oxygenase CarX from *Fusarium fujikuroi*. Eukaryot Cell 6:650–657. https://doi.org/10.1128/ec.00392-06

Promdonkoy P, Watcharawipas A, Bubphasawan S et al (2024) Metabolic engineering of *Saccharomyces cerevisiae* for production of canthaxanthin, zeaxanthin, and astaxanthin. J Fungi Basel Switz 10:433. https://doi.org/10.3390/jof10060433

Przybylska S (2020) Lycopene—a bioactive carotenoid offering multiple health benefits: a review. Int J Food Sci Technol 55:11–32. https://doi.org/10.1111/ijfs.14260

Qi D-D, Jin J, Liu D et al (2020) *In vitro* and *in vivo* recombination of heterologous modules for improving biosynthesis of astaxanthin in yeast. Microb Cell Fact 19:103. https://doi.org/10.1186/s12934-020-01356-7

Qin Z, Liu M, Ren X et al (2024) *De novo* biosynthesis of lutein in *Yarrowia lipolytica*. J Agric Food Chem 72:5348–5357. https://doi.org/10.1021/acs.jafc.3c09080

Rapoport A, Guzhova I, Bernetti L et al (2021) Carotenoids and some other pigments from fungi and yeasts. Metabolites 11:92. https://doi.org/10.3390/metabo11020092

Rebelo BA, Farrona S, Ventura MR, Abranches R (2020) Canthaxanthin, a red-hot carotenoid: applications, synthesis, and biosynthetic evolution. Plants Basel Switz 9:1039. https://doi.org/10.3390/plants9081039

Roche CM, Loros JJ, McCluskey K, Glass NL (2014) *Neurospora crassa*: looking back and looking forward at a model microbe. Am J Bot 101:2022–2035. https://doi.org/10.3732/ajb.1400377

Rodríguez-Ortiz R, Limón MC, Avalos J (2009) Regulation of carotenogenesis and secondary metabolism by nitrogen in wild-type *Fusarium fujikuroi* and carotenoid-overproducing mutants. Appl Environ Microbiol 75:405–413. https://doi.org/10.1128/AEM.01089-08

Rodríguez-Ortiz R, Limón MC, Avalos J (2013) Functional analysis of the *carS* gene of *Fusarium fujikuroi*. Mol Genet Genomics 288:157–173. https://doi.org/10.1007/s00438-013-0739-7

Rodríguez-Sáiz M, Paz B, De La Fuente JL et al (2004) *Blakeslea trispora* genes for carotene biosynthesis. Appl Environ Microbiol 70:5589–5594. https://doi.org/10.1128/AEM.70.9.5589-5594.2004

Rodríguez-Sáiz M, de la Fuente JL, Barredo JL (2010) *Xanthophyllomyces dendrorhous* for the industrial production of astaxanthin. Appl Microbiol Biotechnol 88:645–658. https://doi.org/10.1007/s00253-010-2814-x

Rodríguez-Sáiz M, de la Fuente J-L, Barredo J-L (2012) Metabolic engineering of *Mucor circinelloides* for zeaxanthin production. Methods Mol Biol Clifton NJ 898:133–151. https://doi.org/10.1007/978-1-61779-918-1_8

Ruger-Herreros M, Parra-Rivero O, Pardo-Medina J et al (2019) Comparative transcriptomic analysis unveils interactions between the regulatory CarS protein and light response in *Fusarium*. BMC Genomics 20:67. https://doi.org/10.1186/s12864-019-5430-x

Ruger-Herreros M, Nordzieke S, Vega-Álvarez C et al (2022) Relation between CarS expression and activation of carotenogenesis by stress in *Fusarium fujikuroi*. Front Bioeng Biotechnol 10:1000129. https://doi.org/10.3389/fbioe.2022.1000129

Ruiz-Hidalgo MJ, Benito EP, Sandmann G, Eslava AP (1997) The phytoene dehydrogenase gene of *Phycomyces*: regulation of its expression by blue light and vitamin A. Mol Gen Genet 253:734–744. https://doi.org/10.1007/s004380050378

Saelices L, Youssar L, Holdermann I et al (2007) Identification of the gene responsible for torulene cleavage in the *Neurospora* carotenoid pathway. Mol Genet Genomics 278:527–537. https://doi.org/10.1007/s00438-007-0269-2

Saini RK, Prasad P, Lokesh V et al (2022) Carotenoids: dietary sources, extraction, encapsulation, bioavailability, and health benefits-A review of recent advancements. Antioxid Basel Switz 11:795. https://doi.org/10.3390/antiox11040795

Sakaki H, Nakanishi T, Tada A et al (2001) Activation of torularhodin production by *Rhodotorula glutinis* using weak white light irradiation. J Biosci Bioeng 92:294–297. https://doi.org/10.1016/S1389-1723(01)80265-6

Sakaki H, Nochide H, Komemushi S, Miki W (2002) Effect of active oxygen species on the productivity of torularhodin by *Rhodotorula glutinis* No. 21. J Biosci Bioeng 93:338–340. https://doi.org/10.1016/S1389-1723(02)80040-8

Sandmann G (2009) Evolution of carotene desaturation: The complication of a simple pathway. Arch Biochem Biophys 483:169–174. https://doi.org/10.1016/j.abb.2008.10.004

Sandmann G (2015) Carotenoids of biotechnological importance. Adv Biochem Eng Biotechnol 148:449–467. https://doi.org/10.1007/10_2014_277

Sandmann G (2022) Carotenoids and their biosynthesis in fungi. Mol Basel Switz 27:1431. https://doi.org/10.3390/molecules27041431

Sandmann G, Pollmann H, Gassel S, Breitenbach J (2021) *Xanthophyllomyces dendrorhous*, a versatile platform for the production of carotenoids and other acetyl-CoA-derived compounds. Adv Exp Med Biol 1261:137–151. https://doi.org/10.1007/978-981-15-7360-6_11

Sanz C, Velayos A, Alvarez MI et al (2011) Functional analysis of the *Phycomyces carRA* gene encoding the enzymes phytoene synthase and lycopene cyclase. PLoS One 6:e23102. https://doi.org/10.1371/journal.pone.0023102

Schmidhauser TJ, Lauter FR, Russo VE, Yanofsky C (1990) Cloning, sequence, and photoregulation of *al-1*, a carotenoid biosynthetic gene of *Neurospora crassa*. Mol Cell Biol 10:5064–5070. https://doi.org/10.1128/mcb.10.10.5064-5070.1990

Schmidhauser TJ, Lauter FR, Schumacher M et al (1994) Characterization of *al-2*, the phytoene synthase gene of *Neurospora crassa*. Cloning, sequence analysis, and photoregulation. J Biol Chem 269:12060–12066. https://doi.org/10.1016/S0021-9258(17)32680-7

Schmidt AD, Heinekamp T, Matuschek M et al (2005) Analysis of mating-dependent transcription of *Blakeslea trispora* carotenoid biosynthesis genes *carB* and *carRA* by quantitative real-time PCR. Appl Microbiol Biotechnol 67:549–555. https://doi.org/10.1007/s00253-005-1941-2

Schmidt I, Schewe H, Gassel S et al (2011) Biotechnological production of astaxanthin with *Phaffia rhodozyma/Xanthophyllomyces dendrorhous*. Appl Microbiol Biotechnol 89:555–571. https://doi.org/10.1007/s00253-010-2976-6

Schwartz C, Frogue K, Misa J, Wheeldon I (2017) Host and pathway engineering for enhanced lycopene biosynthesis in *Yarrowia lipolytica*. Front Microbiol 8:2233. https://doi.org/10.3389/fmicb.2017.02233

Schwerdtfeger C, Linden H (2003) VIVID is a flavoprotein and serves as a fungal blue light photoreceptor for photoadaptation. EMBO J 22:4846–4855. https://doi.org/10.1093/emboj/cdg451

Shafe MO, Gumede NM, Nyakudya TT, Chivandi E (2024) Lycopene: a potent antioxidant with multiple health benefits. J Nutr Metab 2024:6252426. https://doi.org/10.1155/2024/6252426

Shah MMR, Liang Y, Cheng JJ, Daroch M (2016) Astaxanthin-producing green microalga *Haematococcus pluvialis*: from single cell to high value commercial products. Front Plant Sci 7:531. https://doi.org/10.3389/fpls.2016.00531

Skibsted LH (2012) Carotenoids in antioxidant networks. Colorants or radical scavengers. J Agric Food Chem 60:2409–2417. https://doi.org/10.1021/jf2051416

So B, Kwon KH (2023) Nutritional approaches of the changing consumer after the pandemic: sustainable potential of phytoene and phytofluene for photoprotection and skin health. Sustainability 15:4416. https://doi.org/10.3390/su15054416

Stachowiak B, Szulc P (2021) Astaxanthin for the food industry. Mol Basel Switz 26:2666. https://doi.org/10.3390/molecules26092666

Stahl W, Sies H (2003) Antioxidant activity of carotenoids. Mol Aspects Med 24:345–351. https://doi.org/10.1016/S0098-2997(03)00030-X

Strobel I, Breitenbach J, Scheckhuber CQ et al (2009) Carotenoids and carotenogenic genes in *Podospora anserina*: engineering of the carotenoid composition extends the life span of the mycelium. Curr Genet 55:175–184. https://doi.org/10.1007/s00294-009-0235-0

Suárez T, Eslava AP (1988) Transformation of *Phycomyces* with a bacterial gene for kanamycin resistance. Mol Gen Genet MGG 212:120–123. https://doi.org/10.1007/BF00322453

Sun N, Lee S, Song KB (2004) Characterization of a carotenoid-hyperproducing yeast mutant isolated by low-dose gamma irradiation. Int J Food Microbiol 94:263–267. https://doi.org/10.1016/S0168-1605(03)00311-8

Sun L, Kwak S, Jin Y-S (2019) Vitamin A production by engineered *Saccharomyces cerevisiae* from xylose via two-phase in situ extraction. ACS Synth Biol 8:2131–2140. https://doi.org/10.1021/acssynbio.9b00217

Sutter RP, Harrison TL, Galasko G (1974) Trisporic acid biosynthesis in *Blakeslea trispora* via mating type-specific precursors. J Biol Chem 249:2282–2284. https://doi.org/10.1016/S0021-9258(19)42829-9

Takemura M, Sahara T, Misawa N (2021) Violaxanthin: natural function and occurrence, biosynthesis, and heterologous production. Appl Microbiol Biotechnol 105:6133–6142. https://doi.org/10.1007/s00253-021-11452-2

Takino J, Kozaki T, Ozaki T et al (2019) Elucidation of biosynthetic pathway of a plant hormone abscisic acid in phytopathogenic fungi. Biosci Biotechnol Biochem 83:1642–1649. https://doi.org/10.1080/09168451.2019.1618700

Tang W, Wang Y, Zhang J et al (2019) Biosynthetic pathway of carotenoids in *Rhodotorula* and strategies for enhanced their production. J Microbiol Biotechnol 29:507–517. https://doi.org/10.4014/jmb.1801.01022

Thewes S, Prado-Cabrero A, Prado MM et al (2005) Characterization of a gene in the *car* cluster of *Fusarium fujikuroi* that codes for a protein of the carotenoid oxygenase family. Mol Genet Genomics 274:217–228. https://doi.org/10.1007/s00438-005-0015-6

Thomas DM, Goodwin TW (1967) Studies on carotenogenesis in *Blakeslea trispora*—I. Phytochemistry 6:355–360. https://doi.org/10.1016/S0031-9422(00)86290-1

Torres-Haro A, Verdín J, Kirchmayr MR, Arellano-Plaza M (2021) Metabolic engineering for high yield synthesis of astaxanthin in *Xanthophyllomyces dendrorhous*. Microb Cell Fact 20:175. https://doi.org/10.1186/s12934-021-01664-6

Torres-Martínez S, Murillo FJ, Cerdá-Olmedo E (1980) Genetics of lycopene cyclization and substrate transfer in β-carotene biosynthesis in *Phycomyces*. Genet Res 36:299–309. https://doi.org/10.1017/S0016672300019911

Ukibe K, Katsuragi T, Tani Y, Takagi H (2008) Efficient screening for astaxanthin-overproducing mutants of the yeast *Xanthophyllomyces dendrorhous* by flow cytometry. FEMS Microbiol Lett 286:241–248. https://doi.org/10.1111/j.1574-6968.2008.01278.x

Valadon LRG, Cooke RC (1963) Carotenoid pigments of the nematode-trapping hyphomycete *Arthrobotrys oligospora* (fres.). Phytochemistry 2:103–105. https://doi.org/10.1016/S0031-9422(00)88022-X

Valadon LR, Heale JB (1965) Carotenoids in mutants of *Verticillium albo-atrum*. J Gen Microbiol 41:341–348. https://doi.org/10.1099/00221287-41-3-341

Van Heeswijck R, Roncero MI (1984) High frequency transformation of *Mucor* with recombinant plasmid DNA. Carlsberg Res Commun 49:691–702. https://doi.org/10.1007/BF02907500

van Poppel G, Goldbohm RA (1995) Epidemiologic evidence for beta-carotene and cancer prevention. Am J Clin Nutr 62:1393S–1402S. https://doi.org/10.1093/ajcn/62.6.1393S

Veiga-Crespo P, Araya-Garay JM, Villa TG (2018) Engineering *Pichia pastoris* for the production of carotenoids. Methods Mol Biol (Clifton, NJ) 1852:311–326. https://doi.org/10.1007/978-1-4939-8742-9_19

Velayos A, López Matas MA, Ruiz-Hidalgo MJ, Eslava AP (1997) Complementation analysis of carotenogenic mutants of *Mucor circinelloides*. Fungal Genet Biol 22:19–27. https://doi.org/10.1006/fgbi.1997.0998

Velayos A, Blasco JL, Álvarez MI et al (2000a) Blue-light regulation of phytoene dehydrogenase (*carB*) gene expression in *Mucor circinelloides*. Planta 210:938–946. https://doi.org/10.1007/s004250050701

Velayos A, Eslava AP, Iturriaga EA (2000b) A bifunctional enzyme with lycopene cyclase and phytoene synthase activities is encoded by the *carRP* gene of *Mucor circinelloides*. Eur J Biochem 267:5509–5519. https://doi.org/10.1046/j.1432-1327.2000.01612.x

Verdoes JC, Krubasik P, Sandmann G, van Ooyen AJJ (1999a) Isolation and functional characterisation of a novel type of carotenoid biosynthetic gene from *Xanthophyllomyces dendrorhous*. Mol Gen Genet 262:453–461. https://doi.org/10.1007/s004380051105

Verdoes JC, Misawa N, van Ooyen AJJ (1999b) Cloning and characterization of the astaxanthin biosynthetic gene encoding phytoene desaturase of *Xanthophyllomyces dendrorhous*. Biotechnol Bioeng 63:750–755. https://doi.org/10.1002/(sici)1097-0290(19990620)63:6<750::aid-bit13>3.0.co;2-7

Wang X, Xu X, Liu J et al (2023) Metabolic engineering of *Saccharomyces cerevisiae* for efficient retinol synthesis. J Fungi Basel Switz 9:512. https://doi.org/10.3390/jof9050512

Wang J, Zhou X, Li K et al (2024) Systems metabolic engineering for efficient violaxanthin production in yeast. J Agric Food Chem 72:10459–10468. https://doi.org/10.1021/acs.jafc.4c01240

Weber D, Grune T (2012) The contribution of β-carotene to vitamin A supply of humans. Mol Nutr Food Res 56:251–258. https://doi.org/10.1002/mnfr.201100230

Wei C, Wu T, Ao H et al (2020) Increased torulene production by the red yeast, *Sporidiobolus pararoseus*, using citrus juice. Prep Biochem Biotechnol 50:66–73. https://doi.org/10.1080/10826068.2019.1663533

Yabuzaki J (2017) Carotenoids database: structures, chemical fingerprints and distribution among organisms. Database (Oxford) 2017:bax004. https://doi.org/10.1093/database/bax004

Yamamoto K, Hara KY, Morita T et al (2016) Enhancement of astaxanthin production in *Xanthophyllomyces dendrorhous* by efficient method for the complete deletion of genes. Microb Cell Fact 15:155. https://doi.org/10.1186/s12934-016-0556-x

Yamano S, Ishii T, Nakagawa M et al (1994) Metabolic engineering for production of β-carotene and lycopene in *Saccharomyces cerevisiae*. Biosci Biotechnol Biochem 58:1112–1114. https://doi.org/10.1271/bbb.58.1112

Youssar L, Avalos J (2007) Genetic basis of the *ovc* phenotype of *Neurospora*: identification and analysis of a 77 kb deletion. Curr Genet 51:19–30. https://doi.org/10.1007/s00294-006-0104-z

Youssar L, Schmidhauser TJ, Avalos J (2005) The *Neurospora crassa* gene responsible for the *cut* and *ovc* phenotypes encodes a protein of the haloacid dehalogenase family. Mol Microbiol 55:828–838. https://doi.org/10.1111/j.1365-2958.2004.04424.x

Yuan JP, Peng J, Yin K, Wang JH (2011) Potential health-promoting effects of astaxanthin: a high-value carotenoid mostly from microalgae. Mol Nutr Food Res 55:150–165. https://doi.org/10.1002/mnfr.201000414

Zare M, Norouzi Roshan Z, Assadpour E, Jafari SM (2021) Improving the cancer prevention/treatment role of carotenoids through various nano-delivery systems. Crit Rev Food Sci Nutr 61:522–534. https://doi.org/10.1080/10408398.2020.1738999

Zhang X-K, Nie M-Y, Chen J et al (2019) Multicopy integrants of *crt* genes and co-expression of AMP deaminase improve lycopene production in *Yarrowia lipolytica*. J Biotechnol 289:46–54. https://doi.org/10.1016/j.jbiotec.2018.11.009

Zhang X, Chen S, Lin Y et al (2023a) Metabolic engineering of *Pichia pastoris* for high-level production of lycopene. ACS Synth Biol 12:2961–2972. https://doi.org/10.1021/acssynbio.3c00294

Zhang Y, Yang J, Na X, Zhao A (2023b) Association between β-carotene supplementation and risk of cancer: a meta-analysis of randomized controlled trials. Nutr Rev 81:1118–1130. https://doi.org/10.1093/nutrit/nuac110

Zhou D, Fei Z, Liu G et al (2024) The bioproduction of astaxanthin: A comprehensive review on the microbial synthesis and downstream extraction. Biotechnol Adv 74:108392. https://doi.org/10.1016/j.biotechadv.2024.108392

Zhu X, Meng C, Sun F et al (2023) Sustainable production of astaxanthin in microorganisms: the past, present, and future. Crit Rev Food Sci Nutr 63:10239–10255. https://doi.org/10.1080/10408398.2022.2080176

Chapter 9
Riboflavin Overproduction in Yeasts and Filamentous Fungi

Justyna Ruchala, Dariya Fedorovych, Dominik Wojdyla, Kostyantyn V. Dmytruk, and Andriy A. Sibirny

Contents

9.1 Introduction 264
9.2 *A. gossypii* 266
 9.2.1 Transformation 266
 9.2.2 Sclectable Markers 267
 9.2.3 Reporters and Promoters 268
 9.2.4 The CRISPR-Cas9 System 269
 9.2.5 Metabolic Engineering to Increase Riboflavin Production 270
 9.2.6 Other Biotechnological Products of *A. gossypii* 272
9.3 *C. famata* 273
 9.3.1 Transformation 273
 9.3.2 Selectable Markers 274
 9.3.3 Insertional Mutagenesis 275
 9.3.4 The Reporter System and Promoters 276
 9.3.5 The CRISPR-Cas9 System 277
 9.3.6 Regulation of Riboflavin Biosynthesis 278
 9.3.7 Development of Strains with Increased Riboflavin Production 279
 9.3.8 Construction of Riboflavin Derivative Producers 281
9.4 Conclusions 283
References 284

J. Ruchala (✉)
University of Rzeszow, Rzeszow, Poland

Institute of Cell Biology, NAS of Ukraine, Lviv, Ukraine
e-mail: jruchala@ur.edu.pl

D. Fedorovych · K. V. Dmytruk
Institute of Cell Biology, NAS of Ukraine, Lviv, Ukraine

D. Wojdyla
University of Rzeszow, Rzeszow, Poland

A. A. Sibirny
Institute of Biotechnology, University of Rzeszow, Rzeszow, Poland

A. A. Sibirny (ed.), *Biotechnology of Yeasts and Filamentous Fungi*,
Grand Challenges in Biology and Biotechnology,
https://doi.org/10.1007/978-3-031-74726-7_9

Abstract Riboflavin (vitamin B_2) is an important nutrient used as a precursor for the synthesis of flavin nucleotides, flavin mononucleotide (FMN), and flavin adenine dinucleotide (FAD), which are used as coenzymes in many enzymatic reactions, mostly oxidoreductive. Riboflavin is widely used in food, agriculture, and medicine. Riboflavin is synthesized by many bacteria and by all fungi and plants, but humans and other animals need to acquire sufficient vitamins with their diet or through supplements to maintain optimal health. The annual market for this vitamin is estimated at US $451.57 million in 2024. Traditionally, riboflavin has been produced chemically, but microbial synthesis is gaining traction due to lower production costs, reduced waste, and decreased energy consumption. Despite these advantages, the genetic stability of microbial overproducers remains a challenge as most yeasts and filamentous fungi naturally produce riboflavin in amounts sufficient only for their metabolic needs. Yeasts and filamentous fungi mostly produce small amounts of riboflavin, which corresponds to their needs. However, there is a group of yeasts and filamentous fungi, known as flavinogenic organisms, which are able to produce larger amounts of riboflavin. Flavinogenic yeasts overproduce riboflavin under iron starvation, and flavinogenic fungi do it during sporulation in the stationary growth phase. Some mutants of the flavinogenic fungus *Ashbya gossypii* and the yeast *Candida famata* belong to the most effective riboflavin overproducers known. This review discusses the development and application of molecular tools, including transformation systems, selectable markers, promoters, and clustered regularly interspaced palindromic repeats-CRISPR-associated protein 9 (CRISPR-Cas9) technology, to enhance riboflavin production in these organisms. The successful genetic engineering of *A. gossypii* and *C. famata* highlights the potential for more efficient and sustainable industrial riboflavin production.

9.1 Introduction

In most cases, primary metabolites are synthesized in optimal amounts; it is thought that otherwise, an organism loses its competitiveness and could be eliminated by natural selection. There are, however, several exceptions to this general picture, an overproduction of glutamic acid by *Corynebacterium glutamicum* being apparently the best-known example (Kimura 2003). Other examples deal with the ability to overproduce riboflavin (vitamin B_2) by several species of filamentous fungi and yeasts. Riboflavin overproducers can be divided into three groups, depending on the amount of riboflavin produced. In this way, organisms that are capable of producing riboflavin at a level of at least 10 mg/l are qualified as riboflavin overproducers. Arnold Demain proposed a classification according to which organisms producing 10 mg/l and less are weak producers of this vitamin, those above 600 mg/l are moderate, and those above 10 g/l are strong (Demain 2007). The current review describes two such natural riboflavin-overproducing organisms: *Ashbya gossypii* and *Candida famata*. *Ashbya gossypii* is a filamentous fungus, evolutionarily related to the yeast

Saccharomyces cerevisiae, isolated as a plant pathogen and characterized as a natural riboflavin (vitamin B_2) overproducer (Wickerham et al. 1946). *A. gossypii* belongs to the group of the most flavinogenic organisms known (Demain 1972). Riboflavin accumulates in the stationary phase (Stahmann et al. 2000), giving mycelium a bright yellow color (Förster et al. 1999). *C. famata* is asporogenic yeast and representative of so-named flavinogenic yeasts—a group of organisms capable of riboflavin oversynthesis under iron limitation (Tanner et al. 1945; Abbas and Sibirny 2011). This group also includes alkane-utilizing yeast *Meyerozyma (Pichia* or *Candida) guilliermondii*, starch-utilizing yeast *Schwanniomyces occidentalis*, pathogenic yeast *Candida albicans*, and some others (Knight et al. 2002; Abbas and Sibirny 2011; Papon et al. 2015). All belong to the so-named CTG clade of yeasts, the organisms that decode the leucine CTG codon to serine. *C. famata* apparently is the most flavinogenic yeast as under iron limitation conditions, it accumulates around 500 μg/ml of riboflavin in the medium, while under optimal concentration of iron ions for growth, the accumulation of riboflavin does not exceed 1–1.5 μg of riboflavin/ml, i.e., similar to that in nonflavinogenic yeasts.

The first microbial riboflavin production processes that were developed in the 1940s used *Eremothecium ashbyi* and *A. gossypii*, two natural riboflavin-overproducing filamentous fungal species, as production strains (Revuelta et al. 2016b; Liu et al. 2020; Zhang et al. 2021).

ADM Company (USA) developed and used a riboflavin production process based on *C. famata* for commercial riboflavin production using hydrolyzed corn starch as feedstock. However, due to the genetic instability of the overproducing strain dep8, resulting in nonoverproducing revertants, the *C. famata*-based production process was stopped, and it is not currently used at the industrial scale. In the early 1970s, an *A. gossypii*-based riboflavin process was developed, which was later implemented on an industrial scale and used in parallel with the traditional chemical process for several years (Revuelta et al. 2016a, b; You et al. 2021; Wang et al. 2021). Since then, the *A. gossypii* process has been continuously improved, with the consequence that the chemical production was shut down in 1996 and the microbial fermentation plant was enlarged to provide sufficient production capacity. *A. gossypii* plays a crucial role in the commercial production of riboflavin due to its high efficiency and the ability to be genetically optimized for even greater yields. Major companies like DSM and BASF are leading the way in utilizing this organism for sustainable riboflavin production. The chemical company BASF has installed a plant in Gunsan (South Korea), which specializes in riboflavin production on an industrial scale using *A. gossypii* with vegetable oil as a substrate (Schwechheimer et al. 2016) (Fig. 9.1).

The one-step fermentation process of *A. gossypii* using highly efficient production strains is economically advantageous to chemical production technologies due to the lower technical complexity and reduced consumption of raw materials, mainly renewable carbon sources, such as vegetable oils or carbohydrates. It should be stressed that microbial riboflavin production is also favorable when it comes to chemical synthesis from an ecological point of view. The biotechnological potential of *C. famata*-producing riboflavin is also very high as some mutants of this species

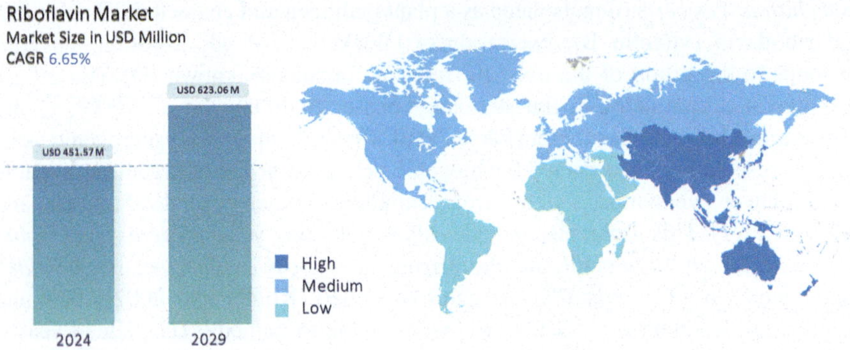

Fig. 9.1 Riboflavin market: market size (%), by region, global, 2021. The riboflavin market size is estimated at US $451.57 million in 2024 and is expected to reach US $623.06 million by 2029, growing at a CAGR of 6.65% during the forecast period (2024–2029) (riboflavin market size & share analysis—growth trends & forecasts (2024–2029), https://www.mordorintelligence.com/industry-reports/riboflavin-market)

are the most flavinogenic organisms known (Dmytruk and Sibirny 2012). The success of *A. gossypii* and *C. famata* in biotechnological processes is based on the development of efficient molecular tools that enable the application of metabolic engineering approaches to improve the productivity of industrial strains. Furthermore, *C. famata* can also be used as a model organism in studying iron metabolism as changes in iron acquisition could be easily detected due to changes in the production of the color substance, riboflavin. Recent studies have underscored the significance of metabolic engineering in enhancing riboflavin production in microbial systems, including both yeasts and filamentous fungi. These advancements have utilized multiomics data to refine metabolic pathways and optimize production conditions (Ding and Ye 2023). This chapter describes the molecular tools developed for *A. gossypii* and *C. famata* and their application to the development of strains overproducing riboflavin. In addition, some features of the regulation of riboflavin synthesis in the flavinogenic yeasts *C. famata*, *M. guilliermondii*, and *C. albicans* are discussed.

9.2 *A. gossypii*

9.2.1 *Transformation*

There are two procedures that achieve the efficient transformation of *A. gossypii*. The first method reported is based on the enzymatic treatment of growing cells to remove their cell walls and the resulting spheroplasts exposed to deoxyribonucleic acid (DNA) in the presence of calcium ions and polyethylene glycol, before being embedded in hypertonic selective agar to facilitate cell wall regeneration and subsequent colony formation. Transformation frequencies of up to 63 transformants per μg of plasmid DNA were obtained (Wright and Philippsen 1991).

Intact germling cells can be transformed by electroporation, which involves using a brief voltage pulse to facilitate the entry of DNA, plasmid, or polymerase chain reaction (PCR) products into the cells (Förster et al. 1999). Transformation by electroporation is simple and at least as efficient as spheroplast transformation (50–100 transformants/μg DNA). Since *A. gossypii* grows as a multinucleate mycelium, transformant colonies are heterokaryotic and contain a mixture of transformed and untransformed nuclei. Therefore, to obtain homokaryotic transformant strains, the primary transformant colonies are subjected to clonal selection. This is achieved by plating the uninucleate spores obtained from the primary transformant colonies onto selectable media.

9.2.2 Selectable Markers

Selectable markers are an important molecular tool because they allow the differentiation of transformed cells from untransformed cells. They can generally be classified into three categories: nutritional markers, resistance markers, and bidirectional or counterselectable markers. In addition, a few reporters (see below) also serve as convenient visual markers to select transformants.

Resistance markers allow the growth of transformants in the presence of inhibitors (antimetabolite or antibiotic). They are routinely used to transform wild-type strains because they circumvent the need to create an auxotrophic host background. The only requirement is that the recipient organism is sensitive to the selection pressure applied. The most commonly used resistance markers in *A. gossypii* gene manipulations (gene knockout, gene over- or underexpression, and gene tagging) are *kan*MX4, *natR*, and *hyg*R, which confer resistance to geneticin (G418®), nourseothricin (Clonat®), and hygromycin B, respectively (Wright and Philippsen 1991; Revuelta et al. 2016a, b; Jiménez et al. 2008; Santos et al. 2005). Whereas antibiotic or nutritional markers are, in general, dominant and provide only a tool for positive selection (the gene imparting a survival phenotype to the recipient on the respective selection medium), counterselectable markers enabling positive or negative selection are powerful molecular tools in genetic engineering because they allow selection for the loss of a genetic marker rather than its presence. Interestingly, *hsv-1 tk* (coding for herpes simplex virus type 1 thymidine kinase) provides a negative selection: its expression is lethal to the host organism when grown in the presence of ganciclovir. Lethality is due to the conversion of ganciclovir into a cytotoxic nucleoside analog. In combination with *kan*MX4, *hsv-1 tk* was used to provide positive and negative selection in *A. gossypii* (Muñoz-Fernández et al. 2021).

The list of markers at our disposal is not extensive, and therefore, recycling/reusing a marker is of essence. This problem is particularly problematic when the same recipient strain has to be transformed repeatedly. A very efficient method for marker recycling has been developed in a wide variety of fungi based on the Cre-*loxP* recombination system adapted from bacteriophage P1. This approach comprises the excision of a DNA fragment that is flanked by loxP sites by the catalytic activity of a Cre recombinase, which needs to be expressed in the fungus. Marker rescue based on the Cre/*loxP* recombination system has also been developed for *A. gossypii* (Aguiar et al. 2014a; Ledesma-Amaro et al. 2014a).

9.2.3 Reporters and Promoters

A. gossypii has been extensively used as a model to study hyphae formation, and therefore several reporters have been developed. Some of them are *LacZ* and fluorescence proteins that can be used fused to target genes/promoters, such as yEGFP, yEmCFP, yEVenus, yECitrine, yEMCitrine, mCherry, mStrawberry, and tdTomoto (Kaufmann 2009).

The wide use of *A. gossypii* in both basic research (hyphae formation) and applied research (the bioproduction of vitamins and other industrially relevant compounds) required the identification and use of different promoters. Some of the constitutive promoters are *GPD1* and *TEF1*. Interestingly, the *AgTEF1* promoter is also widely used to control gene expression in *S. cerevisiae* (Monschau et al. 1998). In contrast, *HIS3*, *TEF1*, *TEF2*, and *PGK1* promoters from *S. cerevisiae* have been used to express genes in *A. gossypii* (Kemper et al. 2011; Ravasio et al. 2014; Ribeiro et al. 2010). *GPD1* is the strongest promoter described so far, and therefore it has been used in most gene overexpressions carried out for metabolic engineering approaches. In addition to the constitutive promoters, inducible and repressible ones, such as *AgICLp*, *AgMET3p*, and *ScTH13p*, have been developed. The *AgICL1* promoter is repressed by glucose, derepressed by glycerol, partially activated by acetate and ethanol, and fully activated by oil (Maeting et al. 1999). The *AgMET3* promoter can be repressed by the presence of methionine (Dünkler and Wendland 2007), while the *ScTH12* promoter is repressed by thiamine (Kaufmann 2009). New promoters for the metabolic engineering of *A. gossypii* were described (Muñoz-Fernández et al. 2021). These promoters cover a wide range of activities with important potential applications for the metabolic engineering of *A. gossypii.* The identification of new promoters was possible due to the method of promoter analysis in *A. gossypii* proposed by Muñoz-Fernández et al. using integration cassettes based on the dual-luciferase reporter (DLR) test in mammals, which enables the sequential quantitative measurement of two luciferase activities (*Renilla* and *firefly* luciferases) in a single protein extract; this ensures the repeatability and accuracy of the system (Muñoz-Fernández et al. 2021).

9.2.4 The CRISPR-Cas9 System

Since the first successful attempts to use the clustered regularly interspaced palindromic repeats-CRISPR-associated protein 9 (CRISPR-Cas9) system as a tool in genetic engineering, the system has been optimized for use in many organisms. It generally consists of a Cas9 endonuclease, guided by a specific ribonucleic acid (RNA), that creates double-stranded breaks (DSBs) at a certain specific locus in the genome, allowing for targeted editing. The RNA associated with Cas9 targets CRISPR RNA (crRNA) and transactivating crRNA (tracrRNA). All of these components, in turn, are required for Cas9 to be catalytically active. In turn, crRNA, as well as tracrRNA, can be combined into a synthetic guide RNA, i.e., sgRNA (*s*ingle *g*uide RNA). To generate a double-stranded cleavage, Cas9 nuclease requires the presence of a PAM (*p*rotospacer *a*djacent *m*otif) sequence at the target site in the genome, which is determined by gRNA (*g*uide RNA). The repair of such double-strand breaks is carried out in the cell by both homologous recombination and non-homologous end joining, but it is the presence of synthetic dDNA (*d*onor DNA) that ensures the possibility of repair carried out by the homologous recombination mechanism (Ran et al. 2013). To ensure the effectiveness of editing using the CRISPR-Cas9 system, it must be optimized for each organism to ensure the effectiveness of gRNA and Cas9 expression. This system was successfully adapted to edit the genome of *A. gossypii* using a single-vector strategy. In this case, a *CAS9* expression module, previously used for the yeast *Saccharomyces cerevisiae*, was proposed, in which a human codon-optimized *Streptococcus pyogenes CAS9* gene was expressed under the control of the yeast *TEF1* gene promoter and the *CYC1* terminator (Dicarlo et al. 2013). Furthermore, the expression of the sgRNA is controlled by regulatory sequences of the *A. gossypii SNR52* gene, which is transcribed by RNA polymerase III. Finally, to ensure the success of editing using this system, sgRNA contained a short sequence targeting a given locus in the genome and a sequence allowing Cas9 binding. According to the assumptions of the system and the induction of DNA repair, the system is completed with dDNA that protects the repair of double-strand breaks by using homologous recombination. The *A. gossypii*

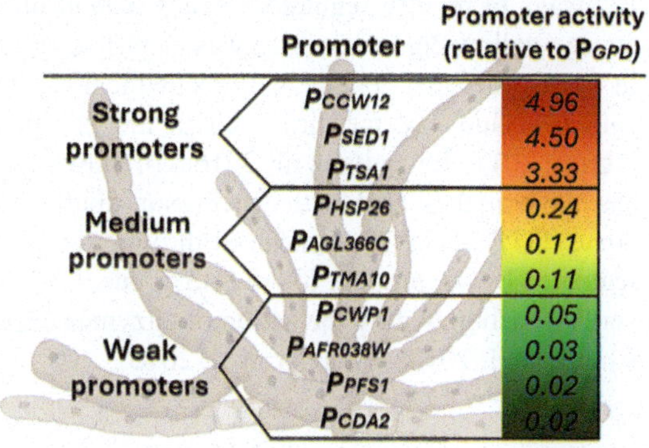

Fig. 9.2 Comparison of the relative activity of selected promoter sequences with respect to the *GPD* promoter. The promoters presented in this work may help implement multiple expression platforms for metabolic engineering in *A. gossypii* to produce industrially relevant metabolites, including riboflavin

genome editing system optimized in this way allows for targeted changes after the appropriate design of a specific sgRNA-dDNA (Muñoz-Fernández et al. 2021) (Fig. 9.2).

A versatile CRISPR/Cpf1 method has been adapted for the efficient genome editing of *A. gossypii*. It complements the CRISPR/Cas9 system, thus minimizing the limitations associated with the existence of specific PAM sequences within the target sites. In addition, a multi-CRISPR/Cpf1 has been designed and validated for multiplex gene deletion of up to four genes using a single multi-CRISPR-Cpf1 plasmid. The existence of two CRISPR/Cas systems adapted for the microbial engineering of *A. gossypii* will contribute significantly to facilitating system metabolic engineering approaches in this industrial fungus (Jiménez et al. 2020).

For the rapid genomic engineering of the industrial fungus *A. gossypii*, the Golden Gate assembly system was adopted. The expression platform contains interchangeable DNA modules, providing flexibility to use different integration loci, selectable markers, and regulatory sequences. The functionality of the system was used to develop strains capable of synthesizing polyunsaturated fatty acids (up to 35% of the total fatty acid content) (Ledesma-Amaro et al. 2018).

9.2.5 Metabolic Engineering to Increase Riboflavin Production

Metabolic engineering involves the reprogramming of cell metabolism to increase the production of specific metabolites or causing the cell to acquire the ability to produce new products. A number of strains overproducing riboflavin have been constructed based on metabolic engineering methods, i.e. targeted genetic changes in metabolic pathways. This was made possible through direct modifications of the riboflavin synthesis pathway or the purine pathway, central carbon metabolism, and glycine synthesis. Such changes are possible due to the increased expression of genes encoding enzymes of riboflavin metabolism pathways or its precursors by introducing additional copies of them or by replacing the native promoter with a strong, constitutive one. Other approaches are based on the disruption of competing pathways to modify regulatory genes (Liu et al. 2020). *A. gossypii* naturally produces a yellow pigment, which was identified as riboflavin (Afroz Toma et al. 2023). In the last decade, genetic and biochemical studies have determined the metabolic pathways and enzymes involved in the production of this vitamin (Abbas and Sibirny 2011; Revuelta et al. 2016a, b; Schwechheimer et al. 2016). The pathway has two branches, each with a precursor, ribulose-5-phosphate or guanosine triphosphate (GTP). GTP is synthesized through the purine nucleotide pathway, which is regulated by the product of the *BAS1* gene. GTP and ribulose-5-phosphate are converted to riboflavin by the action of enzymes encoded by six *RIB* genes (Ledesma-Amaro et al. 2013b; Jimenez et al. 2016).

Deep knowledge of the pathway has allowed its manipulation by rational metabolic engineering in order to increase the production of the vitamin. The overexpression and deregulation of the first two enzymes in the pathway of purine nucleotide synthesis de novo (phosphoribosyl pyrophosphate (PRPP) synthetase and phosphoribosyl pyrophosphate amidotransferase) increased the amount of riboflavin synthesized by *A. gossypii* (Jiménez et al. 2005, 2008). Furthermore, the elimination of a negative regulatory domain in the Bas1 enzyme further increased vitamin production (Mateos et al. 2006). The flux toward GTP and thus riboflavin was also enhanced by the overexpression of inosine 5′-monophosphate (IMP) dehydrogenase (*IMD3*), an intermediate gene of the pathway (Buey et al. 2015a, b).

Recently, the *RIB* genes were found to limit riboflavin production, as well as the competing gene *ADE12*, when analyzed in silico using the genome-scale metabolic model iRL776 (Ledesma-Amaro et al. 2014a, b, c). Both the overexpression of the *RIB* genes and the deletion of the *ADE12* gene lead to increased vitamin production, as predicted by the model (Ledesma-Amaro et al. 2015c).

The combination of such rational engineering approaches with random mutagenesis and selection has created vitamin B_2 overproducer strains, which are currently in use for the large-scale production of this product by the company BASF (Germany).

According to the data obtained by Schlösser et al. on *A. gossypii*, the expression of *RIB3*, *RIB4*, and *RIB5* genes encoding riboflavin biosynthetic enzymes increased during riboflavin overproduction, while the *RIB2* and *RIB7* genes remained constant. This specific transcription profile is observed with simultaneous spore formation and nutrient depletion in the medium (Schlösser et al. 2007). It was shown that oxidative stress and cell aging were involved in the overproduction of riboflavin in an *A. gossypii* riboflavin-overproducing mutant. This finding provides new insights into riboflavin production in *A. gossypii* and demonstrates the usefulness of disparity mutagenesis for the creation of new types of mutants for metabolic engineering (Kato et al. 2020).

It is suggested that the overproduction of riboflavin by *A. gossypii* is an ecological defense mechanism, as confirmed by its production in response to environmental stress. As a result of the interaction of endogenous riboflavin with light, reactive oxygen species (ROS) are generated, resulting in oxidative damage to DNA in mammals; however, it has also been shown that exogenous riboflavin protects *A. gossypii* spores against ultraviolet (UV) radiation. Therefore, Silva's group developed an assay called *Ashbya Comet Assay* to assess the genotoxic effects associated with the overproduction of riboflavin by this organism. This test involved measuring oxidative (hydrogen peroxide/menadione-mediated) and nonoxidative (camptothecin-mediated) DNA damage in *A. gossypii*. Through this experiment, it was determined that exposure to light mimicking sunlight during growth caused a significant increase in DNA damage in *A. gossypii* strains that overproduced riboflavin, but the same effects were not observed in strains that did not overproduce riboflavin. Finally, it was proven that *A. gossypii* strains capable of overproducing riboflavin, exposed to light, accumulated ROS intracellularly, and riboflavin production by these strains increased by 1.5-fold (Silva et al. 2019). Based on the above observations, it can be concluded that strains overproducing riboflavin are more susceptible to light-induced oxidative DNA damage; therefore, its biotechnological production using *A. gossypii* should also be controlled in terms of exposure to light. The industrial process of riboflavin production using the best strains of *A. gossypii* is carried out at a temperature of 30 °C, in 100 m^3 fermenters during fed-batch cultivation, using vegetable oils as a carbon source, as well as complex nitrogen sources, such as soy flour or corn steep liquor. This type of breeding is carried out for 8 days, and the process is divided into two phases, i.e., growth and production. Thanks to process optimization and its control, it is now possible to produce riboflavin by *A. gossypii* up to 20 g/l (Schwechheimer et al. 2016; Wang et al. 2021).

9.2.6 Other Biotechnological Products of A. gossypii

In addition to the production of riboflavin, *A. gossypii* has been engineered for the production of other biotechnological products, such as lipids, nucleosides, folic acid, or proteins. Microbial lipids or single-cell oils are considered an ideal source of biodiesel and chemicals (Ledesma-Amaro 2015; Ledesma-Amaro and Nicaud 2016a, b; Ledesma-Amaro et al. 2016a, b; Aguiar et al. 2017). *A. gossypii* is known to have a very active lipid metabolism, and this characteristic has been exploited to maximize lipid production by metabolic engineering. In one work, the capacity of this fungus to accumulate lipids was increased by blocking beta oxidation and expressing genes from the oleaginous yeast *Yarrowia lipolytica* (Ledesma-Amaro et al. 2014a, b, c, 2015a, b, c). In another work, the profile of the fatty acids produced was adjusted by manipulating the elongase and desaturase pathways (Ledesma-Amaro et al. 2014a, b, c) or the expression of hydroxylase (Lozano-Martínez et al. 2016b). Later on, this host was engineered to enhance lipid production using low-cost carbon sources, such as molasses (Lozano-Martínez et al. 2016a, b) or xyloses (Díaz-Fernández et al. 2017). A lipogenic xylose-utilizing strain of *A. gossypii* was used to apply a metabolic engineering approach to further increase the biosynthesis of lipids. Three genomic manipulations were applied: the overexpression of a feedback inhibition-resistant form of the acetyl coenzyme A carboxylase enzyme; the expression of a truncated form of Mga2, a regulator of the main Δ9 desaturase gene; and the overexpression of an additional copy of *DGA1* that codes for diacylglycerol acyltransferase. The performance of the engineered strain was evaluated in culture media containing mixed formulations of corn-cob hydrolysates, sugarcane molasses, or crude glycerol. This work reports engineered strains of *A. gossypii* able to accumulate 30–40% of cell dry weight (CDW) in lipid content using wastes as raw materials (Díaz-Fernández et al. 2019).

Nucleosides, especially inosine and guanosine, are metabolites with industrial applications, such as flavor enhancers or pharmaceuticals (Ledesma-Amaro et al. 2013a, b). These compounds are synthesized from the purine pathway, which is a very active pathway in *A. gossypii* used for the production of riboflavin, and thus metabolic engineering strategies have redirected the metabolic flux toward inosine and guanosine by the overexpression of purine pathway genes and the deletion of competing pathways, e.g., minimizing riboflavin production (Ledesma-Amaro et al. 2015a, b, c, 2016a, b). These works represent the highest nucleoside titer ever produced in a eukaryote organism, which suggests its potential use as an industrial producer.

A. gossypii also presents the capacity to produce and secrete a high amount of proteins. In this regard, *A. gossypii* has been proposed as a potential producer of recombinant proteins. As pioneers, our laboratory patented the use of this fungus to produce proteins, specifically heterologous cellulases (Althöefer et al. 2001). Subsequently, the expression of *Trichoderma reesei* cellulases (Ribeiro et al. 2010) and a beta-galactosidase of *Aspergillus niger* (Magalhes et al. 2014) has been reported. Furthermore, the natural ability of *A. gossypii* to produce proteins has been improved by both random and targeted mutageneses (Ribeiro et al. 2013). In addition, the characteristics of protein N-glycosylation in this fungus have been studied in detail (Aguiar et al. 2014a, b).

Some work has generated strains of *A. gossypii* capable of producing substantial amounts of folic acid. Folic acid, or vitamin B_9, is an essential compound for the human diet, which, as riboflavin, is synthesized from GTP. These overproducing strains accumulate up to 300 times more folic acid than the wild type, and their production titers are the highest described so far (Serrano-Amatriain et al. 2016).

9.3 *C. famata*

9.3.1 *Transformation*

An efficient transformation method is crucial for the genetic engineering of yeasts as a way to introduce foreign DNA into their cells, which is necessary for the construction of new strains. In the case of *C. famata*, some protocols were adapted that were previously used for other species' transformations (Voronovsky et al. 2002).

C. famata can be modified with both spheroplast transformation and electroporation. The first one requires a number of various buffers, such as lyticase as a cell wall lysing agent, and depends on the quality of all components; thus, electroporation is the main method for *C. famata* transformation. It requires cells in their exponential phase of growth to be briefly incubated with dithiothreitol (DTT), then washed with water and 1 M sucrose prior to electroporation. The frequency of transformation is up to 6.3×10^4 and 1×10^5 transformants/μg DNA, respectively.

9.3.2 Selectable Markers

The development of selectable markers is an important step in genetic engineering because it is usually the main way to select transformants. However, since *C. famata* is still a poorly studied yeast, there are only a few selectable markers available at the moment that significantly limit the research potential of this species.

The first selectable marker adapted for *C. famata* was the *LEU2* gene. Using UV mutagenesis with subsequent growth analysis, the 16 strains unable to grow under leucine-depleted conditions were isolated among more than 40,000 colonies. Further analysis using spheroplast transformation with *Saccharomyces cerevisiae LEU2* harboring plasmids allowed identification among leucine auxotrophs *C. famata leu2* mutants and proved the *S. cerevisiae LEU2* gene as a suitable marker (Voronovsky et al. 2002).

Another auxotrophy-based selectable marker, the *ADE2* gene, was recently developed for *C. famata* using CRISPR-cas9 mutagenesis. The obtained strain has a pink/red colony color and is unable to grow without adenine in the medium. The *ADE2 gene* was isolated from *C. famata* using the basic local alignment search tool (BLAST) to identify its sequence based on homology with the corresponding genes of *C. albicans* and *Debaryomyces hansenii*. An additional benefit of using this marker is the change to white/light-pink color of transformants, which makes selection easier (Lyzak et al. 2016).

The gene *ble* that conferred resistance to the bleomycin family of antibiotics was the first dominant marker used to select the *C. famata* transformants. It was found that *C. famata* growth is inhibited by phleomycin (2.0 mg/l) in a rich yeast extract peptone dextrose (YPD) medium. The *Staphylococcus aureus* version of this gene was used due to the absence of CTG codons in its sequence and was expressed under the control of the *C. famata TEF1* promoter. A transformation frequency obtained was approximately 100 transformants per μg of DNA (Dmytruk et al. 2006).

The *IMH3* gene, which encodes IMP dehydrogenase, can also serve as a selective marker (Köhler et al. 1997). The homolog of the *D. hansenii IMH3* gene and the minimal toxic concentration of mycophenolic acid (15 mg/l) were used for the transformation of *C. famata*. The transformation frequency was approximately 30 transformants per μg of DNA. Furthermore, since *IMH3* participates in purine biosynthesis, it also provides an additional increase in riboflavin synthesis (Dmytruk et al. 2011).

The modified *D. hansenii ARO4* gene encoding 3-deoxy-d-arabino-heptulosonate-7-phosphate (DAHP) synthase not sensitive to retroinhibition with tyrosine, which provides the first step of aromatic amino acid biosynthesis, can also be used as the dominant selectable marker. It confers resistance to fluorophenylalanine (Cebollero and Gonzalez 2004). The transformants can be selected on a mineral medium containing 2.5 g/l fluorophenylalanine and supplemented with 0.8 g/l tyrosine. The transformation frequency for the *ARO4m* marker was around 100 transformants/μg of DNA (Dmytruk et al. 2011). Furthermore, *ARO4m* could confer multicopy

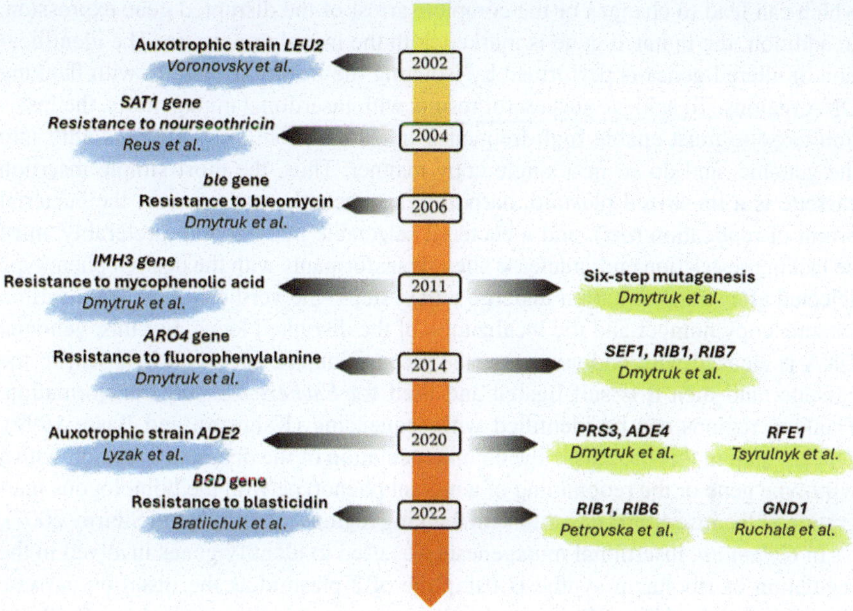

Fig. 9.3 A timeline presenting the most important discoveries that have a direct impact on the increase in the production of riboflavin by the yeast *Candida famata*. The left (blue underline) concerns transformation and selection experiments; the right (yellow underline) shows the main genetic modifications, including mutagenesis and gene overexpression

plasmid integration when the concentration of toxic agents increases (Dmytruk et al. 2014).

The *SAT1 gene* allows yeasts to grow in the medium with nourseothricin as a selective agent. It encodes nourseothricin acetyltransferase, and its sequence was recently adapted for use in the CTG clade of yeasts (Reuss et al. 2004; Millerioux et al. 2011). It is extremely useful as it allows the selection of *C. famata* transformants at low concentration of antibiotic: 2–4 mg/l.

In recent work, Bratiichuk and coauthors proved the effectiveness of blasticidin as a selective agent for the isolation of *C. famata* transformants; they used the *BSD* gene from *Aspergillus terreus* coding for blasticidin S deaminase (Bratiichuk et al. 2020) (Fig. 9.3).

9.3.3 *Insertional Mutagenesis*

Insertional mutagenesis is a method for the identification of genes and their functions and is suitable even for yeasts whose genomes are not yet sequenced. The main idea of this method is the generation of a mutant phenotype library with integration insertion cassette, a small nucleotide sequence, into a genome of a host cell, which can lead to changes or the complete arrest of the disrupted gene expression. In addition, the mutated gene is marked with the insertion cassette. The identification of altered genes is performed by isolating the insertion cassette with flanking DNA regions. To achieve successful results with insertional mutagenesis, the insertion cassette must enable high-frequency transformation, integrate randomly into the genome, and do so in a single-copy manner. Thus, the most simple insertion cassette is a linearized plasmid, harboring a selectable marker gene, the bacterial origin of replication (*ori*), and a bacterial selectable marker, and, preferably, must be lacking restriction endonuclease sites. Transformants with the desired phenotype selected after transformation undergo a next step—the verification of the insertion cassette copy number and the localization of the disrupted locus. For this, genomic DNA is cleaved with endonuclease, the recognition site of which is absent in the cassette, and then it is self-ligated and used for *Escherichia coli* transformation. Flanking regions can be identified with sequencing (Kahmann and Basse 1999). The last step in this process is the complementation of the obtained mutation with a wild-type gene or the reobtaining of a mutant phenotype with the homologous integration of the insertional cassette with flanking regions (Dmytruk and Sibirny 2007).

In *C. famata*, insertional mutagenesis was used to identify genes involved in the regulation of riboflavin synthesis using the pL2 plasmid as the insertion cassette (Dmytruk et al. 2006). Three mutants that were unable to overproduce riboflavin after transformation were obtained and picked for further analysis. Although, in one transformant, the structural gene of riboflavin biosynthesis (*RIB1*) was disrupted, the other two had mutations in the genes *MET2* and *SEF1*.

9.3.4 The Reporter System and Promoters

Gene promoters are the main regulators of gene expression. Constitutive promoters are usually used for the expression of genes, products that are desirable to be in the cell all the time, whereas inducible ones can be regulated using specific chemical compounds. Thus, it is extremely important to have operating promoters from the host cell as they are usually the most efficient. For the evaluation of promoter activity, the report system could be a useful tool.

For *C. famata*, it was proposed to use the gene encoding β-galactosidase as a reporter gene because the activity of this enzyme can be easily detected and measured (Uhl and Johnson 2001; Fu and Xiao 2006). The first step in this work was the isolation of the mutants of *C. famata* that are unable to grow on a lactose-containing medium as the sole carbon source through UV mutagenesis. A mutant, βg6, was used as the parental strain for the report system (Ishchuk et al. 2008). As mentioned above, *C. famata* belongs to the yeast clade with an alternative genetic code (it reassigns the leucine CTG codon as serine). This fact could be one of the obstacles in the search for an active variant of the gene encoding β-galactosidase. *LAC4* from *Kluyveromyces lactis* has only two CTG codons. Therefore, it was a suitable candidate for a report gene. Several promoters of *C. famata* and the closely relative flavinogenic yeast *D. hansenii* have been isolated, which are homologs of that previously used in other yeasts. Only one promoter from *D. hansenii* demonstrated a reasonable level of expression, *DhPGK1*; the others, *DhPGI*, *DhPHO5*, and *DhTPI1*, were quite weak. The promoters of *C. famata*, *CfTEF1* and *CfPGI*, were active; however, the activity of *CfTEF1* was three times higher than that of *CfPGI*. Therefore, the *TEF1* promoter was chosen as the strongest promoter for *C. famata*.

9.3.5 The CRISPR-Cas9 System

The CRISPR-Cas9 system is a recently developed method for genome editing and modification. It is based on one of the bacterial defensive mechanisms against foreign DNA and consists of two components: RNA-guided DNA endonuclease Cas9 and small guide RNA (gRNA). The complex cas9-gRNA can recognize a DNA sequence complementary to 20 nucleotides at the 5′ end of gRNA, followed by NGG. After the recognition of the target site, Cas9 generates a double-strand break (DSB) in this region, approximately three to four nucleotides before NGG. This DSB can be repaired with the host cell via two possible mechanisms: nonhomologous end joining (NHEJ) or homology-directed repair (HDR). In the case of NHEJ, it is possible to generate only small mutations, like mono- or oligo-nucleotide insertions or deletions, which usually result in premature translation stop or frameshift, so the general outcome is gene knockout. Although HDR requires an additional component, a repair template, it has much more wide capabilities for precise gene editing, tagging, or integration. The CRISPR-Cas9 system demonstrates high efficiency and flexibility and might be adapted for various hosts (Gilbert et al. 2013; Xie and Yang 2013; Sander and Joung 2014).

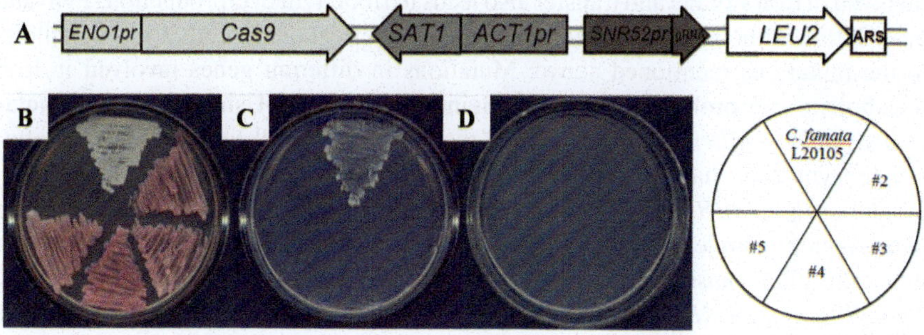

Fig. 9.4 Linear scheme of plasmid based-CRISPR-Cas9 encoding the Cas9 protein under control of *CaENO1* promoter, guide targeting region of *CfADE2* gene controlling with *SNR52* promoter from *C. albicans*, *SAT1* and *ScLEU2* as selective markers and *Cf*ARS-element (**a**). Growth test of parental (*C. famata* L20105, *leu2*) and selected *ade2* strains (#2–5) on YPD (**b**), on YNB with leucine, (**c**) and on YNB without leucine (**d**)

The CRISPR-Cas9 system becomes available for application in *C. famata* after its adaptation from another flavinogenic yeast of the CTG clade, the pathogenic yeast *C. albicans* (Vyas et al. 2015). The CRISPR-Cas9 system has been demonstrated as a system that works in *C. famata* by removing the *ADE2* model gene (Fig. 9.4) (Lyzak et al. 2016). Although the frequency of mutation events is still quite low, this parameter could be significantly improved via promoter modification.

9.3.6 Regulation of Riboflavin Biosynthesis

The regulation of the riboflavin biosynthetic pathway is not completely solved for either the hemiascomycete *A. gossypii* or the yeast *C. famata*. As mentioned earlier, some species of the genera *Pichia*, *Debaryomyces*, *Schwanniomyces*, *Torulopsis*, *and Candida* are able to overproduce riboflavin under cultivation in iron-depleted media. The purpose of the regulation of iron-based riboflavin biosynthesis is unknown. It could be hypothesized that riboflavin secreted by flavinogenic yeasts under iron deficiency conditions could reduce practically insoluble Fe^{3+} ions to more soluble Fe^{2+} ions, in that way providing iron to the cells (Noble 2013).

During riboflavin overproduction, the derepression of riboflavin biosynthetic enzymes (GTP cyclohydrolase, desaminase, riboflavin synthase, etc.) occurs. In *M. guilliermondii*, nine mutant groups, *rib80*, *rib81*, *hit1*, *red1–red6*, have been described that overproduce riboflavin under high iron content in media (Papon et al. 2015). The mutation *rib80* is recessive and has a pleotropic effect. It stimulates not only the expression of riboflavin biosynthesis genes but also the transport and accumulation of iron. Mutations *rib81* and *red6* lead to the constitutive expression of *RIB1*, even in an iron-supplemented condition (Boretsky et al. 2005). Furthermore, the mutations *rib80*, *rib81*, and *hit1*; iron deficiency; and cell exposure to Co^{2+} lead to oxidative stress. The deletion of the *YFH1* gene encoding the frataxin homolog involved in iron storage and transfer also leads to riboflavin overproduction (Pynyaha et al. 2009). Although *YFH* deletion had the same effect on the cell, it was nonallelic to the mutations mentioned above. Mutations in different genes involved in iron metabolism can probably cause riboflavin oversynthesis. Using insertional mutagenesis in *M. guilliermondii*, the role of the *VMA1* gene in the regulation of riboflavin biosynthesis, has been proved (Boretsky et al. 2011). On the other hand, mutations in the *RIB83* gene have been selected, which lead to the loss of riboflavin overproduction under iron deficiency conditions. The *rib83* mutation is epistatic to mutations that cause riboflavin overproduction, such as *rib80* or *rib81*. Later, *M. guilliermondii RIB83* was shown to be homologous to *C. famata SEF1* coding for the transcription activator (Dmytruk and Sibirny 2012).

So far, only one regulatory gene, *SEF1*, involved in the regulation of riboflavin synthesis, has been identified for *C. famata*. Apparently, *SEF1* is involved in a positive regulation of riboflavin biosynthesis. Moreover, its overexpression leads to an increase both in riboflavin production and in the stability of the industrial overproducing strain dep8 (Dmytruk et al. 2006). The product of the *SEF1* gene, the transcriptional regulator Sef1, belongs to the Gal4 family of transcriptional activators and has the zinc finger domain at its N-terminus. To understand the interaction between Sef1 and its putative DNA targets, we developed a hybrid system on the base of the nonflavinogenic yeast *S. cerevisiae* as a useful tool. The one-hybrid system is a derivative of the yeast two-hybrid system. This specific variation is generated to investigate protein-DNA interactions and uses the following scheme: the DNA sequence of interest is fused with the gene, which encodes the reporter protein and the gene of the studied protein under the control of the regulated promoter. Promoters of the structural genes of the riboflavin biosynthesis pathway or of the *SEF1* promoter were fused with the *LAC4* gene encoding β-galactosidase and cotransformed with plasmid, harboring *C. famata SEF1* under the control of the *S. cerevisiae GAL1* promoter. The β-galactosidase activity of the resulting strains grown in a glucose- or galactose-containing medium represented the strength of the binding of the transcription factor to the promoter. The strongest level of activation was observed for the promoters of the *RIB1*, *RIB6*, and *RIB7* genes. These genes encode enzymes of the first and last steps of the riboflavin biosynthetic pathway, so they have to be tightly regulated. Furthermore, the truncation of the *RIB1* promoter allowed the localization of the specific region apparently involved in Sef1 binding. Additionally, evidence of *SEF1* autoregulation has been obtained (Lyzak et al. 2015).

To find out the mechanisms for the upregulation of *SEF1*, the expression level of this gene was tested using qRT-PCR in three yeast species: *C. famata*, *S. cerevisiae*, and *Scheffersomyces stipitis*, cultivated in a medium with high or low iron content. In *S. cerevisiae*, the level of *SEF1* expression in media with low and high iron content was unchanged. In *C. famata*, as expected, an inverse correlation was observed between iron supply and *SEF1* expression; that is, iron deficiency led to an increase in the level of *SEF1* expression. It is interesting to note that in the nonflavinogenic yeast of the CUG clade *S. stipitis*, an inverse correlation was also found between iron supply and *SEF1* expression (though expression level was quite low when it comes to *C. famata*, even in cells cultivated under iron limitation conditions) (O. Lyzak, K. Dmytruk, A. Sibirny, unpublished). It was shown that the *SEF1* promoters from other flavinogenic (*Candida albicans*) and nonflavinogenic (*Candida tropicalis*) yeasts fused with the open reading frame (ORF) of the *SEF1* gene from *C. famata* are able to restore riboflavin oversynthesis in the *sef1Δ* mutant (Andreieva et al. 2020).

9.3.7 *Development of Strains with Increased Riboflavin Production*

A riboflavin overproducing strain has been constructed by the co-overexpression of the genes *SEF1*, *RIB1*, and *RIB7* in a nonreverting riboflavin-producing *C. famata* AF-4 strain (Dmytruk et al. 2014). Under fed-batch fermentation in a 7 l laboratory bioreactor, the constructed strain accumulated up to 16.4 g/l of riboflavin in an optimized medium, representing one of the most known active riboflavin producers. Furthermore, when the two modified genes, *PRS3* (encoding PRPP synthetase) and *ADE4* (encoding PRPP amidotransferase), from *D. hansenii* were introduced and coexpressed in this constructed strain, a twofold increase in riboflavin production is observed (Dmytruk et al. 2020)

Recently, it was found that the overexpression of *RIB1* and *RIB6* genes coding for enzymes GTP cyclohydrolase II and 3,4-dihydroxy2-butanone-4-phosphate synthase, which catalyze the initial steps of riboflavin synthesis, elevated riboflavin production by 13–28% relative to the parental riboflavin-overproducing strains (Petrovska et al. 2022) (Fig. 9.3). It is known that the pentose phosphate pathway influences the formation of substrates that enter the riboflavin production pathway. Data regarding the overproduction of riboflavin as a result of the overexpression of the *GND1* gene turned out to be particularly interesting. This gene encodes a 6-phosphogluconate dehydrogenase involved in the oxidative branch of the pentose phosphate pathway responsible for the synthesis of ribulose-5-phosphate, which serves as one of two precursors of riboflavin synthesis (Fig. 9.5). The results clearly revealed that such overexpression apparently leads to an increase in the intracellular

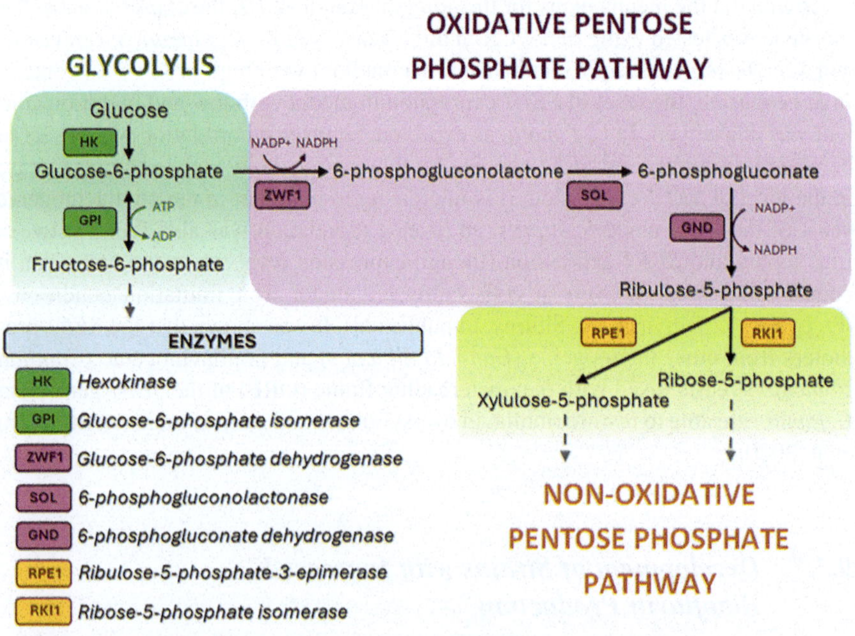

Fig. 9.5 Overview of a fragment of the pentose phosphate pathway (PPP), its enzymes, and connection with glycolysis. The graph shows glycolysis, the oxidative pentose phosphate pathway, and the nonoxidative pentose phosphate pathway, marked with green, purple, and yellow backgrounds, respectively. The end product of oxidative PPP is ribulose-5-phosphate, which is a precursor of riboflavin

pool of ribulose-5-phosphate, which in turn leads to the increased production of riboflavin by the constructed strains also on waste such as whey (Ruchala et al. 2022).

Yeasts of the *Candida* genus, such as *C. famata*, are capable of the internal storage of riboflavin. Studies have shown that genetic modifications of this yeast, such as the overexpression of key riboflavin biosynthetic genes, lead to a significant increase in the amount of this vitamin inside cells (Liu et al. 2020). Nevertheless, overproduced riboflavin accumulates in the cultural medium rather than in the cells, suggesting the existence of the special mechanisms involved in riboflavin excretion. The corresponding protein and gene have not been identified in yeasts. The effects of the overexpression of the yeast homolog of the human riboflavin efflux protein Bcrp (breast cancer resistance protein) on riboflavin production in the flavinogenic yeast have been studied. It was found that the overexpression of the *D. hansenii RFE1* gene in previously constructed riboflavin overproducer *C. famata* (Dmytruk et al. 2011, 2014) led to near twice increase of riboflavin production (Tsyrulnyk et al. 2020). It is known that in the pathogenic flavinogenic yeast *C. albicans*, Sfu1 (GATA-type transcription factor) represses the *SEF1* gene (Chen et al. 2011). It was found that the deletion of the *SFU1* gene in wild-type *C. famata* leads to riboflavin oversynthesis. Moreover, it was shown that the disruption of the *VMA1* gene (coding for vacuolar ATPase subunit A) also results in riboflavin oversynthesis in *C. famata*. The deletion of the *SFU1* gene in wild-type *C. famata* leads to an increase in riboflavin production by the *C. famata* (Andreieva et al. 2020).

Recently, we showed that the overexpression of own transcription activator *SEF1* gene under the control of the promoter of the homologous *LAC4* gene encoding β-galactosidase in *C. famata* overproducers led to an increase in riboflavin production on media with lactose and whey (Ruchala et al. 2022). Thus, whey could be used as a promising substrate for the development of technology for the industrial production of vitamin B_2.

The price of riboflavin significantly depends on the cost of the carbon substrate used (Kato and Park 2012). Using waste as a carbon source would be particularly beneficial. Engineered strains of the yeast *C. famata* have successfully overproduced riboflavin on lignocellulosic hydrolysate. Riboflavin accumulation increased in the recombinant strains *C. famata* overexpressing the *XYL1* and *XYL2* genes, which code for xylose reductase and xylitol dehydrogenase, respectively. The highest riboflavin accumulation during bioreactor cultivation by BRPI/XYL1, using bagasse hydrolysate as the carbon source, reached 1.5 g/l (Dzanaeva et al. 2024).

9.3.8 Construction of Riboflavin Derivative Producers

In contrast to riboflavin biosynthesis, which occurs only in prokaryotes, fungi and plants; flavin nucleotides, i.e. FMN (flavin mononucleotide) and FAD (flavin adenine dinucleotide) are produced by all organisms. This is due to the fact that the systems of these flavin coenzymes are involved in various biochemical processes in the cell, such as mitochondrial electron transport, fatty acid oxidation, and the metabolism of vitamins B6 and B12. Moreover, flavin nucleotides are commonly used in pharmacies and the food industry because they are more soluble in water than riboflavin. Currently, FMN produced by chemical synthesis is available on the market (it constitutes 3% of the total amount of riboflavin produced). Although the production of this flavin nucleotide is cheap, which is an advantage of this process, it contains many flavin impurities that can act as riboflavin antagonists. There are many microorganisms in nature that can overproduce riboflavin but are completely unable to oversynthesize FMN and FAD. The scientific literature contains numerous attempts at the microbiological production of these cofactors through microbial synthesis (Nakagawa et al. 1995; Kitatsuji et al. 1996; Mack et al. 1998; Frago et al. 2008; Lohithakshan et al. 2021), but their main disadvantage is the fact that a mixture of FMN and FAD is produced, as well as the fact that production can only occur in conditions of high adenosine triphosphate (ATP) and riboflavin concentrations in the culture medium for the phosphorylation reaction. Moreover, the advantage of using eukaryotes over prokaryotes is the fact that the former use separate enzymes to produce FMN and FAD, while prokaryotes use a bifunctional enzyme, riboflavin kinase/FAD synthetase. The group of Fedorovych et al. undertook the task of isolating riboflavin overproducers using the flavinogenic yeast *C. famata* as a genetic background. *FMN1* and *FAD1* gene sequences were found in a genomic database and cloned under the control of the native, strong constitutive *TEF1* promoter. As a result of the experiments conducted, it was possible to isolate strains that accumulated 200–250 mg/l of FMN under conditions of high biomass concentration (cell biomass, 5 g of dry matter per 1 kg). After optimizing the culture conditions, 318.2 mg/l FMN was obtained from the best-isolated strains with the overexpression of *FMN1*. Next, a plasmid with an additional copy of *FAD1* was introduced into the strain overproducing FMN, as a result of which the isolated strains were able to produce 3.9–9.6 mg/l of FAD. Moreover, a plasmid was constructed containing, in addition to *FAD1*, an additional copy of *FMN1*, which, interestingly, was able to produce FAD at a level of 30–60 mg/l after 48 h of culture. Nevertheless, the production of FAD did not exceed 18% of the total amount of flavins in the culture medium; therefore, in order to increase the production of this collator, further modifications should be made, e.g., the inactivation of pyrophosphatase genes involved in FAD hydrolysis or the multicopy expression of *FMN1* and *FAD1* in the most efficient producers of riboflavin (Fedorovych et al. 2021).

In addition to the biotechnological usefulness of riboflavin and flavin nucleotides, there are antibiotics, naturally produced by actinomycetes (*Streptomyces davaonensis* and *Streptomyces cinnabarinus*), which are flavins. Studies prove that the biosynthetic precursor of roseoflavin, aminoriboflavin, inhibits the growth of gram-positive bacteria, *Streptococcus aures* (including methicillin-resistant *Streptococcus aures* (MRSA)), *Listeria monocytogenes*, *Clostridium difficile*, *Enterococcus faecalis*, *Bacillus subtilis*, as well as some gram-negative bacteria capable of transporting riboflavin. Thanks to the research of the group of Mack et al., the roseoflavin biosynthesis pathway has been deciphered (Jankowitsch et al. 2012; Mora-Lugo et al. 2019). It turns out that the substrate for its production is FMN. Therefore, having extensive experience in the study and isolation of riboflavin producers, the group of Dmytruk et al. undertook the task of constructing yeast producers for these antibiotics, using as a genetic background the flavinogenic yeast *C. famata* and overproducing riboflavin, as well as classical producers of foreign compounds expressing riboflavin biosynthetic genes, the yeast *Komagataella phaffii* (Dmytruk et al. 2023). A number of plasmid constructs were prepared containing structural genes for roseoflavin production: *rosB* (coding for 8-demethyl-8-amino-riboflavin-5′-phosphate synthase, EC 2.6.1.114), *rosC* (coding for yet unknown phosphatase), and *rosA* (coding for AF-phosphate phosphatase and dimethyltransferase, EC 2.1.1.343), which were codon optimized for the organisms in which they were expressed. As a result of the experiments carried out, it was possible to successfully isolate both roseoflavin and aminoriboflavin producers, but *K. phaffii* synthesized approximately 1.3 times more antibiotics in flasks compared to recombinant *C. famata* strains. Interestingly, despite the correct expression of the *rosA* gene, it was not possible to isolate the roseoflavin producer in the genetic background of the yeast *C. famata*. The authors determined the proper activity of *rosA*, which confirmed that it was caused by the lack of activity of this enzyme. To conclude, it can be stated that despite the fact that *C. famata* is a natural, active overproducer of riboflavin, which is a precursor of flavin antibiotics, it is an inferior organism for the production of these antibiotics compared to *K. phaffii*, which is capable of producing 130 mg/l roseoflavin, thus exceeding the capabilities of natural producers, i.e., the actinomycete *S. davaonensis*, by 9.3 times (Mora-Lugo et al. 2019; Dmytruk et al. 2023).

9.4 Conclusions

A. gossypii and *C. famata* are microorganisms that have been intensively studied. Advanced progress in the development of efficient molecular tools promises a more complete realization of the biotechnological potential of both species. *A. gossypii* is an important industrial riboflavin producer. The application of recently developed tools of molecular research opens possibilities for further improvements. In addition to riboflavin production, *A. gossypii* has been genetically modified for the production of lipids, inosine and guanosine, folic acid (vitamin B_9), and industrially important proteins. Until now, the flavinogenic potential of *C. famata* has not been extensively explored. Industrial riboflavin production based on *C. famata* has been shut down due to the instability of the producer used (Abbas and Sibirny 2011). Much effort is still needed to elucidate the molecular mechanisms responsible for the iron-dependent regulation of riboflavin synthesis in *C. famata*. Genetically stable riboflavin-overproducing strains of *C. famata* are known. Recent molecular studies of *C. famata* permit fast future progress in the construction of the robust competitive biotechnological producers of riboflavin and its derivatives (flavin nucleotides, antibiotic roseoflavin, etc.). Although this yeast has not been used at the industrial scale in recent years, there are still a number of research advances in its genetic engineering, suggesting that it may develop into a new platform for riboflavin and its derivatives (flavin nucleotides, antibiotic roseoflavin, etc.) overproduction.

Acknowledgment This work was supported by the National Research Foundation of Ukraine, grant 2023.03/0035, and the National Science Centre of Poland in the frame of the National Science Centre of Poland (NCN) grant Opus UMO-2021/41/B/NZ1/01224 as well as Preludium 22 grant No. UMO-2023/49/N/NZ1/01040. This work was also partially supported by "Presidential Discretionary-Ukraine Support Grants" from Simons Foundation, Awards No. 1030281 and No. 1290613.

References

Abbas CA, Sibirny AA (2011) Genetic control of biosynthesis and transport of riboflavin and flavin nucleotides and construction of robust biotechnological producers. Microbiol Mol Biol Rev 75(2):321–360

Afroz Toma M, Rahman MH, Rahman MS, Arif M, Nazir KHMNH, Dufossé L (2023) Fungal pigments: carotenoids, riboflavin, and polyketides with diverse applications. J Fungi (Basel) 9(4):454

Aguiar TQ, Dinis C, Domingues L (2014a) Cre-loxP-based system for removal and reuse of selection markers in *Ashbya gossypii* targeted engineering. Fungal Genet Biol 68:1–8

Aguiar TQ, Dinis C, Magalhães F et al (2014b) Molecular and functional characterization of an invertase secreted by *Ashbya gossypii*. Mol Biotechnol 56:524–534

Aguiar TQ, Silva R, Domingues L (2017) New biotechnological applications for *Ashbya gossypii*: Challenges and perspectives. Bioengineered 8(4):309–315

Althöefer H, Pompeus M, Revuelta JL, Santos M, Jiminez A, Benito R, Santamaria R, Fernandez J (2001) Protein production using Ashbya gossypii. Publication date: 05-04-2001. International Filling date: 05-04-2001. Patent WO/2001/023576

Andreieva Y, Petrovska Y, Lyzak O, Liu W, Kang Y, Dmytruk K, Sibirny A (2020) Role of the regulatory genes *SEF1*, *VMA1* and *SFU1* in riboflavin synthesis in the flavinogenic *yeast Candida famata* (*Candida flareri*). Yeast 37(9–10):497–504

Boretsky YR, Kapustyak KY, Fayura LR, Stasyk OV, Stenchuk MM, Bobak YP, Drobot LB, Sibirny AA (2005) Positive selection of mutants defective in transcriptional repression of riboflavin synthesis by iron in the flavinogenic yeast *Pichia guilliermondii*. FEMS Yeast Res 5(9):829–837

Boretsky YR, Pynyaha YV, Boretsky VY, Fedorovych DV, Fayura LR, Protchenko O, Philpott CC, Sibirny AA (2011) Identification of the genes affecting the regulation of riboflavin synthesis in the flavinogenic yeast *Pichia guilliermondii* using insertion mutagenesis. FEMS Yeast Res 11(3):307–314

Bratiichuk D, Kurylenko O, Vasylyshyn R, Zuo M, Kang Y, Dmytruk K, Sibirny A (2020) Development of new dominant selectable markers for the nonconventional yeasts *Ogataea polymorpha* and *Candida famata*. Yeast 9–10:505–513

Buey RM, Ledesma-Amaro R, Balsera M, de Pereda JM, Revuelta JL (2015a) Increased riboflavin production by manipulation of inosine 5′-monophosphate dehydrogenase in *Ashbya gossypii*. Appl Microbiol Biotechnol 99(22):9577–9589

Buey RM, Ledesma-Amaro R, Velázquez-Campoy A, Balsera M, Chagoyen M, De Pereda JM, Revuelta JL (2015b) Guanine nucleotide binding to the Bateman domain mediates the allosteric inhibition of eukaryotic IMP dehydrogenases. Nat Commun 6:8923

Cebollero E, Gonzalez R (2004) Comparison of two alternative dominant selectable markers for wine yeast transformation. Appl Environ Microbiol 70(12):7018–7023

Chen C, Pande K, French SD, Tuch BB, Noble SM (2011) An iron homeostasis regulatory circuit with reciprocal roles in *Candida albicans* commensalism and pathogenesis. Cell Host Microbe 10(2):118–135

Demain AL (1972) Riboflavin oversynthesis. Annu Rev Microbiol 26:369–388

Demain AL (2007) The business of biotechnology. Ind Biotechnol 3:269–283
Díaz-Fernández D, Lozano-Martínez P, Buey RM, Revuelta JL, Jiménez A (2017) Utilization of xylose by engineered strains of *Ashbya gossypii* for the production of microbial oils. Biotechnol Biofuels 10(1):3
Díaz-Fernández D, Aguiarb TQ, Martína VI, Romanib A, Silvab R, Dominguesb L, Revuelta JL, Jiménez A (2019) Microbial lipids from industrial wastes using xylose-utilizing *Ashbya gossypii* strains. Bioresour Technol 293:122054
DiCarlo JE, Norville JE, Mali P, Rios X, Aach J, Church GM (2013) Genome engineering in *Saccharomyces cerevisiae* using CRISPR-Cas systems. Nucleic Acids Res 41(7):4336–4343
Ding Q, Ye C (2023) Microbial cell factories based on filamentous bacteria, yeasts, and fungi. Microb Cell Fact 22:20
Dmytruk KV, Sibirny AA (2007) Molecular mechanisms of insertional mutagenesis in yeasts and mycelium fungi. Russ J Genet 43(8):835–845
Dmytruk KV, Sibirny AA (2012) *Candida famata* (*Candida flareri*). Yeast 29(11):453–458
Dmytruk KV, Voronovsky AY, Sibirny AA (2006) Insertion mutagenesis of the yeast *Candida famata* (*Debaryomyces hansenii*) by random integration of linear DNA fragments. Curr Genet 50(3):183–191
Dmytruk KV, Yatsyshyn VY, Sybirna NO, Fedorovych DV, Sibirny AA (2011) Metabolic engineering and classic selection of the yeast *Candida famata* (*Candida flareri*) for construction of strains with enhanced riboflavin production. Metab Eng 13(1):82–88
Dmytruk K, Lyzak O, Yatsyshyn V, Kluz M, Sibirny V, Puchalski C, Sibirny A (2014) Construction and fed-batch cultivation of *Candida famata* with enhanced riboflavin production. J Biotechnol 172:11–17
Dmytruk KV, Ruchala J, Fedorovych DV, Ostapiv RD, Sibirny AA (2020) Modulation of the purine pathway for riboflavin production in flavinogenic recombinant strain of the yeast *Candida famata*. Biotechnol J 15(7):e1900468
Dmytruk KV, Ruchala J, Fayura LR, Chrzanowski G, Dmytruk OV, Tsyrulnyk AO, Andreieva YA, Fedorovych DV, Motyka OI, Mattanovich D, Marx H, Sibirny AA (2023) Efficient production of bacterial antibiotics aminoriboflavin and roseoflavin in eukaryotic microorganisms, yeasts. Microb Cell Fact 22(1):132. https://doi.org/10.1186/s12934-023-02129-8
Dünkler A, Wendland J (2007) Use of *MET3* promoters for regulated gene expression in *Ashbya gossypii*. Curr Genet 52(1):1–10
Dzanaeva LS, Wojdyła D, Fedorovych DV, Ruchala J, Dmytruk KV, Sibirny AA (2024) Riboflavin overproduction on lignocellulose hydrolysate by the engineered yeast *Candida famata*. FEMS Yeast Res. https://doi.org/10.1093/femsyr/foae020
Fedorovych DV, Dmytruk KV, Sibirny AA. Recent Advances in Construction of the Efficient Producers of Riboflavin and Flavin Nucleotides (FMN, FAD) in the Yeast Candida famata. Methods Mol Biol. 2021;2280:15–30. https://doi.org/10.1007/978-1-0716-1286-6_2
Förster C, Santos MA, Ruffert S, Krämer R, Revuelta JL (1999) Physiological consequence of disruption of the *VMA1* gene in the riboflavin overproducer *Ashbya gossypii*. J Biol Chem 274(14):9442–9448
Frago S, Martínez-Júlvez M, Serrano A, Medina M (2008 Sep) Structural analysis of FAD synthetase from *Corynebacterium ammoniagenes*. BMC Microbiol 23(8):160. https://doi.org/10.1186/1471-2180-8-160
Fu Y, Xiao W (2006) Study of transcriptional regulation using a reporter gene assay. Methods Mol Biol 313:257–264
Gilbert LA, Larson MH, Morsut L, Liu Z, Brar GA, Torres SE, Stern-Ginossar N, Brandman O, Whitehead EH, Doudna JA, Lim WA, Weissman JS, Qi LS (2013) CRISPR-mediated modular RNA-guided regulation of transcription in eukaryotes. Cell 154(2):442–451
Ishchuk OP, Dmytruk KV, Rohulya OV, Voronovsky AY, Abbas CA, Sibirny AA (2008) Development of a promoter assay system for the flavinogenic yeast *Candida famata* based on the *Kluyveromyces lactis* β-galactosidase *LAC4* reporter gene. Enzym Microb Technol 42(3):208–215

Jankowitsch F, Schwarz J, Rückert C, Gust B, Szczepanowski R, Blom J, Pelzer S, Kalinowski J, Mack M (2012) Genome sequence of the bacterium Streptomyces davawensis JCM 4913 and heterologous production of the unique antibiotic roseoflavin. J Bacteriol 194:6818–6827

Jiménez A, Santos MA, Pompejus M, Revuelta JL (2005) Metabolic engineering of the purine pathway for riboflavin production in *Ashbya gossypii*. Appl Environ Microbiol 71(10):5743–5751

Jiménez A, Santos MA, Revuelta JL (2008) Phosphoribosyl pyrophosphate synthetase activity affects growth and riboflavin production in *Ashbya gossypii*. BMC Biotechnol 8(1):67

Jimenez A, Ledesma-Amaro R, Revuelta J (2016) Industrial production of vitamin B_2 by microbial fermentation. In: Vandamme EJ, Revuelta JL (eds) Industrial biotechnology of vitamins, biopigments and antioxidants. Wiley-VCH, Weinheim

Jiménez A, Hoff B, Revuelta JL (2020) Multiplex genome editing in *Ashbya gossypii* using CRISPR-Cpf1. New Biotechnol 57:29–33

Kahmann R, Basse C (1999) REMI (restriction enzyme mediated integration) and its impact on the isolation of pathogenicity genes in fungi attacking plants. Eur J Plant Pathol 105(3):221–229

Kato T, Azegami J, Yokomori A, Dohra H, El Enshasy HA, Park EY (2020) Genomic analysis of a riboflavin-overproducing Ashbya gossypii mutant isolated by disparity mutagenesis. BMC Genomics 21(1):319

Kato T, Park EY (2012) Riboflavin production by Ashbya gossypii Biotechnology Letters 34(4) 611–618. https://doi.org/10.1007/s10529-011-0833-z

Kaufmann A (2009) A plasmid collection for PCR-based gene targeting in the filamentous ascomycete *Ashbya gossypii*. Fungal Genet Biol 46(8):595–603

Kemper M, Mohlzahn L, Lickfeld M, Lang C, Wählisch S, Schmitz HP (2011) A Bnr-like formin links actin to the spindle pole body during sporulation in the filamentous fungus *Ashbya gossypii*. Mol Microbiol 80(5):1276–1295

Kimura E (2003) Metabolic engineering of glutamate production. In: Microbial production of l-amino acids. Springer, Berlin, pp 37–57

Kitatsuji K, Ishino S, Teshiba S, Arimoto M (1996) Method of producing flavine nucleotides. US Patent 5,514,574

Knight SA, Lesuisse E, Stearman R, Klausner RD, Dancis A (2002) Reductive iron uptake by *Candida albicans*: role of copper, iron and the *TUP1* regulator. Microbiology 148:29–40

Köhler GA, White TC, Agabian N (1997) Overexpression of a cloned IMP dehydrogenase gene of *Candida albicans* confers resistance to the specific inhibitor mycophenolic acid. J Bacteriol 179(7):2331–2338

Ledesma-Amaro R (2015) Microbial oils: a customizable feedstock through metabolic engineering. Eur J Lipid Sci Technol 117(2):141–144

Ledesma-Amaro R, Nicaud JM (2016a) Metabolic engineering for expanding the substrate range of *Yarrowia lipolytica*. Trends Biotechnol 34(10):798–809

Ledesma-Amaro R, Nicaud JM (2016b) *Yarrowia lipolytica* as a biotechnological chassis to produce usual and unusual fatty acids. Prog Lipid Res 61:40–50

Ledesma-Amaro R, Jimenez A, Santos M, Revuelta J (2013a) Microbial production of vitamins. In: McNeil B, Archer DB, Giavasis I, Harvey LM (eds) Microbial production of food ingredients, enzymes and nutraceuticals. Woodhead Publishing series in food science, technology and nutrition, Oxford, p 246

Ledesma-Amaro R, Jiménez A, Santos MA, Revuelta JL (2013b) Biotechnological production of feed nucleotides by microbial strain improvement. Process Biochem 48(9):1263–1270

Ledesma-Amaro R, Kerkhoven EJ, Revuelta JL, Nielsen J (2014a) Genome scale metabolic modeling of the riboflavin overproducer *Ashbya gossypii*. Biotechnol Bioeng 111(6):1191–1199

Ledesma-Amaro R, Santos MA, Jiménez A, Revuelta JL (2014b) Strain design of *Ashbya gossypii* for single-cell oil production. Appl Environ Microbiol 80(4):1237–1244

Ledesma-Amaro R, Santos MA, Jiménez A, Revuelta JL (2014c) Tuning single-cell oil production in *Ashbya gossypii* by engineering the elongation and desaturation systems. Biotechnol Bioeng 111(9):1782–1791

Ledesma-Amaro R, Buey RM, Revuelta JL (2015a) Increased production of inosine and guanosine by means of metabolic engineering of the purine pathway in *Ashbya gossypii*. Microb Cell Fact 14(1):58

Ledesma-Amaro R, Lozano-Martínez P, Jiménez A, Revuelta JL (2015b) Engineering *Ashbya gossypii* for efficient biolipid production. Bioengineered 6(2):119–123

Ledesma-Amaro R, Serrano-Amatriain C, Jiménez A, Revuelta JL (2015c) Metabolic engineering of riboflavin production in *Ashbya gossypii* through pathway optimization. Microb Cell Fact 14(1):163

Ledesma-Amaro R, Buey RM, Revuelta JL (2016a) The filamentous fungus *Ashbya gossypii* as a competitive industrial inosine producer. Biotechnol Bioeng 113(9):2060–2063

Ledesma-Amaro R, Lazar Z, Rakicka M, Guo Z, Fouchard F, Crutz-Le Coq AM, Nicaud JM (2016b) Metabolic engineering of *Yarrowia lipolytica* to produce chemicals and fuels from xylose. Metab Eng 38:115–124

Ledesma-Amaro R, Jiménez A, Revuelta JL (2018) Pathway grafting for polyunsaturated fatty acids production in *Ashbya gossypii* through golden gate rapid assembly. ACS Synth Biol 7(10):2340–2347

Liu S, Hu W, Wang Z, Chen T (2020) Production of riboflavin and related cofactors by biotechnological processes. Microb Cell Fact 19(1):31

Lohithakshan A, Narayanasamy R, Potteth US, Keshava S, Nagaraja V, Usharani D, Kumar R (2021) Molecular insights into the mechanism of substrate binding and catalysis of bifunctional FAD synthetase from Staphylococcus aureus. Biochimie 182:217–227. https://doi.org/10.1016/j.biochi.2021.01.013

Lozano-Martínez P, Buey RM, Ledesma-Amaro R, Jiménez A, Revuelta JL (2016a) Engineering *Ashbya gossypii* strains for *de novo* lipid production using industrial by-products. Microb Biotechnol 10(2):425–433

Lozano-Martínez P, Ledesma-Amaro R, Revuelta JL (2016b) Engineering *Ashbya gossypii* for ricinoleic and linoleic acid production. Chem Eng Trans 49:253–258

Lyzak O, Semkiv M, Dmytruk K, Sibirny A (2015) Targets of transcription factor Sef1 in the flavinogenic yeasts *Candida famata*. In: Living organisms and bioanalytical approaches for detoxification and monitoring of toxic compounds

Lyzak O, Dmytruk K, Sibirny A (2016) Generation of mutations in *ADE2* gene using a CRISPR-Cas9 system in the flavinogenic yeast *Candida famata*. Visnyk of L'viv Univ Biol Ser 73:154–159

Mack M, van Loon AP, Hohmann HP (1998) Regulation of riboflavin biosynthesis in Bacillus subtilis is affected by the activity of the flavokinase/flavin adenine dinucleotide synthetase encoded by ribC. J Bacteriol 180:950–955

Maeting I, Schmidt G, Sahm H, Revuelta JL, Stierhof YD, Stahmann KP (1999) Isocitrate lyase of Ashbya gossypii–transcriptional regulation and peroxisomal localization. FEBS Lett 444(1):15–21

Magalhes F, Aguiar TQ, Oliveira C, Domingues L (2014) High-level expression of *Aspergillus niger* β-galactosidase in *Ashbya gossypii*. Biotechnol Prog 30:261

Mateos L, Jiménez A, Revuelta JL, Santos MA (2006) Purine biosynthesis, riboflavin production, and trophic-phase span are controlled by a Myb-related transcription factor in the fungus Ashbya gossypii. Appl Environ Microbiol 72(7):5052–5060

Millerioux Y, Clastre M, Simkin AJ, Courdavault V, Marais E, Sibirny AA, Noel T, Creche J, Giglioli-Guivarc'h N, Papon N (2011) Drug-resistant cassettes for the efficient transformation of *Candida guilliermondii* wild-type strains. FEMS Yeast Res 11(6):457–463

Monschau N, Sahm H, Stahmann K (1998) Threonine aldolase overexpression plus threonine supplementation enhanced riboflavin production in *Ashbya gossypii*. Appl Environ Microbiol 64:4283

Mora-Lugo R, Stegmüller J, Mack M (2019) Metabolic engineering of roseoflavin-overproducing microorganisms. Microb Cell Fact 18:146

Muñoz-Fernández G, Montero-Bullón J-F, Revuelta JL, Jiménez A (2021) New Promoters for Metabolic Engineering of *Ashbya gossypii*. J Fungi 7(11):906

Nakagawa S, Hagihara T, Fujio T, Aisaka K (1995) Metaphosphate-dependent phosphorylation of riboflavin to FMN by Corynebacterium ammoniagenes. Appl Microbiol Biotechnol 43:325–329

Noble SM (2013) *Candida albicans* specializations for iron homeostasis: from commensalism to virulence. Curr Opin Microbiol 16(6):708–715

Papon N, Boretsky YR, Courdavault V, Clastre M, Sibirny AA (2015) Genetic manipulation of *Meyerozyma guilliermondii*. In: van den Berg M, Maruthachalam K (eds) Genetic transformation systems in fungi, vol 2. Springer, Cham, pp 245–262

Petrovska Y, Lyzak O, Ruchala J, Dmytruk K, Sibirny A (2022) Co-Overexpression of *RIB1* and *RIB6* increases riboflavin production in the yeast *Candida famata*. J Ferment 8:141

Pynyaha YV, Boretsky YR, Fedorovych DV, Fayura LR, Levkiv AI, Ubiyvovk VM, Protchenko OV, Philpott CC, Sibirny AA (2009) Deficiency in frataxin homologue *YFH1* in the yeast *Pichia guilliermondii* leads to missregulation of iron acquisition and riboflavin biosynthesis and affects sulfate assimilation. Biometals 22(6):1051–1061

Ran FA, Hsu PD, Wright J, Agarwala V, Scott DA, Zhang F (2013) Genome engineering using the CRISPR-Cas9 system. Nat Protoc 8(11):2281–2308

Ravasio D, Wendland J, Walther A (2014) Major contribution of the Ehrlich pathway for 2-phenylethanol/rose flavor production in Ashbya gossypii. FEMS Yeast Res 14(6):833–844

Reuss O, Vik A, Kolter R, Morschhauser J (2004) The *SAT1* flipper, an optimized tool for gene disruption in *Candida albicans*. Gene 341:119–127

Revuelta JL, Buey RM, Ledesma-Amaro R, Vandamme EJ (2016a) Microbial biotechnology for the synthesis of (pro) vitamins, biopigments and antioxidants: challenges and opportunities. Microb Biotechnol 9(5):564–567

Revuelta JL, Ledesma-Amaro R, Lozano-Martinez P, Díaz-Fernández D, Buey RM, Jiménez A (2016b) Bioproduction of riboflavin: a bright yellow history. J Ind Microbiol Biotechnol 44(4–5):659–665

Ribeiro O, Wiebe M, Ilmén M, Domingues L, Penttilä M (2010) Expression of *Trichoderma reesei* cellulases CBHI and EGI in *Ashbya gossypii*. Appl Microbiol Biotechnol 87(4):1437–1446

Ribeiro O, Magalhães F, Aguiar TQ, Wiebe MG, Penttilä M, Domingues L (2013) Random and direct mutagenesis to enhance protein secretion in *Ashbya gossypii*. Bioengineered 4(5):322–331

Riboflavin market size & share analysis—growth trends & forecasts (2024–2029). https://www.mordorintelligence.com/industry-reports/riboflavin-market

Ruchala J, Andreieva YA, Tsyrulnyk AO, Sobchuk SM, Najdecka A, Wen L, Kang Y, Dmytruk OV, Dmytruk KV, Fedorovych DV, Sibirny AA (2022) Cheese whey supports high riboflavin synthesis by the engineered strains of the flavinogenic yeast *Candida famata*. J Microb Cell Fact 21:161

Sander JD, Joung JK (2014) CRISPR-Cas systems for editing, regulating and targeting genomes. Nat Biotechnol 32(4):347–355

Santos MA, Mateos L, Stahmann KP, Revuelta JL (2005) Insertional mutagenesis in the vitamin B_2 producer fungus *Ashbya gossypii*. In: Barredo JL (ed) Methods in biotechnology: microbial process and products. Humana, Totowa, NJ, pp 283–300

Schlösser T, Wiesenburg A, Gätgens C, Funke A, Viets U, Vijayalakshmi S, Nieland S, Stahmann KP (2007) Growth stress triggers riboflavin overproduction in Ashbya gossypii. Appl Microbiol Biotechnol 76(3):569–578

Schwechheimer SK, Park EY, Revuelta JL, Becker J, Wittmann C (2016) Biotechnology of riboflavin. Appl Microbiol Biotechnol 100(5):2107–2119

Serrano-Amatriain C, Ledesma-Amaro R, López-Nicolás R, Ros G, Jiménez A, Revuelta JL (2016) Folic acid production by engineered *Ashbya gossypii*. Metab Eng 38:473–482

Silva R, Aguiar TQ, Oliveira R, Domingues L (2019) Light exposure during growth increases riboflavin production, reactive oxygen species accumulation and DNA damage in *Ashbya gossypii* riboflavin-overproducing strains. FEMS Yeast Res 19(1)

Stahmann KP, Revuelta JL, Seulberger H (2000) Three biotechnical processes using *Ashbya gossypii*, *Candida famata*, or *Bacillus subtilis* compete with chemical riboflavin production. Appl Microbiol Biotechnol 53(5):509–516

Tanner F Jr, Vojnovich C, Van Lanen JM (1945) Riboflavin production by *Candida* species. Science (Washington) 101:180–161

Tsyrulnyk AO, Andreieva YA, Ruchala J, Fayura LR, Dmytruk KV, Fedorovych DV, Sibirny AA (2020) Expression of yeast homolog of the mammal BCRP gene coding for riboflavin efflux protein activates vitamin B2 production in the flavinogenic yeast *Candida famata*. Yeast 7(9–10):467–473

Uhl MA, Johnson AD (2001) Development of *Streptococcus thermophilus* lacZ as a reporter gene for *Candida albicans*. Microbiology 147(Pt 5):1189–1195

Voronovsky A, Abbas C, Fayura L, Kshanovska B, Dmytruk K, Sybirna K, Sibirny A (2002) Development of a transformation system for the flavinogenic yeast. FEMS Yeast Res 2(3):381–388

Vyas VK, Barrasa MI, Fink GR (2015) A *Candida albicans* CRISPR system permits genetic engineering of essential genes and gene families. Sci Adv 1(3):e1500248

Wang Y, Liu L, Jin Z, Zhang D (2021) Microbial cell factories for green production of vitamins. Front Bioeng Biotechnol 9:661562

Wickerham LJ, Flickinger MH, Johnston RM (1946) The production of riboflavin by Ashbya gossypii. Arch Biochem 9:95–98

Wright MC, Philippsen P (1991) Replicative transformation of the filamentous fungus *Ashbya gossypii* with plasmids containing *Saccharomyces cerevisiae* ARS elements. Gene 109(1):99–105

Xie K, Yang Y (2013) RNA-guided genome editing in plants using a CRISPR-Cas system. Mol Plant 6(6):1975–1983

You J, Pan X, Yang C, Du Y, Osire T, Yang T, Zhang X, Xu M, Xu G, Rao Z (2021) Microbial production of riboflavin: Biotechnological advances and perspectives. Metab Eng 68:46–58

Zhang JR, Ge YY, Liu PH, Wu DT, Liu HY, Li HB, Corke H, Gan R-Y (2021) Biotechnological strategies of riboflavin biosynthesis in microbes. Engineering 12:115–127

Chapter 10
Production of Recombinant Proteins in the Methylotrophic Yeasts

Olena Stasyk and Oleh Stasyk

Contents

10.1 Introduction ... 292
10.2 Overview of Yeast Host Species Commonly Used for Recombinant Protein Production ... 294
10.3 *Pichia pastoris* (Syn. *Komagataella phaffii*) ... 296
10.4 Toolbox of the *P. pastoris* Expression Platform ... 297
10.4.1 Expression Strains ... 297
10.4.2 *P. pastoris* Vectors ... 297
10.4.3 Gene Promoters for Governing Recombinant Protein Expression ... 298
10.4.4 Other Tools ... 300
10.4.5 Industrial, Pharmaceutical, and Other Applications ... 301
10.4.6 The Expression System's Limitations ... 303
10.5 *Hansenula polymorpha* (Syn *Ogataea polymorpha*) ... 303
10.6 Toolbox of the *H. polymorpha* Expression Platform ... 305
10.6.1 Expression Strains ... 305
10.6.2 *H. polymorpha* Vectors ... 305
10.6.3 Gene Promoters for Governing Recombinant Protein Expression ... 306
10.6.4 Other Tools ... 307
10.6.5 Industrial, Pharmaceutical, and Other Applications ... 308
10.7 Other Methylotrophic Yeasts ... 309
10.7.1 *Candida boidinii* ... 309
10.7.2 *Pichia methanolica* ... 309
10.7.3 *Ogataea minuta* ... 310
10.8 Conclusions and Outlook ... 311
References ... 312

O. Stasyk
Department of Biochemistry, Faculty of Biology, Ivan Franko National University of Lviv, Lviv, Ukraine

Department of Cell Signaling, Institute of Cell Biology, NAS of Ukraine, Lviv, Ukraine

O. Stasyk (✉)
Department of Cell Signaling, Institute of Cell Biology, NAS of Ukraine, Lviv, Ukraine

Research Institute of Epidemiology and Hygiene, Danylo Halytsky Lviv National Medical University, Lviv, Ukraine
e-mail: stasyk@cellbiol.lviv.ua

A. A. Sibirny (ed.), *Biotechnology of Yeasts and Filamentous Fungi*, Grand Challenges in Biology and Biotechnology,
https://doi.org/10.1007/978-3-031-74726-7_10

Abstract The production of food and feed, industrial proteins, and biopharmaceuticals requires cost-efficient protocols and rationally chosen heterologous production hosts. Among established production platforms, ranging from bacteria to mammalian cells, yeasts combine the advantages of unicellular organisms, such as fast biomass growth and relatively easy genetic manipulations, with eukaryotic features such as correct posttranslational modifications of recombinant proteins and an efficient secretory pathway. Simultaneously, unlike more complex eukaryotic organisms, yeast expression systems are more economical, and unlike bacterial hosts, they do not contain pyrogenic or viral inclusions. Yeasts also offer a high diversity of metabolic and physiological adaptations that can be exploited in biotechnology, including recombinant protein production. Herein, methylotrophic yeasts, mainly *Pichia pastoris* (syn. *Komagataella phaffii*) and *Hansenula polymorpha* (syn. *Ogataea polymorpha*) (in this review, the traditional nomenclature of yeast species is used), are overviewed as alternative expression platforms for the production of heterologous proteins such as industrial enzymes and biopharmaceuticals.

10.1 Introduction

The production of recombinant proteins is nowadays a sizable multibillion-dollar market. For instance, the biopharmaceutical market is projected to reach more than USD 500 billion by 2026 (https://www.industryarc.com/Report/9586/biopharmaceutical-market.html) and for industrial enzymes, approximately USD 9 billion (https://www.reportbuyer.com/product/870778/global-markets-for-enzymes-in-industrial-applications.html). As natural sources of proteins of interest are often limited or native protein concentration is too low to allow for its efficient extraction, the production of such proteins from genetically engineered heterologous hosts provides an economically justified alternative. This option became feasible with the advent of molecular DNA techniques, and since the 1980s, the number of recombinant proteins expressed and the number of suitable expression hosts have been steadily increasing. Among host organisms, yeasts occupy a special important place because they offer a combination of very useful traits. As eukaryotic unicellular microorganisms, yeasts share with metazoans the core molecular machinery for protein folding, processing, cotranslational and posttranslational modifications, and secretion. This, in most cases, allows for the synthesis of biologically active protein products in yeasts. Importantly, yeast cells are well amendable to genetic manipulations owing to the availability of full-genome sequences and well-developed recombinant DNA techniques, sets of expression vectors, and carbon-source-regulated gene promoters. In addition, a case-specific design of the fermentation process can be implemented in yeast hosts to improve the individual product yield. Conversely, yeasts do not exhibit several characteristic limitations of traditional bacterial expression hosts, such as *Escherichia coli* or *Bacillus subtilis*, namely, contamination with pyrogenic compounds or sensitivity to viral infection (overviewed in Mattanovich et al. 2012; Celik and Calık 2012; Corchero et al. 2013; Meehl and Stadheim 2014; Kim et al. 2015; Rettenbacher et al. 2022; Jayakrishnan et al. 2024).

As a rule, yeasts offer a set of characteristics that make them the host organisms of choice for the production of particular kinds of proteins. For instance, monoclonal antibodies (mABs) and antibody fragments are today one of the most important biopharmaceuticals. Despite the ability of mammalian producer cells to N-glycosylate full-length antibodies, they have important limitations with regard to bioprocessing and scaling up, which lead to elevated production costs. However, antibody fragments that exhibit antigen-binding properties can also be efficiently produced in yeasts, taking into account the availability of several yeast hosts with humanized glycosylation patterns (Spadiut et al. 2014; Li et al. 2022). Some features of particular yeast species make them specifically useful, for instance, for the production of difficult-to-overexpress recombinant membrane proteins for structural studies or those required for food and feed production, lignocellulosic enzymes, or biopharmaceuticals like recombinant human lysosomal enzymes (Mellitzer et al. 2012; Emmerstorfer et al. 2014; Spohner et al. 2015; Bill 2015; Ergün and Calık 2016; Carlesso et al. 2022).

The design of the optimal expression system for a given recombinant protein involves several steps, among which selecting the appropriate host organism and strain is the most crucial. For instance, when proper folding and posttranslational modifications are required for the physiological activity of the protein, the expression host must be chosen among eukaryotic organisms. This is accompanied and followed by the selection of the following:

1. A suitable vector (integrative or episomal) with an appropriate promoter (repressible/inducible or constitutive) and a selectable marker (dominant antibiotic resistance or auxotrophic) that allows either single or multicopy integration of the expression cassette
2. Deciding whether one of the following steps will be necessary:
   - Codon optimization of the recombinant gene
   - Fusion of the gene to an epitope tag necessary for more efficient affinity purification
   - Fusion to the signal sequence to target the protein product to a specific intracellular compartment (e.g., peroxisomes) or to provide secretion into the extracellular milieu

Expression host optimization may also involve introducing such genetic alterations as deletion of the genes encoding vacuolar proteases to prevent the proteolytic cleavage of the desirable product or manipulation with the genes that facilitate secretion, such as the coexpression of certain chaperones. The optimization of an expression system, depending on the chosen host, may further involve designing the optimal fermentation mode and medium composition (as manipulation with carbon and nitrogen sources), optimizing fermentation protocols, and incubating parameters (temperature, pH, aeration conditions, etc.).

Of note, no single yeast expression system can possibly a priori provide all of the desired properties to ensure the successful production of a given recombinant protein. This is especially true when chemical and physical parameters and the chosen feeding strategy may critically influence the yield of the product or its functional characteristics.

10.2 Overview of Yeast Host Species Commonly Used for Recombinant Protein Production

Saccharomyces cerevisiae, a traditional baker's yeast, is the first and best-characterized yeast expression system, developed in the 1980s. It was also the first eukaryote for which a full-genome sequence was established. However, its limitations as a protein expression host, such as the hyperglycosylation of proteins and glycosylation terminated via α-1,3-linked mannose residues, which are considered to be allergenic, and low protein yields, along with cases of plasmid instability, soon became apparent and have limited the number of marketed protein products from *S. cerevisiae* (Gellissen et al. 2005a; Darby et al. 2012). The current popularity of *S. cerevisiae* is mostly due to the accumulated in-depth knowledge of its physiology, genetics, and fermentation techniques and the fact that it has been generally recognized as a safe (GRAS) species. Also, *S. cerevisiae* is possibly the organism with the most comprehensive experimental dataset available (Petranovic et al. 2010).

Nevertheless, in the last years, the number of published reports on the utilization of alternative yeasts for recombinant protein production exceeds that of *S. cerevisiae*. Also, in the few systematic comparisons between different yeast production systems, the clear advantages of the so-called nonconventional species over *S. cerevisiae*, especially for the production of secreted proteins, have been found (Mack et al. 2009; Dragosits et al. 2011; Darby et al. 2012). For instance, the production of the anticoagulant peptide hirudin in different expression systems demonstrated the significantly higher comparative productivities achieved with *P. pastoris* and *H. polymorpha* (Demain and Vaishnav 2009). Thus, the folding and secretion pathways (Mattanovich et al. 2012; Unver and Dagci 2024), glycosylation patterns (Li et al. 2022), higher yields of membrane-bound proteins (Carlesso et al. 2022), and strong regulatable promoters (Tsuda and Nonaka 2024) are the major advantages of alternative yeasts as expression systems.

Several nonconventional yeast species, including the methylotrophs *H. polymorpha* and *P. pastoris*, the two dimorphic yeasts *Arxula adeninivorans* and *Yarrowia lipolytica*, the budding yeast *Kluyveromyces lactis*, and the fission yeast *Schizosaccharomyces pombe* have been proposed and used, to a different extent, as recombinant protein expression platforms.

In this review, we focus on the characterization of the alternative methylotrophic yeast species as hosts for recombinant protein production. In contrast to *S. cerevisiae*, they belong to the group of so-called Crabtree-negative yeasts, which do not ferment and produce ethanol under fully aerobic conditions (Ata et al. 2021). This enables these yeasts to grow to very high cell densities in bioreactor cultures, which results, in turn, in higher product yields (Cereghino and Cregg 2000). Methylotrophic yeasts are also characterized by a unique enzymatic machinery involved in the utilization of toxic to other microbes one-carbon compound methanol. Many enzymes, especially those of a peroxisomal nature, are highly overexpressed and are strictly regulated by carbon sources, and their gene promoters have been used, therefore, to drive heterologous protein production (Fig. 10.1). For comparative characterization of nonmethylotrophic alternative yeast expression platforms, readers are referred to reviews on the subject (Gellissen et al. 2005b; Celik and Calık 2012; Corchero et al. 2013; Rettenbacher et al. 2022; Jayakrishnan et al. 2024).

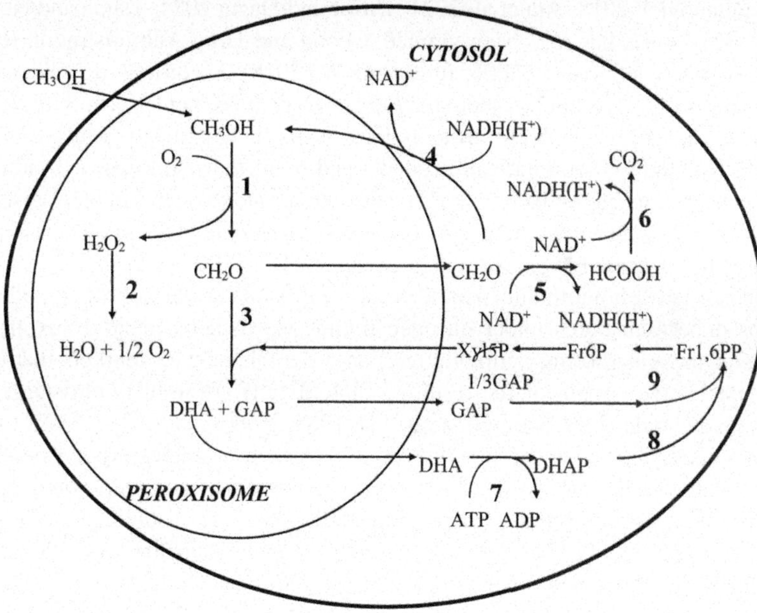

Fig. 10.1 Simplified schematic representation of the metabolic pathways of methanol oxidation and assimilation in the methylotrophic yeast cell. Main enzymes are numbered, and the names of the corresponding gene promoters previously used to drive recombinant protein expression in different methylotrophic yeasts are listed in the legend: 1) alcohol (methanol) oxidase (*H. polymorpha MOX*, *P. pastoris AOX1*, *P. methanolica MOD1/AUG1*, *C. boidinii AOD1*), 2) catalase, 3) dihydroxyacetone synthase, 4) formaldehyde reductase, 5) formaldehyde dehydrogenase (*H. polymorpha FLD*, *P. pastoris FLD1*), 6) formate dehydrogenase (*H. polymorpha FMD*), 7) dihydroxyacetone kinase, 8) fructose-1,6-diphosphate aldolase, 9) fructose-1,6-diphosphatase. Abbreviations for intermediates: GAP—glyceraldehyde, DHAP—dihydroxyacetone (phosphate), Fr1,6PP—fructose-1,6-diphosphate, Fr6P—fructose-6-phosphate, Xyl5P—xylulose-5-phosphate

10.3 *Pichia pastoris* (Syn. *Komagataella phaffii*)

Of note, *Pichia pastoris* has been reclassified (Yamada et al. 1995), and new species names, such as *Komagataella pastoris* and *Komagataella phaffii*, can be found in more recent literature (for further details, please see Kurtzman 2005, 2009; Sturmberger et al. 2016). Most industrially exploited strains, formerly called *P. pastoris*, belong, in fact, to the newly denoted species of *K. phaffii* (Gasser and Mattanovich 2018).

Since the unsuccessful introduction by Phillips Petroleum of a single-cell protein (SCP) for commercial production more than 50 years ago, the methylotrophic yeast *P. pastoris* has become nowadays the most frequently used yeast host for heterologous protein production. This yeast is distinguished for its growth to very high cell densities, the availability of strong and tightly regulated gene promoters, very efficient protein secretion, and the ability to produce recombinant proteins in amounts of grams per liter of culture broth (for reviews, see Cereghino and Cregg 2000; Daly and Hearn 2005; Gasser et al. 2013; Ahmad et al. 2014; Spohner et al. 2015; Gasser and Mattanovich 2018; Ata et al. 2021; Unver and Dagci 2024; Tsuda and Nonaka 2024). *P. pastoris* has also been granted a Food and Drug Administration (FDA) GRAS status (Ciofalo et al. 2006; Juturu and Wu 2018). According to Pichia.com, a web platform for information about the *P. pastoris* expression platform (http://www.pichia.com), more than 5000 proteins have been successfully produced, among which more than 70 commercial products and more than 300 licensed industrial processes are carried out using the *P. pastoris* expression system. In basic research, *P. pastoris* has been a very popular yeast model in research on the molecular mechanisms of peroxisome biogenesis and degradation (Dunn Jr et al. 2005).

P. pastoris is a homothallic haploid yeast species able to mate, form diploid cells, and produce ascospores under nitrogen limitation. Usually, haploid auxotrophic strains or those harboring antibiotic resistance complementary markers that allow for the selective growth of the resulting diploid cells are used in classic genetic crossings in strain construction (Cregg et al. 1985, 2009).

As a methylotrophic organism, *P. pastoris* can use single-carbon compound methanol as its sole carbon and energy source in the absence of a repressing carbon sources, such as glucose or glycerol. Since *P. pastoris* is an obligate aerobe and prefers a respiratory mode of growth (Ata et al. 2021), fermentation products such as ethanol and acetic acid do not build up in the culture, enabling to reach very high cell densities of up to 200 g L^{-1} dry weight (Cereghino et al. 2002). Simultaneously, *P. pastoris* is an efficient secreting host for recombinant proteins because a few of its own proteins are found in extracellular media, which allows simple purification of the secreted products. Another advantage of *P. pastoris* is its less extensive hypermannosylation as compared to *S. cerevisiae* and the absence of α-1,3-linked mannosyltransferase, which produces highly immunogenic terminal mannosyl epitopes in *S. cerevisiae* (Gemmill and Trimble 1999).

10.4 Toolbox of the *P. pastoris* Expression Platform

10.4.1 Expression Strains

All *P. pastoris* expression strains are derivative of the wild-type strains NRRL-Y 11430 from the Agriculture Research Service Culture Collection (Peoria IL, USA) and NRRL Y-48124 (X-33, Invitrogen expression kit strains, Carlsbad CA, USA) (Kurtzman 2009) and are now denoted as *K. phaffii* (Sturmberger et al. 2016). In particular, the GS115 *his4* auxotrophic mutant lacking the gene encoding histidinol dehydrogenase (Cereghino and Cregg 2000), the KM71 Mut S strain that exhibits slow growth on methanol (see below), and protease-deficient strains (e.g., SMD1163, SMD1165, SMD1168) are commonly used as classical hosts for heterologous protein expression. The SMD1163 strain (*his4, pep4, prb1*) lacks both proteinase A, which activates other endogenous *P. pastoris* proteases (*pep4*), and proteinase B (*prb1*) activities, whereas SMD1165 and SMD1168 are deficient in *prb1* and *pep4*, respectively. Therefore, strain SMD1163 has been demonstrated to be the most efficient host for the synthesis of recombinant proteins sensitive to proteolytic degradation (Cereghino and Cregg 2000). Depending on the protein of interest and further predicted modifications of the protein production process, all the abovementioned strains provide certain useful features.

10.4.2 **P. pastoris** *Vectors*

The standard *P. pastoris* vectors used in biotechnological applications are shuttle vectors, which can be replicated in *E. coli* and maintained in *P. pastoris* using a selectable marker, such as auxotrophic genetic markers (e.g., *HIS4*, *ARG4*, *MET2*, *ADE1*, *URA3*). A set of plasmids for intracellular and secretory protein expression in *P. pastoris* based on different auxotrophic genetic markers has been developed (Lin Cereghino et al. 2001; Fickers 2014; Unver and Dagci 2024). Selection in *P. pastoris* can also be provided by the heterologous genes conferring resistance to certain drugs, e.g., geneticin (G418) and Zeocin.

The distinctive feature of the *P. pastoris* expression system (patented by Research Corporation Technologies (Tucson, AZ, USA)) is that the basic expression kit with standard vector systems for intracellular and secretory expression is commercially available for research use from Invitrogen Corporation (Carlsbad, CA, USA). It also provides *P. pastoris* expression vectors and strains for the humanized glycosylation of target proteins (Vogl and Glieder 2013) and can be purchased from BioGrammatics (Carlsbad, CA, USA).

Of note, unlike in *S. cerevisiae*, where homologous recombination predominates, in *P. pastoris*, nonhomologous recombination usually occurs upon transformation with a linearized vector. Also, homologous recombination in *P. pastoris* requires elongated homologous regions flanking heterologous DNA in the expression cassettes (Cereghino and Cregg 2000). Several different transformation methods and protocols can be applied in *P. pastoris* to integrate vector DNA into its genome. The most commonly used transformation method is electroporation, yielding up to 10^5 transformants/μg of DNA (Cregg and Russell 1998). For other up-to-date tools of the *P. pastoris* expression system, readers are referred to comprehensive reviews on the subject (Gasser et al. 2013; Unver and Dagci 2024).

10.4.3 Gene Promoters for Governing Recombinant Protein Expression

The promoter of the gene of alcohol oxidase (*AOX1*) encoding a key peroxisomal enzyme of the methanol utilization pathway was exploited as a component of the classical *P. pastoris* expression system (Cregg et al. 1985). The *AOX1* promoter is repressed in the presence of glucose or glycerol but is strongly upregulated in the presence of methanol, making it tightly controlled by the carbon source expression system. The advantage of the catabolite regulation of the *AOX1* promoter, especially important in case a toxic recombinant product is to be expressed, is that high biomass growth can be achieved under repressing conditions, and afterward, the expression can be initiated. The transcriptional regulation of *pAOX1* has been extensively studied, and cis-acting elements and transacting factors were characterized (Hartner and Glieder 2006). A wide range of proteins have been expressed from the control of this promoter, with claimed yields of recombinant protein of up to 20–30 g L^{1} (Spohner et al. 2015). Of note, one of the studies suggested that protein expression from under *pAOX1* may occur during the preinduction phase in cultures grown in bioreactors, contrary to shake flask cultures, tentatively indicating that the corresponding promoter may be partially leaky under certain conditions (Bawa et al. 2014).

However, methanol, the *pAOX1* inducer, is a potential fire hazard and a toxic compound, which may pose risks under large-scale industrial fermentation conditions. Therefore, methanol nonutilizing strains have been constructed that utilize alternative carbon sources for growth but can also be cultivated on a controlled methanol concentration at a low growth rate or under mixed feed protocols (Cereghino et al. 2002). *P. pastoris* harbors in its genome two genes encoding, for peroxisomal alcohol oxidase, major *AOX1* and auxiliary *AOX2*. Therefore, depending on these genes' status, there are three phenotypes of *P. pastoris* with respect to methanol utilization: Mut^+ (wild-type or intact methanol utilization), Mut^S (slow methanol utilization, disrupted *AOX1*, intact *AOX2*), and Mut^- (deficient methanol utilization, both *AOX1* and *AOX2* disrupted) (Cereghino and Cregg 2000).

Several transcription factors that govern *pAOX1* were identified, namely, Mit1, Mxr1, and Prm1 (Wang et al. 2016a, b). The overexpression of *MXR1* led to the inhibition of the glycerol transporter and the simultaneous upregulation of *AOX1* in the glycerol medium or in the glycerol/methanol mixture (Zhan et al. 2017). Also, via *AOX1* promoter engineering, several of its variants have been isolated, some of which exceeded the strength of the wild-type promoter (Yang et al. 2018).

Besides, a number of alternative promoters have also been identified in *P. pastoris* and characterized with regard to their effectiveness for protein production. These included several constitutive promoters from the housekeeping genes (e.g., *GAP1, TEF1, PGK1*, TPI1, *GCW*) and regulatable (inducible) promoters from particular biochemical pathways (e.g., *FLD1, PHO89, THI11, AOD*) (Prielhofer et al. 2013). Interestingly, the application of the *P. pastoris FLD1* promoter of the formaldehyde dehydrogenase gene allows the induction of recombinant protein expression with methylamine instead of methanol and its repression by glucose (Resina et al. 2005). Recently, the formate dehydrogenase (*FDH*) gene promoter has been proposed for a regulated by carbon sources expression of foreign proteins in *P. pastoris*, which exhibited comparable, or even superior under certain conditions, strength relative to *pAOX1* or *pGAP* (Shen et al. 2024).

The *GAP* promoter provides a high-level constitutive expression of foreign proteins on glycerol- and glucose-containing media. This feature is particularly useful for protein production in large-scale fermenters since it does not require flammable and therefore hazardous methanol in the process. In some cases, a constitutive expression from under *pGAP* yields even higher protein levels relative to that controlled by *pAOX1* (Waterham et al. 1997). Therefore, when recombinant proteins are not toxic for the yeast host, as in the case of most enzymes of the food industry, constitutive promoters are applicable and are even favorable because they support more productive continuous modes of fermentation.

A methanol-free *P. pastoris* expression system was proposed that involves a strong methanol-inducible promoter of dihydroxyacetone synthase (*DAS1*) and the overexpression of *TRM1*, encoding a positive transcription regulator of the methylotrophic genes. It was shown by Takagi et al. (2019) that in such a combination, a *pDAS1*-driven expression of the recombinant phytase did not require methanol for induction. Another possible alternative would be the promoter of the peroxisomal catalase gene *CAT*, which can be induced by oleic acid in the absence of methanol (Vogl et al. 2018).

For a detailed and comprehensive description of the available *P. pastoris* promoters, readers are referred to the reviews by Vogl and Glieder 2013 and Unver and Dagci 2024. Several *P. pastoris* transcriptional factors have been identified and studied with respect to their control of host physiology and fermentation-relevant traits. These include Hac1 involved in the control of unfolded protein response (UPR), Fep1 in iron uptake, and Yap1 in oxidative stress response (see Gasser et al. 2013; Puxbaum et al. 2015; and references thereof). The understanding of these transcriptional regulatory networks, together with synthetic biology and metabolic engineering approaches, forms a solid basis that enables the design of tailor-made *P. pastoris* hosts for the production of recombinant proteins.

10.4.4 Other Tools

The genes of interest are often codon and guanine and cytosine (G + C) content optimized for expression in *P. pastoris*. Since the comparison with the non-codon-optimized gene is not usually reported, what difference this generally makes to the product yield is not clear. However, in the case of membrane proteins, as demonstrated in one of the studies with mouse P-glycoprotein, expression from a codon-optimized *P. pastoris* gene not only substantially yielded more protein than expression from the wild-type gene but also concomitantly correlated with better product stability (Bai et al. 2011). A positive effect of codon optimization in *P. pastoris* has also been reported for endochitinase (Yu et al. 2013), inulinase (He et al. 2014), keratinase (Hu et al. 2013), lignocellulolytic enzymes (Mellitzer et al. 2012), and some other enzymes.

The *P. pastoris* vectors usually integrate into the host cell genome, producing stably expressing clones. The number of copies (up to 30 in *P. pastoris* (Cregg et al. 2009)) that are integrated cannot be precisely controlled, so the optimal clone must be experimentally identified. The copy number can be roughly estimated via colony screening in the presence of increasing concentrations of antibiotics when a corresponding resistance gene is included in the vector as a selectable marker or by other methods, such as real-time polymerase chain reaction (RT-PCR). It has to be stressed that higher copy numbers of the integrated vector do not necessarily correlate with higher expression of a protein product, so it is usually necessary to follow copy number analysis with a direct assessment of protein expression levels. The main approaches for how to achieve high-expression cassette copy numbers are summarized in Gasser et al. (2013).

An advanced highly efficient cloning system for *P. pastoris*, called Golden*Pi*CS, has been recently proposed. It is capable of integrating up to eight expression cassettes in one plasmid and allows a flexible exchange of the promoter and integration and secretion signals in multigene constructs for expression optimization (Prielhofer et al. 2017). Also, a relatively recently developed clustered regularly interspaced palindromic repeats/CRISPR-associated protein 9 (CRISPR/Cas9) platform appeared as an efficient and accurate gene editing technique in *P. pastoris* (Weninger et al. 2016).

Although untagged recombinant proteins have been successfully isolated from *P. pastoris* based on their individual properties, it is more usual to produce tagged versions of the desirable proteins to facilitate their detection and isolation. A range of different tags is compatible with expression in *P. pastoris*. The vast majority of recombinant proteins can and have been expressed as His-tagged constructs to allow one-step purification by affinity chromatography. Besides, FLAG tags have also been used for both protein analysis and their isolation. Green fluorescent protein (GFP) tags can be incorporated into expression vectors to assist in protein visualization. A number of different protease-cleavage sites have also been employed to successfully remove tags after product purification if required (Gasser et al. 2013).

For efficient secretion, *P. pastoris* expression vectors can incorporate the *S. cerevisiae* alpha-mating factor (α-MF) pre-pro leader sequence signal upstream of the expressed heterologous gene. The α-MF should be cleaved following expression by the yeast Golgi-associated Kex2 protease. *P. pastoris* strain with an increased Kex2 copy number, which alters the Kex2 endogenous level and increases the production of a secreted protein, has been described (Yang et al. 2013). Also, some modifications of the α-MF pre-pro leader to further enhance secretion in *P. pastoris* have been proposed (Lin-Cereghino et al. 2013). A hybrid secretion signal composed of the *S. cerevisiae OST1* signal sequence and α-MF proregion was suggested as facilitating cotranslational translocation into the endoplasmic reticulum (ER) (Barrero et al. 2018). The coexpression of several chaperones has been demonstrated to facilitate ER translocation in *P. pastoris* (Duan et al. 2019; Raschmanová et al. 2021).

A better understanding of *P. pastoris* genetics and promoter regulation and physiology was greatly facilitated after full-genome sequences became available (Küberl et al. 2011). Readers are referred to the publication by Sturmberger et al. (2016) on the *P. pastoris* genome for more details. The Genome Browser database for *P. pastoris* genome sequences is publicly available (Sturmberger et al. 2016). Also, achievements in strain optimization, which concerned intracellular trafficking, protein folding, glycosylation, and proteolytic degradation, were reported (Kim et al. 2004; Puxbaum et al. 2015; Yu et al. 2015). The toolbox for genetic elements and *P. pastoris* strains for the improvement of recombinant protein production has been steadily expanding and includes novel promoters, gene copy-number enhancing vectors, and host strains adapted for efficient secretion (Gasser et al. 2013; Ata et al. 2021; Unver and Dagci 2024; Tsuda and Nonaka 2024).

Several genome-scale metabolic models for *P. pastoris* have been created and have been updated to a consensus model (Toma's-Gamisans et al. 2016). This provides an efficient tool for manipulating *P. pastoris* metabolic networks, constructing a new generation of host strains, and in silico simulations for optimizing fermentation protocols for recombinant proteins (Nocon et al. 2014). Other applications of yeast synthetic biology for the production of therapeutic proteins are summarized in the review by Kim et al. (2015).

10.4.5 Industrial, Pharmaceutical, and Other Applications

A number of recombinant products based on the *P. pastoris* platform have obtained approval as biopharmaceuticals (reviewed in Meehl and Stadheim 2014). Several products, e.g., human serum albumin, insulin, interferon alpha, and hepatitis B vaccine, are marketed in India and Japan (Shekhar 2008). The first therapeutic polypeptide, ecallantide, expressed in *P. pastoris* and used for the treatment of hereditary angioedema, received approval from the FDA in 2009 (Thompson 2010) and is produced by Dyax Corporation (Cambridge, MA). Successful genetic engineering to humanize N-glycosylation in *P. pastoris* has led to the further improvement of this host as a producer of pharmaceutical proteins dependent on glycosylation mode for their activity (Puxbaum et al. 2015 and references thereof). This permitted the efficient production of fully active human erythropoietin in *P. pastoris* (Hamilton et al. 2006). These features made *P. pastoris* a remarkably cost-efficient expression platform. The most important event for the *P. pastoris* expression system in food technology was receiving the GRAS status from the FDA and the FDA approval of a number of recombinant products (Ciofalo et al. 2006; Thompson 2010).

One important feature of the *P. pastoris* system is that strains-producers are highly suitable for large-scale growth and cultivation in bioreactors (Cereghino et al. 2002). Bioreactors provide precise control over culture parameters, including temperature, pH, and feed rate, and allow the real-time monitoring of changes in cell optical density and dissolved oxygen. This enables ultrahigh cell densities (>100 g/L dry cell weight, >500 OD600 units/mL) to be achieved (Cereghino et al. 2002). However, usually, medium cell-density cultures are utilized in bioreactors since they may reduce cellular stress and proteolysis associated with high-density cultures (Mattanovich et al. 2004). This feature combined with inexpensive growth media makes *P. pastoris* a highly cost-effective expression system.

Depending on which promoter is used, different fermentation techniques are applied in *P. pastoris*. The standard fermentation procedure for the expression system based on *pAOX1* is a three-stage fed-batch fermentation (Zhang et al. 2000). Usually, the first stage, the so-called "growth phase," consists of batch fermentation with glycerol as a carbon source to get sufficient initial cell biomass. The second "transition" stage starts right after all glycerol is exhausted and is a fed-batch phase aimed at the production of more biomass and at metabolic transition from glycerol to methanol metabolism. In this phase, a sole glycerol feed or a mixed feed of glycerol and methanol, where the glycerol feed progressively decreases and the methanol feed progressively increases, can be applied. For the "production phase," different feeding profiles can be applied, e.g., constant oxygen feeding, constant methanol concentration feeding, constant specific growth rate feeding, or oxygen- or temperature-limited fed-batch technique (Potvin et al. 2012).

The fermentation strategy for recombinant protein production under the control of constitutive *pGAP*, or other similar promoters, differs significantly from the ones described above. In this case, no separate transition and production phases nor methanol induction is necessary, and the amount of produced protein depends solely on the cell biomass. Therefore, fed-batch fermentation and continuous fermentation are the modes of choice for *GAP* promoters. An example of an efficient large-scale fed-batch protocol using *pGAP* is the production of lipase from *Candida rugosa*, generating up to 500 g/L wet cell weight and stable lipase activity of 14,000 U/mL of culture broth (Zhao et al. 2008). For a review of the industrial expression of food and feed enzymes in *P. pastoris*, see Spohner et al. (2015). For more details on fermentation process optimization and vector design for recombinant protein production, readers are referred to specialized reviews on the subjects (Celik and Calık 2012; Potvin et al. 2012; Looser et al. 2015; Zahrl et al. 2017; Unver and Dagci 2024).

P. pastoris has been particularly successful as an expression platform for various eukaryotic membrane proteins, especially those required for structural biology studies. One of the key issues associated with such proteins is the difficulty in expressing sufficiently large amounts of the product. This is due to the fact that eukaryotic membrane proteins require specific co- and posttranslational processing and membrane lipid composition, rendering bacterial expression systems, as a rule, not suitable as production hosts. Although higher eukaryotic systems (e.g., based on insect cells) have a good track record for producing membrane proteins, they are much more expensive and time-consuming as compared to microbial hosts. Although the baker's yeast *S. cerevisiae* has a good track record in the production of membrane proteins, much better success has been achieved in *P. pastoris*. Several reviews summarize successful examples of eukaryotic membrane proteins produced for structural studies in *P. pastoris* (Byrne 2015; Emmerstorfer et al. 2014; Routledge et al. 2016; Carlesso et al. 2022).

10.4.6 The Expression System's Limitations

Not every protein of interest is produced, especially secreted, in *P. pastoris* at high titers. For complex proteins that are membrane bound, hetero-oligomers, or those prone to proteolysis, yields are often lower than usually achieved. Protein folding, glycosylation, and secretion efficiency have been identified among the major bottlenecks and limiting factors for this host, pinpointing major targets for strain optimization. It is known that *P. pastoris* can hyperglycosylate recombinant products, although to a lesser extent than observed in *S. cerevisiae*. A comparative analysis of 25 nonglycosylated recombinant G-protein-coupled receptors (GPCRs) expressed in *P. pastoris* revealed that the loss of glycosylation resulted, in many cases, in reduced functional expression as compared to a glycosylated counterpart (Yurugi-Kobayashi et al. 2009). The data indicated that the effects of deviant glycosylation need to be assessed for each individual heterologous protein. For a comprehensive review on the subject of the challenges of recombinant protein folding, glycosylation, and secretion in the *P. pastoris* expression platform, see Puxbaum et al. (2015).

10.5 *Hansenula polymorpha* (Syn *Ogataea polymorpha*)

(Of note, *Hansenula polymorpha* has been reclassified several times, and species names such as *Pichia angusta* and *Ogataea polymorpha* can be found in the recent literature and should be considered synonyms (for further details, please see Yamada et al. 1994, 1995; Kurtzman 2009).)

H. polymorpha is another methylotrophic yeast successfully utilized as an expression platform for recombinant proteins. The peculiarity of this species is its remarkable thermotolerance and ability to grow up to 45 C and even higher as well as its profound resistance to heavy metals and oxidative stress (Gellissen 2002).

Therefore, *H. polymorpha* is well suited for the production of thermostable enzymes and proteins. Another peculiarity is that it harbors genes of the nitrate assimilation pathway horizontally introgressed from bacteria and not present in other yeasts, including methylotrophs (Morales and Dujon 2012; Morales et al. 2013). The *H. polymorpha* expression system is also known for the remarkable genome stability of strains-producers, a feature very important for large-scale fermentation (Kang and Gellissen 2004). The presence of excess peroxisomes and their membranes in *H. polymorpha* provides a convenient site for targeting heterologous proteins when fused with an appropriate targeting signal peptide (Leao and Kiel 2003) (Fig. 10.2). Similarly to *P. pastoris*, *H. polymorpha* is a Crabtree-negative yeast, allowing for high dilution rates and high biomass yields in bioreactors. *H. polymorpha* efficiently secretes foreign proteins and exhibits a lesser extent of hypermannosylation and no terminal α-1,3-linked mannose residues considered to be allergenic, as compared to *S. cerevisiae* (Kang et al. 1998; Gellissen 2000; Kim et al. 2004; Tsuda and Nonaka 2024). Several recombinant products derived from the *H. polymorpha* expression platform have reached the market, including the hepatitis B vaccine Hepavax-Gene (Crucell, NL), insulin, and interferon alpha-2 (Rhein-Biotech, DE). In addition, *H. polymorpha* has received a GRAS status as a producer of food enzymes (Gellissen et al. 2005a). Additional information on the *H. polymorpha* expression system can be found in several reviews on the subject (Sudbery et al. 1988; Gellissen 2000; Gellissen et al. 2005b; Mattanovich et al. 2012; Kim et al. 2014; Tsuda and Nonaka 2024).

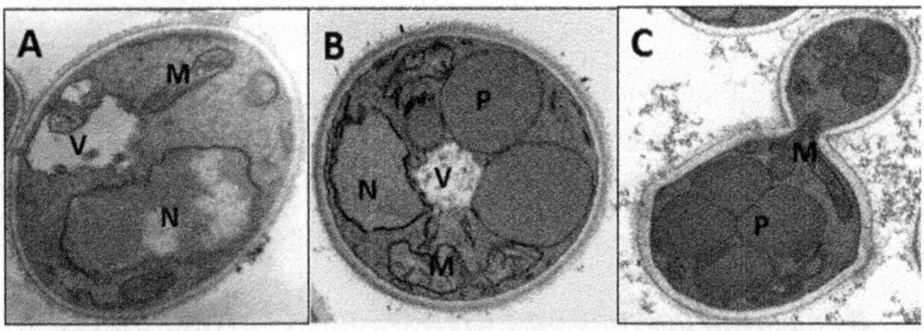

Fig. 10.2 Induction of peroxisomes in the cells of *H. polymorpha* grown in (**a**) glucose-supplemented batch culture, (**b**) methanol medium in batch culture, (**c**) chemostat culture on methanol to maximally induce the alcohol oxidase (MOX) promoter. Big crystalloids of alcohol oxidase protein tightly packed in peroxisomes and the intracellular volume they occupy inform the strength of the MOX promoter. Abbreviations: N—nucleus, M—mitochondrion, P—peroxisome, V—vacuole (Photo "**c**" adapted from Gellissen et al. 2005b)]

10.6 Toolbox of the *H. polymorpha* Expression Platform

10.6.1 Expression Strains

There are three known main strains of *H. polymorpha* that have independent origins. For biotechnological applications, mostly strains derivative of CBS4732 and DL-1 are used, whereas strain NCYC495 is used primarily in fundamental studies. Some time ago, strain DL-1 has been reclassified as *Hansenula (Ogataea) polymorpha's* sister species—*Ogataea parapolymorpha* (Kurtzman 2011; Ravin et al. 2013). All *H. polymorpha* isolates are homothallic, and reproduction occurs vegetatively by budding. Methods of classic and molecular genetics are well developed for this yeast (Gellissen 2002; Kang and Gellissen 2004).

10.6.2 H. polymorpha *Vectors*

There are many variants of *H. polymorpha* vectors that provide either a replicative or an integrative expression of heterologous genes (Mattanovich et al. 2012). Among auxotrophic marker genes, the native and heterologous *LEU2* and *URA3* are the most frequently used. Previously, we successfully applied a *P. pastoris* heterologous gene marker, *ADE1*, for the isolation of multicopy transformants and *H. polymorpha* as well as the native *FLD1* gene under a shortened *GAP* promoter as novel selectable markers (Krasovska et al. 2007, 2013). A wide-range yeast vector system (CoMedTM) has been constructed that provides a versatile transformation/expression toolkit for several hosts, namely, *H. polymorpha*, *S. cerevisiae*, *P. pastoris*, *A. adeninivorans*, and *K. lactis*. In its basic form, the vector is composed of genetic modules that are functional in all the mentioned yeasts, namely, an rDNA targeting sequence, an appropriate selection marker, and an expression cassette under the control of a *TEF1* promoter from various sources (Steinborn et al. 2006).

Efficient strategies for gene replacements and marker recycling and for introducing multiple heterologous genes are essential for the maximal efficacy of the expression platforms. In addition to the counterselection strategy using the *URA3* marker, more advanced approaches such as the Cre/loxP and Flp/FRT recombinase-based systems have also been developed. These systems act via active excision of the marker gene fragments to recycle them and allow for sequential rounds of deletions in several yeast species, including *H. polymorpha* (Qian et al. 2009).

Of note, the frequency of homologous versus nonhomologous recombination upon the integration of foreign DNA fragments into the *H. polymorpha* genome is much lower than that in *S. cerevisiae* (Kang and Gellissen 2004). It was shown in several yeasts, including *H. polymorpha*, that after the deletion of KU70 or KU80 homologs involved in nonhomologous end joining, the frequency of homologous recombination was significantly enhanced (Kim et al. 2015).

Positive selection markers such as geneticin- or Zeocin-resistance genes are routinely applied in *H. polymorpha* for the construction of multicopy transformants with different vectors harboring expression cassettes. However, it has become apparent that there is not always a linear correlation between an expression cassette copy number and the resulting protein expression level, particularly for secretory recombinant proteins (Aw and Polizzi 2013). For example, a study involving the secretory production of human serum albumin in *H. polymorpha* showed that the effect of gene dosage on expression was not apparent, and a single copy of the optimized expression cassette integrated into the *MOX* locus was sufficient for maximal expression (Kang et al. 2001).

10.6.3 Gene Promoters for Governing Recombinant Protein Expression

The *H. polymorpha* expression platform is based on strong, strictly regulated promoters of the genes of the methanol utilization pathway, encoding for methanol oxidase (MOX) and formate dehydrogenase (FMD). Unlike the *P. pastoris pAOX1*-based expression system, which strictly requires methanol for induction, *H. polymorpha pMOX* is highly derepressed under glucose exhaustion, thus supporting a methanol-free fermentation process. *H. polymorpha* is a rather exceptional host among methylotrophic yeasts as its *FMD* promoter is more frequently employed for governing recombinant protein production (Kang and Gellissen 2004). Also, the promoter of methylotrophic cytosolic dihydroxyacetone synthase (*DAS*), exhibiting tight carbon-source regulation, has been proposed to drive the expression of heterologous proteins (Manfrão-Netto et al. 2019). Interestingly, heterologous *H. polymorpha MOX* and *FMD* promoters were shown to be active and even exceed the activity of the native counterparts in *P. pastoris* (Vogl et al. 2020).

For more details on the regulation of the promoters of the methanol utilization pathway in this yeast, see the review of Hartner and Glieder (2006). Besides methylotrophic promoters, native and heterologous constitutive *GAP* promoters are used in *H. polymorpha* (Kang and Gellissen 2004). Recently, a new constitutive promoter, *HpGCW*, has been described in *H. polymorpha*, which exhibited 1.3-fold higher activity than *pGAP* in a glucose-containing medium and stable activity on various carbon substrates (Yan et al. 2022).

We described the mutations in the *H. polymorpha* putative glucose transceptor Gcr1 that affect the catabolite repression of *pMOX* (Stasyk and Stasyk 2019; Stasyk et al. 2004, 2018). The mutants deficient in *GCR1* exhibited defects in *pMOX* regulation only by hexoses and xylose but not by disaccharides or ethanol. With the derivatives of *gcr1* mutant strains as hosts, we developed a modified two-carbon-source-mode expression platform that utilizes convenient sugar substrates for growth (such as sucrose) and the induction of recombinant protein expression (glucose or xylose). Efficient and regulatable by sugar carbon sources, the expression of three recombinant proteins—secreted glucose oxidase from the fungus *A. niger*, secreted human mini-proinsulin, and membrane-bound hepatitis B virus surface antigen—was demonstrated in these mutant hosts (Krasovska et al. 2007, 2013). Thus, this mutant expression platform preserved the favorable regulatable nature of *pMOX* without the addition of methanol, making it a convenient alternative to the traditional system.

10.6.4 Other Tools

H. polymorpha full-genome sequence has been established and is available online at http://genome.jgi.doe.gov/Hanpo1/Hanpo1.home.html for the strain NCYC495, which allows convenient screening for gene homologs and the application of different system biology approaches using this species.

It was recently found that *H. polymorpha* differs from *S. cerevisiae* or *P. pastoris* in that it has a unique codon usage frequency. Even "yeast-likewise" codon-optimized genes that are expressed well in *S. cerevisiae* and *P. pastoris* may be inefficiently expressed in *H. polymorpha.* Thus, rare codons must be avoided when universally optimized gene versions are designed for expression in the host panel that includes *H. polymorpha* (Liu et al. 2014). Similarly to *P. pastoris*, the CRISPR/Cas9 gene editing technique has been recently adopted for *H. polymorpha* (Manfrão-Netto et al. 2019).

In addition, similarly to *P. pastoris*, studies have been conducted for host strain engineering in *H. polymorpha* to perform humanized recombinant protein glycosylation (Kim et al. 2004; Oh et al. 2008). These glycoengineering modifications have yielded glycoproteins mainly containing trimannosyl core N-glycan (Man3GlcNAc2) (Oh et al. 2008). Also, an alternative strategy for reconstructing the biosynthetic pathway of initial lipid-linked oligosaccharide synthesis was developed based on the deletion of the *H. polymorpha ALG3* gene coding for dolichyl-phosphate-mannose (Dol-P-Man)-dependent α-1,3 mannosyltransferase (Cheon et al. 2012).

It is known that correct protein folding in the ER is very important as it determines whether the protein enters the secretory pathway or is targeted for ER-associated degradation (ERAD), thus affecting the recombinant protein yield. It was reported that an increase in the dosage of the calnexin gene, which controls the folding and quality of glycoproteins, boosts the secretion of recombinant proteins in *H. polymorpha* (Klabunde et al. 2007). Also, an improved processing of secretory proteins in *H. polymorpha* by sequence variation near the processing site of α-MF pre-pro sequence was reported (Eilert et al. 2013).

H. polymorpha, similarly to *P. pastoris*, can reach high cell densities and efficiently secretes recombinant proteins, as exemplified by phytase production yield of 13.5 g L^{-1} (Mayer et al. 1999). Fermentation process development in *H. polymorpha* was comprehensively reviewed by Stockmann et al. (2009) and Celik and Calık (2012). Similarly to *P. pastoris*, a defined minimal mineral medium composed of salts, vitamins, and trace elements is used to support growth. Based on the selected promoter (*FMD* or *MOX*), fermentation strategies are developed using either glucose or glycerol at the beginning of fermentation, followed by carbon-source limitation in the second phase (Stockmann et al. 2009). Of note, one interesting example of the comparative analysis of rotavirus VP6 production has been reported where *H. polymorpha* outperformed both *P. pastoris* and *E. coli* during bioreactor cultivation but not in shake flask cultures (Bredell et al. 2016).

10.6.5 Industrial, Pharmaceutical, and Other Applications

Although not as popular as *P. pastoris* has been, *H. polymorpha* was exploited for the production of a number of recombinant proteins, e.g., enzymes, such as glucose oxidase (Hodgkins et al. 1993; Krasovska et al. 2007), phytase (Mayer et al. 1999), glycolate oxidase and catalase (Gellissen et al. 1996), isopenicillin N synthase (Gidijala et al. 2008), and novel fibrinolytic recombinant staphylokinase (Moussa et al. 2012), and pharmaceuticals: IFNα-2a (Müller et al. 2002), interleukin-6 (Boer et al. 2007), and recombinant human serum albumin (Heo et al. 2008).

As mentioned, the recombinant hepatitis B vaccine produced in *H. polymorpha* has been commercialized (Seo et al. 2008). The *H. polymorpha*-produced hepatitis B virus S antigen (HbsAg) was found to be assembled into pseudoviral particles in yeast membranes. Other studies have indicated that this lipoprotein particle structure is essential for the antigenicity of HBsAg (Petre et al. 1992). Methanol induction, which is routinely used for the high-level production of recombinant proteins in *H. polymorpha*, is a favorable condition for membrane formation (e.g., peroxisomal membranes) and is thus advantageous for the production of the recombinant hepatitis B vaccine with desirable antigenicity. A codon-adapted HPV 52 L1 gene encoding for recombinant human papillomavirus protein was recently successfully expressed in *H. polymorpha*. It was found that the recombinant proteins, similarly to HBsAg, could form virus-like particles in this host (Liu et al. 2014, 2015).

H. polymorpha has recently become a new promising yeast factory for the production of biofuel ethanol from pentose xylose (see Chap. 1 of this book). Several heterologous proteins have been successfully expressed in *H. polymorpha* and improved xylose fermentation, e.g., bacterial xylose isomerase (Voronovsky et al. 2005; Dmytruk et al. 2008), amylolytic enzymes alpha-amylase and glucoamylase from the yeast *Schwanniomyces occidentalis*, endoxylanase of the fungus *Trichoderma reesei*, beta-xylosidase of the fungus *A. niger* (Voronovsky et al. 2009), and others.

H. polymorpha can also be exploited as a yeast model of certain human diseases, e.g., neurodegenerative disorders (Menezes et al. 2015; Tenreiro et al. 2017). We have recently constructed and investigated *H. polymorpha* strains expressing recombinant human alpha-synuclein, a well-known trigger or protein reporter of Parkinson's disease (Kim et al. 2022). It was observed that alpha-synuclein cytotoxicity interferes with redox balance in the yeast cell and is affected by ions of divalent metals, suggesting the physiological validity of this novel *H. polymorpha* disease model (Denega et al. 2014; Hrushanyk et al. 2020; Grushanyk et al. 2020).

10.7 Other Methylotrophic Yeasts

10.7.1 Candida boidinii

Using the *Candida boidinii* expression system, several cytosolic, peroxisomal, and extracellular recombinant proteins have been produced. They included cytosolic human α1-antitrypsin (Yurimoto and Sakai 2009), *A. niger* pectin methylesterase (Kawaguchi et al. 2014), secretory glucoamylase from *Rhizopus oryzae* (Sakai et al. 1996), bovine cathepsin C (Komeda et al. 2002a), and peroxisome-targeted fructosyl amino acid oxidase (FAOD) from *Penicillium janthinellum* (Sakai et al. 1999). Thus, based on *C. boidinii*, a platform has been elaborated for recombinant toxic proteins which are targeted to peroxisomes. It was also observed that the propeptide of transglutaminase (TGase) from the actinomycete *Streptomyces mobaraensis* efficiently drives secretion in *C. boidinii* (Yurimoto et al. 2004). The *C. boidinii* expression system also exhibits several characteristics that may be beneficial for certain application as compared with other methylotrophic yeasts because different carbon sources provide three different expression levels in the same *C. boidinii* system exploiting a repressible alcohol oxidase (*AOD1*) promoter. *pAOD1* exhibits a high level of expression in cells grown on methanol or methanol supplemented with glycerol, an intermediate expression level in cells grown on glycerol as a sole carbon source, and a repressed expression in cells grown on glucose or ethanol (Sakai et al. 1995). As in other methylotrophs, high recombinant protein expression levels can be obtained in *C. boidinii* using cheap synthetic salt-based media. In addition, like in *P. pastoris*, a *C. boidinii* recipient host strain is available in which vacuolar protease genes, encoding the proteinase A (*PEP4*) and proteinase B (*PRB1*), are disrupted (Komeda et al. 2002b).

10.7.2 Pichia methanolica

Only a few recombinant proteins have been expressed in the methylotrophic yeast *Pichia methanolica* under the control of the promoter of its alcohol oxidase gene, *AUG1*. They included an isoform of human glutamate decarboxylase (Raymond et al. 1998), hepatitis B virus polymerase (HBV P protein) (Choi et al. 2002), secreted lignin peroxidase of *Phanerochaete chrysosporium* (Wang et al. 2004), bovine lactoferricin (Wang et al. 2007), and recombinant laccase from *Trametes versicolor* (Guo et al. 2006). In Mayson et al. (2003), various methanol concentrations were investigated in mixed glucose/methanol fed-batch cultures of *P.*

methanolica expressing the human transferrin N-lobe protein. Of note, lignin peroxidase expressed under the control of *pAUG1* adjoined to either the lignin peroxidase leader peptide of *P. chrysosporium* or the *S. cerevisiae* α-MF signal peptide was equally efficiently secreted (Wang et al. 2004).

It is known that *P. pastoris* and *P. methanolica* share the structure of the methanol utilization pathway and have two alcohol oxidase genes: *AOX1* and *AOX2* in *P. pastoris* (Cereghino and Cregg 2000) and *AUG1* and *AUG2* in *P. methanolica* (Raymond et al. 1998). Some studies suggested that Mut^S strains may produce higher levels of recombinant proteins than wild-type (Mut^+) strains in certain applications while utilizing substantially less methanol (Cereghino and Cregg 2000). Therefore, the system analogous to Mut^S and Mut^- phenotypes has also been developed in *P. methanolica* (Raymond et al. 1998). The *AUG1* gene disruptions resulted in a slow methanol utilization phenotype (Mut^S); in contrast, a disruption of the *AUG2* did not markedly affect methanol-dependent growth, whereas the double mutant strain was totally deficient in methanol utilization (Mut^-) (Raymond et al. 1998). Of note, *P. methanolica* and *P. pastoris* have been shown to produce similar levels of a recombinant protein when transformed with the same expression cassettes (Raymond et al. 1998). For additional details on *P. methanolica* as well as *C. boidinii* expression systems, please see the review by Gellissen (2000).

10.7.3 Ogataea minuta

Ogataea minuta is another and less known methylotrophic yeast closely related to *H. polymorpha,* recently adopted as a host for the production of recombinant proteins, human antibodies in particular (Yamada et al. 1994; Tsuda and Nonaka 2024). In yeast glycosylation studies, the collection of vectors and host strains of this species was constructed, and the strain producing a Man5GlcNAc2 high mannose-type sugar chain as a prototype of a humanized sugar chain was obtained by disrupting the *OCH1* gene and by introducing the α-1,2-mannosidase gene (Kuroda et al. 2006).

In addition, some strategies based on the partial suppression of O-glycosylation by the addition of specific inhibitors of this pathway (e.g., rhodanine-3-acetic acid) have been successfully tested in the *O. minuta*-producing immunoglobulin G (IgG) (Kuroda et al. 2008). Strains bearing the *OCH1* disruption (thus eliminating the alpha-1,6-mannosyltransferase initiating the synthesis of the hypermannosylated glycan structures in yeast) and co-overexpressing one of the genes involved in the mannosyl phosphorylation of N-glycans (e.g., *MNN4*, *MNN6*, and *PNO1*) have been constructed in *O. minuta* (Akeboshi et al. 2009). The usefulness of protease-deficient strains in the production of antibodies in *O. minuta* has been demonstrated (Kuroda et al. 2007). Also, the production of recombinant beta-hexosaminidase A, a potential enzyme for replacement therapy for Tay-Sachs and Sandhoff diseases, has been achieved (Akeboshi et al. 2007). The therapeutic effect in mice of a recombinant human lysosomal β-hexosaminidase A (HexA) with terminal mannose-6-phosphate residues on their N-glycans produced in the glycoengineered *O. minuta* strain has been shown as well (Tsuji et al. 2011). It was also demonstrated that co-expression of certain chaperones increased recombinant antibody productivity in this species (Suzuki et al. 2017). Interestingly, Yoko-o et al. (2021) demonstrated that, although *O. minuta* was more closely taxonomically related to *H. polymorpha*, the *OmAOX* promoter in *O. minuta* was suppressed by either glucose or glycerol, being in this respect more similar to *P. pastoris pAOX1* (Cereghino and Cregg 2000).

10.8 Conclusions and Outlook

For the production of recombinant proteins, including biopharmaceuticals, food and feed enzymes, and proteins for structural studies, various yeast expression systems have been successfully used over the years. While *S. cerevisiae* is possibly the best genetically characterized eukaryotic organism out of all, for quite some time, the most frequently used yeast species for heterologous protein expression has been *P. pastoris* (Spohner et al. 2015). For *P. pastoris*, as well as other methylotrophic yeasts, the availability of commercial expression kits and large strain collections, wide variety of integrative vectors and auxotrophic or dominant selectable markers that allow stable integration into the genomes in a controlled copy number, strictly regulated by carbon sources inducible and constitutive promoters, and efficient transformation protocols, combined with the well-elaborated large-scale fermentation techniques, make these expression platforms very attractive for biotechnological exploitation. In addition to such common yeast features as the ability to perform complex posttranslational modifications and being neither pyrogenic nor pathogenic, which is relevant for pharmaceutical proteins, methylotrophic yeasts are also known to secrete recombinant proteins more efficiently than others. Some methylotrophic yeasts already possess a status of generally recognized as safe (GRAS) for application in the food industry. Efforts to humanize the glycosylation pathway similar to that in *P. pastoris* will continue with other glycoengineered yeast hosts to further develop them as producers of biopharmaceuticals. The recent revolutionary advances in next-generation DNA sequencing, systems and synthetic biology, and comprehensive mathematical genome-scale metabolic models (GEMs) are predicted to greatly intensify the works in methylotrophic yeasts on new host construction and the optimization of high cell-density fermentation in the near future. It remains to be seen whether the application spectrum of alternative methylotrophic yeast expression platforms will also be broadened.

Acknowledgment This study was partially supported by "Presidential Discretionary-Ukraine Support Grants" from Simons Foundation, Award No 1030281. This work was also supported in part by the NRFU project N 2023.04/0048, State registration number: 0123U103586, "Development of a broad-spectrum antimicrobial agent based on the extract of the leguminous plant *Indigofera spicata*". The authors want to acknowledge the Armed Forces of Ukraine and the Allies for the opportunity to work on this manuscript in Ukraine.

References

Ahmad M, Hirz M, Pichler H, Schwab H (2014) Protein expression in *Pichia pastoris*: recent achievements and perspectives for heterologous protein production. Appl Microbiol Biotechnol 98:5301–5317

Akeboshi H, Chiba Y, Kasahara Y, Takashiba M, Takaoka Y, Ohsawa M, Tajima Y, Kawashima I, Tsuji D, Itoh K, Sakuraba H, Jigami Y (2007) Production of recombinant beta-hexosaminidase A, a potential enzyme for replacement therapy for Tay-Sachs and Sandhoff diseases, in the methylotrophic yeast *Ogataea minuta*. Appl Environ Microbiol 73:4805–4812

Akeboshi H, Kasahara Y, Tsuji D, Itoh K, Sakuraba H, Chiba Y, Jigami Y (2009) Production of human beta-hexosaminidase A with highly phosphorylated N-glycans by the overexpression of the *Ogataea minuta* MNN4 gene. Glycobiology 19:1002–1009

Ata Ö, Ergün BG, Fickers P, Heistinger L, Mattanovich D, Rebnegger C, Gasser B (2021) What makes *Komagataella phaffii* non-conventional? FEMS Yeast Res 21(8):foab059

Aw R, Polizzi KM (2013) Can too many copies spoil the broth? Microb Cell Factories 12:128

Bai J, Swartz DJ, Protasevich II, Brouillette CG, Harrell PM, Hildebrandt E, Gasser B, Mattanovich D, Ward A, Chang G et al (2011) A gene optimization strategy that enhances production of fully functional P-glycoprotein in *Pichia pastoris*. PLoS One 6:e22577

Barrero JJ, Casler JC, Valero F, Ferrer P, Glick BS (2018) An improved secretion signal enhances the secretion of model proteins from *Pichia pastoris*. Microb Cell Factories 17:161

Bawa Z, Routledge SJ, Jamshad M, Clare M, Sarkar D, Dickerson I, Ganzlin M, Poyner DR, Bill RM (2014) Functional recombinant protein is present in the pre-induction phases of *Pichia pastoris* cultures when grown in bioreactors, but not shake-flasks. Microb Cell Factories 13:561

Bill RM (2015) Recombinant protein subunit vaccine synthesis in microbes: a role for yeast? J Pharm Pharmacol 67:319–328

Boer E, Steinborn G, Gellissen G, Kunze G (2007) Production of interleukin-6 in *Arxula adeninivorans*, *Hansenula polymorpha* and *Saccharomyces cerevisiae* by applying a wide-range yeast vector (CoMed™) system to simultaneous comparative assessment. FEMS Yeast Res 7:1181–1187

Bredell H, Smith JJ, Prins WA, Gorgens JF, van Zyl WH (2016) Expression of rotavirus VP6 protein: a comparison amongst *Escherichia coli*, *Pichia pastoris* and *Hansenula polymorpha*. FEMS Yeast Res 16:fow001

Byrne B (2015) *Pichia pastoris* as an expression host for membrane protein structural biology. Curr Opin Struct Biol 32:9–17

Carlesso A, Delgado R, Ruiz Isant O, Uwangue O, Valli D, Bill RM, Hedfalk K (2022) Yeast as a tool for membrane protein production and structure determination. FEMS Yeast Res 22:foac047

Celik E, Calık P (2012) Production of recombinant proteins by yeast cells. Biotechnol Adv 30:1108–1118

Cereghino JL, Cregg JM (2000) Heterologous protein expression in the methylotrophic yeast *Pichia pastoris*. FEMS Microbiol Rev 24:45–66

Cereghino GPL, Cereghino JL, Ilgen C, Cregg JM (2002) Production of recombinant proteins in fermenter cultures of the yeast *Pichia pastoris*. Curr Opin Biotechnol 13:329–332

Cheon SA, Kim H, Oh DB, Kwon O, Kang HA (2012) Remodeling of the glycosylation pathway in the methylotrophic yeast *Hansenula polymorpha* to produce human hybrid-type N-glycans. J Microbiol 50:341–348

Choi J, Kim EE, Park YI, Han YS (2002) Expression of the active human and duck hepatitis B virus polymerases in heterologous system of *Pichia methanolica*. Antivir Res 55:279–290

Ciofalo V, Barton N, Kreps J, Coats I, Shanahan D (2006) Safety evaluation of a lipase enzyme preparation, expressed in *Pichia pastoris*, intended for use in the degumming of edible vegetable oil. Regul Toxicol Pharmacol (ISRTP) 45:1–8

Corchero JL, Gasser B, Resina D, Smith W, Parrilli E, Va'zquez F, Abasolo I, Giuliani M, Jantti J, Ferrer P, Saloheimo M, Mattanovich D, Schwartz S Jr, Tutino ML, Villaverde A (2013)

Unconventional microbial systems for the cost-efficient production of high-quality protein therapeutics. Biotechnol Adv 31:140–153

Cregg JM, Russell KA (1998) Transformation. Methods Mol Biol 103:27–39

Cregg JM, Barringer KJ, Hessler AY, Madden KR (1985) *Pichia pastoris* as a host system for transformations. Mol Cell Biol 5:3376–3385

Cregg JM, Tolstorukov I, Kusari A, Sunga J, Madden K, Chappell T (2009) Expression in the yeast Pichia pastoris. In: Burgess RR, Deutscher MP (eds) Methods in enzymology. Guide to protein purification, vol 463, 2nd edn. Academic, pp 169–189

Daly R, Hearn MT (2005) Expression of heterologous proteins in *Pichia pastoris*: a useful experimental tool in protein engineering and production. J Mol Recognit 18:119–138

Darby RA, Cartwright SP, Dilworth MV, Bill RM (2012) Which yeast species shall I choose? *Saccharomyces cerevisiae* versus *Pichia pastoris*. Methods Mol Biol 866:11–23

Demain AL, Vaishnav P (2009) Production of recombinant proteins by microbes and higher organisms. Biotechnol Adv 27:297–306

Denega IO, Klymyshyn NI, Sybirna NO, Stasyk OV, Stasyk OG (2014) Modeling of molecular processes underlying Parkinson's disease in cells of methylotrophic yeast *Hansenula polymorpha*. Stud Biol 8:5–16

Dmytruk OV, Dmytruk KV, Abbas CA, Voronovsky AY, Sibirny AA (2008) Engineering of xylose reductase and overexpression of xylitol dehydrogenase and xylulokinase improves xylose alcoholic fermentation in the thermotolerant yeast *Hansenula polymorpha*. Microb Cell Factories 7:21

Dragosits M, Frascotti G, Bernard-Granger L, Vazquez F, Giuliani M, Baumann K et al (2011) Influence of growth temperature on the production of antibody Fab fragments in different microbes: a host comparative analysis. Biotechnol Prog 27:38–46

Duan G, Ding L, Wei D, Zhou H, Chu J, Zhang S, Qian J (2019) Screening endogenous signal peptides and protein folding factors to promote the secretory expression of heterologous proteins in Pichia pastoris. J Biotechnol 306:193–202

Dunn WA Jr, Cregg JM, Kiel JA, van der Klei IJ, Oku M, Sakai Y, Sibirny AA, Stasyk OV, Veenhuis M (2005) Pexophagy: the selective autophagy of peroxisomes. Autophagy 1:75–83

Eilert E, Rolf T, Heumaier A, Hollenberg CP, Piontek M, Suckow M (2013) Improved processing of secretory proteins in *Hansenula polymorpha* by sequence variation near the processing site of the alpha mating factor prepro sequence. J Biotechnol 167:94–100

Emmerstorfer A, Wriessnegger T, Hirz M, Pichler H (2014) Overexpression of membrane proteins from higher eukaryotes in yeasts. Appl Microbiol Biotechnol 98:7671–7698

Ergün BG, Calık P (2016) Lignocellulose degrading extremozymes produced by *Pichia pastoris*: current status and future prospects. Bioprocess Biosyst Eng 39:1–36

Fickers P (2014) *Pichia pastoris*: a workhorse for recombinant protein production. Curr Res Microbiol Biotechnol 2:354–363

Gasser B, Mattanovich D (2018) A yeast for all seasons - is *Pichia pastoris* a suitable chassis organism for future bioproduction? FEMS Microbiol Lett 365:fny181

Gasser B, Prielhofer R, Marx H, Maurer M, Nocon J, Steiger M, Puxbaum V, Sauer M, Mattanovich D (2013) *Pichia pastoris*: protein production host and model organism for biomedical research. Future Microbiol 8:191–208

Gellissen G (2000) Heterologous protein production in methylotrophic yeasts. Appl Microbiol Biotechnol 54:741–750

Gellissen G (ed) (2002) *Hansenula polymorpha* – biology and applications. Wiley-VCH, Weinheim

Gellissen G, Piontek M, Dahlems U, Jenzelewski V, Gavagan JE, DiCosimo R (1996) Recombinant *Hansenula polymorpha* as a biocatalyst: coexpression of the spinach glycolate oxidase (GO) and the *S. cerevisiae* catalase T (CTT1) gene. Appl Microbiol Biotechnol 46:46–54

Gellissen G, Strasser AWM, Suckow M (2005a) Key and criteria to the selection of an expression platform. In: Gellissen G (ed) Production of recombinant proteins — novel microbial and eukaryotic expression systems. Wiley-VCH, Weinheim, pp 1–5

Gellissen G, Kunze G, Gaillardin C, Cregg JM, Berardi E, Veenhuis M, van der Klei I (2005b) New yeast expression platforms based on methylotrophic *Hansenula polymorpha* and *Pichia pastoris* and on dimorphic *Arxula adeninivorans* and *Yarrowia lipolytica* – a comparison. FEMS Yeast Res 5:1079–1096

Gemmill TR, Trimble RB (1999) Overview of N- and O-linked oligosaccharide structures found in various yeast species. BBA-Gen Subjects 1426:227–237

Gidijala L, Bovenberg RAL, Klaassen P, van der Klei IJ, Veenhuis M, Kiel JAKW (2008) Production of functionally active *Penicillium chrysogenum* isopenicillin N synthase in the yeast *Hansenula polymorpha*. BMC Biotechnol 8:29

Grushanyk NV, Sarai IA, Stasyk OV, Stasyk OG (2020) Prooxidant-antioxidant balance in the methylotrophic yeast *Ogataea polymorpha* exposed to spermidine. Stud Biol 14:13–28

Guo M, Lu F, Du L, Pu J, Bai D (2006) Optimization of the expression of a laccase gene from *Trametes versicolor* in *Pichia methanolica*. Appl Microbiol Biotechnol 71:848–852

Hamilton SR, Davidson RC, Sethuraman N, Nett JH, Jiang Y, Rios S, Bobrowicz P, Stadheim TA, Li H, Choi B-K, Hopkins D, Wischnewski H, Roser J, Mitchell T, Strawbridge RR, Hoopes J, Wildt S, Gerngross TU (2006) Humanization of yeast to produce complex terminally sialylated glycoproteins. Science 313:1441–1443

Hartner FS, Glieder A (2006) Regulation of methanol utilisation pathway genes in yeasts. Microb Cell Factories 5:3

He M, Wu D, Wu J, Chen J (2014) Enhanced expression of endoinulinase from *Aspergillus niger* by codon optimization in *Pichia pastoris* and its application in inulooligosaccharide production. J Ind Microbiol Biotechnol 41:105–114

Heo JH, Ananin V, Kang HA, Rhee SK, Kim CH (2008) Feeding strategies for the enhanced production of recombinant human serum albumin in the fed-batch cultivation of *Hansenula polymorpha*. Process Biochem 43:918–924

Hodgkins M, Mead D, Ballance DJ, Goodey A, Sudbery P (1993) Expression of the glucose oxidase gene from *Aspergillus niger* in *Hansenula polymorpha* and its use as a reporter gene to isolate regulatory mutations. Yeast 9:625–635

Hrushanyk NV, Stasyk OV, Stasyk OG (2020) Oxidative stress regulation in the yeast *Ogataea polymorpha* producer of human α-synuclein. Ukrainian Biochem J 92:120–133

Hu H, Gao J, He J, Yu B, Zheng P, Huang Z, Mao X, Yu J, Han G, Chen D (2013) Codon optimization significantly improves the expression level of a keratinase gene in *Pichia pastoris*. PLoS One 8(3):e58393

Jayakrishnan A, Wan Rosli WR, Tahir ARM, Razak FSA, Kee PE, Ng HS, Chew Y-L, Lee S-K, Ramasamy M, Tan CS, Liew KB (2024) Evolving paradigms of recombinant protein production in pharmaceutical industry: a rigorous review. Sci 6:9

Juturu V, Wu JC (2018) Heterologous protein expression in *Pichia pastoris*: latest research progress and applications. Chembiochem Eur J Chem Biol 19:7–21

Kang HA, Gellissen G (2004) Hansenula polymorpha. In: Gellissen G (ed) Production of recombinant proteins, novel microbial and eukaryotic expression systems, 1st edn. Wiley-VCH, Weinheim, pp 111–142

Kang HA, Sohn JH, Choi ES, Chung BH, Yu MH, Rhee SK (1998) Glycosylation of human alpha 1-antitrypsin in *Saccharomyces cerevisiae* and methylotrophic yeasts. Yeast 14:371–381

Kang HA, Kang W, Hong WK, Kim MW, Kim JY, Sohn JH, Choi ES, Choe KB, Rhee SK (2001) Development of expression systems for the production of recombinant human serum albumin using the MOX promoter in *Hansenula polymorpha* DL-1. Biotechnol Bioeng 76:175–185

Kawaguchi K, Yurimoto H, Sakai Y (2014) Expression of a codon-optimized *Aspergillus niger* pectin methylesterase gene in the methylotrophic yeast *Candida boidinii*. Biosci Biotechnol Biochem 78:718–721

Kim MW, Rhee SK, Kim JY, Shimma Y, Chiba Y, Jigami Y, Kang HA (2004) Characterization of N-linked oligosaccharides assembled on secretory recombinant glucose oxidase and cell wall mannoproteins from the methylotrophic yeast *Hansenula polymorpha*. Glycobiology 14:243–251

Kim H, Yoo SJ, Kang HA (2015) Yeast synthetic biology for the production of recombinant therapeutic proteins. FEMS Yeast Res 15:1–16

Kim H, Jeong J, Lee C (2022) A method to study α-synuclein toxicity and aggregation using a humanized yeast model. J Vis Exp 189. https://doi.org/10.3791/64418

Klabunde J, Kleebank S, Piontek M, Hollenberg CP, Hellwig S, Degelmann A (2007) Increase of calnexin gene dosage boosts the secretion of heterologous proteins by *Hansenula polymorpha*. FEMS Yeast Res 7:1168–1180

Komeda T, Tazumi K, Shimada H, Kano K, Hayashi T, Saito H, Tsumura H, Kato N, Sakai Y, Kondo K (2002a) Production of active bovine cathepsin C (dipeptidyl aminopeptidase I) in the methylotrophic yeast *Candida boidinii*. Appl Microbiol Biotechnol 59:252–258

Komeda T, Sakai Y, Kato N, Kondo K (2002b) Construction of protease-deficient *Candida boidinii* strains useful for recombinant protein production: cloning and disruption of proteinase A gene (PEP4) and proteinase B gene (PRBI). Biosci Biotechnol Biochem 66:628–631

Krasovska OS, Stasyk OV, Nahorny VO, Stasyk OV, Granovski N, Kordium VA, Vozianov OF, Sibirny AA (2007) Glucose-induced production of recombinant proteins in *Hansenula polymorpha* mutants deficient in catabolite repression. Biotechnol Bioeng 97:858–870

Krasovska OS, Stasyk OV, Sibirny AA (2013) Stable overproducer of hepatitis B surface antigen in the methylotrophic yeast *Hansenula polymorpha* due to multiple integration of heterologous auxotrophic selective markers and defect in peroxisome biogenesis. Appl Microbiol Biotechnol 97:9969–9979

Küberl A, Schneider J, Thallinger GG, Anderl I, Wibberg D, Hajek T, Jaenicke S, Brinkrolf K, Goesmann A, Szczepanowski R, Pühler A, Schwab H, Glieder A, Pichler H (2011) High-quality genome sequence of *Pichia pastoris* CBS7435. J Biotechnol 154:312

Kuroda K, Kobayashi K, Tsumura H, Komeda T, Chiba Y, Jigami Y (2006) Production of Man5GlcNAc2-type sugar chain by the methylotrophic yeast *Ogataea minuta*. FEMS Yeast Res 6:1052–1062

Kuroda K, Kitagawa Y, Kobayashi K, Tsumura H, Komeda T, Mori E, Motoki K, Kataoka S, Chiba Y, Jigami Y (2007) Antibody expression in protease-deficient strains of the methylotrophic yeast *Ogataea minuta*. FEMS Yeast Res 7:1307–1316

Kuroda K, Kobayashi K, Kitagawa Y, Nakagawa T, Tsumura H, Komeda T, Shinmi D, Mori E, Motoki K, Fuju K, Sakai T, Nonaka K, Suzuki T, Ichikawa K, Chiba Y, Jigami Y (2008) Efficient antibody production upon suppression of *O mannosylation* in the yeast *Ogataea minuta*. Appl Environ Microbiol 74:446–453

Kurtzman CP (2005) Description of *Komagataella phaffii* sp. nov. and the transfer of *Pichia pseudopastoris* to the methylotrophic yeast genus *Komagataella*. Int J Syst Evol Microbiol 55:973–976

Kurtzman CP (2009) Biotechnological strains of *Komagataella* (*Pichia*) pastoris are *Komagataella phaffii* as determined from multigene sequence analysis. J Ind Microbiol Biotechnol 36:1435–1438

Kurtzman CP (2011) A new methanol assimilating yeast, *Ogataea parapolymorpha*, the ascosporic state of *Candida parapolymorpha*. Antonie Van Leeuwenhoek 100:455–462

Leao AN, Kiel JA (2003) Peroxisome homeostasis in *Hansenula polymorpha*. FEMS Yeast Res 4:131–139

Li X, Shen J, Chen X, Chen L, Wan S, Qiu X, Chen K, Chen C, Tan H (2022) Humanization of yeasts for glycan-type end-products. Front Microbiol 13:930658

Lin Cereghino GP, Lin Cereghino J, Sunga AJ, Johnson MA, Lim M, Gleeson MA, Cregg JM (2001) New selectable marker/auxotrophic host strain combinations for molecular genetic manipulation of *Pichia pastoris*. Gene 263:159–169

Lin-Cereghino GP, Stark CM, Kim D, Chang J, Shaheen N, Poerwanto H, Agari K, Moua P, Low LK, Tran N, Huang AD, Nattestad M, Oshiro KT, Chang JW, Chavan A, Tsai JW, Lin-Cereghino J (2013) The effect of alpha-mating factor secretion signal mutations on recombinant protein expression in *Pichia pastoris*. Gene 519:311–317

Liu C, Yang X, Yao Y, Huang W, Sun W, Ma Y (2014) Diverse expression levels of two codon-optimized genes that encode human papilloma virus type 16 major protein L1 in *Hansenula polymorpha*. Biotechnol Lett 36:937–945

Liu C, Yao Y, Yang X, Bai H, Huang W, Xia Y, Ma Y (2015) Production of recombinant human papillomavirus type 52 L1 protein in *Hansenula polymorpha* formed virus-like particles. J Microbiol Biotechnol 25:936–940

Looser V, Bruhlmann B, Bumbak F, Stenger C, Costa M, Camattari A, Fotiadis D, Kovar K (2015) Cultivation strategies to enhance productivity of *Pichia pastoris*: a review. Biotechnol Adv 33:1177–1193

Mack M, Wannemacher M, Hobl B, Pietschmann P, Hock B (2009) Comparison of two expression platforms in respect to protein yield and quality: *Pichia pastoris* versus *Pichia angusta*. Protein Expr Purif 66:165–171

Manfrão-Netto JHC, Gomes AMV, Parachin NS (2019) Advances in using *Hansenula polymorpha* as chassis for recombinant protein production. Front Bioeng Biotechnol 7:94

Mattanovich D, Gasser B, Hohenblum H, Sauer M (2004) Stress in recombinant protein producing yeasts. J Biotechnol 113:121–135

Mattanovich D, Branduardi P, Dato L, Gasser B, Sauer M, Porro D (2012) Recombinant protein production in yeasts. Methods Mol Biol 824:329–358

Mayer AF, Hellmuth K, Schlieker H, Lopez-Ulibarri R, Oertel S, Dahlems U et al (1999) An expression system matures: a highly efficient and cost-effective process for phytase production by recombinant strains of *Hansenula polymorpha*. Biotechnol Bioeng 63:373–381

Mayson BE, Kilburn DG, Zamost BL, Raymond CK, Lesnicki GJ (2003) Effects of methanol concentration on expression levels of recombinant protein in fed-batch cultures of *Pichia methanolica*. Biotechnol Bioeng 81:291–298

Meehl MA, Stadheim TA (2014) Biopharmaceutical discovery and production in yeast. Curr Opin Biotechnol 30:120–127

Mellitzer A, Weis R, Glieder A, Flicker K (2012) Expression of lignocellulolytic enzymes in *Pichia pastoris*. Microb Cell Factories 11:61

Menezes R, Tenreiro S, Macedo D, Santos CN, Outeiro TF (2015) From the baker to the bedside: yeast models of Parkinson's disease. Microb Cell 2:262–279

Morales L, Dujon B (2012) Evolutionary role of interspecies hybridization and genetic exchanges in yeasts. Microbiol Mol Biol Rev 76:721–739

Morales L, Noel B, Porcel B, Marcet-Houben M, Hullo MF, Sacerdot C, Tekaia F, Leh-Louis V, Despons L, Khanna V, Aury JM, Barbe V, Couloux A, Labadie K, Pelletier E, Souciet JL, Boekhout T, Gabaldon T, Wincker P, Dujon B (2013) Complete DNA sequence of *Kuraishia capsulata* illustrates novel genomic features among budding yeasts (*Saccharomycotina*). Genome Biol Evol 5:2524–2539

Moussa M, Ibrahim M, El Ghazaly M, Rohde J, Gnoth S, Anton A, Kensy F, Mueller F (2012) Expression of recombinant staphylokinase in the methylotrophic yeast *Hansenula polymorpha*. BMC Biotechnol 12:96

Müller IIF, Tieke A, Waschk D, Mühle C, Müller FI, Seigelchifer M et al (2002) Production of IFN α -2a in *Hansenula polymorpha*. Process Biochem 38:15–25

Nocon J, Steiger MG, Pfeffer M, Sohn SB, Kim TY, Maurer M, Rußmayer H, Pflügl S, Ask M, Haberhauer-Troyer C, Ortmayr K, Hann S, Koellensperger G, Gasser B, Lee SY, Mattanovich D (2014) Model based engineering of *Pichia pastoris* central metabolism enhances recombinant protein production. Metab Eng 24:129–138

Oh DB, Park JS, Kim MW, Cheon SA, Kim EJ, Moon HY et al (2008) Glycoengineering of the methylotrophic yeast *Hansenula polymorpha* for the production of glycoproteins with trimannosyl core N-glycan by blocking core oligosaccharide assembly. Biotechnol J 3(5):659–668

Petranovic D, Tyo K, Vemuri G, Nielsen J (2010) Prospects of yeast systems biology for human health: integrating lipid, protein and energy metabolism. FEMS Yeast Res 10:1046–1059

Petre J, Rutgers T, Hauser P (1992) Properties of a recombinant yeast-derived hepatitis B surface antigen containing S, preS2 and preS1 antigenic domains. Arch Virol Suppl 4:137–141

Potvin G, Ahmad A, Zhang Z (2012) Bioprocess engineering aspects of heterologous protein production in *Pichia pastoris*: a review. Biochem Eng J 64:91–105

Prielhofer R, Maurer M, Klein J, Wenger J, Kiziak C, Gasser B, Mattanovich D (2013) Induction without methanol: novel regulated promoters enable high-level expression in *Pichia pastoris*. Microb Cell Factories 12:5

Prielhofer R, Barrero JJ, Steuer S, Gassler T, Zahrl R, Baumann K, Sauer M, Mattanovich D, Gasser B, Marx H (2017) Golden*Pi*CS: a Golden Gate-derived modular cloning system for applied synthetic biology in the yeast *Pichia pastoris*. BMC Syst Biol 11:1–14

Puxbaum V, Mattanovich D, Gasser B (2015) Quo vadis? The challenges of recombinant protein folding and secretion in *Pichia pastoris*. Appl Microbiol Biotechnol 99:2925–2938

Qian W, Song H, Liu Y, Zhang C, Niu Z, Wang H, Qiu B (2009) Improved gene disruption method and Cre-loxP mutant system for multiple gene disruptions in *Hansenula polymorpha*. J Microbiol Methods 79:253–259

Raschmanová H, Weninger A, Knejzlik Z, Melzoch K, Kovai K (2021) Engineering of the unfolded protein response pathway in *Pichia pastoris*: enhancing production of secreted recombinant protein. Appl Microbiol Biotechnol 105:4397–4414

Ravin NV, Eldarov MA, Kadnikov VV, Beletsky AV, Schneider J, Mardanova ES, Smekalova EM, Zvereva MI, Dontsova OA, Mardanov AV, Skryabin KG (2013) Genome sequence and analysis of methylotrophic yeast *Hansenula polymorpha* DL1. BMC Genomics 14:837

Raymond CK, Bukowski T, Holderman SD, Ching AFT, Vanaja E, Stamm MR (1998) Development of the methylotrophic yeast, *Pichia methanolica*, for the expression of the 65-kilodalton isoform of human glutamate decarboxylase. Yeast 14:11–23

Resina D, Cos O, Ferrer P, Valero F (2005) Developing high cell density fed-batch cultivation strategies for heterologous protein production in *Pichia pastoris* using the nitrogen source-regulated FLD1 promoter. Biotechnol Bioeng 91:760–767

Rettenbacher LA, Arauzo-Aguilera K, Buscajoni L, Castillo-Corujo A, Ferrero-Bordera B, Kostopoulou A, Moran-Torres R, Núñez-Nepomuceno D, Öktem A, Palma A, Pisent B, Puricelli M, Schilling T, Tungekar AA, Walgraeve J, Humphreys D, von der Haar T, Gasser B, Mattanovich D, Ruddock L, van Dijl JM (2022) Microbial protein cell factories fight back? Trends Biotechnol 40:576–590

Routledge SJ, Mikaliunaite L, Patel A, Clare M, Cartwright SP, Bawa Z, Wilks MD, Low F, Hardy D, Rothnie AJ, Bill RM (2016) The synthesis of recombinant membrane proteins in yeast for structural studies. Methods 95:26–37

Sakai Y, Rogi T, Takeuchi R, Kato N, Tani Y (1995) Expression of *Saccharomyces* adenylate kinase gene in *Candida boidinii* under the regulation of its alcohol oxidase promoter. Appl Microbiol Biotechnol 42:860–864

Sakai Y, Akiyama M, Kondoh H, Shibano Y, Kato N (1996) High-level secretion of fungal glucoamylase using the *Candida boidinii* gene expression system. Biochim Biophys Acta 1308:81–87

Sakai Y, Yoshida H, Yurimoto H, Yoshida N, Fukuya H, Takabe K, Kato N (1999) Production of fungal fructosyl amino acid oxidase useful for diabetic diagnosis in the peroxisome of *Candida boidinii*. FEBS Lett 459:233–237

Seo HS, Park JS, Han KY, Bae KD, Ahn SJ, Kang HA, Lee J (2008) Analysis and characterization of hepatitis B vaccine particles synthesized from *Hansenula polymorpha*. Vaccine 26:4138–4144

Shekhar C (2008) Pichia power: India's biotech industry puts unconventional yeast to work. Chem Biol 15:201–202

Shen Q, Cui J, Wang Y, Hu ZC, Xue YP, Zheng YG (2024) Identification of a novel growth-associated promoter for biphasic expression of heterogenous proteins in *Pichia pastoris*. Appl Environ Microbiol 90:e0174023

Spadiut O, Capone S, Krainer F, Glieder A, Herwig C (2014) Microbials for the production of monoclonal antibodies and antibody fragments. Trends Biotechnol 32:54–60

Spohner SC, Müller H, Quitmann H, Czermak P (2015) Expression of enzymes for the usage in food and feed industry with *Pichia pastoris*. J Biotechnol 202:118–134

Stasyk OG, Stasyk OV (2019) Glucose sensing and regulation in yeasts. In: Sibirny AA (ed) Non-conventional yeasts: from basic research to application. Springer International Publishing AG, pp 477–519

Stasyk OV, Stasyk OG, Komduur J, Veenhuis M, Cregg JM, Sibirny AA (2004) A hexose transporter homologue controls glucose repression in the methylotrophic yeast *Hansenula polymorpha*. J Biol Chem 279:8116–8125

Stasyk OG, Denega IO, Padhorny D, Dmytruk KV, Kozakov D, Abbas C, Stasyk OV (2018) Glucose regulation in the methylotrophic yeast *Hansenula (Ogataea) polymorpha* is mediated by a putative transceptor Gcr1. Int J Biochem Cell Biol 103:25–34

Steinborn G, Boer E, Scholz A, Tag K, Kunze G, Gellissen G (2006) Application of a wide-range yeast vector (Comed (TM)) system to recombinant protein production in dimorphic *Arxula adeninivorans*, methylotrophic *Hansenula polymorpha* and other yeasts. Microb Cell Factories 5:33

Stockmann C, Scheidle M, Klee D, Dittrich B, Merckelbach A, Hehmann G, Melmer G, Buchs J, Kang HA, Gellissen G (2009) Process development in *Hansenula polymorpha* and *Arxula adeninivorans*, a re-assessment. Microb Cell Factories 8:22–31

Sturmberger L, Chappell T, Geier M, Krainer F, Day KJ, Vide U, Trstenjak S, Schiefer A, Richardson T, Soriaga L, Darnhofer B, Birner-Gruenberger R, Glick BS, Tolstorukov I, Cregg J, Madden K, Glieder A (2016) Refined *Pichia pastoris* reference genome sequence. J Biotechnol 235:121–131

Sudbery PE, Gleeson MA, Veale RA, Ledeboer AM, Zoetmulder MC (1988) *Hansenula polymorpha* as a novel yeast system for the expression of heterologous genes. Biochem Soc Trans 16:1081–1083

Suzuki T, Baba S, Ono M, Nonaka K, Ichikawa K, Yabuta M, Ito R, Chiba Y (2017) Efficient antibody production in the methylotrophic yeast *Ogataea minuta* by overexpression of chaperones. J Biosci Bioeng 124:156–163

Takagi S, Tsutsumi N, Terui Y, Kong X, Yurimoto H, Sakai Y (2019) Engineering the expression system for *Komagataella phaffii* (*Pichia pastoris*): an attempt to develop a methanol-free expression system. FEMS Yeast Res 19:foz059

Tenreiro S, Franssens V, Winderickx J, Outeiro TF (2017) Yeast models of Parkinson's disease-associated molecular pathologies. Curr Opin Genet Dev 44:74–83

Thompson CA (2010) FDA approves kallikrein inhibitor to treat hereditary angioedema. Am J Health Syst Pharm 67:93

Toma's-Gamisans M, Ferrer P, Albiol J (2016) Integration and validation of the genome-scale metabolic models of *Pichia pastoris*: a comprehensive update of protein glycosylation pathways, lipid and energy metabolism. PLoS One 11(1):e0148031

Tsuda M, Nonaka K (2024) Recent progress on heterologous protein production in methylotrophic yeast systems. World J Microbiol Biotechnol 40:200

Tsuji D, Akeboshi H, Matsuoka K, Yasuoka H, Miyasaki E, Kasahara Y, Kawashima I, Chiba Y, Jigami Y, Taki T, Sakuraba H, Itoh K (2011) Highly phosphomannosylated enzyme replacement therapy for GM2 gangliosidosis. Ann Neurol 69:691–701

Unver Y, Dagci I (2024) Komagataella phaffii (Pichia pastoris) as a powerful yeast expression system for biologics production. Front Biosci (Elite Ed) 16:19

Vogl T, Glieder A (2013) Regulation of *Pichia pastoris* promoters and its consequences for protein production. New Biotechnol 30:385–404

Vogl T, Kickenweiz T, Pitzer J, Sturmberger L, Weninger A, Biggs BW, Köhler EM, Baumschlager A, Fischer JE, Hyden P, Wagner M, Baumann M, Borth N, Geier M, Ajikumar PK, Glieder A (2018) Engineered bidirectional promoters enable rapid multi-gene coexpression optimization. Nat Commun 12:1287

Vogl T, Fischer JE, Hyden P, Wasmayer R, Sturmberger L, Glieder A (2020) Orthologous promoters from related methylotrophic yeasts surpass expression of endogenous promoters of *Pichia pastoris*. AMB Express 10:38

Voronovsky AY, Ryabova OB, Verba OV, Ishchuk OP, Dmytruk KV, Sibirny AA (2005) Expression of xylA genes encoding xylose isomerases from *Escherichia coli* and *Streptomyces coelicolor* in the methylotrophic yeast *Hansenula polymorpha*. FEMS Yeast Res 5:1055–1062

Voronovsky AY, Rohulya OV, Abbas CA, Sibirny AA (2009) Development of strains of the thermotolerant yeast *Hansenula polymorpha* capable of alcoholic fermentation of starch and xylan. Metab Eng 11:234–242

Wang H, Lu F, Sun Y, Du L (2004) Heterologous expression of lignin peroxidase of *Phanerochaete chrysosporium* in *Pichia methanolica*. Biotechnol Lett 26:1569–1573

Wang H, Zhao X, Lu F (2007) Heterologous expression of bovine lactoferricin in *Pichia methanolica*. Biochemistry (Mosc) 72:640–643

Wang M, Jiang S, Wang Y (2016a) Recent advances in the production of recombinant subunit vaccines in *Pichia pastoris*. Bioengineered 7:155–165

Wang X, Wang Q, Wang J, Bai P, Shi L, Shen W, Zhou M, Zhou X, Zhang Y, Cai M (2016b) Mit1 transcription factor mediates methanol signaling and regulates the alcohol oxidase 1 (AOX1) promoter in *Pichia pastoris*. J Biol Chem 291:6245–6261

Waterham HR, Digan ME, Koutz PJ, Lair SV, Cregg JM (1997) Isolation of the *Pichia pastoris* glyceraldehyde-3-phosphate dehydrogenase gene and regulation and use of its promoter. Gene 186:37–44

Weninger A, Hatzl AM, Schmid C, Vogl T, Glieder A (2016) Combinatorial optimization of CRISPR/Cas9 expression enables precision genome engineering in the methylotrophic yeast *Pichia pastoris*. J Biotechnol 235:139–149

Yamada Y, Maeda K, Mikata K (1994) The phylogenetic relationships of the hat-shaped ascospore-forming, nitrate-assimilating Pichia species, formerly classified in the genus *Hansenula Sydow et Sydow*, based on the partial sequences of 18S and 26Sribosomal RNAs (*Saccharomycetaceae*): the proposals of three new genera, *Ogataea*, *Kuraishia*, and *Nakazawaea*. Biosci Biotechnol Biochem 58:1245–1257

Yamada Y, Matsuda M, Maeda K, Mikata K (1995) The phylogenetic relationships of methanol-assimilating yeasts based on the partial sequences of 18S and 26S ribosomal RNAs: the proposal of *Komagataella* gen. nov. (*Saccharomycetaceae*). Biosci Biotechnol Biochem 59:439–444

Yan C, Yu W, Zhai X, Yao L, Guo X, Gao J, Zhou YJ (2022) Characterizing and engineering promoters for metabolic engineering of *Ogataea polymorpha*. Synth Syst Biotechnol 7:498–505

Yang S, Kuang Y, Li H, Liu Y, Hui X, Li P, Jiang Z, Zhou Y, Wang Y, Xu A et al (2013) Enhanced production of recombinant secretory proteins in *Pichia pastoris* by optimizing Kex2 P10 site. PLoS One 8:e75347

Yang J, Cai H, Liu J, Zeng M, Chen J, Cheng Q, Zhang L (2018) Controlling AOX1 promoter strength in *Pichia pastoris* by manipulating poly (dA:dT) tracts. Sci Rep 8:1401

Yoko-o T, Komatsuzaki A, Yoshihara E, Zhao S, Umemura M, Gao XD, Chiba Y (2021) Regulation of alcohol oxidase gene expression in methylotrophic yeast *Ogataea minuta*. J Biosci Bioeng 132:437–444

Yu P, Yan Y, Gu Q, Wang X (2013) Codon optimisation improves the expression of *Trichoderma viride sp.* endochitinase in *Pichia pastoris*. Sci Rep 3:3043

Yu P, Zhu Q, Chen K, Lv X (2015) Improving the secretory production of the heterologous protein in *Pichia pastoris* by focusing on protein folding. Appl Biochem Biotechnol 175:535–548

Yurimoto H, Sakai Y (2009) Methanol-inducible gene expression and heterologous protein production in the methylotrophic yeast *Candida boidinii*. Biotechnol Appl Biochem 53:85–92

Yurimoto H, Yamane M, Kikuchi Y, Matsui H, Kato N, Sakai Y (2004) The pro-peptide of *Streptomyces mobaraensis* transglutaminase functions in cis and in trans to mediate efficient secretion of active enzyme from methylotrophic yeasts. Biosci Biotechnol Biochem 68:2058–2069

Yurugi-Kobayashi T, Asada H, Shiroishi M, Shimamura T, Funamoto S, Katsuta N, Ito K, Sugawara T, Tokuda N, Tsujimoto H, Murata T, Nomura N, Haga K, Haga T, Iwata S, Kobayashi T (2009) Comparison of functional non-glycosylated GPCRs expression in *Pichia pastoris*. Biochem Biophys Res Commun 380:271–276

Zahrl RJ, Peña DA, Mattanovich D, Gasser B (2017) Systems biotechnology for protein production in *Pichia pastoris*. FEMS Yeast Res 17:1–31

Zhan C, Yang Y, Zhang Z, Li X, Liu X, Bai Z (2017) Transcription factor Mxr1 promotes the expression of Aox1 by repressing glycerol transporter 1 in Pichia pastoris. FEMS Yeast Res 17(4):fox015. https://doi.org/10.1093/femsyr/fox015

Zhang W, Inan M, Meagher MM (2000) Fermentation strategies for recombinant protein expression in the methylotrophic yeast *Pichia pastoris*. Biotechnol Bioprocess Eng 5:275–287

Zhao W, Wang J, Deng R, Wang X (2008) Scale-up fermentation of recombinant *Candida rugosa* lipase expressed in *Pichia pastoris* using the GAP promoter. J Ind Microbiol Biotechnol 35:189–195

Chapter 11
Mrakia spp. Yeasts: Extremophilic Organisms with Underexplored Biotechnological Potential

B. Turchetti, G. Mugnai, D. Andreani, C. Sannino, and P. Buzzini

Contents

11.1 Introduction 322
11.2 Taxonomy and Ecology 323
11.3 Cold Adaptation Mechanisms 329
11.4 Biotechnological Applications 330
11.4.1 Low-Alcoholic Beer Production 330
11.4.2 Ethanol Production from Biomasses 331
11.4.3 Production of Cold-Active Enzymes 331
11.4.4 Miscellaneous 333
11.5 Future Perspectives and Challenges 333
References 334

Abstract Studies on the yeast genus *Mrakia* can be considered crucial to understanding the physiological and molecular bases of its adaptation to cold and also to selecting novel strains useful for food, chemical, cosmetic, and pharmaceutical applications. The present chapter summarizes the current state of the art on the taxonomy, ecology, physiology, and biotechnology of *Mrakia* spp. yeasts, with special emphasis on their role as starter cultures for fermentation processes and as producers of many valuable molecules of biotechnological interest. The future implications of its exploitation, also considering the most recent genomic studies, are discussed. Based on these, the genus *Mrakia* undoubtedly deserves special attention in view of its great, although still understudied, biotechnological potential.

B. Turchetti · G. Mugnai · D. Andreani · C. Sannino · P. Buzzini (✉)
Department of Agricultural, Food and Environmental Sciences, University of Perugia, Perugia, Italy
e-mail: pietro.buzzini@unipg.it

A. A. Sibirny (ed.), *Biotechnology of Yeasts and Filamentous Fungi*,
Grand Challenges in Biology and Biotechnology,
https://doi.org/10.1007/978-3-031-74726-7_11

11.1 Introduction

Yeasts are adaptable unicellular fungi, which are able to colonize all environments worldwide (Rosa and Péter 2006; Buzzini et al. 2017a, b). Although they are not often seen as extremophilic organisms when compared to prokaryotes, some species grow under conditions commonly regarded as extreme (Raspor and Zupan 2006; Buzzini et al. 2018). Following the current definition, the term extremophilic includes both true extremophiles (i.e., requiring extreme conditions for life) and extremotolerant organisms (i.e., tolerating some limiting factors to a certain extent) (Rothschild and Mancinelli 2001; Péter et al. 2017). Raspor and Zupan (2006) suggested that the boundary between extremophilic and extremotolerant organisms is sometimes very thin, and very often, the decision to include a certain life form in one of the two categories must take into consideration its biological and ecological features.

Low temperatures have a strong impact on microbial life, both directly (through a reduction in the activity rates of enzymes) and indirectly (through a change in the physical state of water and water-associated habitats) (Poindexter 2009). Overall, despite the influence of other environmental parameters, the key attribute of cold-adapted microorganisms is their ability to successfully overcome the negative effects of low temperatures through the development of a range of structural and functional adaptations (Deming 2002; Siddiqui and Cavicchioli 2006; Feller 2007; Margesin et al. 2008; Deming 2009; Shivaji and Prasad 2009; Margesin and Miteva 2011; Maggi et al. 2013; Buzzini and Margesin 2014). Among them, the basidiomycetous yeast genus *Mrakia* has been known for many decades as one of the dominant (and most adaptive) fungal taxa colonizing cold environments worldwide (Di Menna 1966a, b; Tsuji et al. 2019b). *Mrakia* spp. strains have been isolated from polar (Arctic and Antarctic areas), subpolar (e.g., Siberia and Alaska), and nonpolar (e.g., Alps, Patagonia, China) areas (Margesin et al. 2005; Margesin and Fell 2008; Thomas-Hall et al. 2010; Singh and Singh 2012; de García et al. 2012; Panikov and Sizova 2007; Buzzini et al. 2017c; Sannino et al. 2017; Tsuji et al. 2018; Zhang et al. 2020). The presence of this yeast genus has also been sometimes detected in some food where the use of low temperatures represents a preservation strategy (Moreira et al. 2001).

Some studies reported the ability of members of the genus *Mrakia* to express some physiological abilities, which could be successfully used for biotechnological scopes, both for traditional (e.g., fermented beverages) and for more innovative exploitations (e.g., the production of enzymes, bioethanol, etc.) (Tsuji et al. 2019a). However, despite the growing number of studies on their possible biotechnological exploitation, there are still significant opportunities to expand knowledge on these cold-adapted extremophilic yeasts in order to further increase the number of *Mrakia* spp. strains expressing novel and attractive biotechnological properties.

11.2 Taxonomy and Ecology

The genus *Mrakia* belongs to the phylum Basidiomycota, subphylum Agaricomycotina, class Tremellomycetes, order Cystofilobasidiales, and family Mrakiaceae (Fell 2011). The first isolation of *Mrakia* strains can be dated to 1965, when Japanese scientists isolated a few yeast strains from frozen food and identified them as the new yeast species *Candida curiosa* (Komagata and Nakase 1965). One year later, New Zealand scientist Margaret E. di Menna described three new species of *Candida* isolated from Ross Island (Scott Base, Antarctica) as *Candida frigida*, *Candida gelida*, and *Candida nivalis* (di Menna 1966a, b). Lastly, Fell et al. (1969) proposed the establishment of the new genus *Leucosporidium* and the new species *Leucosporidium stokesii* and the reassignment of the above *Candida* species as *Leucosporidium nivalis*, *Leucosporidium gelidum*, and *Leucosporidium frigidum* due to their ability to form teliospores in the absence of mating (self-sporulation) (Fell et al. 1969).

The genus *Mrakia* was formally established in 1987 in honor of the American scientist Emil M. Mrak. Following this, *Leucosporidium* species were accommodated into the new genus as *M. frigida*, *M. gelida* (Figs. 11.1 and 11.2), *M. nivalis*, and *M. stokesii* due to their CoQ-8 system and xylose presence in cell hydrolysates (Sugiyama et al. 1985; Yamada and Komagata 1987; Tsuji et al. 2019a). In recent years, the use of more advanced taxonomic approaches led to a general reassignment of *Mrakia* species: *M. nivalis* and *M. stokesii* were indicated as synonyms of *M. frigida* and *M. gelida*, respectively (Yamada and Matsumoto 1988; Diaz and Fell 2000). Subsequently, Margesin and Fell (2008) proposed the establishment of the new genus *Mrakiella* to accommodate two species closely related to *Mrakia* but lacking teliospores production: *Mrakiella cryoconiti* and *Mrakiella aquatica*. Finally, in 2015, in the framework of a general reorganization of Tremellomycetes,

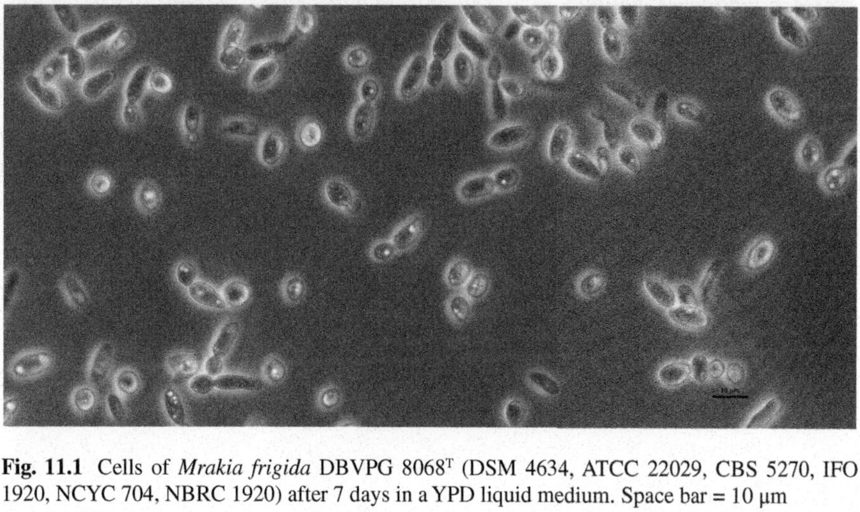

Fig. 11.1 Cells of *Mrakia frigida* DBVPG 8068^T (DSM 4634, ATCC 22029, CBS 5270, IFO 1920, NCYC 704, NBRC 1920) after 7 days in a YPD liquid medium. Space bar = 10 µm

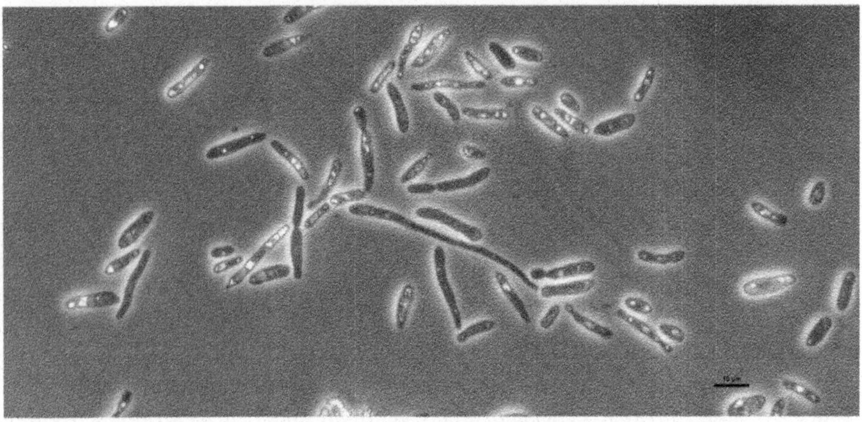

Fig. 11.2 Cells of *Mrakia gelida* DBVPG 8069^T (DSM 70882, TCC 22030, CBS 5272, IFO 1921, NCYC 706, NBRC 1921, PYCC 3898) after 7 days in a YPD liquid medium. Space bar = 10 µm

the genus *Mrakiella* was incorporated into *Mrakia* in agreement with the taxonomic code "one fungus = one name" (Liu et al. 2015).

The recent description of new *Mrakia* species, mostly isolated from cold environments, has led to the expansion of this genus. At the time this chapter was written, the genus encompasses 16 species: *M. aquatica*, *M. arctica*, *M. blollopis*, *M. cryoconiti*, *M. fibulata*, *M. frigida*, *M. gelida*, *M. hoshinonis*, *M. montana*, *M. niccombsii*, *M. panshiensis*, *M. psychrophila*, *M. robertii*, *M. soli*, *M. stelviica*, and *M. terrae* (Park et al. 2021; Tsuji et al. 2019a, b; Turchetti et al. 2020; Yurkov et al. 2020; Zhang et al. 2020). Phylogenetic analyses revealed three primary clusters within the *Mrakia* genus: the *M. gelida* cluster (including *M. blollopis*, *M. frigida*, *M. gelida*, *M. psychrophila*, and *M. robertii*), the *M. cryoconiti* cluster (*M. arctica*, *M. cryoconiti*, *M. montana*, and *M. stelviica*), and the *M. aquatica* cluster (*M. aquatica*, *M. fibulata*, *M. hoshinonis*, *M. niccombsii*, *M. panshiensis*, *M. soli*, and *M. terrae*) (Park et al. 2021; Turchetti et al. 2023).

Except for the species *M. fibulata*, *M. terrae*, and *M. soli*, which display a maximum growth temperature of 25 °C, all other *Mrakia* species exhibit a maximum growth temperature of 20 °C. Therefore, *Mrakia* species have been primarily isolated from cold environments worldwide (Table 11.1). Additionally, metabarcoding analysis documented the presence of DNA sequences assigned to the genus *Mrakia* in different substrates (Table 11.1).

From an ecological viewpoint, *Mrakia* species have the potential to exert a significant influence on biogeochemical cycles in cold ecosystems, mainly through the secretion of a number of extracellular cold-active hydrolytic enzymes, empowering *Mrakia* species to decompose a diverse array of organic materials across a broad temperature spectrum (Pathan et al. 2010; de García et al. 2012; Turchetti et al. 2008; Tsuji et al. 2013b, 2014a, b, 2015b).

Table 11.1 Mrakia species found in cold environments worldwide

Species	Source (locality)	References
Mrakia aquatica	Fragments of *Deschampsia antarctica* (Antarctic Peninsula)	Ferreira et al. (2019)
	[a]Frozen soil (Iceland)	Thomas-Hall et al. (2010)
	[a]Frozen soil, plant leaves, and branches (Heidmörk, Iceland)	Birgisson et al. (2003)
	[a]Glacial meltwater, sediment, and ice (Miage glacier, Alps, Italy)	Turchetti et al. (2013)
	[a]Meltwaters and ice cores (Calderone glacier, Apennines, Italy)	Branda et al. (2010)
	[a]Spring waters (Podlasie, Poland)	Kiziewicz (2004)
	[a]Water scum (Malham Tarn lake, Yorkshire Dales, England)	Jones and Slooff (1966)
	[a]Water scum (Yorkshire, Malham Tarn, UK)	Thomas-Hall et al. (2010)
	Soil from a high-altitude volcano crater (Xinantécatl volcano, Mexico)	Tapia-Vázquez et al. (2020)
Mrakia arctica	Ice island microbial mat (Disraeli Fiord, Ellesmere Island, Canada)	Tsuji et al. (2018)
	[b]Sediment (Walker Glacier, Ellesmere Island, Canada)	Tsuji et al. (2022)
Mrakia blollopis	Algal mat, Naga Ike lake (Skarvsnes, East Antarctica)	Tsuji et al. (2013d)
	[b]Fermented yak milk (Qinghai-Tibet Plateau, China)	Ding et al. (2022)
	Fragments of *Deschampsia antarctica* (Antarctic Peninsula)	Ferreira et al. (2019)
	Meltwaters and ice cores (Calderone glacier, Apennines, Italy)	Branda et al. (2010)
	Soil (King George Island, Shetland South Archipelago, Antarctica)	Carrasco et al. (2012)
	Soil (Marine Plain, Antarctica)	Thomas-Hall et al. (2010)
	Soil, benthic, algal mat (Moss Cirque, Antarctica)	Thomas-Hall et al. (2010)
	Spring flux of *Betula pendula* (Braunschweig, Germany)	Yurkov et al. (2020)
Mrakia cryoconiti	[a]Cryoconite hole (Stubaier glacier, Alps, Austria)	Margesin and Fell (2008)
	[a]Glacial meltwater, sediment, and ice (Miage glacier, Alps, Italy)	Turchetti et al. (2013)
	[a]Meltwaters and ice cores (Calderone glacier, Apennines, Italy)	Branda et al. (2010)
	[a]Mud, spring water, and moss (Gyda Peninsula, North Siberia, Russia)	Gounot (2001)
	[a]Snow (Glacier du Géant, Alps, France)	Turchetti et al. (2013)

(continued)

Table 11.1 (continued)

Species	Source (locality)	References
Mrakia fibulata	Spring flux of *Betula pendula* (Braunschweig, Germany)	Yurkov et al. (2020)
Mrakia frigida	[b]Air samples from waste treatment sites (Quebec, Canada)	Mbareche et al. (2020)
	Fragments of bryophytes (Antarctic Peninsula)	Ferreira et al. (2019)
	Frozen *Sillago japonica* (Japan)	Komagata and Nakase (1965)
	Glacial meltwater rivers (Patagonia, Argentina)	De García et al. (2007)
	[b]Orange-colored protrusions from frozen muscles of *Lobatus gigas* (Colombia)	Cuartas et al. (2018)
	Sea sediments (Antarctica)	Hua et al. (2010)
	[b]Soil (Anshan, Liaoning province, and Taizhou, Zhejiang province, China)	Yan et al. (2023)
	Snow and soil (Scott Base, Antarctica)	Di Menna (1966a, b)
	Snow petrel (Lichen Valley, Antarctica)	Thomas-Hall et al. (2010)
	[b]Vascular plants (Ny-Ålesund region, Svalbard archipelago, Norway)	Zhang and Yao (2015)
Mrakia gelida	[b]Forest soil (South Tyrol, Alps, Italy)	França et al. (2016)
	Fragments of *Deschampsia antarctica*, *Colobanthus quitensis*, bryophytes, and biofilm of puddles (Antarctic Peninsula)	Ferreira et al. (2019)
	Glacial meltwater, sediment, and ice (Miage glacier, Alps, Italy)	Turchetti et al. (2013)
	Groundwater of natural springs, and wells (Moscow, Russia)	Kurakov et al. (2024)
	[b]Ice and cryoconite (South Greenland, Denmark)	Jaarsma et al. (2023)
	Meltwater and soil (King George Island, Shetland South Archipelago, Antarctica)	Carrasco et al. (2012)
	Meltwaters and ice cores (Calderone glacier, Apennines, Italy)	Branda et al. (2010)
	[b]Sediment (Walker glacier, northern Ellesmere Island, Canada)	Tsuji et al. (2022)
	Soil and lichens (Moss Cirque, Antarctica)	Thomas-Hall et al. (2010)
	[c]Cave ice (Scărişoara, Bihor Mountains, Romania)	Brad et al. (2018)
	Spring flux of *Betula pendula* and *Fagus sylvatica* (Braunschweig, Germany)	Yurkov et al. (2020)
	Subglacial sediment and meltwater (Forni Glacier, Alps, Italy)	Turchetti et al. (2008)
Mrakia hoshinonis	[b]Sediment (Walker Glacier, northern Ellesmere Island, Canada)	Tsuji et al. (2022)
	Walker Glacier (northern Ellesmere Island, Canada)	Tsuji et al. (2019b)

Mrakia montana	Glacial meltwater (Miage glacier, Alps, Italy)	Turchetti et al. (2013)
	Soil and Alpine grassland (Stelvio Pass, Alps, Italy)	Turchetti et al. (2020)
	Supraglacial sediment (Calderone glacier, Apennines, Italy)	Turchetti et al. (2020)
Mrakia niccombsii	[a]Soil and lichens (Vestfold Hills area, Antarctica)	Thomas-Hall et al. (2010)
	[a]Snow (Glacier du Géant, Alps, France)	Turchetti et al. (2013)
Mrakia panshiensis	Soil (Lianhuashan, National Forest Park, Jilin province, China)	Zhang et al. (2020)
Mrakia psychrophila	Glacial meltwater, sediment, and ice (Miage glacier, Alps, Italy)	Turchetti et al. (2013)
	Fragments of *Colobanthus quitensis* (Antarctic Peninsula)	Ferreira et al. (2019)
	Soil (Fildes, Antarctic Peninsula)	Xin and Zhou (2007)
	Soil (King George Island, Shetland South Archipelago, Antarctica)	Carrasco et al. (2012)
Mrakia robertii	Cryoconite hole (Hamtah glacier, Himalaya, India)	Dhume et al. (2019)
	Cryoconite hole (Larsemann Hills, Antarctica)	Sanyal et al. (2020)
	Cryoconite hole (Sutri Dhaka glacier, Himalaya, India)	Sanyal et al. (2020)
	Fragments of *Deschampsia antarctica*, *Colobanthus quitensis*, and biofilm of puddles (Antarctic Peninsula)	Ferreira et al. (2019)
	Glacial meltwater, sediment, and ice (Miage glacier, Alps, Italy)	Turchetti et al. (2013)
	[b]Sediment (Walker Glacier, northern Ellesmere Island, Canada)	Tsuji et al. (2022)
	Soil (King George Island, Shetland South Archipelago, Antarctica)	Carrasco et al. (2012)
	Soil and lichens (Mossel Lake, Antarctica)	Thomas-Hall et al. (2010)
	Soil (Mossel Cirque, Antarctica)	Thomas-Hall et al. (2010)
	Soil and lichens (Lichen Valley, Antarctica)	Thomas-Hall et al. (2010)
	Subglacial sediment and meltwater, Forni and Sforzellina glaciers, Italian Alps, Italy	Thomas-Hall et al. (2010)
Mrakia soli	Soil (Pocheon city, Gyeonggi province, South Korea)	Park et al. (2021)
Mrakia stelviica	Soil, nival belt, and alpine grassland (Stelvio Pass, Alps, Italy)	Turchetti et al. (2020)
Mrakia terrae	Soil (Pocheon city, Gyeonggi province, South Korea)	Park et al. (2021)

(continued)

Table 11.1 (continued)

Species	Source (locality)	References
Mrakia spp.	[b]Aged cigars (Cigar aging warehouse, Sichuan, China)	Xue et al. (2023)
	Boreal bog (western Siberia, Russia)	Poliakova et al. (2001)
	[b]Coastal waters (King George Island, Antarctica)	Garmendia et al. (2021)
	Glacial meltwater rivers (Patagonia, Argentina)	De García et al. (2007)
	[b]Grapevine plant (Alentejo region, Portugal)	Patanita et al. (2022)
	[b]Hypersaline brines (Tarn Flat, northern Victoria Land, Antarctica)	Borruso et al. (2018)
	[b]Ice (Adélie Cove, northern Victoria Land, Antarctica)	Sannino et al. (2023)
	Lake bottom sediment (Lake Baikal, Russia)	Fedorova and Kurakov (2023)
	[b]Litter sample (Bayinbuluk alpine wetland, Xinjiang Uygur, China)	Chen et al. (2021)
	[b]Maple sap and maple syrup (Québec, Canada)	Filteau et al. (2012)
	[b]Permafrost and hypersaline brines (Boulder Clay, northern Victoria Land, Antarctica)	Sannino et al. (2020)
	[b]Prepackaged fennels (Campania, Italy)	Sequino et al. (2022)
	[b]Soil (Hanzhong, Shaanxi province, China)	Jiang et al. 2019
	[b]Soil (Hubei province, China)	Yu et al. (2016)
	[b]Soil and permafrost (mountain ridge Muot da Barba Peider, Alps, Switzerland)	Frey et al. (2016)
	[b]Spontaneous fermented ice grape must (Liaoning province, China)	Li et al. (2018)
	Tundra wetland soil (Alaska, USA)	Poliakova et al. (2001)
	[b]Vascular plants (Ny-Ålesund region, Svalbard archipelago, Norway)	Zhang and Yao (2015)
	[b]Weaned piglets caecal and colonic digesta (China)	Li et al. (2020b)

[a]Species previously described using an old taxonomic designation then reassigned to the genus *Mrakia* (Fell 2011)
[b]DNA sequences assigned to the genus *Mrakia* by metabarcoding analysis
[c]Strains previously identified as *M. stokesii* then reassigned to *M. gelida* (Diaz and Fell 2000)

11.3 Cold Adaptation Mechanisms

Cold environments cover a large percentage of planet surface, with many ecosystems permanently exposed to temperatures below 5 °C (Feller and Gerday 2003). Cold-adapted microorganisms (including yeasts) are able to grow at temperatures close (or even below) 0 °C due to the development of some physiological adaptations capable of ensuring cell survival and development under these harsh environmental conditions, e.g., the production of cold-active enzymes, ice-binding proteins, polyunsaturated lipids, etc. (Buzzini et al. 2012; Maggi et al. 2013; Tiquia-Arashiro and Rodrigues 2016; Tsuji et al. 2019a). Although the number of studies conducted so far on *Mrakia* spp. yeasts is very limited, their ability to grow and metabolize at subzero temperatures have been demonstrated. Therefore, this genus is supposed to play a pivotal role in nutrient cycling in cold ecosystems worldwide, especially in decomposing organic matter, although some inter- and intraspecies differences were found (Panikov and Sizova 2007; Bakermans 2008). Tsuji (2016) reported that the growth of *M. blollopis* under subzero conditions was strain-dependent and associated to the accumulation of high levels of tricarboxylic acid (TCA)-cycle metabolites, lactic acid, aromatic amino acids, and polyamines. Interestingly, *M. frigida* and *M. gelida* have also been evaluated for their heat shock response: when incubated at 20 °C and 25 °C for 3 h, both species induced the production of heat-shock proteins (Deegenaars and Watson 1998).

In some *Mrakia* spp. Strains, the percentage of unsaturated fatty acids was found to account from 50 to 90% of the total fatty acid profile of triacylglycerols (TAGs) (Watson 1984; Chintalapati et al. 2004). Tsuji et al. (2013a) observed that some *Mrakia* species exhibited the ability to accumulate high amounts of unsaturated TAGs, suggesting that members of this genus have adapted to cold environments primarily by increasing their membrane fluidity, while they did not demonstrate any antifreeze activity and exhibited only limited secretion of extracellular polysaccharides. Likewise, the species *M. psychrophila* and *M. gelida* were found to produce TAGs containing high amounts of oleic and linoleic acids, while the amount of branched fatty acids and the unsaturation degree increased at low temperatures (Pathan et al. 2010). More recently, a transcriptomic analysis of *M. psychrophila* revealed an upregulation at 4 °C of genes encoding the synthesis of specific desaturases and dehydrogenases involved in the biosynthetic pathways of unsaturated fatty acid and glycerol, associated both to the maintenance of membrane fluidity and to the cryoprotectant role expressed by glycerol. Besides, the proteomic analysis suggested that *M. psychrophila* grown at 4 °C upregulated energy metabolism, ribosome activity, glycerol production, and the synthesis of enzymes involved in the unsaturation of fatty acids (i.e., desaturase and glycerol 3-phosphate dehydrogenase), while genes involved in protein processing at the endoplasmic reticulum level; proteasome, spliceosome, and mRNA surveillance; and unfolded protein binding are upregulated at 20 °C. The accumulation of unfolded proteins in the endoplasmic reticulum might be involved in the death of *M. psychrophile*, which was observed above 20 °C (Su et al. 2016). Interestingly, the whole genome analysis of an Antarctic strain of *M. blollopis* showed that it lacks the protease K gene and ice-binding gene (Tsuji et al. 2015a). Besides, draft genome comparison among cold-adapted yeasts showing different optimal growth temperatures exhibited a large number of putative genes associated to the synthesis of enzymes involved in oxidative and cold stress responses and suggested that the speed of enzyme activity in cold conditions depends on the growth rate of the strain (Baeza et al. 2021).

11.4 Biotechnological Applications

11.4.1 Low-Alcoholic Beer Production

Despite the fact that the regulations of the EU Food Safety Authority still represent a bottlenecked hampering of the commercial release of novel healthier and safer beers into the market, the use of non-*Saccharomyces* yeasts as alternative starters for brewing has received increasing attention in recent years (Turchetti et al. 2023). Due to the increasing consumer demands, the production of low alcoholic beer is the new goal of beer producers. Accordingly, the use of *Mrakia* spp. strains as brewing starters could open novel perspectives for the development of nonconventional beers (De Francesco et al. 2018; Burini et al. 2021; Miguel et al. 2022; Simões et al. 2023). The fermentation of some C sources (including glucose and sucrose) has been found in almost all *Mrakia* species (Fell 2011), although the amount of wort carbohydrates converted into ethanol is considered a species-specific character (Thomas-Hall et al. 2010; Tsuji et al. 2019a; Linnakoski et al. 2023; Turchetti et al. 2023).

De Francesco et al. (2018) investigated the ethanol content and the volatile profile of a beer obtained using *M. gelida*, in comparison with a commercial starter of *Saccharomycodes ludwigii*. Both beers exhibited low content of alcohol and diacetyl (1.40% and 1.32%; 5.04 and 5.20 μg L^{-1}, respectively), while the sensorial characters of the beer obtained using *M. gelida* were more appreciated by the panelists. A more recent study found that members of *M. gelida* clade showed a superior ability to produce ethanol, higher alcohols, esters and carbohydrate conversion compared to the members of *M. cryoconiti* and *M. aquatica* clades. Among the *M. gelida* clade, the species *M. blollopis* exhibited higher tolerance to ethanol and to iso-α-acids, a considerable production of lactic and acetic acids and glycerol, as well as an inverse relationship between fermentative performances and incubation temperature (Turchetti et al. 2023). Linnakoski et al. (2023) selected some *M. gelida* strains (isolated from Finnish forest environments) for their capacity to produce low-alcoholic beers in small-scale fermentation and the subsequent level up to pilot scale (40 L). After maturation, filtration, carbonation, and bottling, the beer fermented by *M. gelida* was evaluated for ethanol and sensorial profiles, comparable with that produced by the commercial low-alcohol brewing yeast *S'codes ludwigii.*

11.4.2 Ethanol Production from Biomasses

Ethanol from lignocelluloses is one of the new frontiers of biofuel production for decreasing fossil oil consumption and pollution (El-Sheekh et al. 2023). Although fermentation abilities were uncommon features of basidiomycetous yeast (Kurtzman et al. 2011), some studies have investigated the use of *Mrakia* spp. for producing ethanol from different C sources. Thomas-Hall et al. (2010) reported that some *Mrakia* species were unable to completely convert sucrose to ethanol, probably due to the growth inhibition observed when alcohol concentration exceeded 2%. Some *Mrakia* strains have shown superior fermentation performances: an Antarctic strain of *M. blollopis* exhibited the ability to convert 12% glucose to ethanol (48.2 g L^{-1}) at 10 °C and within a wide pH range (from 5 to 10, with optimum at 8–10) (Tsuji et al. 2013b, c).

Direct ethanol fermentation (DEF) from cellulosic biomass is a technique that includes enzymatic hydrolysis followed by simultaneous ethanol fermentation. This approach allows minimizing the feed-back inhibition of cellulase, thanks to the rapid conversion of free glucose to ethanol by yeasts (Takagi et al. 1977). In addition, this approach has proven to reduce the concentration of furfural and 5-methylfolate (Thomsen et al. 2009). DEF of cellulosic biomass catalyzed by *M. blollopis* strains obtained approximately from 30 to 65% of the theoretical ethanol yield thanks to simultaneous use of nonionic surfactants (Tsuji et al. 2014a, 2019a).

11.4.3 Production of Cold-Active Enzymes

Although cold-adapted basidiomycetous yeasts (including *Mrakia* spp.) are generally considered less attractive than bacteria and filamentous fungi as enzyme producers, they have been taken into consideration in the last decades as putative sources of cold-active enzymes for eco-friendly biotechnologies. Such enzymes may provide significant financial and technical advantages as energy savers for developing heat-sensitive manufacturing processes requiring easy and fast postcatalysis inactivation at moderate temperatures (Feller and Gerday 2003; Gerday et al. 1997, 2000; Buzzini et al. 2012; Gerday 2014; Santiago et al. 2016; Javed and Qazi 2016; Baeza et al. 2017; Tasselli et al. 2017).

Regarding bioprospecting studies on *Mrakia* spp. strains isolated from cold environments, most of them have so far been performed using qualitative/semiquantitative tests (often carried out in Petri dishes). A number of *Mrakia* spp. strains exhibited the ability to release extracellular cold-active hydrolases catalyzing the degradation of a variety of organic polymers over a wide range of temperatures (Table 11.2) (Nakagawa et al. 2004; Turchetti et al. 2008; Pathan et al. 2010; de García et al. 2012; Tsuji et al. 2013d, 2014b; Carrasco et al. 2016; Martorell et al. 2017; Yuivar et al. 2019; Sanyal et al. 2020; Da Silva et al. 2022; De Souza et al. 2023).

Table 11.2 Production of cold-active enzymes from *Mrakia* spp.

Species	Degree of purification and characterization	Cold-active enzymes	References
Mrakia blollopis	Characterization of the activity of partially purified enzymes	Lipase	Tsuji et al. (2013d)
	Characterization of the activity of cell-free extracts	Cellulase	Carrasco et al. (2016)
	Screening of enzymatic activity on Petri dishes	Cellulase, pectic enzymes	de Souza et al. (2023)
Mrakia frigida	Characterization of the activity of cell-free extracts	Pectate lyase	Margesin et al. (2005)
	Characterization of the activity of cell-free extracts	Pectic enzymes	Nakagawa et al. (2004)
	Screening of enzymatic activity on Petri dishes	Amylase, cellulase, esterase, lipase, pectic enzymes, protease, xylanase	Martorell et al. (2017)
Mrakia gelida	Screening of enzymatic activity on Petri dishes	Protease	Da Silva et al. (2022)
Mrakia niccombsii	Screening of enzymatic activity on Petri dishes	Amylase	Da Silva et al. (2022)
Mrakia robertii	Characterization of the activity of partially purified enzymes	Endoglucanase	Dhume et al. (2019)
	Screening of enzymatic activity on Petri dishes	Amylase, cellulase, esterase, lignin-modifying enzymes, pectic enzymes, phosphate-solubilizing enzymes, protease, silicate-solubilizing enzymes, tannase	Sanyal et al. (2020)
Mrakia niccombsii	Screening of enzymatic activity on Petri dishes	Cellulase, esterase, protease	Da Silva et al. (2022)
Mrakia sp.	Characterization of the activity of partially purified enzymes	Gelatinase	Yuivar et al. (2019)
	Screening of enzymatic activity on Petri dishes	Amylase, lipase, pectic enzymes, protease, urease	Pathan et al. (2010)
	Screening of enzymatic activity on Petri dishes	Cellulase, pectic enzymes	Da Silva et al. (2022)

Quantitative determination of enzyme activity in *Mrakia* spp. cell-free extracts has been performed less often: only a few studies concerned the purification and characterization of cold-active enzymes (Table 11.2). Two *M. frigida* strains were found to produce, at 1–10 °C, extracellular cold-active alkaline pectate lyases resistant to repeated freezing and thawing (Margesin et al. 2005). More recently, a psychrotolerant strain of *M. robertii* (isolated from Himalayan cryoconites) was studied for its ability to produce cold-active endoglucanases. After partial purification, the endoglucanase exhibited a molecular weight around 57 kDa and showed its higher activity at 15 °C and pH 11. Divalent cations played the role of activators of hydrolytic activity (Dhume et al. 2019).

A lipase extracted and purified from *M. blollopis* cells exhibited stability under a wide range of temperatures and pH, even in the presence of various metal ions and organic solvents (Tsuji et al. 2013d), while an extracellular gelatinase was purified from cultures of *Mrakia* spp. After purification, the enzyme exhibited a molecular weight around 37 kDa and displayed its highest activity at 36 °C, pH 7, in the presence of Ca^{2+} and Zn^{2+} salts (Yuivar et al. 2019).

11.4.4 Miscellaneous

Dairy plant drainage produced during the cleaning of transport pipes and milking tanks can pollute rivers and groundwater with detergents, disinfectants, mucus, and milk fat residues, which increases the biochemical oxygen demand (BOD) of drainage systems (Healy et al. 2007). Tsuji et al. (2015b) investigated the bioremediation aptitude of some Antarctic *Mrakia* spp. strains to decompose milk fat at temperatures ranging from 4 to 10 °C. Interestingly, a strain of *M. blollopis* exhibited a BOD removal rate of 1.25-fold higher than that of the control due to its superior lipolytic activity (Tsuji et al. 2013d).

The presence of *Mrakia* spp. strains has also been found in some natural fermentation of special wines carried out at low temperatures. Studies on yeast diversity and dynamics in the spontaneous fermentation of Vidal blanc icewine (produced in some Chinese regions) revealed that non-*Saccharomyces* yeasts were dominant in the initial stages: among them, *Mrakia* spp. strains were found (Li et al. 2018). Likewise, *Mrakia* spp. strains were among the most abundant yeasts found in the spontaneous fermentation of raspberry wine (Li et al. 2020a).

An Antarctic strain of *M. frigida* showed the ability to produce a cold-active killer toxin lethal to the pathogenic yeasts *Metschnikowia bicuspidata*, *Candida tropicalis*, and *Candida albicans* (Hua et al. 2010). After purification, this toxin exhibited a molecular weight of around 55.6 kDa and 35% sequence homology with a protein kinase. The optimal temperature and pH for activity were 16 °C and 4.5, respectively. The maximum killing activity was observed in the presence of 3% NaCl, while protein was stable at temperatures below 25 °C (Liu et al. 2012).

11.5 Future Perspectives and Challenges

Although cold ecosystems are among the largest biospheres on Earth, the physiology (and possible biotechnological implications) of the prokaryotic and eukaryotic organisms that colonize such extreme habitats are still poorly studied. Among cold-adapted yeasts, the genus *Mrakia* can be considered one of the most attractive life forms isolated from cold habitats worldwide (Margesin et al. 2002; Brizzio et al. 2007; Turchetti et al. 2008; Singh and Singh 2012; Tsuji et al. 2019a).

The most recent research on the taxonomy, ecology, physiology, and biotechnology of members of the genus *Mrakia* and the availability of already deposited whole genomes (Tsuji et al. 2015a, 2023), together with those that will be hopefully deposited in the future, will probably lead to new insights on the physiological and molecular bases of cold adaptation, thus providing the scientific tools for developing future biotechnological exploitations, especially for processes carried out at low temperatures. The most advanced biotechnologies involving strains of *Mrakia* spp. are related to their use as starter cultures for low-alcoholic brewing and as producers of cold-active enzymes catalyzing the hydrolysis of polymeric organic compounds. The production of killer toxins active against pathogenic yeasts can also be considered one more interesting prospect for the pharmaceutical industry.

All the above applications could make the genus *Mrakia* an attractive producer of innovative beverages, enzymes, and new molecules. However, to the best of the author's knowledge, none of the products potentially obtainable from *Mrakia* spp. strains are currently available on the market, whereas some possible chemicals produced by these yeasts are still confined to the laboratory scale (Shivaji and Prasad 2009; Margesin and Feller 2010; Buzzini et al. 2012; Buzzini and Margesin 2014). Therefore, the biotechnological potential expressed by *Mrakia* spp. strains concerns, at present, only possible future applications. At the time of writing this chapter, only two Japanese patents claiming the use of *Mrakia* spp. strains (the first for producing ethanol and the second for decomposing milk fat-based wastes) are available on the website of the World Intellectual Property Organization (www.wipo.int).

In conformity with Joseph et al. (2008), the small number of patented *Mrakia* spp. strains could be a consequence of the high risks and costs associated with studying the little unexplored field of cold-adapted organisms and the difficulties associated with the development of low-temperature biotechnologies. The use of recombinant DNA technologies could probably have a profound effect on developing new and efficient heterologous expressions of *Mrakia* spp. genes encoding the production of tailored recombinant enzymes in suitable hosts by expanding the portfolio of relevant products obtainable from this yeast genus in the future.

References

Baeza M, Alcaíno J, Cifuentes V, Turchetti B, Buzzini P (2017) Cold-active enzymes from cold-adapted yeasts. In: Sibirny A (ed) Biotechnology of yeasts and filamentous fungi. Springer International Publishing, pp 297–324

Baeza M, Zúñiga S, Peragallo V, Barahona S, Alcaino J, Cifuentes V (2021) Identification of stress-related genes and a comparative analysis of the amino acid compositions of translated coding sequences based on draft genome sequences of Antarctic yeasts. Front Microbiol 12:623171

Bakermans C (2008) Limits for microbial life at subzero temperatures. In: Margesin R, Schinner F, Marx JC, Gerday C (eds) Psychrophiles: from biodiversity to biotechnology. Springer, Berlin Heidelberg, pp 17–28

Birgisson H, Delgado O, García Arroyo L, Hatti-Kaul R, Mattiasson B (2003) Cold-adapted yeasts as producers of cold-active polygalacturonases. Extremophiles 7:185–193

Borruso L, Sannino C, Selbmann L, Battistel D, Zucconi L, Azzaro M, Turchetti B, Buzzini P, Guglielmin M (2018) A thin ice layer segregates two distinct fungal communities in Antarctic brines from Tarn Flat (Northern Victoria Land). Sci Rep 8:6582

Brad T, Itcus C, Pascu MD, Perşoiu A, Hillebrand-Voiculescu A, Iancu L, Purcarea C (2018) Fungi in perennial ice from Scărişoara Ice Cave (Romania). Sci Rep 8:10096

Branda E, Turchetti B, Diolaiuti G, Pecci M, Smiraglia C, Buzzini P (2010) Yeast and yeast-like diversity in the southernmost glacier of Europe (Calderone Glacier, Apennines, Italy). FEMS Microbiol Ecol 7:354–369

Brizzio S, Turchetti B, de García V, Libkind D, Buzzini P, van Broock M (2007) Extracellular enzymatic activities (EEA) of basidiomycetous yeasts isolated from glacial and subglacial waters of Northwest Patagonia (Argentina). Can J Microbiol 53:519–525

Burini JA, Eizaguirre JI, Loviso C, Libkind D (2021) Levaduras no convencionales como herramientas de innovación y diferenciación en la producción de cerveza. Rev Argentina Microbiol 53:359–377

Buzzini P, Margesin R (2014) Cold-adapted yeasts: a lesson from the cold and a challenge for the XXI century. In: Buzzini P, Margesin R (eds) Cold-adapted yeasts. Springer International Publishing, pp 3–22

Buzzini P, Branda E, Goretti M, Turchetti B (2012) Psychrophilic yeasts from worldwide glacial habitats: diversity, adaptation strategies and biotechnological potential. FEMS Microbiol Ecol 82:217–241

Buzzini P, Lachance MA, Yurkov AM (2017a) Yeasts in natural ecosystems: diversity. Springer International Publishing, Cham

Buzzini P, Lachance MA, Yurkov AM (2017b) Yeasts in natural ecosystems: ecology. Springer International Publishing

Buzzini P, Turk M, Perini L, Turchetti B, Gunde-Cimerman (2017c) Yeasts in polar and subpolar habitats. In: Buzzini P, Lachance MA, Yurkov A (eds) Yeasts in natural ecosystems: diversity. Springer International Publishing, pp. 330–365

Buzzini P, Turchetti B, Yurkov A (2018) Extremophilic yeasts: the toughest yeasts around? Yeast 35:487–497

Carrasco M, Rozas JM, Barahona S, Alcaíno J, Cifuentes V, Baeza M (2012) Diversity and extracellular enzymatic activities of yeasts isolated from King George Island, the sub-Antarctic region. BMC Microbiol 12:1–9

Carrasco M, Villarreal P, Barahona S, Alcaíno J, Cifuentes V, Baeza M (2016) Screening and characterization of amylase and cellulase activities in psychrotolerant yeasts. BMC Microbiol 16:21

Chen M, Zhu X, Zhao C, Yu P, Abulaizi M, Jia H (2021) Rapid microbial community evolution in initial Carex litter decomposition stages in Bayinbuluk alpine wetland during the freeze–thaw period. Ecol Indic 121:107180

Chintalapati S, Kiran MD, Shivaji S (2004) Role of membrane lipid fatty acids in cold adaptation. Cell Mol Biol 50:631–642

Cuartas JH, Alzate JF, Moreno-Herrera CX, Marquez EJ (2018) Metagenomic analysis of orange colored protrusions from the muscle of Queen Conch *Lobatus gigas* (Linnaeus, 1758). PeerJ 6:e4307

Da Silva MK, Da Silva AV, Fernandez PM, Montone RC, Alves RP, De Queiroz AC, De Oliveira VM, Dos Santos VP, Putzke J, Rosa LH, Duarte AWF (2022) Extracellular hydrolytic enzymes produced by yeasts from Antarctic lichens. An Acad Bras Cienc 94:e20210540

De Francesco G, Sannino C, Sileoni V, Marconi O, Filippucci S, Tasselli G, Turchetti B (2018) *Mrakia gelida* in brewing process: an innovative production of low alcohol beer using a psychrophilic yeast strain. Food Microbiol 76:354–362

De García V, Brizzio S, Libkind D, Buzzini P, Van Broock M (2007) Biodiversity of cold-adapted yeasts from glacial meltwater rivers in Patagonia, Argentina. FEMS Microbiol Ecol 59:331–341 de García V, Brizzio S, Broock MR (2012) Yeasts from glacial ice of Patagonian Andes, Argentina. FEMS Microbiol Ecol 82:540–550

de Souza LMD, Ogaki MB, Teixeira EAA, de Menezes GCA, Convey P, Rosa CA, Rosa LH (2023) Communities of culturable freshwater fungi present in Antarctic lakes and detection of their low-temperature-active enzymes. Braz J Microbiol 54:1923–1933

Deegenaars ML, Watson K (1998) Heat shock response in psychrophilic and psychrotrophic yeast from Antarctica. Extremophiles 2:41–49

Deming JW (2002) Psychrophiles and polar regions. Curr Opin Microbiol 5:301–309

Deming JW (2009) Extremophiles: cold environments. In: Schaechter M (ed) Encyclopedia of microbiology. Elsevier, Oxford, pp 147–158

Dhume GM, Maharana AK, Tsuji M, Srivastava AK, Singh SM (2019) Cold-tolerant endoglucanase producing ability of *Mrakia robertii* A2-3 isolated from cryoconites, Hamtha glacier, Himalaya. J Basic Microbiol 59:667–679

Di Menna ME (1966a) Three new yeasts from Antarctic soils: *Candida nivalis*, *Candida gelida* and *Candida frigida* spp. n. A van Leeuwenhoek 32:25–28

Di Menna ME (1966b) Yeasts in Antarctic soils. A van Leeuwenhoek 32:29–38

Diaz MR, Fell JW (2000) Molecular analyses of the IGS & ITS regions of rDNA of the psychrophilic yeasts in the genus *Mrakia*. A van Leeuwenhoek 77:7–12

Ding Z, Mi J, Zhou J, Ke W, Guo X (2022) Characterization of the microbial community compositions and diversities in the traditional fermented yak milk from different ecotopes on the Qinghai-Tibetan plateau. LWT 156:113027

El-Sheekh MM, El-Nagar AA, El Kelawy M, Bastawissi HA (2023) Bioethanol from wheat straw hydrolysate solubility and stability in waste cooking oil biodiesel/diesel and gasoline fuel at different blends ratio. Biotechnol Biofuels 16:15

Fedorova MD, Kurakov AV (2023) Microbiota of bottom sediments in the coastal zone of lake Baikal. Contemp Probl Ecol 16:492–508

Fell JW (2011) Mrakia Y. Yamada & Komagata (1987). In: Kurtzman CP, Fell JW, Boekhout T (eds) The yeast, a taxonomic study, 5th edn. Elsevier, Amsterdam, pp 1503–1510

Fell JW, Statzell AC, Hunter IL, Phaff HJ (1969) *Leucosporidium* gen.n., the heterobasidiomycetous stage of several yeasts of the genus *Candida*. A van Leeuwenhoek 35:433–462

Feller G (2007) Life at low temperatures: is disorder the driving force? Extremophiles 11:211–216

Feller G, Gerday C (2003) Psychrophilic enzymes: hot topics in cold adaptation. Nature Rev Microbiol 1:200–208

Ferreira EMS, de Sousa FMP, Rosa LH, Pimenta RS (2019) Taxonomy and richness of yeasts associated with angiosperms, bryophytes, and meltwater biofilms collected in the Antarctic Peninsula. Extremophiles 23:151–159

Filteau M, Lagacé L, LaPointe G, Roy D (2012) Maple sap predominant microbial contaminants are correlated with the physicochemical and sensorial properties of maple syrup. Int J Food Microbiol 154:30–36

França L, Sannino C, Turchetti B, Buzzini P, Margesin R (2016) Seasonal and altitudinal changes of culturable bacterial and yeast diversity in Alpine forest soils. Extremophiles 20:855–873

Frey B, Rime T, Phillips M, Stierli B, Hajdas I, Widmer F, Hartmann M (2016) Microbial diversity in European alpine permafrost and active layers. FEMS Microbiol Ecol 92:fiw018

Garmendia G, Alvarez A, Villarreal R, Martínez-Silveira A, Wisniewski M, Vero S (2021) Fungal diversity in the coastal waters of King George Island (maritime Antarctica). World J Microbiol Biotechnol 37:1–12

Gerday C (2014) Fundamentals of cold-adapted yeasts. In: Buzzini P, Margesin R (eds) Cold-adapted yeasts. Springer International Publishing, pp 325–350

Gerday C, Aittaleb M, Arpigny JL, Baise E, Chessa JP, Garsoux G, Petrescu I, Feller G (1997) Psychrophilic enzymes: a thermodynamic challenge. Biochim Biophys Acta 1342:119–131

Gerday C, Aittaleb M, Bentahir M, Chessa JP, Claverie P, Collins T, D'Amico S, Dumont J, Garsoux G, Georlette D, Hoyoux A, Lonhienne T, Meuwis MA, Feller G (2000) Cold-adapted enzymes: from fundamentals to biotechnology. Trend Biotechnol 18:103–107

Gounot AM (2001) Ecology of psychrophilic and psychrotrophic micro-organisms in cold and frozen soils. In: Paepe R, Melnikov VP, Van Overloop E, Gorokhov VD (eds) Permafrost response on economic development, environmental security and natural resources, NATO science series, vol 76. Springer Netherlands, Dordrecht, pp 543–551

Healy MG, Rodgers M, Mulqueen J (2007) Treatment of dairy wastewater using constructed wetland and intermittent sand filter. Bioresour Technol 98:2268–2281

Hua MX, Chi Z, Liu GL, Buzdar MA, Chi ZM (2010) Production of a novel and cold-active killer toxin by *Mrakia frigida* 2E00797 isolated from sea sediment in Antarctica. Extremophiles 14:515–521

Jaarsma AH, Sipes K, Zervas A, Jiménez FC, Ellegaard-Jensen L, Thøgersen MS, Stougaard P, Benning LG, Tranter M, Anesio AM (2023) Exploring microbial diversity in Greenland ice sheet supraglacial habitats through culturing-dependent and-independent approaches. FEMS Microbiol Ecol 99:fiad119

Javed A, Qazi JI (2016) Psychrophilic microbial enzymes implications in coming biotechnological processes. ASRJETS 23:2313–4410

Jiang J, Yu M, Hou R, Li L, Ren X, Jiao C, Yang L, Xu H (2019) Changes in the soil microbial community are associated with the occurrence of *Panax quinquefolius* L. root rot diseases. Plant Soil 438:143–156

Jones EBG, Slooff WC (1966) *Candida aquatica* sp. n. isolated from water scums. A van Leeuwenhoek 32:223–228

Joseph B, Ramteke PW, Thomas G (2008) Cold active microbial lipases: some hot issues and recent developments. Biotechnol Adv 26:457–470

Kiziewicz B (2004) *Candida albicans*, *Candida aquatica*, *Candida krusei* and *Candida tropicalis* strains isolated from springs of Podlasie Province. Wiadomosci Parazytologiczne 50:535–544

Komagata K, Nakase T (1965) New species of the genus *Candida* isolated from frozen foods. J Gen Appl Microbiol 11:255–267

Kurakov AV, Shumakova AA, Fedorova MD (2024) Fungi in groundwater springs and wells in Moscow region. Biol Bull 51:78–90

Kurtzman CP, Fell JW, Boekhout T (2011) The yeast, a taxonomic study, 5th edn. Elsevier, Amsterdam

Li J, Hu W, Huang X, Xu Y (2018) Investigation of yeast population diversity and dynamics in spontaneous fermentation of Vidal blanc icewine by traditional culture-dependent and high-throughput sequencing methods. Food Res Int 112:66–77

Li H, Jiang D, Liu W, Yang Y, Zhang Y, Jin C, Sun S (2020a) Comparison of fermentation behaviors and properties of raspberry wines by spontaneous and controlled alcoholic fermentations. Food Res Int 128:108801

Li J, Luo Y, Chen D, Yu B, He J, Huang Z, Mao X, Zheng P, Yu J, Luo J, Tian G, Yan H, Wang Q, Wang H (2020b) The fungal community and its interaction with the concentration of short-chain fatty acids in the caecum and colon of weaned piglets. J Anim Physiol Anim Nutr 104:616–628

Linnakoski R, Jyske T, Eerikäinen R, Veteli P, Cortina-Escribano M, Magalhães F, Järvenpää E, Heikkilä L, Hutzler M, Gibson B (2023) Brewing potential of strains of the boreal wild yeast *Mrakia gelida*. Front Microbiol 14:1108961

Liu GL, Wang K, Hua MX, Buzdar MA, Chi ZM (2012) Purification and characterization of the cold-active killer toxin from the psychrotolerant yeast *Mrakia frigida* isolated from sea sediments in Antarctica. Process Biochem 47:822–827

Liu XZ, Wang QM, Göker M, Groenewald M, Kachalkin AV, Lumbsch HT, Millanes AM, Wedin M, Yurkov A, Boekhout T, Bai FY (2015) Towards an integrated phylogenetic classification of the Tremellomycetes. Stud Mycol 81:85–147

Maggi O, Tosi S, Angelova M, Lagostina E, Fabbri AA, Pecoraro L, Altobelli E, Picco AM, Savino E, Branda E, Turchetti B, Zotti M, Vizzini A, Buzzini P (2013) Adaptation of fungi, including yeasts, to cold environments. Plant Biosyst 147:247–258

Margesin R, Fell JW (2008) *Mrakiella cryoconiti* gen. nov., sp. nov., a psychrophilic, anamorphic, basidiomycetous yeast from alpine and arctic habitats. Int J Syst Evol Microbiol 58:2977–2982

Margesin R, Feller G (2010) Biotechnological applications of psychrophiles. Environ Technol 31:835–844
Margesin R, Miteva V (2011) Diversity and ecology of psychrophilic microorganisms. Res Microbiol 162:346–361
Margesin R, Zacke G, Schinner F (2002) Characterization of heterotrophic microorganisms in alpine glacier cryoconite. Arct Antarct Alp Res 34:88–93
Margesin R, Fauster V, Fonteyne PA (2005) Characterization of cold-active pectate lyases from psychrophilic *Mrakia frigida*. Lett Appl Microbiol 40:453–459
Margesin R, Schinner F, Marx JC, Gerday C (2008) Psychrophiles. From biodiversity to biotechnology. Springer International Publishing
Martorell MM, Ruberto LAM, Fernández PM, Castellanos de Figueroa LI, Mac Cormack WP (2017) Bioprospection of cold-adapted yeasts with biotechnological potential from Antarctica. J Basic Microbiol 57:504–516
Mbareche H, Veillette M, Bilodeau G, Duchaine C (2020) Comparison of the performance of ITS1 and ITS2 as barcodes in amplicon-based sequencing of bioaerosols. PeerJ 8:e8523
Miguel GA, Carlsen S, Arneborg N, Saerens SMG, Laulund S, Knudsen GM (2022) Non-*Saccharomyces* yeasts for beer production: insights into safety aspects and considerations. Int J Food Microbiol 383:109951
Moreira SR, Schwan RF, de Carvalho P, Wheals AE (2001) Isolation and identification of yeasts and filamentous fungi from yoghurts in Brazil. Braz J Microbiol 32:117–122
Nakagawa T, Nagaoka T, Taniguchi S, Miyaji T, Tomizuka N (2004) Isolation and characterization of psychrophilic yeasts producing cold-adapted pectinolytic enzymes. Lett Appl Microbiol 38:383–387
Panikov NS, Sizova M (2007) Growth kinetics of microorganisms isolated from Alaskan soil and permafrost in solid media frozen down to −35 °C. FEMS Microbiol Ecol 59:500–512
Park Y, Maeng S, Oh J, Sung GH, Srinivasan S (2021) *Mrakia terrae* sp. nov. and *Mrakia soli* sp. nov., two novel basidiomycetous yeast species isolated from soil in Korea. Mycobiology 49:469–475
Patanita M, Albuquerque A, Campos MD, Materatski P, Varanda CM, Ribeiro JA, Féli MDR (2022) Metagenomic assessment unravels fungal microbiota associated to grapevine trunk diseases. Horticulturae 8:288
Pathan AA, Bhadra B, Begum Z, Shivaji S (2010) Diversity of yeasts from puddles in the vicinity of midre lovénbreen glacier, arctic and bioprospecting for enzymes and fatty acids. Curr Microbiol 60:307–314
Péter G, Takashima M, Čadež N (2017) Yeast habitats: different but global. In: Buzzini P, Lachance MA, Yurkov A (eds) Yeasts in natural ecosystems: ecology. Springer International Publishing, pp 38–71
Poindexter JS (2009) Encyclopedia of microbiology. Elsevier, Amsterdam
Poliakova AV, Ilu C, Panikov NS (2001) Yeast biodiversity in hydromorphic soils with reference to grass-Sphagnum swamp in Western Siberia and the hammocky tundra region (Barrow, Alaska). Mikrobiologiia 70:714–720
Raspor P, Zupan J (2006) Yeasts in extreme environments. In: Rosa CA, Péter G (eds) Biodiversity and ecophysiology of yeasts. Springer International Publishing, pp 371–417
Rosa CA, Péter G (2006) Biodiversity and ecophysiology of yeasts. Springer International Publishing
Rothschild LJ, Mancinelli RL (2001) Life in extreme environments. Nature 409:1092–1101
Sannino C, Tasselli G, Filippucci S, Turchetti B, Buzzini P (2017) Yeasts in nonpolar cold habitats. In: Buzzini P, Lachance MA, Yurkov A (eds) Yeasts in natural ecosystems: diversity. Springer International Publishing, pp 366–396
Sannino C, Borruso L, Mezzasoma A, Battistel D, Zucconi L, Selbmann L, Azzaro M, Onofri S, Turchetti B, Buzzini P, Guglielmin M (2020) Intra-and inter-cores fungal diversity suggests interconnection of different habitats in an Antarctic frozen lake (Boulder Clay, Northern Victoria Land). Environ Microbiol 22:3463–3477

Sannino C, Borruso L, Mezzasoma A, Turchetti B, Ponti S, Buzzini P, Mimmo T, Guglielmin M (2023) The unusual dominance of the yeast genus *Glaciozyma* in the deeper layer in an Antarctic permafrost core (Adélie Cove, Northern Victoria Land) is driven by elemental composition. J Fungi 9:435

Santiago M, Ramìrez-Sarmiento CA, Zamora RA, Parra LP (2016) Discovery, molecular mechanisms, and industrial applications of cold-active enzymes. Front Microbiol 7:1408

Sanyal A, Antony R, Ganesan P, Thamban M (2020) Metabolic activity and bioweathering properties of yeasts isolated from different supraglacial environments of Antarctica and Himalaya. A van Leeuwenhoek 113:2243–2258

Sequino G, Valentino V, Torrieri E, De Filippis F (2022) Specific microbial communities are selected in minimally-processed fruit and vegetables according to the type of product. Foods 11:2164

Shivaji S, Prasad GS (2009) Antarctic yeasts: biodiversity and potential applications. In: Satyanarayana T, Kunze G (eds) Yeast biotechnology: diversity and applications. Springer International Publishing, pp 3–16

Siddiqui KS, Cavicchioli R (2006) Cold-adapted enzymes. Annu Rev Biochem 75:403–433

Simões J, Coelho E, Magalhães P, Brandão T, Rodrigues P, Teixeira JA, Domingues L (2023) Exploiting non-conventional yeasts for low-alcohol beer production. Microorganisms 11:316

Singh P, Singh SM (2012) Characterization of yeast and filamentous fungi isolated from cryoconite holes of Svalbard, Arctic. Polar Biol 35:575–583

Su Y, Jiang X, Wu W, Wang M, Hamid MI, Xiang M, Liu X (2016) Genomic, transcriptomic, and proteomic analysis provide insights into the cold adaptation mechanism of the obligate psychrophilic fungus *Mrakia psychrophila*. G3 Genes 6:3603–3613

Sugiyama J, Fukagawa M, Chiu SW, Komagata K (1985) Cellular carbohydrate composition, DNA base composition, ubiquinone systems, and diazonium blue B color test in the genera *Rhodosporidium*, *Leucosporidium*, *Rhodotorula* and related basidiomycetous yeasts. J Gen Appl Microbiol 31:519–550

Takagi M, Abe S, Suzuki S, Emert GH, Yata N (1977) A method for production of alcohol directly from cellulose using cellulose and yeast. In: Proc Bioconversion Symp IIT, Delhi. pp 551–571

Tapia-Vázquez I, Sánchez-Cruz R, Arroyo-Domínguez M, Lira-Ruan V, Sánchez-Reyes A, Del Rayo S-CM, Padilla-Chacón D, Batista-García RA, Folch-Mallol JL (2020) Isolation and characterization of psychrophilic and psychrotolerant plant-growth promoting microorganisms from a high-altitude volcano crater in Mexico. Microbiol Res 232:126394

Tasselli G, Filippucci S, Sannino C, Turchetti B, Buzzini P (2017) Cold-adapted basidiomycetous yeasts as a source of biochemicals. In: Margesin R (ed) Psychrophiles: from biodiversity to biotechnology. Springer International Publishing, pp 5556–5584

Thomas-Hall SR, Turchetti B, Buzzini P, Branda E, Boekhout T, Threelen B, Watson K (2010) Cold-adapted yeasts from Antarctica and Italian alps-description of three novel species: *Mrakia robertii* sp. nov., *Mrakia blollopis* sp. nov. and *Mrakiella niccombsii* sp. nov. Extremophiles 14:47–59

Thomsen MH, Thygesen A, Thomsen AB (2009) Identification and characterization of fermentation inhibitors formed during hydrothermal treatment and following SSF of wheat straw. Appl Microbiol Biotechnol 83:447–455

Tiquia-Arashiro SM, Rodrigues D (2016) Extremophiles: applications in nanotechnology. Springer International Publishing

Tsuji M (2016) Cold-stress responses in the Antarctic basidiomycetous yeast *Mrakia blollopis*. R Soc Open Sci 3:160106

Tsuji M, Fujiu S, Xiao N, Hanada Y, Kudoh S, Kondo H, Tsuda S, Hoshino T (2013a) Cold adaptation of fungi obtained from soil and lake sediment in the Skarvsnes ice-free area, Antarctic. FEMS Microbiol Lett 346:121–130

Tsuji M, Goshima T, Matsushika A, Kudoh S, Hoshino T (2013b) Direct ethanol fermentation from lignocellulosic biomass by Antarctic basidiomycetous yeast *Mrakia blollopis* under a low temperature condition. Cryobiology 67:241–243

Tsuji M, Singh SM, Yokota Y, Kudoh S, Hoshino T (2013c) Influence of initial pH on ethanol production by the Antarctic basidiomycetous yeast *Mrakia blollopis*. Biosci Biotechnol Biochem 77:2483–2485

Tsuji M, Yokota Y, Shimohara K, Kudoh S, Hoshino T (2013d) An application of wastewater treatment in a cold environment and stable lipase production of Antarctic basidiomycetous yeast *Mrakia blollopis*. PLoS One 8:e59376

Tsuji M, Yokota Y, Kudoh S, Hoshino T (2014a) Improvement of direct ethanol fermentation from woody biomasses by the Antarctic basidiomycetous yeast, *Mrakia blollopis*, under a low temperature condition. Cryobiology 68:303–305

Tsuji M, Yokota Y, Kudoh S, Hoshino T (2014b) Effects of nitrogen concentration and culturing temperature on lipase secretion and morphology of the Antarctic basidiomycetous yeast *Mrakia blollopis*. Int J Res Eng Sci 2:49–54

Tsuji M, Kudoh S, Hoshino T (2015a) Draft genome sequence of cryophilic basidiomycetous yeast *Mrakia blollopis* SK-4, isolated from an algal mat of Naga-ike Lake in the Skarvsnes ice-free area, East Antarctica. Genome Announc 3:e01454–e01414

Tsuji M, Yokota Y, Kudoh S, Hoshino T (2015b) Comparative analysis of milk fat decomposition activity by *Mrakia* spp. isolated from Skarvsnes ice-free area, East Antarctica. Cryobiology 70:293–296

Tsuji M, Tanabe Y, Vincent WF, Uchida M (2018) *Mrakia arctica* sp. nov., a new psychrophilic yeast isolated from an ice Island in the Canadian High Arctic. Mycoscience 59:54–58

Tsuji M, Kudoh S, Tanabe Y, Hoshino T (2019a) Basidiomycetous yeast of the genus *Mrakia*. In: Tiquia-Arashiro S, Grube M (eds) Fungi in extreme environments. Springer International Publishing, pp 145–156

Tsuji M, Tanabe Y, Vincent WF, Uchida M (2019b) *Mrakia hoshinonis* sp. nov., a novel psychrophilic yeast isolated from a retreating glacier on Ellesmere Island in the Canadian High Arctic. Int J Syst Evol Microbiol 69:944–948

Tsuji M, Vincent WF, Tanabe Y, Uchida M (2022) Glacier retreat results in loss of fungal diversity. Sustainability 14:1617

Tsuji M, Ishihara JI, Toyoda A, Takahashi H, Kudoh S (2023) Genome sequence of basidiomycetous yeast *Mrakia gelida* MGH-2, isolated from Skarvsnes ice-free area, East Antarctica. Microbiol Resour Announc 12:e0106422

Turchetti B, Buzzini P, Goretti M, Branda E, Diolaiuti G, D'Agata C, Smiraglia C, Vaughan-Martini A (2008) Psychrophilic yeasts in glacial environments of Alpine glaciers. FEMS Microbiol Ecol 63:73–83

Turchetti B, Goretti M, Branda E, Diolaiuti G, D'Agata C, Smiraglia C, Onofri A, Buzzini P (2013) Influence of abiotic variables on culturable yeast diversity in two distinct Alpine glaciers. FEMS Microbiol Ecol 86:327–340

Turchetti B, Sannino C, Mezzasoma A, Zucconi L, Onofri S, Buzzini P (2020) *Mrakia stelviica* sp. nov. and *Mrakia montana* sp. nov., two novel basidiomycetous yeast species isolated from cold environments. Int J Syst Evol Microbiol 70:4704–4713

Turchetti B, De Francesco G, Mugnai G, Sileoni V, Alfeo V, Buzzini P, Yurkov A, Marconi O (2023) Species and temperature-dependent fermentative aptitudes of *Mrakia* genus for innovative brewing. Food Res Int 170:113004

Watson K (1984) Membrane lipid adaptation in yeast. In: Kates M, Manson L (eds) Membrane fluidity, Biomembranes, vol 12. Plenum, New York, pp 517–542

Xin MX, Zhou PJ (2007) *Mrakia psychrophila* sp. nov., a new species isolated from Antarctic soil. J Zhejiang Univ Sci B 8:260–265

Xue F, Yang J, Luo C, Li D, Shi G, Song G, Li Y (2023) Metagenomic insight into the biodegradation of biomass and alkaloids in the aging process of cigar. Bioresour Bioprocess 10:45

Yamada Y, Komagata K (1987) *Mrakia* gen. nov., a heterobasidiomycetous yeast genus for the Q-equipped, self-sporulating organisms which produce a unicellular metabasidium, formerly classified in the genus *Leucosporidium*. J Gen Appl Microbiol 33:455–457

Yamada Y, Matsumoto A (1988) An electrophoretic comparison of enzymes in strains of species in the genus *Mrakia* Yamada et Komagata. J Gen Appl Microbiol 34:201–208

Yan K, Zhou J, Feng C, Wang S, Haegeman B, Zhang W, Chen J, Zhao S, Zhou J, Xu J, Wang H (2023) Abundant fungi dominate the complexity of microbial networks in soil of contaminated site: high-precision community analysis by full-length sequencing. Sci Total Environ 861:160563

Yu C, Hu X, Deng W, Li Y, Han G, Ye C (2016) Soil fungal community comparison of different mulberry genotypes and the relationship with mulberry fruit sclerotiniosis. Sci Rep 6:28365

Yuivar Y, Alcaino J, Cifuentes V, Baeza M (2019) Characterization of gelatinase produced by Antarctic *Mrakia* sp. J Basic Microbiol 59:846–852

Yurkov AM, Sannino C, Turchetti B (2020) *Mrakia fibulata* sp. nov., a psychrotolerant yeast from temperate and cold habitats. A van Leeuwenhoek 113:499–510

Zhang T, Yao YF (2015) Endophytic fungal communities associated with vascular plants in the high arctic zone are highly diverse and host-plant specific. PLoS One 10:e0130051

Zhang KH, Shi CF, Chai CY, Hui FL (2020) *Mrakia panshiensis* sp. nov. a new member of the Cystofilobasidiales from soil in China, and description of the teleomorphic-stage of *M. arctica*. MycoKeys 74:75–90

Chapter 12
Dehydration of Yeasts and Anhydrobiosis

Alexander Rapoport

Contents

12.1 Introduction 344
12.2 Influence of Dehydration Treatment on Yeast Cells 347
12.2.1 Yeast as a Eukaryotic Cell Model 347
12.2.2 Cell Wall 348
12.2.3 The Plasma Membrane 349
12.2.4 Mitochondria 352
12.2.5 The Endoplasmic Reticulum 353
12.2.6 Nucleus 354
12.2.7 Other Intracellular Organelles 355
12.2.8 Intracellular Protective Reactions 356
12.3 Some Aspects of Industrial Drying and the Use of Dry Yeasts 358
12.3.1 Fluidized Bed Drying 359
12.3.2 Spray Drying 360
12.3.3 Air Drying of Yeast 362
12.3.4 Rehydration Stage 364
12.3.5 Construction of New Strains 365
12.3.6 Application of Active Dry Yeasts 367
12.4 Conclusions 368
References 369

A. Rapoport (✉)
Laboratory of Cell Biology, Institute of Microbiology and Biotechnology, University of Latvia, Riga, Latvia
e-mail: rapoport@mail.eunet.lv

A. A. Sibirny (ed.), *Biotechnology of Yeasts and Filamentous Fungi*,
Grand Challenges in Biology and Biotechnology,
https://doi.org/10.1007/978-3-031-74726-7_12

Abstract Anhydrobiosis is a unique phenomenon of nature that gives the possibility for various live organisms to survive during very hot and dry seasons of the year. It is a state in which all processes of metabolism are temporarily reversibly suspended as a result of strong dehydration of the cells. The state of anhydrobiosis is used for the industrial production of active dry yeasts. The results of detailed studies of the mechanisms of anhydrobiosis and of changes in the cells that take place during their drying by different methods help improve the dehydration resistance of yeast strains necessary in traditional and nonconventional biotechnologies. The progress that has been made in these research studies within the last decades has led to the formulation of the major factors influencing the maintenance of yeast viability during their drying. It was revealed that all intracellular organelles are subjected to definite changes during dehydration and rehydration procedures. A number of earlier unknown intracellular protective reactions that can be artificially activated for the improvement of cells' resistance have been found. These results are also used in the construction of new biotechnologically efficient yeast strains that are very resistant to the dehydration process.

12.1 Introduction

Since the last century, the use of active dry yeasts has become more and more popular in various branches of biotechnology. Active dry yeasts are yeasts in the state of anhydrobiosis, in which their metabolism is temporarily reversibly suspended (with the maintenance of the yeast cells' viability) as a result of their strong dehydration. The increased interest by industries in the use of this state for practical goals is linked to many reasons, the main one being economic grounds. It is much cheaper and more convenient to transport dry yeasts containing only 5–8% water than fresh ones, where water content can reach 75–80%. In this case, the same amount of dry yeasts takes a considerably smaller volume, but refrigeration is not necessary for the transportation of dry yeasts. Using active dry yeasts is also much more convenient for customers because refrigerators are not necessary for their storage, and they can be stored for very long periods of time, which can reach 2–3 years if the quality of the active dry yeasts is good enough. To achieve such a high quality of active dry yeasts, first of all, it is necessary to clarify the major factors promoting the preservation of their viability at dehydration. It means we have to understand what happens with yeast cells during their dehydration and the subsequent rehydration/reactivation and to reveal the effects of the various factors in different industrial dehydration methods. Then it is necessary to find the possibilities for improving the resistance of biotechnologically important yeast strains to dehydration-rehydration treatments and to diminish the negative effects of the damaging factors of industrial drying procedures. Our current knowledge of these problems is the main topic of this review.

The history of research on anhydrobiosis started at the beginning of the eighteenth century in 1701, when great Dutch naturalist Anthony van Leeuwenhoek discovered this state using his very own simple hand-made microscopes. He studied sand taken from the roof gutter of his house after one very hot summer and added a drop of water to the sample of this sand. Unexpectedly, he found in this preparation a number of moving organisms, which he called "animalcules." He was very surprised that there were live organisms in such a dry sample. He repeated this procedure a lot of times, and every time, he saw the restoration of the active life of these small organisms when water was added. Now we know that these organisms are rotifers. In 1743, a new paper by T. Needham appeared that revealed a similar effect on other organisms—nematodes—under other conditions. Then a number of well-known researchers in the eighteenth century, including H. Backer, L. Spallanzani, and others, continued to experiment in this direction. In the twentieth century, results of the experiments of W. Preyer were published, which once more evidenced the maintenance of viability by dry rotifers, tardigrades, and nematodes, giving the term "anabiosis" for this phenomenon. From this time, studies on anabiosis have actively been continued in different countries. Another term, a synonym for anabiosis—"cryptobiosis"—was proposed by D. Keilin (1959), describing it as "the state of an organism when it shows no visible signs of life and when its metabolic activity becomes hardly measurable or comes reversibly to a standstill" (Keilin 1959). In the second half of the twentieth century, a new stage of studies on anabiosis (cryptobiosis) started because of the development of powerful techniques for cytological, biophysical, and biochemical research and the accumulation of new, more-or-less modern knowledge. This new stage in the study of anabiosis (cryptobiosis) was started mainly through experimentation by John Crowe in the USA, Folkert Hoekstra in the Netherlands, and some research groups in Russia and Latvia.

Today, following the proposal by D. Keilin (1959), we distinguish between forms of anabiosis (cryptobiosis), corresponding to the various ways in which live organisms can reach this state:

- Anhydrobiosis—when metabolism is suspended through water evaporation from the cells
- Osmobiosis—caused by the extraction of water by solutions with high osmotic pressure
- Cryobiosis—when liquid water, the medium for metabolic reactions, is eliminated by freezing
- Anoxybiosis—caused by the decrease of oxygen concentration below critical limits

On the basis of the results of active studies by G. El-Registan's group, it is now possible to discuss one more form, which conditionally can be called "autochemo-anabiosis." It can be reached under the influence of chemical factors produced by the cells themselves, which are the autoinducers of anabiosis—alkylhydroxybenzenes (Mulyukin et al. 1996, 2001; Bespalov et al. 2000; Stepanenko et al. 2004; El-Registan et al. 2005; Nikolaev et al. 2006 etc.).

There are a number of live organisms belonging to completely different groups that are able to transition into a state of anhydrobiosis at unfavorable conditions of the environment. These are seeds of various plants; mosses; cysts of crustaceans;

microscopic animals such as rotifers, tardigrades, and nematodes; as well as the majority of microorganisms.

There are a number of different factors that may determine microorganisms' viability at their dehydration. One of the attempts to summarize these factors was made by C.A. Morgan et al. (2006). In accordance with this classification, the main factors are:

- Resistance of microorganisms to dehydration treatment
- The growth media taken for the cultivation of microorganisms
- The growth phase of a microbial culture
- The concentration of cells taken for drying
- Protective substances that are used in the dehydration stage
- The methods for dehydration used
- Conditions at the rehydration/reactivation stage

All these factors are rather conditional ones. Here are some data that may evidence such a postulate: if we think, for example, of the resistance of microorganisms to dehydration, usually it is supposed that yeasts are more resistant to dehydration treatments compared to bacteria. However, a comparison of genetically closely related yeasts—*Saccharomices cerevisiae*, *Saccharomyces mikatae*, and *Naumovia castellii*—showed that they differ very significantly in their resistance to dehydration treatment. The viability of the cells after dehydration-rehydration was 84, 20, and 40%, correspondingly, but when these yeasts were dehydrated in the presence of trehalose, the viability of each of the three species increased to approximately 20% (Lopez-Martinez et al. 2014). Also, the resistance to drying of the various strains of the same species can be rather different. What is more, the resistance of the same strain differs rather considerably depending on the conditions for cultivation, rehydration, and reactivation. For example, the viability of yeast strain *Saccharomyces cerevisiae* 14 grown in a molasses-containing medium after dehydration-rehydration may reach 95% and higher, whereas after growth in an ethanol-containing medium, viability reaches only 30% (Beker and Rapoport 1987). At the same time, the viability of *Lactobacillus reuterii* dehydrated after their growth at pH 5 was at the level of 90%, whereas after their growth at pH 6, it was about 65% (Palmfeldt and Hahn-Hagerdal 2000). Another example may be linked to the influence of the microbial culture growth phase on resistance to dehydration-rehydration. Usually, microbial cells (both bacteria and yeasts) are the most resistant to dehydration in the stationary growth phase, but in the exponential growth phase, their viability after dehydration-rehydration, in the best-case scenario, can reach only 30% (Beker and Rapoport 1987). Some recent studies showed that this postulate is also not the common rule. Very high resistance to dehydration-rehydration treatment was shown for the yeast *Debaryomyces hansenii* in the exponential growth phase, too (Khroustalyova et al. 2001). High resistance to dehydration in the exponential growth phase was found for *S. cerevisiae* thermo-resistant strains (Rapoport et al. 2014). If optimum cell concentration for the highest viability at dehydration-rehydration is to be considered, it depends on the protective medium used. If the optimum protective medium is to be taken into account, the optimum cell concentration depends on the concentration of the protective medium (Morgan et al. 2006). The conclusion made on the basis of an analysis of all these factors was that the maintenance of high viability for microorganisms at their drying "appears to be still a science based on empirical testing rather than facts and tested theories. Major problem to overcome seems to be the lack of generic theories" (Morgan et al. 2006). The current review is devoted to an attempt to make an additional step toward the generation of such theories. It is linked to only one kind of anabiosis (cryptobiosis)—to anhydrobiosis—the form discovered by Anthony van Leeuwenhoek, and only information received for one group of microorganisms—for yeasts—will be discussed.

12.2 Influence of Dehydration Treatment on Yeast Cells

Main studies on the structural and functional changes in yeast cells during their dehydration were performed in 1970–1987 and were summarized in a review published 30 years ago (Beker and Rapoport 1987). They will be described rather briefly in this review as the main attention will be paid to the results of later studies in this direction.

12.2.1 Yeast as a Eukaryotic Cell Model

In recent decades, yeast has been widely used as a model for eukaryotic cells in various medical, pharmacological, and molecular biological studies (Botstein et al. 1997; Bolotin-Fukuhara et al. 2010; Nielsen 2019; Vanderwaeren et al. 2022). This is due to the fact that they have a short life cycle, a well-annotated genome, and an expansive molecular toolbox. Beyond that, they are easy to manipulate genetically, with strong conservation of basic eukaryotic biological and biochemical pathways. Usually, yeast cells are used in the active physiological state. Recently, American scientists have used desiccated yeast as a model in their experiments devoted to the understanding of the deep space environment's influence on eukaryotic systems. In these experiments, dry yeast cells that are in a state of anhydrobiosis are necessary to test the long-term treatment of highly ionizing radiation, composed of high-energy protons and heavy ions, and solar particle events, composed of primarily energetic protons (Cucinotta 2014). Such treatment is similar to the one astronauts experience in long-term deep space missions. Yeasts are used as a robust model for human cells in this study. The experiments that were performed in preparation for the BioSentinel mission revealed the best conditions for yeast cell dehydration and storage for 6–12 months in the state of anhydrobiosis. This study shows that dehydrated and stored yeast cells may survive during a Mars mission for relevant periods under flight-like conditions. Yeast cells in a desiccated state were irradiated with 0, 10, or 250 cGy of 1 GeV protons at the NASA Space Radiation Laboratory (NSRL);

both doses and particle species are relevant to the deep space environment. Importantly, this study revealed a dose-dependent decrease in yeast cell growth and metabolism after their rehydration and reactivation from anhydrobiosis in response to space-like radiation, and this effect was significantly more pronounced in a strain of yeast that was deficient in deoxyribonucleic acid (DNA) damage repair compared with a wild-type strain. These preliminary experiments showed that desiccated yeast cells are sensitive to deep-space-relevant ionizing radiation (Liddell et al. 2023; Santa Maria et al. 2023).

One more research using yeast in the state of anhydrobiosis as a model was recently performed with the goal of understanding the biological effects of very low-dose radiation. These experiments were performed in a surface control laboratory (radiation level: 68.0 nGy h^{-1}) and at the SNOLAB underground laboratory—located 2 km underground—with a radiation level of 10.1 nGy h^{-1} for up to 48 wk. The results of this study showed that the absence of natural radiation negatively influenced the survival and metabolism of desiccated yeast, especially in the case of a strain with a defective DNA repair system. An interesting and important conclusion was made on the potential importance of natural radiation exposure in maintaining the homeostasis of living organisms (Lapointe et al. 2024).

12.2.2 Cell Wall

Studies of this surface structure response to dehydration treatment which have been performed using different methodological approaches has showed that in these conditions changes both main its components—mannoproteins and glucans. In the case of mannoproteins, the changes were revealed using scanning electron microscopy—there were fibrils on the surface of the cells instead of smooth cell walls, like in native cells (Rapoport et al. 1983; Ventina et al. 1984). These changes were accompanied by an increase in the negative and positive charges of the cell walls (Rapoport and Beker 1985; Rapoport and Muter 1995), which can be explained by the disruption of certain connections between the molecules. As a result, "free ends" of mannoprotein molecules with negative and positive charges may be formed and may appear on cell wall surfaces. The conclusion on the changes in the glucan component of the cell walls was made on the basis of the revealed changes in the shape of the cells (Rapoport et al. 1986a), which mainly depend on the structure and stability of glucans. It is important to mention that all these changes in the cell walls were revealed for yeast cells after being viably dehydrated and rehydrated after dehydration. So a conclusion was made that these changes are reversible and do not influence cell viability. More recent studies showed that mannoproteins' importance for the stability of yeast cells at dehydration was definitely underestimated. It was shown recently that the absence of some mannoproteins (glycosylphosphatidylinositol (GPI)-anchored cell wall protein Ccw12) in the cell wall may unexpectedly have very positive effects on the higher resistance of yeast to dehydration. At the same time, mutants lacking proteins belonging to two structurally and functionally unrelated groups (proteins noncovalently attached to the cell wall and Pir proteins) possessed significantly lower cell resistance to dehydration-rehydration than the mother wild-type strains (Borovikova et al. 2016). On the basis of these results, a conclusion can be made that the cell wall structure and mannoprotein composition are characteristics that may be very important for the resistance of yeasts to drying. It simultaneously means that the possible damage of cell wall proteins during yeast dehydration can be an unfavorable factor for cell resistance and viability. Besides that, a hypothesis was proposed that increased amounts of chitin may have positive effects on the stabilization of yeast cell walls under conditions of dehydration-rehydration (Borovikova et al. 2016). Recently, additional, rather interesting, indirect information was received specifying the definite changes in the glucan component in the cell walls during the dehydration of the cells. It appeared that β-glucans isolated from dry yeast cells possess higher immunogenic activity compared with β-glucans isolated from native yeast biomass (Liepins et al. 2015). The reasons for this effect of yeast cells' dehydration process, of course, have to be discovered in future studies. As one of the working hypotheses, the idea can be proposed that dehydration treatment leads to a critical weakening of the chemical bonds between mannoproteins and glucans in yeast cell walls. As a result, it facilitates the easier isolation of β-glucans from dry yeast biomass, and the obtained preparations of β-glucans have a higher level of purity.

12.2.3 The Plasma Membrane

Essential changes were also found in the plasma membrane (PM) of yeast cells at their dehydration-rehydration. Because of the significant decrease in cell size during this treatment, to maintain its integrity, the plasma membrane becomes strongly folded (Fig. 12.1) (Biryusova and Rapoport 1978). These structural changes are accompanied by an essential increase in its permeability characteristics at the rehydration stage (Beker et al. 1984; Novichkova and Rapoport 1984). It is supposed that the changes in the molecular organization of the plasma membrane, which lead to an increase in membrane permeability and significant leakage from the cells of various intracellular compounds at their rehydration, may be one of the main reasons for their loss of viability (Beker and Rapoport 1987). A model describing the changes in membranes at the dehydration and subsequent rehydration of the cells was developed (Crowe et al. 1989). In accordance with this model, dehydration leads to the transition of membrane lipids from the liquid-crystalline phase to the gel phase. If the reverse phase transition during rehydration of dry cells occurs in an aqueous medium, this can be quite dangerous for the condition of the membrane and accompanied at least by its high permeability. This model postulates that to decrease such damage, it would be necessary to increase the rehydration temperature above the phase transition temperature or to reach this phase transition in waterless conditions, for example, in water vapors. The correctness of these postulates was confirmed also for dehydrated yeast cells (Leslie et al. 1994: Rapoport et al.

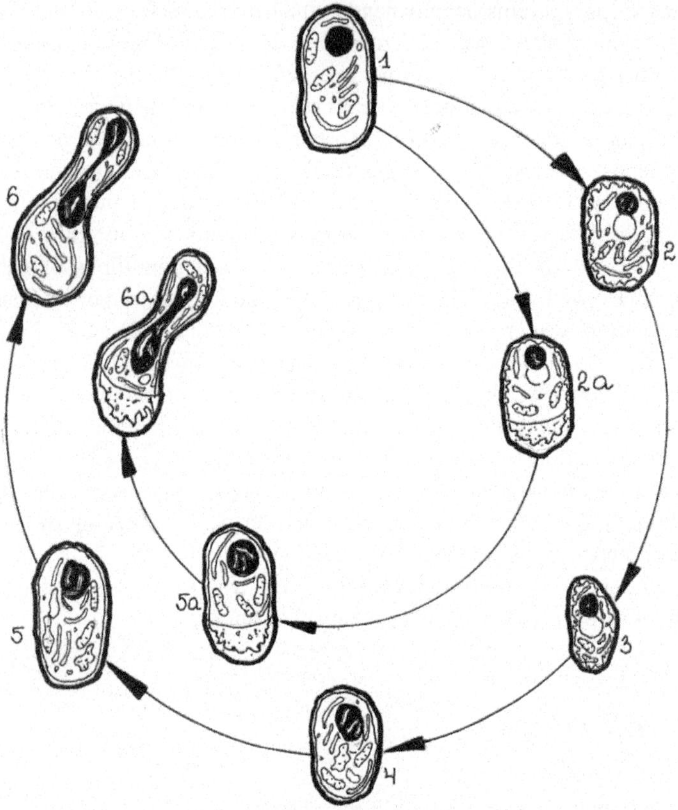

Fig. 12.1 A scheme of several structural transformations in a yeast cell during its dehydration (1–3) and the subsequent rehydration/reactivation (4-6). 1—initial cell; 2–3—formation of plasma membrane invaginations (folds) during dehydration, chromatin condensation, its localization at one of the poles of the nucleus, and separation of the chromosome containing a part of the nucleus; 2a—formation of a new intracellular plasma membrane during yeast dehydration, which separates the irreversibly damaged part of the cell from the viable one; 4-6—rehydration and reactivation of the dry cell, further degradation of the part of the nucleus to be separated, and its elimination by phagosomes during autophagy; 5a—autolysis in the separated irreversibly damaged part of the cell; 6, 6a—division of a reactivated cell (Beker and Rapoport 1987, modified)

1995, 2009). It is already known for many years that to have resistance to dehydration, yeast cells have to accumulate rather significant amounts of trehalose (Panek 1985, 1995; Eleutherio et al. 2015; etc.). The water replacement hypothesis was proposed to explain its role in the protection of the cells (Crowe 2008; Crowe et al. 1992). In accordance with developments in this hypothesis, trehalose hydrogen bonds to polar residues in dry phospholipids (mainly the phosphate of the headgroup) and maintains its lateral spacing, similar to substituted water, thereby preserving membrane native structures (Crowe 2008). It is supposed also that besides interaction with lipids, trehalose can also form a glassy matrix with a high glass-transition temperature (Crowe et al. 1998), which prevents phase separation in membranes during the drying of live organisms as well as maintains existing phases or rafts (Ricker et al. 2003). Various papers appeared during the last years that show that yeast cells can survive desiccation treatment also in the absence of trehalose (Ratnakumar and Tunnacliffe 2006; Petitjean et al. 2015; etc.). It is possible that, in this case, the trehalose for the membrane and for macromolecule protection may be substituted by other compounds, for example, sugar alcohols (Rapoport et al. 1988: Krallish et al. 1997). Recently, the water replacement hypothesis has been additionally developed on the basis of molecular dynamic simulation results. Up to these new ideas, trehalose interaction with membrane phospholipids at the dehydration of cells results in increased spacing between lipids and the disordering of the hydrocarbon chains in a concentration-dependent manner (Golovina et al. 2009, 2010; Rapoport et al. 2019). It was shown recently that the kinetics of both processes—dehydration and rehydration—may very significantly influence the level of possible damage in the plasma membrane. The lowering of both processes facilitates a more safe reorganization of its lipid components (Dupont et al. 2010; Lemetais et al. 2012). It was also shown that essential importance in the maintenance of plasma membrane integrity during yeast cells dehydration has ergosterol (Dupont et al. 2011, 2012). It provides mechanical and structural stability to the membrane and also protects yeast cells from oxidation damage (Dupont et al. 2011, 2012). Another new direction in the study of plasma membrane changes is the understanding of the role of membrane proteins in the maintenance of cell viability in dehydration-rehydration treatments. First experiments have been performed using a collection of deletion mutants, strains lacking definite membrane proteins—transporters of potassium and glycerol. Their results clearly showed the importance of potassium transporter Trk2, which is responsible for the accumulation by cells of relatively high intracellular potassium content, and it is crucial for the maintenance of yeast cells' viability during their transfer into the anhydrobiosis state and their subsequent reactivation to active life (Borovikova et al. 2014c). Similar results were also received regarding glycerol transporters Stl1p (glycerol importer) and Fps1p (glycerol exporter). It appears that both of them contribute to cells' ability to survive during the transfer into the state of anhydrobiosis (Duskova et al. 2015). It means that damage to at least these membrane proteins at the dehydration-rehydration of the cells may be very dangerous for their viability. In continuation of this work direction, researchers studied the changes in another plasma membrane protein—α--glucoside transporter Agt1. This protein is involved in maltose accumulation in *S. cerevisiae*. The obtained results showed, for the first time, that some changes in membrane protein functionality (most probably of a conformational character) take place already at the stage of elimination from the cells of only a small portion of the total amount of water (Kulikova-Borovikova et al. 2018). These results are the first indication that not only lipid components of the plasma membrane but also its protein components play an important role in yeast transfer to the state of anhydrobiosis and the subsequent restoration and return to active life.

12.2.4 Mitochondria

In spite of the importance of mitochondria to the life of any eukaryotic organism, there is still too little information about these organelles' role and the possible changes during the cells' transition into the state of anhydrobiosis and the subsequent restoration of their normal physiological activities. At the same time, it is necessary to keep in mind that besides the production of energy necessary for the reactivation of cells and the repair processes during oxidative phosphorylation, there also appears reactive oxygen species (ROS) as by-products. ROS can be rather dangerous for the cells, especially when they are in such a weakened state after anhydrobiosis. It is well known that oxidative stress processes are one of the reasons for cell death during dehydration-rehydration treatments (Espindola et al. 2003; Franca et al. 2005; Dupont et al. 2014). It means that any information about these organelles is not only interesting but also important for the further understanding of the ways to increase yeast resistance to dehydration. Nevertheless, our knowledge about their changes is, at the moment, too scarce. It is known that mitochondria are organelles whose recovery from an inactive state in anhydrobiosis can be detected earlier than of any other organelles. Their multiplication by dividing was found via electron microscopy at the very early stages of yeast reactivation from the state of anhydrobiosis (Rapoport and Kostrikina 1973). Studies on possible mitochondrial DNA (mtDNA) changes as a result of cell dehydration have not revealed them till now. No increase in petite mutant occurrence after dehydration treatment was revealed, as well as mtDNA copy number changes, sequence rearrangements, and resistance to mutagen challenge (Finn and Stewart 2002; Jenkins et al. 2010). Meanwhile, it was shown that yeast cell viability after dehydration-rehydration correlates with mtDNA resistance (Picazo et al. 2015). This fact was revealed by comparing the viability of hybrid strain clones that contained the mtDNA of *S. cerevisiae* or *S. uvarum*. In the first case, the viability of yeast after dehydration was significantly higher (Picazo et al. 2015). Electron microscopy studies also showed that during yeast dehydration, the condensation of the mitochondrial genome takes place (Rapoport and Kostrikina 1973). It definitely specifies the attempts of cells to protect mtDNA from possible damage during this treatment and to transfer into the state of anhydrobiosis. The first attempt to understand whether the chemical structure of mitochondrial membranes may be important for the resistance of yeast cells to dehydration was performed through the external introduction of lithocholic acid. This compound is an agent that induces changes in glycerophospholipids in mitochondrial membranes. Results of these experiments showed that lipid changes in mitochondrial membranes influence the resistance of cells to dehydration-rehydration treatments and may improve the survival rate of dehydrated cells (Khroustalyova and Rapoport 2019). Mitochondria are dynamic organelles, and during cells' life, they continuously undergo fission and fusion. These processes provide mitochondrial network morphology as well as regulate their physiological activity (Chen et al. 2019; Giacomello et al. 2020). Recent results revealed that disrupting mitochondrial dynamics severely hampered the resistance of yeast cells at dehydration/rehydration stress. A well-maintained dynamic balance of fusion and fission supports mitochondrial genome integrity and functions that are necessary for the survival of yeast cells during their dehydration-rehydration (Chen et al. 2021). A recent study analyzed changes in the energy status of yeast cells during dehydration and rehydration that are associated with mitochondrial status. A comparison between strains of the yeast *S. cerevisiae* that are semi-resistant and very resistant to dehydration-rehydration led to a conclusion that at least two interconnected factors are responsible for the resistance of cells to dehydration–rehydration: the stability of mitochondria and the intracellular content of adenosine triphosphate (ATP). The higher resistance of cells could be linked to the better maintenance of mitochondria during cell dehydration and, correspondingly, to more efficient ATP synthesis by oxidative phosphorylation during yeast reactivation from anhydrobiosis (Kuliešiene et al. 2021). Additional studies in this direction are undoubtedly necessary.

12.2.5 The Endoplasmic Reticulum

In recent decades, serious attention has been paid to one more intracellular structure—the endoplasmic reticulum (ER). The ER in eukaryotic cells is considered a major membrane source. The main activities of ER include the folding, modification, and assembly of newly synthesized proteins, which may be used by the cells themselves or excreted from the cells (Hijazi et al. 2020). Yeast ER consists of cortical (cell peripheral) ER, cytoplasmic ER, and perinuclear ER (Voeltz et al. 2002; Nakatogawa 2020). First information recently appeared about ER reactions and the changes during dehydration-rehydration treatments. Electron microscopy studies revealed the appearance, in these conditions, of ER whorls, which look like multilayered, ring-shaped structures (Ren et al. 2020). In addition, these researchers found that desiccation causes the expansion and disorganization of the ER tangled around the nuclear membrane in yeast cells (Ren et al. 2020). The ER that is close to the plasma membrane may be broken into small pieces, forming circular structures (Ren et al. 2020).

First studies of ER's role in the stability of yeast cells at their dehydration-rehydration revealed a number of very important results. Comparison of viability of cells from two dehydrated strains *S. cerevisiae* showed low viability of the mutant strain lacking the IST2 gene and correspondingly cortical ER protein Ist2. Further experiments led to very important conclusions. The first was that not only protein components of the plasma membrane but also at least one ER membrane protein (Ist2) play an important role in the maintenance of the viability of yeast cells during their dehydration and subsequent rehydration. It appeared that the plasma membrane of cells where ER membrane protein Ist2 was absent was not able to restore its molecular organization during reactivation from the state of anhydrobiosis. It means that this ER protein plays an important role in ensuring the stability of the molecular organization and functionality of the PM during yeast transition into anhydrobiosis and the subsequent restoration of cells' normal functional activities. The second conclusion is also the important role of ER-PM interactions during cells' transition into the state of anhydrobiosis (Dauss et al. 2021). Of course, these studies have to be continued.

12.2.6 Nucleus

Studies on the changes in yeasts' nucleus during yeast transition into the state of anhydrobiosis, first of all, revealed the condensation of chromatin, which makes it inactive and more resistant to extreme factors in the environment (Rapoport et al. 1973). This is now assumed as the main protective reaction in cells, and absent it, cells cannot maintain their viability during dehydration-rehydration processes. Recent developments in research in this direction was performed with yeast *S. cerevisiae* cells taken from a culture in the exponential growth phase (Zhang and Zhang 2022). These cells are usually not resistant during the drying process, and in the best-case scenario, only up to 30% of dehydrated cells can maintain their viability (Beker and Rapoport 1987; Rapoport et al. 2019; Sęk et al. 2023). Evaluation of the chromosome state in surviving yeast cells led to a conclusion that also yeast cells from the exponential growth phase may survive during dehydration if they are in the G1 phase and have condensed chromosomes (Zhang and Zhang 2022). Electron microscopy investigations also showed changes in the nucleus' shape as well the existence of opened areas of the nucleus that are not covered by a nuclear membrane (Rapoport et al. 1973). Such pictures are interpreted now as evidence of considerable changes in the lipid components of the nuclear membrane, which are targets for fixing agents used in the preparation of samples for electron microscopy studies (Dupont et al. 2014). Usually in these areas, mitochondria and lipid droplets are located, which, as sources of energy and lipids, are most probably necessary for the restoration of changed areas of nuclear membranes during the reactivation of cells from anhydrobiosis (Rapoport et al. 1973). Further studies have not revealed any changes in the size and number of chromosomes, as well as any sequence changes in the DNA of brewer's yeasts at their dehydration-rehydration (Jenkins et al. 2010). It was revealed that all the genes needed for desiccation tolerance are already transcribed and translated prior to desiccation (Calahan et al. 2011). Studies on gene expression after the rehydration of dry yeast revealed the activation of some genes linked to the fermentation pathway and the nonoxidative branch of the pentose phosphate pathway. The induction of genes linked to ribosomal biogenesis and protein synthesis was shown, too (Novo et al. 2007). Observations made at transcriptomic and proteomic levels showed that after the rehydration of dry yeast cells, there can be revealed very quickly the start of protein synthesis and the expression of various genes, including those that are responsible for the assimilation of various nutrient compounds (Singh et al. 2005; Rossignol et al. 2006; Novo et al. 2007; Vaudano et al. 2009, 2010; Noti et al. 2015). All these results lead to the conclusion that if the transition of yeast cells into the state of anhydrobiosis takes place in adequate conditions, there can be no serious damage to the cellular genome and nucleus.

12.2.7 Other Intracellular Organelles

A rather limited amount of information has been accumulated at the moment regarding the reactions of other intracellular organelles during dehydration-rehydration as well as on their changes during these processes, in spite of some of them having a very important role in the maintenance of yeast cells' physiological activities. One such important organelles are *vacuoles*, which have many functions, including the regulation of cellular homeostasis. Besides, it is a storage organelle, and, what can be very important for cells' transfer into the state of anhydrobiosis, is linked to degradation, the turnover of various intracellular molecules, and the process of autophagy. It was revealed that at dehydration-rehydration, yeast cell vacuoles could diminish in their sizes as well as fragment. There takes place also the redistribution of protein molecules at least at some limited areas of the vacuolar membrane, which specifies changes in the molecular organization of the membrane (Rapoport et al. 1986b). It is clear that such changes can be rather dangerous for the cell if extrapolate our knowledge about plasma membrane changes and strong increase of plasma membrane permeability. Vacuoles contain a lot of lytic enzymes, whose leakage from these organelles can lead to the death of the cells. An increase in the permeability of vacuolar membranes has already been shown at least for Mg^{2+} and Ca^{2+} ions (Rapoport et al. 1982). Electron microscopy studies also revealed the activation of autophagy during the rehydration and reactivation of yeast cells from the state of anhydrobiosis (Beker and Rapoport 1987). It definitely can be supposed that autophagy is an important protective reaction of the cells at these processes because it gives the possibility of isolating from actively functioning organelles in the rest of the cells and subject to degradation (and further recycling) irreversibly damaged molecules and structures which can be dangerous for the cell.

The only information we have at the moment about the changes in *peroxisomes* at yeast cell dehydration-rehydration was obtained in experiments with the yeast *Candida boidinii*. Their cells contain packets of peroxisomes. Electron microscopy studies showed obvious changes in the lipid components of their membranes. Such a conclusion was made because in dehydrated-rehydrated cells, it was not possible to visualize peroxisomal membranes inside the packet of peroxisomes. Similar observations were made in studies on the changes in the membranes of nuclei (which have been discussed above). Such effects, revealed by electron microscopy studies, are associated with poor chemical fixation of these membranes when preparing samples for research by standard methods used also for samples of native (not dehydrated) yeast. Because the fixation procedure is linked to the chemical interaction of the fixing agent with the lipid membrane, the differences in fixation results may be explained only by the changes in the lipid components of the membranes during the dehydration of yeast cells (Rapoport et al., unpublished).

One more interesting organelle is *lipid droplets* (Klug and Daum 2014). Lipids which are located in lipid droplets are linked to energy production via β-oxidation, intracellular membrane synthesis, protein modification, signaling, secretion within lipoproteins, and cell reactions to stress treatments (Klug and Daum 2014; Barbosa et al. 2015; Wang 2016). In spite of the recently essentially increased attention by researchers to these organelles, we still have too scarce information on their changes at yeast dehydration-rehydration. Electron microscopy studies revealed that dehydration treatment may result in the merging of lipid droplets in baker's yeast cells. At the reactivation stage, sometimes the existence of direct contact between lipid droplets and vacuoles can be found, as well as the intrusion of merged lipid droplets into the vacuole in yeast cells (Ventina et al. 1986). These observations give the possibility to think about the active participation of lipid droplets in the processes of repair and restoration of the membrane structure. Experiments performed with yeast species that are genetically closely related to *S. cerevisiae*—*S. paradoxus* and *S. bayanus*—revealed the correlation between a higher amount of phosphatidylcholine accumulated by cells in lipid droplets as a result of special cultivation conditions and increased resistance in these cells at dehydration treatment (Rodrıguez-Porrata et al. 2011). It was also shown that lipid droplets may be linked to the storage or inactivation of misfolded or aggregated proteins, which can appear during dehydration-rehydration (Wang 2016).

Unfortunately, we still have no information about the possible changes in the *Golgi complex*.

12.2.8 Intracellular Protective Reactions

To survive in the constantly changing conditions of the environment, under the influence of its various unfavorable physical and chemical factors, all live organisms, including yeasts, have worked out a lot of special protective reactions during their long evolution period. In principle, the suspension of metabolism and transferring into a state of anhydrobiosis are also protective reactions of live organisms that help them maintain their viability during very dry periods of the year. Sometimes, it is rather difficult to separate the protective reactions of organisms or cells from the mere changes in them under the influence of extreme conditions in the environment. One example is changes in the plasma membrane, which becomes strongly folded during the dehydration of cells (Fig. 12.1). Taking into account the fact that cells during the dehydration process considerably diminish in volume, to maintain its integrity, the plasma membrane has to become strongly folded, and this is realized by the formation of folded nets from small folds, which exist in the membranes of native cells. So it can be assumed that such changes are also one of their protective reactions. A similar situation is autophagy, already mentioned above. It can also be supposed as a protective reaction because its role is definitely associated with the necessity to eliminate damaged molecules and structures, such as proteins, membranes, or, in some cases, organelles, which can be toxic to actively functioning cells (Ratnakumar et al. 2011). In addition, electron microscopy studies revealed some other protective reactions. A main one is most probably the condensation of chromatin in the nucleus and of mtDNA in mitochondria, which makes DNA molecules inactive, protecting them from damage. For nuclei, the possibility of being diminished significantly in size was found, with the formation of a new nuclear membrane covering the condensed chromatin-containing part of the nucleus (Fig. 12.1) (Rapoport et al. 1973). In some cases, the possibility of the formation of a new intracellular plasma membrane was discovered, which separate two parts of the cell—dead and viable (Fig. 12.1) (Rapoport 1973). A number of chemical compounds have protective functions. The most well known of them is trehalose, which can protect the membranes and macromolecules by substitution of water, which is necessary for the maintenance of their structure and conformation (Panek 1995; Eleutherio et al. 2015). Despite the fact that a number of researchers have shown the ability of mutant yeast cells to maintain their viability also in the absence of trehalose, most likely, this only indicates that the cells have alternatives how to survive in these conditions. It was found that one of the alternatives to trehalose can be the group of sugar alcohols (Rapoport et al. 1988, 2009; Krallish et al. 1997). A number of substances, including catalase, glutathione, and proline, protect cells from oxidative stress during dehydration-rehydration (Takagi et al. 2000, 2016; Takagi 2008; Espindola et al. 2003; Franca et al. 2005). Information on the protective function of proline is discussed in detail by H. Takagi in the chapter of this book titled "Construction of Baker's Yeast Strains with Enhanced Tolerance to Baking-Associated Stresses." It was revealed that higher amounts of thioredoxin enhance oxidative stress resistance in wine yeast strains. The thioredoxin and also the glutathione/glutaredoxin systems constitute the most important defense against oxidation. Both these antioxidant systems are very important at various stages of active dry yeast production, starting from biomass propagation (Grant 2001; Herrero et al. 2008; Gómez-Pastor et al. 2010). Proteins, such as thioredoxins and glutaredoxins, also participate in the protection of protein activity against oxidative damage by repairing chemically modified proteins or by modulating the redox state of protein sulfhydryl groups (Grant 2001). In accordance with these effects of the protection mechanisms against oxidative stress, lower levels of molecular damage were revealed, and both lipid peroxidation and protein carbonylation were diminished at yeast dehydration if the cells contained higher amounts of these protective proteins (Gómez-Pastor et al. 2010, 2012). Some researchers showed the protective role of heat shock proteins and, especially, of Hsp70 (Guzhova et al. 2008; Dupont et al. 2014; Capece et al. 2016), which can be important in the facilitation of endoplasmic-reticulum-associated degradation of damaged proteins. Recently, it was shown that ergosterol is an extremely important compound, which, as believed, can increase the mechanical resistance of the plasma membrane (Dupont et al. 2011, 2012, 2014). The most recent discovery in this direction is the function of one more important protein—trehalose-6P synthase—which, in addition to its catalytic function in trehalose synthesis, is a sensing/signaling intermediate that has regulatory function(s), at least in energy homeostasis. Such function in preventing energy depletion may definitely be very essential to withstand adverse conditions (Petitjean et al. 2015). This short list of different intracellular protective reactions is not complete, of course, but it definitely shows how complex and various reactions of cells are under unfavorable conditions of the environment. Such a conclusion can be supported by recent research results showing that mutants defective in trehalose biosynthesis, hydrophilins, responses to hyperosmolarity and hypersalinity, reactive oxygen species (ROS) scavenging, and DNA damage repair nevertheless retain this strain desiccation tolerance (Calahan et al. 2011). It means that there are a lot of various alternatives that give the possibility to substitute one protective reaction with the other. It is also clear that further studies on the mechanisms of anhydrobiosis in yeasts will lead to new discoveries on intracellular protective reactions that were earlier unknown to us. Some of them may appear very important not only for basic biology and industrial biotechnology but also for the goals of medicine. It can be expected that this new knowledge will provide the possibility for developing new approaches to the protection of human organisms and the artificial increase in their resistance, which can be necessary, for example, in the case of complicated medical procedures and treatments.

12.3 Some Aspects of Industrial Drying and the Use of Dry Yeasts

The history of the commercial production of active dry yeasts was perfectly described in the review by P. Gelinas, who made a detailed analysis of the patent literature on the subject (Gelinas 2019). The main objective of this review was to determine the key challenges in the production of active dry yeast for fermented food. The first commercial active dry yeast product, called Florylin, was obtained in 1911 in Berlin by Dauerhefe G.m.b.H. Unfortunately, it was not too efficient, was too expensive, and had a number of other disadvantages (Gelinas 2019). Essential progress in yeast-drying processes and formulations was reached starting from the 1950s. In this review, the differences in the drying processes and formulations that were proposed during the further years are described in detail (Gelinas 2019). This extremely interesting and important review can provide a lot of comprehensive information for yeast technologists interested in detail in the efficient production of high-quality active dry yeast (Gelinas 2019).

Different methods for dehydration are used to obtain active dry yeast preparations as well as for the production of dry starter cultures for various biotechnological processes. These are freeze drying (Lodato et al. 1999; Abadias et al. 2001; Ale et al. 2015), drum dryers, fluidized bed drying (Bayrock and Ingledew 1997), spray drying (Luna-Solano et al. 2005; Isono et al. 1995), convective dehydration, and air-blast processes between these approaches. All these approaches have their advantages and disadvantages. For example, the disadvantage of freeze drying is linked to the rather high expenses necessary for the application of this method. Fluidized bed drying and spray drying can sometimes give rather low viability of yeast cells in the preparations obtained after dehydration (Lee et al. 2016). Therefore, a lot of studies during the last years were devoted to attempts to improve and optimize these drying methods to have the possibility of using them rather efficiently. The sections below will discuss several results obtained during the last 15–20 years using some drying approaches—fluidized bed drying, spray drying, and air drying.

12.3.1 Fluidized Bed Drying

Fluidized bed technology has already been used in industries for many years. Between different products that were dehydrated using this technology, there are various wet solid products, including yeasts (Turker et al. 2006). The main advantages of this technology are linked to the large contact surface area between solids and gas, a good degree of solids mixing, and the rapid transfer of heat and moisture between solids and gas, which shortens the drying time as well as the level of damage to heat-sensitive biological materials. At the same time, the main problem in industrial fluidized bed dryers is the scaling up of the processes developed in laboratory conditions. The most important issues linked to the application of this technology and the construction of various possible fluidized bed dryer models are discussed in detail by Daud (2008). It is reported that a multistage fluidized bed drying process can be used as a commercially effective method for yeast drying (Akbari et al. 2012).

Some researchers suppose that fluidized bed technologies have significant advantages compared to other ways of industrial yeast dehydration (Grabowski et al. 1997; Bayrock and Ingledew 1997). Bayrock and Ingledew (1997) showed that the viability of yeasts in fluidized bed dryers was not significantly affected by the drying temperature used during the warming-up and constant-rate drying periods (when the moisture content was greater than 15%). On the basis of the obtained results, these researchers stated that the mechanisms of death for fluidized bed drying are very different from those for spray drying. It is suggested that factors other than thermal destruction influence the viability seen during fluidized bed drying.

Studies directed to the understanding of the possible role of oxidative stress in the case of the fluidized bed drying of yeasts showed that it is a critical one (Grant 2001; Herrero et al. 2008; Gómez-Pastor et al. 2010). It was shown in these studies that one of the factors that provide yeast strain resistance at their dehydration in fluidized bed dryers is the accumulation of glutathione. In experiments with simulation of industrial growth conditions, yeast cells accumulated more than 1% of glutathione, and this percentage increased during the first period of drying, reaching approximately 2%. Such results led to a low level of lipid peroxidation in dry biomass (Garre et al. 2010), and it was five- to tenfold lower than in other experiments with laboratory dehydration conditions (Espindola et al. 2003). Other research investigated if total glutathione and trehalose accumulation may increase the viability of yeast during fluid bed drying. The effect of glutathione accumulation after yeast cultivation in a relatively nutrient-rich medium on cell resistance to mild stress (45 °C—s90 min) was obvious. These growing conditions also improved the performance of cells after their subsequent rehydration. It was also revealed that the highest amount of trehalose stored by yeasts cultivated in nutrient-poor media led to cells being more resistant to severe stress (60 °C—60 min) (Câmara et al. 2018; Camara and Santana 2021).

Mathematical modeling has been used for the analysis of drying processes, and a number of more or less complex empirical or theoretical models are available (see Debaste et al. 2008). Operating conditions such as temperature, the loading rate of compressed yeast granules, and hot air humidity had direct effects on both yeast activity and viability. The most important factors that affected the quality of the product were the loading rate and the operational temperature in each zone of the bed. Compared to the normal operating conditions at the plant, the optimization resulted in more than 12% and 27% improvement in yeast activity and viability, respectively (Akbari et al. 2012). A model has been developed for the drying of granular baker's yeast in fluidized bed dryers. The model was applied to the drying of granular baker's yeast with different granular dimensions. Good correspondence was found between model predictions and experimental measurements for small granule sizes. As a result, based on the opinion of this model's authors, it may be successfully used in technologies linked to the drying of relatively small particles where diffusion is not the limiting step (Turker et al. 2006). Current models take into account several simultaneous mechanisms of heat, liquid, and vapor diffusion (Turker et al. 2006). Comparatively, recent modeling of yeast drying in a fluidized bed led to the construction of a mathematical model in which deep physical meaning was developed, using Fick's law, to model the molecular diffusion of vapor inside a tortuous porous medium (Debaste et al. 2008).

The important aspects in the control and optimization of drying processes are energy cost, drying time, product quality, and the moisture content in the material at the end of the process (Koni et al. 2009). The optimal control algorithm is proposed to improve the drying process of baker's yeast in a batch fluidized bed. The developed model helps to reach smaller energy consumption and quality loss at the end of the drying process using proposed optimization procedure (Koni et al. 2009).

12.3.2 Spray Drying

Spray drying is another process used for the production of active dry microbial preparations, including yeasts, at the industrial scale (Luna-Solano et al. 2005; Aponte et al. 2016). This approach renders the possibility for high production rates with low operating costs. It is commonly used for the production of various stable dry food products (Silva et al. 2011; Aponte et al. 2016). The following advantages of this process are usually mentioned: 1) heat-sensitive materials can be dried under atmospheric pressure and at low temperatures; 2) this allows pilot and large-scale production in continuous operation with relatively simple equipment; 3) spray-dried products are usually homogeneous; 4) different types of carriers and protective compounds can be used to enhance microbial survival (Fu and Chen 2011; Aponte et al. 2016). The process of spray drying may be divided into three main steps. During the first step, the liquid feed is atomized and fine droplets form. At the next stage, the droplets interact with drying gas at an appropriate temperature, and the solvent is vaporized, generating solid particles. In the third step, the dried particles are separated from the drying gas and collected (Casas-Godoy et al. 2021).

Similar to other industrial processes related to the dehydration of yeast, spray drying may also lead to rather serious changes in as well as damage to DNA, ribonucleic acid (RNA), cell walls, the plasma membrane, and other organelles (Labuza et al. 1970; Santivarangkna et al. 2008; Aponte et al. 2016). Therefore, some protective compounds, such as sugars, polyols, and amino acids, are recommended to be incorporated into coating materials to increase yeast survival during spray drying. Sucrose, sorbitol, and trehalose showed a protective effect against osmosis, temperature, and salt and oxidative stress (Suryabhan et al. 2019).

In a number of studies, the optimal conditions for the spray drying of several kinds of yeast were investigated. For example, to obtain dry brewer's yeast with rather high viability, it was recommended to add 10% maltodextrins and to use the following temperatures: inlet—145 °C, outlet—60 °C (Luna-Solano et al. (2000, 2003, 2005). Another study resulted in obtaining high-quality dry yeast when a soluble starch was used as a protective compound but inlet temperature was lowered to 120 °C with the same outlet temperature (60 °C). The further decrease of inlet/outlet temperatures to 110/45 °C also gave rather high viability of dry preparations, but the level of remaining water in these powders was higher than is usually recommended; as result, it can be unfavorable for the long-term storage of such dry preparations (Gomez-Pastor et al. 2011). In some papers, experiments on spray-drying process optimization for yeasts used as biocontrol agents are described. For example, the optimal values for the drying process of *Meyerozyma caribbica* were trehalose at 7.75% w/v and an inlet temperature of 112.5 °C, which reached 95.4% viability (Aguirre-Güitrón et al. 2018). The optimal conditions found were 113 °C dryer inlet temperature, 3.8% soy protein isolates, and 0.06% sodium alginate, giving viability of 97.8%. The optimal conditions found for another biocontrol yeast, *Meyerozyma guilliermondii*, were 113 °C dryer inlet temperature, 3.8% soy protein isolates, and 0.06% sodium alginate, resulting in viability of 97.8% (López-Cruz et al. 2020).

Some researchers separate thermal and dehydration inactivation in their studies on the main factors affecting yeast damage during the drying procedure. They showed that yeast fermentative activity was lost very sharply when drying temperature was over 40 °C, but when it was at the 30 °C level, thermal inactivation was negligible. Dehydration inactivation also occurred at such a low temperature. It was also concluded that, although it is difficult to control the drying rate for spray drying, low dehydration rates result in higher activity of dried yeast preparations (Fujii et al. 2011).

The possibilities for the optimization of the spray-drying process are discussed in a number of papers (Luna-Solano et al. 2000; Luna-Solano et al. 2003, 2005). A combination of simulation with experimental treatments of spray drying process was developed in order to minimize the energy resources required to obtain a production of spray dried brewer's yeast at given viability and moisture. Cost function, viability, output moisture and production were related with process variables, both empirically and by simulation with a nonlinear space state. This developed methodology is not specific to a definite technical construction of the dryers but is rather general for yeast spray drying (Luna-Solano et al. 2005). An approach to determine the kinetics of degradation of the substances, as a function of the current state of the material (temperature, residual moisture content and rates of their changes) during the spray drying process of yeasts is described by Gordienko et al. (2015). The obtained kinetic equations were used to predict the quality of yeast biomass dried using pilot-scale spray equipment (Gordienko et al. 2015).

Studies on the efficient storage of dry yeast preparations produced using the spray-drying method showed that there can be losses in viability from 10 to 25% per year at 20 °C, but at the same time, there was no revealed significant changes in the viability of dry yeast preparations for 3 years when they were stored at 4 °C in a dry atmosphere (Gomez-Pastor et al. 2011).

Spray drying is used in the food industry for microencapsulation purposes. Microencapsulation is necessary to cover bioactive compounds with protective materials, which can protect the bioactive core from various environmental stress factors (Huq et al. 2013). In the case of the yeast *Saccharomyces boulardii*, this process provides a possibility for their safe passing through the gastrointestinal system of humans and animals and for the use of this yeast as an efficient probiotic. It can be used also as a preventative and therapeutic agent for the curing of diarrhea (Holzapfel and Schillinger 2002; Kuhle et al. 2005; Czerucka et al. 2007; McFarland 2010; Rajkowska et al. 2012). *S. boulardii* was microencapsulated by spray drying in a number of studies (Joshi and Thorat 2011; Duongthingoc et al. 2013a, b). Gelatin and gum Arabic are recommended as promising protective "wall" materials for *S. boulardii* microencapsulation. An inlet temperature of 125 °C should be preferred in this spray-drying microencapsulation process (Arslan et al. 2015). The microencapsulation of epiphytic coffee yeasts by spray drying was shown as promising for use in coffee fermentation. In these studies, it was revealed that maltodextrin was efficient as a wall material in the microencapsulation of the yeasts *S. cerevisiae* and *Torulaspora delbrueckii*. Maltose incorporation increased the cell viability of *S. cerevisiae* during storage (Martins et al. 2022a, b).

12.3.3 Air Drying of Yeast

Some researchers prefer to use air-blast drying, which, according to their findings, can lower the cost fivefold, decrease cell damage during the drying procedure, as well as provide easier control of the moisture of yeast preparations compared with other drying methods (Santivarangkna et al. 2007; Lee et al. 2016). In spite of these advantages, a rather small number of studies related to air-blast dried yeast have been conducted. One of them was devoted to the optimization of the production of *Saccharomyces cerevisiae* and non-*Saccharomyces* (*Hanseniaspora uvarum* and *Issatchenkia orientalis*) yeast starters at the industrial level using this method (Lee et al. 2016). The objective of this work also included the selection of efficient excipients and protective compounds for the treatment of dehydration, which are very important because the intracellular accumulation of appropriate solutes can significantly improve microbial cell survival during and after the drying process (Kets et al. 1996; Champagne et al. 2012; Lee et al. 2016 etc.). Seven types of sugars were studied in this work with the goal of improving the viability of air-dried yeast cells. The best results were obtained with the use of 10% trehalose, 10% sucrose, and 10% glucose. The viability of these dehydrated yeasts was at 80–97%. Also, after 3 months of storage, *S. cerevisiae* and *H. uvarum* still had good survival rates, whereas the viability of *I. orientalis* decreased rather significantly. Also, after their storage at 4 °C, dried *S. cerevisiae* and *H. uvarum* showed metabolic activities similar to those of nondried yeast cells (Lee et al. 2016).

Air drying or convective drying was used for the development of a novel approach to yeast cell immobilization (Rapoport et al. 2011). As a result of this technology stable in technological processes, biotechnologically active and economically efficient immobilized yeast preparations were obtained. It was shown that their use might also be more efficient than the utilization of free microbial preparations, for example, in such biotechnological processes as the purification of wastewater from heavy metals (Rapoport et al. 2011). Yeast preparations immobilized on hydroxylapatite or chamotte using this method with drying procedure were tested also in bioethanol production. In these experiments, it was shown that immobilized preparations produced the same amount of ethanol as free cells after 36 hours of fermentation. At the same time, during the early stages of fermentation, immobilized yeast cells produced ethanol at a higher rate than free cells. Yeast preparations immobilized on both supports (hydroxylapatite and chamotte) were successfully used in at least six sequential batch fermentations without any loss of biotechnological activity (Borovikova et al. 2014a). A special study was performed with the goal of determining whether the dehydration procedure used in this new immobilization method noticeably decreased the viability of yeast cells in immobilized preparations. Various yeasts were used in this research: *S. cerevisiae* cells that were rather sensitive to dehydration and had been aerobically grown in an ethanol-containing medium, a recombinant strain of *S. cerevisiae* grown in aerobic conditions which was completely nonresistant to dehydration, and an anaerobically grown bakers' yeast strain of *S. cerevisiae*, as well as a fairly resistant *Pichia pastoris* strain. The performed experiments showed that the immobilization of all these strains essentially increased their resistance to dehydration-rehydration treatment. The increase in cell viability (compared with control cells dehydrated in similar conditions) was from 30 to 60%. It is concluded that a new immobilization method, which includes a dehydration stage, does not lead to a critical loss of yeast cell viability. Correspondingly, there is no risk of losing the biotechnological activities of immobilized preparations. At the same time, this research showed a new and interesting possibility of producing dry, active yeast preparations also for those strains that are very sensitive to dehydration and that can be used in biotechnology in an immobilized form. Finally, immobilization with the drying approach can also be used for the development of efficient methods for the long-term storage of biotechnologically active recombinant yeast strains (Borovikova et al. 2014b).

A similar approach to yeast immobilization with a dehydration stage was recently used in other studies, and it was found that five yeast strains representing *S. cerevisiae*, *Hanseniaspora uvarum*, and *Issatchenkia orientalis* entrapped in 2% air-blast-dried Ca-alginate beads with protective agents had rather high survival rates after storage at 4 °C for 3 months. The protective agents efficiently used in this research were 10% skim milk containing 10% sucrose or 10% raffinose or 10% trehalose or 10% glucose. In all these cases, the initial viability of immobilized dried yeasts was at least 87% and higher. After storage for 3 months at 4 °C, the survival rate of these yeast preparations was maintained at 51% at least (Kim et al. 2017).

12.3.4 Rehydration Stage

It is well known that the rehydration process is one of the critical stages in the reactivation of dehydrated organisms, which can help restore their active state but, in inadequate conditions, can increase damage to their cells, which can also lead to their death. Therefore, it is important to find ways to promote the restoration of the conformation of macromolecules, the state of membrane lipids and proteins, the molecular organization of membranes, which, in turn, will lead to a decrease in membrane permeability and loss of intracellular compounds during the rehydration stage (Crowe et al. 1984, 1989, 1992; Beker and Rapoport 1987). Such studies were performed in laboratory conditions also with dehydrated yeast cells (Beker and Rapoport 1987; Beker et al. 1984; Rapoport et al. 1988). To understand what should be recommended for the industrial use of active dry yeast preparations for the production of wine, the best yeast rehydration conditions for ensuring good cell viability and vitality before their inoculation into the must were revealed. In this study, the biochemical and biophysical behavior of dehydrated yeast after their rehydration in various media under oenological conditions (incubation at 37 °C for 30 min) were investigated. It was shown that the addition of magnesium to the rehydration medium provided the highest vitality rate as well as significantly diminished the period necessary for the reactivation of the cells and the beginning of their biotechnological activity (approximately 60%) (Rodríguez-Porrata et al. 2008). The magnesium rehydrating conditions revealed in this study may definitely have a broad industrial application.

There are also some other additives to the rehydration media that can influence the restoration of high viability and vitality, which, according to some recent data (Vaudano et al. 2014), better correspond to the fermentation activities of active dry yeast preparations. They include ergosterol (Soubeyrand et al. 2005); inactive dry yeast, which contains a lot of active compounds (Vaudano et al. 2014); ammonium (Vaudano et al. 2014); and ascorbic acid, which is most probably efficient due to its antioxidant activity (Díaz-Hellín et al. 2013).

It is known that there are a lot of factors that can influence the viability of cells at the rehydration stage. They include the duration of the rehydration period, the rehydration temperature (Poirier et al. 1999), and the pH of the medium (Zikmanis et al. 1984). These factors probably affect the structure of the yeast membrane by altering the permeability and flow of molecules and ions, resulting in higher or lower vitality/viability of the rehydrated cells (Attfield 1997; Poirier et al. 1999). Based on some results, the mechanical stirring of dry yeast preparations during rehydration may lead to a decrease in the number of vital cells by 5–7% and the consumption of sugars in fermentation by 18–26%. Other studies revealed the possibility of a weak positive effect of slight stirring during rehydration (reviewed by Ferrarini et al. 2007), but the experiments of Attfield et al. (2000) showed that an increase in the number of viable cells may be reached without presoaking and stirring. A detailed study performed by Ferrarini et al. (2007) showed that dispersion procedures play an important role in the recovery of dry yeast preparations. The authors of this research suppose that stirring is necessary for breaking the clusters of yeast cells, which in turn would facilitate easier water absorption, positively influence the decrease of osmotic pressure in cells' cytoplasm, and avoid the leakage of cell material. The other proposal or recommendation of these researchers is extending the usual duration (30 min) for the rehydration procedure to 120 min at 38 °C. At the same time, it is necessary to keep in mind that these conditions may be suitable only for the strains used in this study; therefore, such conclusions should most probably be confirmed also in other conditions and with other strains of dry wine yeast preparations.

Some researchers suppose that the additional procedure of "propagation" or "conditioning" of active dry yeast may reduce lag times for fermentation and, correspondingly, the overall fermentation duration. This issue was studied by Bellissimi and Ingledew (2005), who also introduced a new term—metabolic acclimatization—to describe the events taking place during the process of conditioning. At the same time, their research showed that metabolic acclimatization was not necessary, at least prior to alcoholic fermentation. Lag times were reduced, but overall fermentation times were unaffected (Bellissimi and Ingledew 2005).

12.3.5 Construction of New Strains

Another direction in the studies related to the improvement of the quality of industrially produced active dry yeast preparations is the construction of new biotechnologically efficient yeast strains that are very resistant to the dehydration process. One successful example of this approach is the recent development of a new strain for the production of bioethanol (Zheng et al. 2013). In this work, researchers applied the method of whole-genome shuffling, in which gene recombination is achieved at different positions throughout the whole genome through recursive protoplast fusion or crossing (Gong et al. 2009; Zheng et al. 2013). In this process, genetic diversity is generated, and resultant mutants with desired phenotypes can be selected using specific screening methods (Patnaik et al. 2002; Zhang et al. 2002; Zheng et al. 2013). The application of this approach made it possible for the authors of this work to obtain a novel yeast strain with high resistance to dehydration treatment. During the subsequent analysis of this novel strain and its differences from the initial strain, a lot of proteomics and physiological changes were revealed in this new strain. In total, 479 genes had different expression levels. Further studies showed that 299 upregulated genes in the new strain are linked to ribosome biosynthesis, proteasomes, glycolysis/gluconeogenesis, and secondary metabolite biosynthesis. At the same time, 180 downregulated genes were enriched in pathways for oxidative phosphorylation, endocytosis, RNA transport, and vitamin B6 metabolism (Zheng et al. 2013). Besides that, it was also revealed that this new strain accumulates higher amounts of trehalose, which can protect macromolecule conformation and the molecular organization of various intracellular membranes, including the plasma membrane. As was shown earlier in this review, it is one of the main factors that may determine the stability of cells at dehydration-rehydration. Very interesting features of the new strain were revealed during a comparison of fatty acid content in the initial and new strains. It appeared that in normal conditions, there were no distinct differences in the four main fatty acid ratios between these two strains. At the same time, after treatment, there was found an increased content of unsaturated fatty acids (palmitoleic acid, oleic acid, and linoleic acid) in both strains, resulting in 17.6% and 24.6% increase in the unsaturation ratios of the initial and new strains, respectively. The new strain had a higher proportion of C16:1 and a lower proportion of C18:1. It is supposed that a higher amount of palmitoleic acid (C16:1) content may be important for the maintenance of membrane fluidity (Zheng et al. 2013). The new strain also accumulated 14.6% more ergosterol (Zheng et al. 2013), which, as shown earlier, stabilizes the molecular organization of the plasma membrane at the dehydration-rehydration of yeast cells (Dupont et al. 2011).

One more interesting attempt to construct a new efficient and dehydration-resistant strain was performed by Kotarska and colleagues (Kotarska et al. 2011). These researchers, by electrofusion between *S. cerevisiae* and *S. diastaticus*, constructed a new strain with rather high amylolytic activities for industrial use in ethanol production. They checked the technological efficiency and stability of the biotechnological application of this strain after its dehydration. Very good results were obtained in such experiments. This dry distillery yeast showed high fermentation activity and high ethanol production. It was concluded that the application of this new strain in the form of active dry yeast in distillery practice significantly lowered the costs of spirit production by saving 30% of glucoamylase preparations used for starch saccharification. Besides that, the authors of this research confirmed that the use of active dry yeast preparation considerably simplifies the technological process in distillery. The application of this strain in the form of dry preparation makes the start of fermentation at an optional time easier as well as eliminates the possibility of mash infection, which sometimes takes place when conventional yeast cultures are applied (Kotarska et al. 2011).

12.3.6 Application of Active Dry Yeasts

The most traditional application of active or instant dry yeasts is in classic technologies and biotechnologies. These include the production of bread, beer, ethanol, and wine (Fig. 12.2). The simplicity of its transportation and storage and its long shelf life are the main reasons that dry yeast has becomes more and more popular in these industries. Several reviews describe different issues relating to these classical technologies (Strumillo and Kudra 1986; Adamiec et al. 1995; Beker and Rapoport 1987; Schmidt and Henschke 2015; etc.).

During the last two decades, there have been a number of new and interesting ideas and proposals for the additional and nonconventional use of dry yeast preparations (Rapoport et al. 2016). They include the biomedical, pharmaceutical, agricultural, environmental, and possible chemical applications of dry yeasts (Fig. 12.2). One such biomedical application, already mentioned above, is the use of the dry microencapsulated yeast *S. boulardii* as an efficient probiotic in medicine (Arslan et al. 2015). Another biomedical application of dry yeast is its use as a drug delivery system. This idea is based on the use of dry recombinant yeasts, which can either carry out a bioconversion reaction or produce compounds of interest directly in the

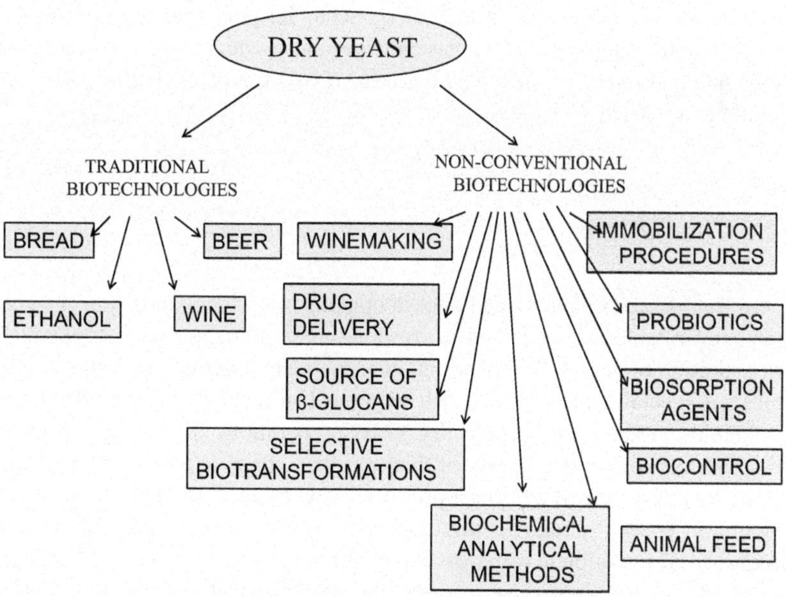

Fig. 12.2 Use of dry yeasts in traditional and nonconventional biotechnologies (Rapoport et al. 2016, modified)

digestive environment (Blanquet et al. 2005). One more interesting example is the possibility of using dry yeast as a source of efficient immunomodulators—β-glucans (Liepins et al. 2015). Dry yeasts are already used in animal feeds because the yeast biomass has a good composition of amino acids and has high levels of lysine, threonine, leucine, and B-complex vitamins (Rapoport et al. 2016). Because of the changes in cell wall surface characteristics during yeast dehydration, it appears that they become much more active in the sorption of various pollutants, including such dangerous substances as heavy metals. This makes various dry yeasts very promising agents for the protection and/or purification of the environment (Muter et al. 2001a, b, 2002; Muter and Rapoport 2014). The use of the drying method to obtain very stable and active immobilized preparations of dry yeasts was already mentioned above (Rapoport et al. 2011; Kim et al. 2017). Some yeasts possess biocontrol activities against various fruit and cereal grain diseases. Dry preparations of these yeasts allow for their long-term storage without the loss of activity and for a decrease in the cost of their transportation to various countries and regions (Melin et al. 2007, 2011; Rapoport et al. 2016). An interesting application of dry yeasts was recently proposed during the development of a new efficient yeast model system for the rapid evaluation of the possible efficiency of various chemical compounds (Borovikova et al. 2015). Dry yeasts can also be used in some biochemical analytical methods and in selective biotransformations (Mowshowitz 1976; Buzzini et al. 2012; Tsuda et al. 2012; Forti et al. 2015; etc.). Then in winemaking, besides the already classical use of dry yeasts as starter cultures, they can also be successfully applied for other goals. This nonconventional application can improve the sensorial characteristics of wine and may be efficient for the removal of the unpleasant smell of volatile phenols. They have effects on wine fermentation, composition, and organoleptic characteristics, as well as on wine safety, which depends on the various antioxidant compounds produced by the yeast (Pozo-Bayon et al. 2009; Rodriguez-Bencomo et al. 2016).

12.4 Conclusions

The last four decades have helped a lot in opening our understanding of the general mechanisms of anhydrobiosis and, correspondingly, of the events that take place at the dehydration of yeasts, during which they can be transferred into a state of anhydrobiosis. Especially, big progress has been reached recently. Accumulated knowledge enables producers of active dry yeast preparations to find new interesting approaches to improve the quality of their products. In spite of very high quality of baker's active (or instant) dry yeast produced now by leading yeast companies this problem still is very actual for the case of brewer's, ethanol and wine yeasts. In addition to the undoubted value of the acquired information for the purposes of practical biotechnology it is very important also for basic science. We know now much more than earlier about the role of water for the maintenance of biological systems, structures, and macromolecules. Studies of anhydrobiosis considerably developed our knowledge on the molecular organization of biological membranes and on the intracellular protective reactions which sometimes are rather unexpected for us. We understood that there are a lot of various alternatives for the protection of the most important functions of live organisms in the unfavorable conditions and that the losses of some protective compounds are not critical because they may be substituted by the rapid synthesis of other ones which will perform necessary protection. What is more, theoretical ideas obtained from studies on anhydrobiosis in yeast (which is the optimal model for any eukaryotic cell) can be applied to higher organisms, including animals and people. They can be used, for example, for medical purposes for the special artificial increase or improvement of organisms' resistance in the case of various diseases or complicated therapeutic treatments and surgical procedures. At the same time, it is necessary to underline once more that we are now still at the early stages of the understanding of the mechanisms of anhydrobiosis and potential of live organisms in the unfavorable conditions of the environment or physical and chemical treatments. John H. Crowe, the most well-known world leader in the studies of anhydrobiosis, in 1971 and in 2015 published two big reviews with similar titles—"Anhydrobiosis: An Unsolved Problem" indicating a number of problems which still have to be solved by the researchers of this unique phenomenon of live nature (Crowe 1971, 2015). And it is clear that already in the next few years, we have to expect very rapid progress in this direction.

Acknowledgments This work was supported by the grant from University of Latvia No. Y5-AZ20-ZF-N-270.

References

Abadias M, Benabarre A, Teixido N, Usall J, Vinas I (2001) Effect of freeze-drying and protectants on viability of the biocontrol yeast *Candida sake*. Int J Food Microbiol 65:173–182

Adamiec J, Kaminski W, Markowski AS, Strumillo C (1995) Drying of biotechnological products. In: Mujumdar AS (ed) Handbook of industrial drying. Marcel Dekker, New York, pp 775–808

Aguirre-Güitrón L, Calderón-Santoyo M, Ortiz-Basurto RI, Bautista-Rosales PU, Ragazzo-Sánchez JA (2018) Optimisation of the spray drying process of formulating the post-harvest biocontrol agent *Meyerozyma caribbica*. Biocontrol Sci Tech 28:574–590

Akbari H, Karimia K, Lundin M, Taherzadeh MJ (2012) Optimization of baker's yeast drying in industrial continuous fluidized bed dryer. Food Bioprod Process 90:52–57

Ale CE, Otero MC, Pasteris SE (2015) Freeze-drying of wine yeasts and *Oenococcus oeni* and selection of the inoculation conditions after storage. J Bioprocess Biotech 5:248

Aponte M, Troianiello GD, Di Capua M, Romano R, Blaiotta G (2016) Impact of different spray-drying conditions on the viability of wine *Saccharomyces cerevisiae* strains. World J Microbiol Biotechnol 32:13

Arslan S, Erbas M, Tontul I, Topuz A (2015) Microencapsulation of probiotic *Saccharomyces cerevisiae* var. *boulardii* with different wall materials by spray drying. LWT – Food Sci Technol 63:685–690

Attfield PV (1997) Stress tolerance: the key to effective strains of industrial baker's yeast. Nat Biotechnol 15:1351–1357

Attfield PV, Kletsas S, Veal DA, Van Rooijen R, Bell PJL (2000) Use of flow cytometry to monitor cell damage and predict fermentation activity of dried yeasts. J Appl Microbiol 89:207–214

Barbosa AD, Savage DB, Siniossoglou S (2015) Lipid droplet–organelle interactions: emerging roles in lipid metabolism. Curr Opin Cell Biol 35:91–97

Bayrock D, Ingledew WM (1997) Fluidized bed drying of baker's yeast: moisture levels, drying rates, and viability changes during drying. Food Res Int 30:407–415

Beker MJ, Rapoport AI (1987) Conservation of yeasts by dehydration. Adv Biochem Engin/Biotechnol 35:127–171

Beker MJ, Blumbergs JE, Ventina EJ, Rapoport AI (1984) Characteristics of cellular membranes at rehydration of dehydrated yeast Saccharomyces cerevisiae. Eur J Appl Microbiol Biotechnol 19:347–352

Bellissimi E, Ingledew WM (2005) Metabolic acclimatization: preparing active dry yeast for fuel ethanol production. Process Biochem 40:2205–2213

Bespalov MM, Kolpako AI, Loiko NG, Doroshenko EV, Mulyukin AL, Kozlova AN, Varlamova EA, Kurganov BI, El-Registan GI (2000) The role of microbial dormancy autoinducers in metabolism blockade. Microbiology 69:217–223

Biryusova VI, Rapoport AI (1978) Cryofractographic investigation of the structure of yeast cells in an anabiotic state. Microbiology 47:245–251

Blanquet S, Garrait G, Beyssac E, Perrier C, Denis S, Hebrard G, Alric M (2005) Effects of cryoprotectants on the viability and activity of freeze dried recombinant yeasts as novel oral drug delivery systems assessed by an artificial digestive system. Eur J Pharm Biopharm 61:32–39

Bolotin-Fukuhara M, Dumas B, Gaillardin C (2010) Yeasts as a model for human diseases. FEMS Yeast Res 10:959–960

Borovikova D, Scherbaka R, Patmalnieks A, Rapoport A (2014a) Effects of yeast immobilisation on bioethanol production. Biotechnol Appl Biochem 61:33–39

Borovikova D, Rozenfelde L, Pavlovska I, Rapoport A (2014b) Immobilisation increases yeast cells resistance to dehydration-rehydration treatment. J Biotechnol 184:169–171

Borovikova D, Herynkova P, Rapoport A, Sychrova H (2014c) Potassium uptake system Trk2 is crucial for yeast cell viability during anhydrobiosis. FEMS Microbiol Lett 350:28–33

Borovikova D, Muiznieks I, Rapoport A (2015) New test-system based on the evaluation of yeast cells resistance to dehydration-rehydration stress. Open Biotechnol J 9:49–53

Borovikova D, Teparic R, Mrsa V, Rapoport A (2016) Anhydrobiosis in yeast: cell wall mannoproteins are important for yeast *Saccharomyces cerevisiae* resistance to dehydration. Yeast 33:247–353

Botstein D, Chervitz SA, Cherry JM (1997) Yeast as a model organism. Science 277:1259–1260

Buzzini P, Goretti M, Ponzoni C, Caselli E, Branda E, Cramarossa MR, Turchetti B, Forti L (2012) Asymmetric reduction of (4S)-(+)-carvone catalyzed by enoate reductases (ERs) expressed by non-conventional yeast (NCY) whole-cells. In: Whittall J, Sutton P (eds) Practical methods for biocatalysis and biotransformations. Wiley, New York, pp 104–108

Calahan D, Dunham M, DeSevo C, Koshland DE (2011) Genetic analysis of desiccation tolerance in *Saccharomyces cerevisiae*. Genetics 189:507–519

Camara AA, Santana AS (2021) Advances in yeast preservation: physiological aspects for cell perpetuation. Cur Opin Food Sci 38:62–70

Câmara AA, Nguyen TD, Jossier A, Endrizzi A, Saurel R, Simonin H, Husson F (2018) Improving total glutathione and trehalose contents in *Saccharomyces cerevisiae* cells to enhance their resistance to fluidized bed drying. Process Biochem 69:45–51

Capece A, Votta S, Guaragnella N, Zambuto M, Romaniello R, Romano P (2016) Comparative study of *Saccharomyces cerevisiae* wine strains to identify potential marker genes correlated to desiccation stress tolerance. FEMS Yeast Res 16:fow015

Casas-Godoy L, Arellano-Plaza M, Kirchmayr M, Barrera-Martínez I, Gschaedler-Mathis A (2021) Preservation of non-*Saccharomyces* yeasts: current technologies and challenges. Compr Rev Food Sci Food Saf 20:3464–3503

Champagne CP, Raymond Y, Simon JP (2012) Effect of water activity and protective solutes on growth and subsequent survival to air-drying of *Lactobacillus* and *Bifidobacterium* cultures. Appl Microbiol Biotechnol 95:745–756

Chen YC, Cheng TH, Lin WL, Chen CL, Yang WY, Blackstone C, Chang CR (2019) Srv2 is a pro-fission factor that modulates yeast mitochondrial morphology and respiration by regulating actin assembly. iScience 11:305–317

Chen CL, Chen YC, Huang WL, Lin S, Daugelavicius R, Rapoport A, Chang CR (2021) A crucial role of mitochondrial dynamics in dehydration resistance in *Saccharomyces cerevisiae*. Int J Mol Sci 22:4607

Crowe JH (1971) Anhydrobiosis: an unsolved problem. Am Nat 105:563–574

Crowe JH (2008) Trehalose as a "chemical chaperone": fact and fantasy. Adv Exp Med Biol 594:143–158

Crowe JH (2015) Anhydrobiosis: an unsolved problem with applications in human welfare. Subcell Biochem 71:263–280

Crowe JH, Crowe LM, Chapman D (1984) Preservation of membranes in anhydrobiotic organisms: the role of trehalose. Science 223:701–703

Crowe JH, Hoekstra FA, Crowe LM (1989) Membrane phase transitions are responsible for imbibitional damage in dry pollen. Proc Natl Acad Sci USA 86:520–523

Crowe JH, Hockstra FA, Crowe LM (1992) Anhydrobiosis. Annu Rev Physiol 54:579–599

Crowe JH, Carpenter JF, Crowe LM (1998) The role of vitrification in anhydrobiosis. Annu Rev Physiol 60:73–103

Cucinotta FA (2014) Space radiation risks for astronauts on multiple international space station missions. PLoS One 9:16–23

Czerucka D, Piche T, Rampal P (2007) Yeast as probiotics – *Saccharomyces boulardii*. Aliment Pharmacol Ther 26:767–778

Daud WRW (2008) Fluidized bed dryers - recent advances. Adv Powder Technol 19:403–418

Dauss E, Papouškova K, Sychrova H, Rapoport A (2021) Anhydrobiosis in yeast: role of cortical endoplasmic reticulum protein Ist2 in *Saccharomyces cerevisiae* cells during dehydration and subsequent rehydration. Antonie Van Leeuwenhoek 114:1069–1077

Debaste F, Halloin V, Bossart L, Haut B (2008) A new modeling approach for the prediction of yeast drying rates in fluidized beds. J Food Eng 84:335–347

Díaz-Hellín P, Úbeda J, Briones A (2013) Improving alcoholic fermentation by activation of *Saccharomyces* species during the rehydration stage. LWT – Food Sci Technol 50:126–131

Duongthingoc D, George P, Gorczyca E, Kasapis S (2013a) Studies on the viability of *Saccharomyces boulardii* within microcapsules in relation to the thermomechanical properties of whey protein. Food Hydrocoll 42:232–238

Duongthingoc D, George P, Katopo L, Gorczyca E, Kasapis S (2013b) Effect of whey protein agglomeration on spray dried microcapsules containing *Saccharomyces boulardii*. Food Chem 141:1782–1788

Dupont S, Beney L, Ritt JF, Lherminier J, Gervais P (2010) Lateral reorganization of plasma membrane is involved in the yeast resistance to severe dehydration. Biochim Biophys Acta 1798:975–985

Dupont S, Beney L, Ferreira T, Gervais P (2011) Nature of sterols affects plasma membrane behavior and yeast survival during dehydration. Biochim Biophys Acta 1808:1520–1528

Dupont S, Lemetais G, Ferreira T, Cayot P, Gervais P, Beney L (2012) Ergosterol biosynthesis: a fungal pathway for life on land? Evolution 66:2961–2968

Dupont S, Rapoport A, Gervais P, Beney L (2014) The survival kit of *Saccharomyces cerevisiae* for anhydrobiosis. Appl Microbiol Biotechnol 98:8821–8834

Duskova M, Borovikova D, Herynkova P, Rapoport A, Sychrova H (2015) The role of glycerol transporters in yeast cells in various physiological and stress conditions. FEMS Microbiol Lett 362:1–8

Eleutherio E, Panek A, De Mesquita JF, Trevisol D, Magalhães R (2015) Revisiting yeast trehalose metabolism. Curr Genet 61:263–274

El-Registan GI, Mulyukin AL, Nikolaev YA, Stepanenko IY, Kozlova AN, Martirosova EI, Shanenko EF, Strakhovskaya MG, Revina AA (2005) The role of microbial low-molecular-weight autoregulatory factors (alkylhydroxybenzenes) in resistance of microorganisms to radiation and heat shock. J Adv Space Res 36:1718–1728

Espindola AS, Gomes DS, Panek AD, Eleutherio EC (2003) The role of glutathione in yeast dehydration tolerance. Cryobiology 47:236–241

Ferrarini R, Bocca E, Cavazza A (2007) Mechanical dispersion procedures improve the rehydration of active dry yeast. Enz Microb Technol 40:1251–1255

Finn DA, Stewart GG (2002) Fermentation characteristics of dried brewers yeast: effect of drying on flocculation and fermentation. J Am Soc Brew Chem 60:135–139

Forti L, Di Mauro S, Cramarossa MR, Filippucci S, Turchetti B, Buzzini P (2015) Non-conventional yeasts whole cells as efficient biocatalysts for the production of flavors and fragrances. Molecules 20:10377–10398

Franca MB, Panek AD, Eleutherio ECA (2005) The role of cytoplasmic catalase in dehydration tolerance of *Saccharomyces cerevisiae*. Cell Stress Chaperones 10:167–170

Fu N, Chen XD (2011) Towards a maximal cell survival in convective thermal drying processes. Food Res Int 44:1127–1149

Fujii S, Sakamoto Y, Aso T, Aktas T, Yoshimoto N, Yamamoto S (2011) Drying of yeasts—factors affecting inactivation during drying. Drying Technol Int J 29:1981–1985

Garre E, Raginel F, Palacios A, Julien A, Matallana E (2010) Oxidative stress responses and lipid peroxidation damage are induced during dehydration in the production of dry active wine yeasts. Int J Food Microbiol 136:295–303

Gelinas P (2019) Active dry yeast: lessons from patents and science. Compr Rev Food Sci Food Saf 18:1227–1255

Giacomello M, Pyakurel A, Glytsou C, Scorrano L (2020) The cell biology of mitochondrial membrane dynamics. Nat Rev Mol Cell Biol 21:204–224

Golovina EA, Golovin AV, Hoekstra FA, Faller R (2009) Water replacement hypothesis in atomic detail—factors determining the structure of dehydrated bilayer stacks. Biophys J 97:490–499

Golovina E, Golovin A, Hoekstra F, Faller R (2010) Water replacement hypothesis in atomic details: effect of trehalose on the structure of single dehydrated POPC bilayers. Langmuir 26:11118–11126

Gómez-Pastor R, Pérez-Torrado R, Cabiscol E, Ros J, Matallana E (2010) Reduction of oxidative cellular damage by overexpression of the thioredoxin *TRX2* gene improves yield and quality of wine yeast dry active biomass. Microb Cell Fact 9:9

Gomez-Pastor R, Perez-Torrado R, Garre E, Matallana E (2011) Recent advances in yeast biomass production. In: Matovic MD (ed) Biomass-detection, production and usage. InTech Publisher, Rijeka

Gomez-Pastor R, Perez-Torrado R, Cabiscol E, Ros J, Matallana E (2012) Engineered Trx2p industrial yeast strain protects glycolysis and fermentation proteins from oxidative carbonylation during biomass propagation. Microb Cell Factories 11:4

Gong JX, Zheng HJ, Wu ZJ, Chen T, Zhao XM (2009) Genome shuffling: progress and applications for phenotype improvement. Biotechnol Adv 27:996–1005

Gordienko MG, Voynovskiy AA, Menshutina NV (2015) Design of a kinetic model for degradation of substances during spray drying using yeast biosuspension, oil-in-water emulsion, and alumosilicate suspension. Dry Technol 33:24–36

Grabowski S, Mujumdar AS, Ramaswamy HS, Strumillo C (1997) Evaluation of fluidized versus spouted bed drying of baker's yeast. Dry Technol 15:625–634

Grant CM (2001) Role of the glutathione/glutaredoxin and thioredoxin systems in yeast growth and response to stress conditions. Mol Microbiol 39:533–541

Guzhova I, Krallish I, Khroustalyova G, Margulis B, Rapoport A (2008) Dehydration of yeast: changes in the intracellular content of Hsp 70 family proteins. Process Biochem 43:1138–1141

Herrero E, Ros J, Belli G, Cabiscol E (2008) Redox control and oxidative stress in yeast cells. Biochim Biophys Acta 1780:1217–1235

Hijazi I, Knupp J, Chang A (2020) Retrograde signaling mediates an adaptive survival response to endoplasmic reticulum stress in *Saccharomyces cerevisiae*. J Cell Sci 133:jcs241539

Holzapfel WH, Schillinger U (2002) Introduction to pre- and probiotics. Food Res Int 35:109–116

Huq T, Khan A, Khan RA, Riedl B, Lacroix M (2013) Encapsulation of probiotic bacteria in biopolymeric system. Critic Rev Food Sci Nutrit 53:909–916

Isono Y, Araya G, Hoshino A (1995) Immobilization of *Saccharomyces cerevisiae* for ethanol fermentation on γ-alumina particles using a spray-dryer. Process Biochem 30:743–746

Jenkins DM, Powell CD, Smart KA (2010) Dried yeast: impact of dehydration and rehydration on brewing yeast DNA integrity. J Am Soc Brew Chem 68:132–138

Joshi VS, Thorat BN (2011) Formulation and cost-effective drying of probiotic yeast. Dry Technol 29:749–757

Keilin D (1959) The problem of anabiosis or latent life: history and current concept. The Leeuwenhoek lecture. Proc Roy Soc London B150:148–191

Kets EPW, Teunissen PJM, de Bont JAM (1996) Effect of compatible solutes on survival of lactic acid bacteria subjected to drying. Appl Environ Microbiol 62:259–261

Khroustalyova G, Rapoport A (2019) Anhydrobiosis in yeasts: changes in mitochondrial membranes improve the resistance of *Saccharomyces cerevisiae* cells to dehydration-rehydration. Fermentation 5:82

Khroustalyova GM, Adler L, Rapoport AI (2001) Exponential growth phase cells of the osmotolerant yeast *Debaryomyces hansenii* are extremely resistant to dehydration stress. Process Biochem 36:1163–1166

Kim D-H, Lee S-B, Park H-D (2017) Effect of air-blast drying and the presence of protectants on the viability of yeast entrapped in calcium alginate beads with an aim to improve the survival rate. Appl Microbiol Biotechnol 101:93–102

Klug L, Daum G (2014) Yeast lipid metabolism at a glance. Yeast FEMS Res 14:369–388

Koni M, Yuzgec U, Turker M, Dincer H (2009) Optimal quality control of baker's yeast drying in large scale batch fluidized bed. Chem Eng Process 48:1361–1370

Kotarska K, Kłosowski G, Czuprynski B (2011) Characterization of technological features of dry yeast (strain I-7-43) preparation, product of electrofusion between *Saccharomyces cerevisiae* and *Saccharomyces diastaticus*, in industrial application. Enzyme Microb Technol 49:38–43

Krallish I, Jeppsson H, Rapoport A, Hahn-Hagerdal B (1997) Effect of xylitol and trehalose on dry resistance of yeasts. Appl Microbiol Biotechnol 47:447–451

Kuhle AVDA, Skovgaard K, Jespersen L (2005) In vitro screening of probiotic properties of *Saccharomyces cerevisiae* var. *boulardii* and food-borne *Saccharomyces cerevisiae* strains. Int J Food Microbiol 101:29–39

Kuliešiene N, Žukiene R, Khroustalyova G, Chang CR, Rapoport A, Daugelavicius R (2021) Changes in energy status of *Saccharomyces cerevisiae* cells during dehydration and rehydration. Microorganisms 9:444

Kulikova-Borovikova D, Lisi S, Dauss E, Alamae T, Buzzini P, Hallsworth JE, Rapoport A (2018) Activity of the a-glucoside transporter Agt1 in *Saccharomyces cerevisiae* cells during dehydration-rehydration events. Fung Biol 122:613–620

Labuza TP, Le Roux JP, Fan TS, Tannenbaum SR (1970) Engineering factors in single-cell protein production. Spray drying and cell viability. Biotechnol Bioeng 12:135–140

Lapointe MR, Laframboise T, Pirkkanen J, Tai TC, Lees SJ, Santa Maria SR, Tharmalingam S, Boreham DR, Thome C (2024) Protracted exposure to a sub-background radiation environment negatively impacts the anhydrobiotic recovery of desiccated yeast sentinels. Health Phys 126:397–404

Lee S-B, Choi W-S, Jo H-J, Yeo S-H, Park H-D (2016) Optimization of air-blast drying process for manufacturing *Saccharomyces cerevisiae* and non-*Saccharomyces* yeast as industrial wine starters. AMB Express 6:105

Lemetais G, Dupont S, Beney L, Gervais P (2012) Air-drying kinetics affect yeast membrane organization and survival. Appl Microbiol Biotechnol 96:471–480

Leslie SB, Teter SA, Crowe LM, Crowe JH (1994) Trehalose lowers membrane phase transitions in dry yeast cells. Biochim Biophys Acta 1192:7–13
Liddell CL, Gentry DM, Gilbert R, Marina D, Massaro Tieze S, Padgen MR, Akiyama K, Keenan K, Bhattacharya S, Santa Maria SR (2023) BioSentinel: validating sensitivity of yeast biosensors to deep space relevant radiation. Astrobiology 23:648–656
Liepins J, Kovacova E, Shvirksts K, Grube M, Rapoport A, Kogan G (2015) Drying enhances immunoactivity of spent brewer's yeast cell wall beta-D-glucans. J Biotechnol 206:12–16
Lodato P, Segovia de Huergo M, Buera MP (1999) Viability and thermal stability of a strain of *Saccharomyces cerevisiae* freeze-dried in different sugar and polymer matrices. Appl Microbiol Biotechnol 52:215–220
López-Cruz RL, Ragazzo-Sánchez JA, Calderón-Santoyo M (2020) Microencapsulation of *Meyerozyma guilliermondii* by spray drying using sodium alginate and soy protein isolate as wall materials: a biocontrol formulation for anthracnose disease of mango. Biocontrol Sci Tech 30:1116–1132
Lopez-Martinez G, Borull A, Poblet M, Rozes N, Cordero-Otero R (2014) Metabolomic characterization of yeast cells after dehydration stress. Int Microbiol 17:131–139
Luna-Solano G, Salgado-Cervantes MA, Garcia-Alvarado MA, Rodriguez-Jimenes G (2000) Improved viability of spray dried brewer's yeast by using starch (grits) and maltodextrin as processing aids. J Food Process Eng 23:453–462
Luna-Solano G, Salgado-Cervantes MA, Ramirez-Lepe M, Garcia-Alvarado MA, Rodriguez-Jimenes GC (2003) Effect of drying type and drying conditions over fermentative ability of brewer's yeast. J Food Process Eng 26:135–147
Luna-Solano G, Salgado-Cervantes MA, Rodríguez-Jimenes GC, García-Alvarado MA (2005) Optimization of brewer's yeast spray drying process. J Food Eng 68:9–18
Martins PMM, Batista NN, Santos LD, Ribeiro Dias D, Schwan RF (2022a) Microencapsulation of epiphytic coffee yeasts by spray drying using different wall materials: implementation in coffee medium. Int J Food Microbiol 379:109839
Martins PMM, Batista NN, Santos LD, Ribeiro Dias D, Schwan RF (2022b) Microencapsulation by spray drying of coffee epiphytic yeasts *Saccharomyces cerevisiae* CCMA 0543 and *Torulaspora delbrueckii* CCMA 0684. Braz J Microbiol 53:1565–1576
McFarland LV (2010) Systematic review and meta-analysis of *Saccharomyces boulardii* in adult patients. World J Gastroenterol 16:2202–2222
Melin P, Hakansson S, Schnurer J (2007) Optimization of liquid and dry formulations of the biocontrol yeast *Pichia anomala*. Appl Microbiol Biotechnol 73:1008–1016
Melin P, Schnurer J, Hakansson S (2011) Formulation and stabilization of the biocontrol yeast *Pichia anomala*. Antonie Van Leeuwenhoek 99:107–112
Morgan CA, Herman N, White PA, Vesey G (2006) Preservation of micro-organisms by drying; a review. J Microbiol Methods 66:183–193
Mowshowitz DB (1976) Permeabilization of yeast for enzyme assays: an extremely simple method for small samples. Anal Biochem 70:94–99
Mulyukin AL, Lusta KA, Gryaznova MN, Kozlova AN, Duzha MV, Duda VI, El-Registan GI (1996) Formation of resting cells by Bacillus cereus and Micrococcus luteus. Mikrobiologiia 65:683–689
Mulyukin AL, Demkina EV, Kozlova AN, Soina VS, El-Registan GI (2001) Synthesis of anabiosis autoinducers by non-spore-forming bacteria as a mechanism regulating their activity in soil and subsoil sedimentary rocks. Microbiology 70:535–541
Muter O, Rapoport A (2014) Some physiological aspects of chromium sorption by yeasts. Lambert Academic Publishing, Saarbrucken
Muter O, Millers D, Grigorjeva L, Ventina E, Rapoport A (2001a) Cr(VI) sorption by intact and dehydrated *Candida utilis* cells: differences in mechanisms. Process Biochem 37:505–511
Muter O, Patmalnieks A, Rapoport A (2001b) Interrelations of the yeast *Candida utilis* and Cr(VI): metal reduction and its distribution in the cell and medium. Process Biochem 36:963–970

Muter O, Lubinya I, Millers D, Grigorjeva L, Ventinya E, Rapoport A (2002) Cr (VI) sorption by intact and dehydrated *Candida utilis* cells in the presence of other metals. Process Biochem 38:123–131

Nakatogawa H (2020) Autophagic degradation of the endoplasmic reticulum. Proc Jpn Acad Ser B Phys Biol Sci 96:1–9

Nielsen J (2019) Yeast systems biology: model organism and cell factory. Biotechnol J 14:1800421

Nikolaev YA, Mulyukin AL, Stepanenko IY, El-Registan GI (2006) Autoregulation of stress response in microorganisms. Microbiology 75:420–426

Noti O, Vaudano E, Pessione E, Garcia-Moruno E (2015) Short-term response of different *Saccharomyces cerevisiae* strains to hyperosmotic stress caused by inoculation in grape must: RT-qPCR study and metabolite analysis. Food Microbiol 52:49–58

Novichkova AT, Rapoport AI (1984) State of the intracellular pool of free amino acids in dehydrated yeast organisms. Microbiology 53:1–4

Novo M, Beltran G, Rozes N, Guillamon J-S, Sokol S, Leberre V, Francois J, Mas A (2007) Early transcriptional response of wine yeast after rehydration: osmotic shock and metabolic activation. FEMS Yeast Res 7:304–316

Palmfeldt J, Hahn-Hagerdal B (2000) Influence of culture pH on survival of *Lactobacillus reuteri* subjected to freeze-drying. Int J Food Microbiol 55:235–238

Panek AD (1985) Trehalosc metabolism and its role in *Saccharomyces cerevisiae.* J Biotechnol 3:121–130

Panek AD (1995) Trehalose metabolism - new horizons in technological applications. Braz J Med Biol Res 28:169–181

Patnaik R, Louie S, Gavrilovic V, Perry K, Stemmer WPC, Ryan CM, Cardayre D (2002) Genome shuffling of *Lactobacillus* for improved acid tolerance. Nat Biotechnol 20:707–712

Petitjean M, Teste M-A, Francois JM, Parrou J-L (2015) Yeast tolerance to various stresses relies on the trehalose- 6P synthase (Tps1) protein, not on trehalose. J Biol Chem 290:16177–16190

Picazo C, Gamero-Sandemetrio E, Orozco H, Albertin W, Marullo P, Matallana E, Aranda A (2015) Mitochondria inheritance is a key factor for tolerance to dehydration in wine yeast production. Lett Appl Microbiol 60:217–222

Poirier I, Maréchal PA, Richard S, Gervais P (1999) Saccharomyces cerevisiae viability is strongly dependant on rehydration kinetics and the temperature of dried cells. J Appl Microbiol 86:87–92

Pozo-Bayon MA, Andujar-Ortiz I, Moreno-Arribas MV (2009) Scientific evidences beyond the application of inactive dry yeast preparations in winemaking. Food Res Int 42:754–761

Rajkowska K, Kunicka-Styczynska A, Rygala A (2012) Probiotic activity of *Saccharomyces cerevisiae* var. *boulardii* against human pathogens. Food Technol Biotechnol 50:230–236

Rapoport AI (1973) Rejection of areas of damaged cytoplasm by microorganisms in a state of anabiosis. Microbiology 42:317–318

Rapoport AI, Beker ME (1985) Changes in the surface charge of yeast cells during their dehydration and rehydration. Microbiology 54:362–365

Rapoport AI, Kostrikina NA (1973) Cytological investigation of an anabiosis state of yeast organisms. Proc Acad Sci USSR Biol 5:770–773. (in Russian)

Rapoport A, Muter O (1995) Biosorption of hexavalent chromium by yeasts. Process Biochem 30:145–149

Rapoport AI, Birjusova VI, Meissel MN (1973) The structure of the yeast cell nucleus in an anabiotic state. Rep Acad Sci USSR 213:708–710. (in Russian)

Rapoport AI, Markovskii AB, Beker ME (1982) Increase in the permeability of intracellular membranes on dehydration and rehydration of *Saccharomyces cerevisiae* yeast. Microbiology 51:707–710

Rapoport AI, Biryuzova VI, Beker ME (1983) Effect of dehydration on the surface structure of the yeast cell wall. Microbiology 52:204–207

Rapoport AI, Berestennikova ND, Yanovsky KA, Beker MJ (1986a) Changes in the shape and size of yeast cells upon their dehydration followed by reactivation. Microbiologiya 55:881–882. (in Russian)

Rapoport AI, Biryusova VI, Svetlichnaya TP (1986b) Freeze-fracture study of changes in vacuoles during dehydration of *Saccharomyces cerevisiae* cells. Microbiology 55:835–839

Rapoport AI, Puzyrevskaya OM, Saubenova MG (1988) Polyols and resistance of yeasts to dehydration. Microbiology 57:269–271

Rapoport AI, Khroustalyova GM, Camanis GJ, Beker MJ (1995) Yeast anhydrobiosis: permeability of the plasma membrane. Mikrobiologiya 64:275–278

Rapoport AI, Khroustalyova GM, Crowe LM, Crowe JH (2009) Anhydrobiosis in yeast: stabilization by exogenous lactose. Microbiology 78:624–629

Rapoport A, Borovikova D, Kokina A, Patmalnieks A, Polyak N, Pavlovska I, Mezinskis G, Dekhtyar Y (2011) Immobilisation of yeast cells on the surface of hydroxyapatite ceramics. Process Biochem 46:665–670

Rapoport A, Rusakova A, Khroustalyova G, Walker G (2014) Thermotolerance in *Saccharomyces cerevisiae* is linked to resistance to anhydrobiosis. Process Biochem 49:1889–1892

Rapoport A, Turchetti B, Buzzini P (2016) Application of anhydrobiosis and dehydration of yeasts for non-conventional biotechnological goals. World J Microbiol Biotechnol 32:104

Rapoport A, Golovina EA, Gervais P, Dupont S, Beney L (2019) Anhydrobiosis: inside yeast cells. Biotechnol Adv 37:51–67

Ratnakumar S, Tunnacliffe A (2006) Intracellular trehalose is neither necessary nor sufficient for desiccation tolerance in yeast. FEMS Yeast Res 6:902–913

Ratnakumar S, Hesketh A, Gkargkas K, Wilson M, Rash BM, Hayes A, Tunnacliffe A, Oliver SG (2011) Phenomic and transcriptomic analyses reveal that autophagy plays a major role in desiccation tolerance in *Saccharomyces cerevisiae*. Mol BioSyst 7:139–149

Ren Q, Brenner R, Boothby TC, Zhang Z (2020) Membrane and lipid metabolism plays an important role in desiccation resistance in the yeast *Saccharomyces cerevisiae*. BMC Microbiol 20:338

Ricker JV, Tsvetkova NM, Wolkers WF, Leidy C, Tablin F, Longo M, Crowe JH (2003) Trehalose maintains phase separation in an air-dried binary lipid mixture. Biophys J 84:3045–3051

Rodriguez-Bencomo JJ, Andujar-Ortiz I, Sanchez-Patan F, Moreno-Arribas MV, Pozo-Bayon MA (2016) Fate of the glutathione released from inactive dry yeast preparations during the alcoholic fermentation of white musts. Aust J Grape Wine Res 22:46–51

Rodríguez-Porrata B, Novo M, Guillamon J, Rozes N, Mas A, Cordero-Otero R (2008) Vitality enhancement of the rehydrated active dry wine yeast. Int J Food Microbiol 126:116–122

Rodrıguez-Porrata B, Lopez-Martinez G, Redon M, Sancho M, Mas A, Rozes N, Cordero-Otero R (2011) Enhancing yeast cell viability after dehydration by modification of the lipid profile. World J Microbiol Biotechnol 27:75–83

Rossignol T, Postaire O, Storaı J, Blondin B (2006) Analysis of the genomic response of a wine yeast to rehydration and inoculation. Appl Microbiol Biotechnol 71:699–712

Santa Maria SR, Marina DB, Massaro Tieze S, Liddell LC, Bhattacharya S (2023) BioSentinel: long-term *Saccharomyces cerevisiae* preservation for a deep space biosensor mission. Astrobiology 23:617–630

Santivarangkna C, Kulozik U, Foerst P (2007) Alternative drying processes for the industrial preservation of lactic acid starter cultures. Biotechnol Prog 23:302–315

Santivarangkna C, Kulozik U, Foerst P (2008) Inactivation mechanisms of lactic starter cultures preserved by drying processes. J Appl Microbiol 105:1–13

Schmidt SA, Henschke PA (2015) Production, reactivation and nutrient requirements of active dried yeast in winemaking: theory and practice. Aust J Grape Wine Res 21:651–662

Sęk W, Kot AM, Rapoport A, Kieliszek M (2023) Physiological and genetic regulation of anhydrobiosis in yeast cells. Arch Microbiol 205:348

Silva J, Freixo R, Gibbs P, Teixeira P (2011) Spray-drying for the production of dried cultures. Int J Dairy Technol 64:321–335

Singh J, Kumar D, Ramakrishnan N, Singhal V, Jervis J, Garst J, Slaughter S, DeSantis A, Potts M, Helm R (2005) Transcriptional response of *Saccharomyces cerevisiae* to desiccation and rehydration. Appl Environ Microbiol 17:8752–8763

Soubeyrand V, Luparia V, Williams P, Doco T, Vernhet A, Ortiz-Julien A, Salmon JM (2005) Formation of micella containing solubilized sterols during rehydration of active dry yeasts improves their fermenting capacity. J Agric Food Chem 53:8025–8032

Stepanenko IY, Strakhovskaya MG, Belenikina NS, Nikolaev YA, Mulyukin AL, Kozlova AN, Revina AA, El-Registan GI (2004) Protection of *Saccharomyces cerevisiae* against oxidative and radiation-caused damage by alkylhydroxybenzenes. Microbiology 73:163–169

Strumillo C, Kudra T (1986) Drying: principles applications and design. Gordon and Breach Science Publishers, New York

Suryabhan P, Lohith K, Anu-Appaiah KA (2019) Sucrose and sorbitol supplementation on maltodextrin encapsulation enhance the potential probiotic yeast survival by spray drying. LWT – Food Sci Technol 107:243–248

Takagi H (2008) Proline as a stress protectant in yeast: physiological functions, metabolic regulations, and biotechnological applications. Appl Microbiol Biotechnol 81:211–223

Takagi H, Sakai K, Morida K, Nakamori S (2000) Proline accumulation by mutation or disruption of the proline oxidase gene improves resistance to freezing and desiccation stresses in *Saccharomyces cerevisiae*. FEMS Microbiol Lett 184:103–108

Takagi H, Taguchi J, Kaino T (2016) Proline accumulation protects *Saccharomyces cerevisiae* cells in stationary phase from ethanol stress by reducing reactive oxygen species levels. Yeast 33:355–363

Tsuda H, Matsumoto T, Ishimi Y (2012) Time reduction of vitamin B6 and inositol assay by using lyophilized *Saccharomyces cerevisiae* ATCC 9080. J Nutr Sci Vitaminol 58:149–151

Turker M, Kanarya A, Yuzgec U, Kapucu H, Senalp Z (2006) Drying of baker's yeast in batch fluidized bed. Chem Eng Process 45:1019–1028

Vanderwaeren L, Dok R, Voordeckers K, Nuyts S, Verstrepen KJ (2022) Saccharomyces cerevisiae as a model system for eukaryotic cell biology, from cell cycle control to DNA damage response. Int J Mol Sci 23:11665

Vaudano E, Costantini A, Cersosimo M, Del Prete V, Garcia-Moruno E (2009) Application of real-time RT–PCR to study gene expression in active dry yeast (ADY) during the rehydration phase. Int J Food Microbiol 129:30–36

Vaudano E, Costantini A, Noti O, Garcia-Moruno E (2010) An RT-qPCR approach to study the expression of genes responsible for sugar assimilation during rehydration of active dry yeast. Food Microbiol 27:802–808

Vaudano E, Noti O, Costantini A, Garcia-Moruno E (2014) Effect of additives on the rehydration of *Saccharomyces cerevisiae* wine strains in active dry form: influence on viability and performance in the early fermentation phase. J Inst Brew 120:71–76

Ventina EJ, Saulite LA, Rapoport AI, Beker ME (1984) Electron-microscopic study of yeasts in a state of anabiosis and reactivated from this state. Microbiology 53:536–541

Ventina EJ, Saulite LA, Rapoport AI, Beker ME (1986) Lipid inclusions in cells and changes in them during dehydration and reactivation of yeasts. Microbiology 55:99–104

Voeltz GK, Rolls MM, Rapoport TA (2002) Structural organization of the endoplasmic reticulum. EMBO Rep 3:944–950

Wang C-W (2016) Lipid droplets, lipophagy and beyond. Biochim Biophys Acta 1861:793–805

Zhang Z, Zhang GR (2022) Chromosome-condensed G1 phase yeast cells are tolerant to desiccation stress. Microbial Cell 9:42–51

Zhang YX, Perry K, Vinci VA, Powell K, Stemmer WPC, Del Cardayre SB (2002) Genome shuffling leads to rapid phenotypic improvement in bacteria. Nature 415:644–646

Zheng D, Zhang K, Gao K, Liu Z, Zhang X, Li O, Sun J, Zhang X, Du F, Sun P, Qu A, Wu X (2013) Construction of novel *Saccharomyces cerevisiae* strains for bioethanol active dry yeast (ADY) production. PLoS One 8:e85022

Zikmanis PB, Auzan SI, Kruche RV, Zikmane MA, Auzinia LP (1984) Rehydration conditions for dehydrated *Saccharomyces cerevisiae* yeasts and cell viability. Mikrobiologiia 53:208–212

Chapter 13
Biosensors Based on Yeast/Fungal Cells

Alexandre Chamas, Martin Jähne, Uwe Wegner, Ha Thi Minh Pham, Kirsten Simon, Steffen Uhlig, and Gotthard Kunze

Contents

13.1 Introduction 380
13.2 Utilization of Receptors 382
13.2.1 Steroid Receptors 382
13.2.2 Other Nuclear Receptors 385
13.2.3 GPRC 385
13.3 Transduction of Biological Signal: Reporter Proteins 386
13.3.1 Utilization of Enzymes as Reporter Proteins 387
13.3.2 Direct Measurement of Emitted Light 389
13.3.3 Direct Measurement of Pigments 391
13.4 Applications of EDC Targeting Yeast Biosensors 391
13.4.1 Sensitivity of the Biosensors 391
13.4.2 Protocols 391
13.4.3 Applications with Real Samples 394

A. Chamas
Department of Microbial Pathogenicity Mechanisms, Leibniz Institute for Natural Product Research and Infection Biology, Jena, Germany

Cluster of Excellence Balance of the Microverse, Friedrich Schiller University Jena, Jena, Germany

M. Jähne · K. Simon · S. Uhlig
QuoData GmbH, Dresden, Germany

U. Wegner · G. Kunze (✉)
Leibniz Institute of Plant Genetics and Crop Plant Research (IPK), Seeland, OT Gatersleben, Germany
e-mail: kunzeg@ipk-gatersleben.de

H. T. M. Pham
INM – Leibniz-Institute for New Materials gGmbH, Saarbrücken, Germany

A. A. Sibirny (ed.), *Biotechnology of Yeasts and Filamentous Fungi*, Grand Challenges in Biology and Biotechnology,
https://doi.org/10.1007/978-3-031-74726-7_13

13.5 Emerging Biosensors Strategies. 395
13.6 Applications of *Blastobotrys raffinosifermentans* for the Effect-Based Determination of Hormonal Activity. 398
13.7 Conclusions. 399
References. 400

Abstract Yeast cells can be utilized as whole-cell biosensors for the detection of an array of target analytes. This technology is derived from the latest molecular biology techniques enabling the modification of the yeast genome to enhance the cells' metabolic properties. Current methods to produce correctly folded recombinant proteins, particularly receptors in yeast, are now highly successful, and the implementation of diverse reporting strategies is well documented. A myriad of yeast whole-cell biosensors are now utilized in various laboratories, and some are also commercially available. A major target of these biosensors is endocrine disruptors, a class of organic molecules that can disturb the vertebrate endocrine system and therefore pose a potential threat to the environment. Recent studies focused on building new biosensors using receptors such as G-protein-coupled receptors and the thyroid receptor. In this chapter, we will present a selection of newly developed biosensors and the diverse reporting strategies used to transduce the binding event into a measurable signal.

13.1 Introduction

A multitude of studies aim to measure the concentrations of molecules from the simplest gas to complicated organic molecules. These results are applicable to diverse areas such as environmental analytics, food production, and medical science. For example, the composition analysis of rivers, the atmosphere, food, or bodily fluids and effective threshold concentrations enabled the discovery of several compounds with negative impacts on the ecosystem, animals, and human health. This data was used by governments to establish recommendations related to the production, utilization, storage, and disposal of these compounds (European Commission 2014). The detection of natural or artificial compounds impacting biological processes increasingly involves the use of biosensors. A biosensor has the particular ability to use a biological partner, which can be strictly identical or derived from a naturally occurring molecule. The second part of a biosensor is the transducer, which transforms this biological interaction into physically detectable information. This information can be provided in several states (e.g., light, current, heat) to a detector, which will transform, amplify, and process this information to give a numerical value.

In this chapter, only the latest advances in the development of transcription-related whole-cell yeast biosensors will be presented. Since the 1980s, other biosensors relying on nontranscriptional events have been presented—biosensors allowing the detection of sugar, alcohol, or lactate, for example (Adeniran et al. 2015; Riedel et al. 2006). These take advantage of the yeasts' diverse metabolic pathways to transform these hardly detectable molecules into simple compounds. By monitoring the parameters of the yeast culture such as changes in the amount of dissolved oxygen (biological oxygen demand (BOD)) or the milieu pH, these biosensors can indirectly assess the concentration of the target analytes. However, a few new biosensors relying on these strategies have been presented in the last decade, probably because analytical detection methods now provide enhanced sensitivity and reliability in the results. On the contrary, during the same period, publications concerning transcription-based yeast cell biosensors increased dramatically, and some of these biosensors have been commercialized. Their success is dependent on the fact that biosensors can detect the biological activity of compounds that are highly specific, even at very low concentrations, a characteristic that cannot be obtained by analytical detection methods.

In the field of cell-based biosensors, yeast biosensors also present decisive advantages in comparison to bacteria-based biosensors or mammal cell suspensions. Firstly, most of the target compounds of these biosensors interact with eukaryote organisms by binding to specific receptors or disturbing a specific signal pathway. Therefore yeasts, as eukaryotic organisms, are more likely to mimic what will happen in higher eukaryotes than bacteria. Additionally, the latest advances in molecular biology tools allow a precise and easy modification of the yeast genome to either modulate the yeast's own metabolism or insert new coding sequences for recombinant proteins. This reduces the historical advantage of bacteria-based biosensors in terms of genetic modification easiness. Advanced genome-editing techniques are nowadays routine and allow greater control for knockouts or targeted gene integration. In comparison to mammalian cell culture, yeast-based biosensors present mostly practical advantages, such as the fast and cheap propagation of cells and the possibility of lyophilizing cells for storage or shipment. However, there are applications in pathology where yeast-based biosensors cannot replicate the responses of mammalian cell cultures because of the large difference between the metabolism of lower and higher eukaryotes.

Besides *S. cerevisiae*, other microorganisms are used; especially, *Blastobotrys raffinosifermentans* is well established as a nonconventional yeast biosensor. This yeast is able to synthesize inducer and reporter genes in high concentrations.

Up to the year 2019, this yeast was classified as *Arxula adeninivorans*. Seven wild-type strains of this species were isolated: LS3 isolated from wood hydrolysates in Siberia (Russia); CSIR 1136 and CSIR 1138 isolated from maize silage in the Netherlands; CRIR1147, CSIR 1148, and CSIR 1149 isolated from soil in South Africa; and CBS 8244T isolated from soil in the Netherlands. The nucleic and mitochondrial DNA of only one wild-type strain (LS3 from Russia) has completely been sequenced so far (Kunze et al. 2014). The genome of this strain contains 11,787,453 nucleotides (chromosome 1—1,659,399 bp; chromosome 2—2,016,785 bp; chromosome 3—3,827,910 bp; chromosome 4—4,251,697 bp; mitochondrion—31,662; number of open reading frames (ORFs)—6083). The other six wild-type sequences also include four chromosomes; however, their chromosomes differ in the base number per chromosome, especially for chromosome 4 (Kunze and Kunze 1994; Samsonova et al. 1996). In the year 2020, all *A. adeninivorans* strains were reclassified, and only the wild-type LS3 strain was classified as *B. raffinosifermentans* (FEMS Yeast Research, 20, 2020).

13.2 Utilization of Receptors

A key element for the development of a whole-cell yeast biosensor is the intracellular production of a molecule, generally a protein, able to bind the target analyte. Even though new *in silico* techniques strive to make the design of an artificial binding protein for every potential ligand theoretically possible, an easier way is to take advantage of the millions of years of vertebrates' evolution and the existence of specific binding proteins called receptors. These proteins can generally be produced recombinantly in yeasts, a topic that is discussed in other chapters of this book. Since the analyte is often presumed to have an impact on human health, the utilization of recombinant human receptors appears as the natural choice, although other vertebrate receptors are preferred in some biosensors. Among the receptors, a certain class—the nuclear receptors—is of particular interest because of its simple signaling pathway, which does not necessitate extra partner proteins to transform the binding into a transcriptional response. These nuclear receptors contain a ligand-binding domain and act as a transcription factor by directly binding to their specific deoxyribonucleic acid (DNA) domain in the nucleus. Another large class of receptors, which is utilized for yeast-based biosensors, is membrane-bound receptors, known as G-protein-coupled receptors. Since these receptors cannot directly translocate into the nucleus in order to induce the transcription of a reporter protein, the presence of a specific mobile G protein or of a specific kinase in the cytosol is mandatory. While in some cases the existing endogenous yeast G protein or kinases can be used for signal transduction, the majority of these biosensors require the additional production of recombinant signaling proteins.

13.2.1 Steroid Receptors

The detection of molecules binding to steroid hormone receptors gained increased importance since the discovery that molecules called endocrine disruptors (EDCs) can interfere with vertebrate endocrine systems and can lead to environmental or health hazards (Fent 2015; Kabir et al. 2015). Because of animal welfare and cost considerations in the determination of the potential endocrine-disrupting activity of a molecule, pharmaceutical and cosmetic industries are attempting to replace classical animal testing. Many companies are now using yeast-cell biosensors to determine at a very early developmental stage if a new compound is worth further effort or if its endocrine-disrupting properties make it unsuitable to continue development. Initially receptor-based biosensors focused on the detection of estrogens, but new biosensors for the detection of androgens (Bhattacharjee and Khurana 2014; Gerlach et al. 2014), progesterone (Viswanath et al. 2008; Chamas et al. 2015), glucocorticoids (Bovee et al. 2011; Pham et al. 2016), and mineralocorticoids (Miller et al. 2010) were recently presented. As a large number of research articles, describing continuous optimization of biosensors, have been published in the last years, this chapter will only focus on the latest version of selected biosensors. Therefore, fundamental work, which paved the way for the development of actual biosensors, will not be discussed. For a more complete overview of the steroid hormone yeast biosensors history, readers are invited to consider the selected references (Tag et al. 2006).

Two organisms, *S. cerevisiae* and *B. raffinosifermentans*, have been mainly used for the development of steroid hormone yeast biosensors. However, a hormone-receptor-based biosensor dedicated to the detection of endocrine disruptors has been recently engineered in *Y. lipolytica* (Cho et al. 2010). In all organisms, the integration of a human steroid receptor under the control of a constitutive promoter is the first step in biosensor design. If a target ligand is present in the cultivation medium, it can enter the cell by passive diffusion through the membrane and bind to the receptor. This binding event induces a conformational change that allows the homodimerization of the receptor in the cytoplasm. The nuclear localization signal of the receptor will then allow the translocation of the receptor homodimer into the nucleus to bind to its DNA response element, thus initiating transcription factor activity. This mechanism of hormone recognition is presented schematically in Fig. 13.1.

To obtain an inducible reporter gene, the promoter region upstream of this gene was modified. This construct consists of a modified glucoamylase promoter (*GAA*) containing two 15 bp sequences acting as the hormone response element (*HRE*) at position -107 of the promoter. *HRE* is a conserved DNA-binding site for the dimerized hormone receptor, and only when this dimerized receptor is bound to the *HRE* can the expression of the downstream gene occur. In the case of the estrogen-related assays, *HRE* was replaced by the slightly different estrogen response element (*ERE*) sequence as it was shown to have a higher affinity to the estrogen receptor than *HRE* (Hahn et al. 2006). In *S. cerevisiae* biosensors, *PGK*, *GPD*, or *ADH1* promoters containing *HRE* or *ERE* sequences have been described (Leskinen et al. 2005; Sanseverino et al. 2005; Routledge and Sumpter 1996), while *ALK1*, *ICL1*, *RPS7*, or *TEF1* promoters have been used in *Y. lipolytica* (Cho et al. 2010). The reporter protein utilized will then lead to different variations of the assays with different sensitivity or practical needs; these aspects will be considered in the section "Reporter Proteins" of this chapter.

Besides the mechanism described above, additional research to enhance the performance of biosensors has been reported, notably the expression of accessory proteins and the modification of the native hormone receptor. For example, the coexpression of the steroid receptor cofactor I (SRC1) has been implemented in the *S. cerevisiae* YES assay in order to increase the sensitivity of the biosensor (Chu et al. 2009). In humans, SRC1 is known to bind to a number of nuclear hormone receptors in the nucleus and help stimulate transcriptional activity. The successful use of the β isoform of the estrogen receptor (Lee et al. 2007) is also reported.

Recently, Zhou et al. (2022) described efforts to enhance estrogen-biosensing performances by both the addition of flanking activator regions to the nuclear receptor and promoter engineering.

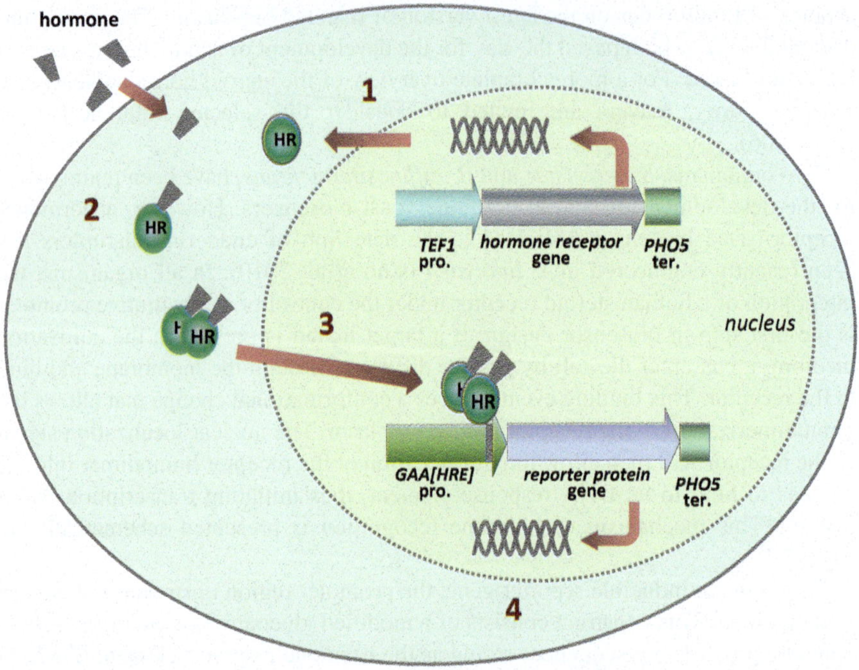

Fig. 13.1 Schematic representation of a modified yeast cell in a steroid-receptor-based biosensor. In the nucleus, a *human hormone receptor* gene is constitutively expressed, and the receptor monomers are transferred to the cytoplasm (1). Receptor ligands in the extracellular medium pass through the yeast cell wall and the membrane and bind to the receptor monomers (2). After binding-induced receptor dimerization, this dimer can relocate into the nucleus and bind to the *hormone response element* (*HRE*) (3), thus activating a promoter and allowing the expression of the reporter protein gene (4). Modified after Chamas et al. (2017)

Another interesting research is the one performed by Rajasärkkä et al. in 2013 to develop a bisphenol-A biosensor. Bisphenol-A is an artificial organic compound that possesses endocrine-disrupting properties and which is extensively used in industry, although it has been banned in products such as baby bottles in several countries. But despite its endocrine-disrupting property, the yeast estrogen biosensor cannot detect low concentrations of bisphenol-A, probably due to the low affinity of the human estrogen receptor to this compound. Since no specific receptor for this pollutant has been found in nature, the bisphenol-A biosensor used a human estrogen receptor with a sequence modified by targeted mutations to decrease its affinity toward most estrogens and to increase its affinity toward bisphenol-A.

Very recently, a novel *S. cerevisiae* biosensor for the detection of S-adenosylmethionine (SAM)—a well-known metabolite with interesting therapeutic potential—was developed (Chen et al. 2023). Sensing is based on the binding site of the nuclear estrogen receptor, subsequently modified to bind SAM in addition to 17β-estradiol. The detection of the two molecules is mediated through the expression of a fluorescent reporter protein.

13.2.2 Other Nuclear Receptors

Newly developed yeast biosensors also utilize other nuclear receptors, which are not steroid receptors, such as the human aryl hydrocarbon receptor (hAhR), the retinoid X receptor (RXR), and the thyroid receptor (TR). AhR-based biosensors were designed to detect dioxin-like compounds, which can accumulate in the environment, and some of their members are known to cause health hazards in humans. As for the development of steroid-receptor-based yeast biosensors, most of the research in the field of AhR yeast biosensors were conducted in *S. cerevisiae* (Miller 1999; Kawanishi et al. 2013), although *B. raffinosifermentans* was also used to build a similar biosensor (Pham et al. 2015). The strategy to detect dioxin-like compounds is similar to the strategy used to detect steroid hormones, albeit with some modifications. To detect dioxin-like compounds, two genes have to be integrated into the yeast genome to express constitutively hAhR and the human aryl hydrocarbon receptor nuclear translocator (hARNT). Only the hAhR protein can bind to dioxin-like compounds, but it has to form a heterodimer with hARNT in order to translocate into the nucleus. Additionally, in *B. raffinosifermentans*, the cyp1A1-derived core sequence replaced the *HRE* sequence in the *GAA* promoter, which allowed the interaction of the promoter and the hAhR-hARNT heterodimer.

The construction of a yeast biosensor using the binding properties of RXR is described in *S. cerevisiae* by Kabiersch et al. (2013). RXR can bind organotin compounds, a large family of molecules shown to have a negative impact on gastropod reproductive organs and on mammal endocrine systems. Kabiersch et al. proposed a hybrid receptor consisting of the ligand-binding domain of the RXR and the DNA-binding domain of the estrogen receptor α to function as a transcription factor.

Similarly, yeast biosensors dedicated to the detection of compounds interfering with the thyroid receptor (TR) were also presented (Li et al. 2008; Shiizaki et al. 2010). TR acts as a transcriptional factor and can be found as a homodimer or a heterodimer, together with RXR. Several molecules present in our environment, such as polychlorinated biphenyls (PCBs), pesticides, or dioxin, are believed to bind to the thyroid receptor and have a negative impact on an organism's development.

13.2.3 GPRC

In vertebrates, the largest family of receptors is the family of G-protein-coupled receptors (GPCRs). These membrane-bound receptors specifically bind a diverse group of molecules from neurotransmitters to hormones and large proteins, but contrary to nuclear receptors, they cannot directly act as transcription factors. The binding event induces a conformational change of the receptor, which will release a multimeric protein known as G protein, which will then be able to activate downstream effectors. The native GPCR pathway in yeast is dedicated to pheromone sensing via the recognition of the α or a-factor by the STE2 and STE3 receptors, an important mechanism for mating but often not of interest for biosensing applications. Additionally, the knockout of these two endogenous GPCRs is often required to avoid interference and improve biosensor performance. This leads to the fact that heterologous GPCRs of vertebrate origin have to be integrated into the yeast genome in order to detect the molecule of interest. In some cases, these heterologous receptors can interact with the endogenous yeast mating pheromone pathway (Bardwell 2005). In other cases, a heterologous expression of the mammalian G protein is mandatory, thus adding more complexity to the system.

Although not yet available commercially, several yeast biosensors based on GPCRs have been described. Ostrov et al. (2017) presented a biosensor for point-of-care (POC) diagnostics to detect human and plant pathogens based on the modification of native *S. cerevisiae* GPCR receptors. The readout of their assay is based on lycopene production on a dipstick, a method that does not necessitate heavy laboratory equipment. Similarly, Adeniran et al. (2018) performed a directed evolution of the native *S. cerevisiae* GPCR receptor to detect cystatin C, a marker for chronic kidney disease (Ault & Broach 2006). The reporter gene for their POC biosensor was the green fluorescent protein (GFP). Very recently, studies described the complete refactoring of the GPCR signaling pathway by deleting several endogenous genes and expressing human GPCR to detect cannabinoids, a class of molecules with increasing importance in the medical field (Shaw et al. 2019; Miettinen et al. 2022).

Given the large number of GPCRs and the diversity of ligands they target, it is expected that yeast-based GPCR biosensors will rapidly be commercialized. Knockouts of dozens of genes and the expression of human membrane proteins are not a major hindrance anymore, thus allowing for fast and smooth development of such biosensors. Although most of the work with GPCR biosensors was done in *S. cerevisiae*, other yeasts with different membrane compositions or wider sample tolerance may also prove to be better suited for the construction of a GPCR biosensor.

13.3 Transduction of Biological Signal: Reporter Proteins

The second fundamental part of a biosensor is the part responsible for the transformation of a biological interaction, the binding of a ligand to its receptor, into a measurable signal. For this, reporter proteins have to be integrated into the yeast cells.

Two main types of reporter proteins are used in yeast biosensors: enzymes and fluorescent proteins. Enzymes were the first to be integrated and are still part of a large number of yeast biosensors. However, their main drawback is that these proteins have to be transported in the extracellular milieu in order to perform enzymatic reactions. This implies either the breakdown of the cells or the implementation of an extracellular secretion signal sequence in the reporter gene coding sequence.

Fluorescent proteins as reporters are relatively new and allow direct observation of reporter production without extracellular localization or cell rupture.

13.3.1 Utilization of Enzymes as Reporter Proteins

β-galactosidase, phytase, and tannase are the most common reporter enzymes used in yeast biosensors. In *S. cerevisiae*, the use of *lacZ* and *luc* reporter genes, responsible for the production of β-galactosidase and the firefly luciferase, respectively, is often reported (Leskinen et al. 2005; Bovee et al. 2004; Balsiger et al. 2010). In *B. raffinosifermentans*, the *Klebsiella*-derived *phytase K* gene (*phyK*) and the *B. raffinosifermentans*-derived *tannase* gene (*ATAN1*) were used (Kaiser et al. 2010; Gerlach et al. 2014). Both *phyK* and *ATAN1* genes produce an extracellular-located enzyme that can then be used for an enzymatic reaction. The transformation of the substrate into a reaction product can be followed by either colorimetry, amperometry, or fluorescence.

It is notable that *ATAN1* was successfully used in a biosensor for estrogen detection (Kaiser et al. 2010). The enzyme phytase K was solely used in other *B. raffinosifermentans* biosensors, e.g., for the detection of androgens (Gerlach et al. 2014).

Colorimetric Detection

Colorimetric detection is generally considered the easiest and least expensive method to follow an enzymatic reaction. Chromogenic substrates, such as red-β-D-galactopyranoside (CPRG) for β-galactosidase or p-nitrophenyl phosphate (p-NPP) for phytase, are used, and the appearance of the colored products is monitored using a spectrophotometer. To measure enzyme activity, the classical protocol necessitates either collecting the extracellular milieu or lysing the cells in order to obtain the reporter enzyme. The enzymatic reaction itself generally takes place in a second plate and is followed in a spectrophotometer. The increase of the reaction product is then directly linked to the concentration of the ligand, for example, a hormone, in the cultivation medium. This reporting system is used in the *S. cerevisiae* YES/YAS/YPS/YGS biosensors and in the *B. raffinosifermentans* A-YES, A-YAS, A-YPS, and A-YGS biosensors.

Fluorescence, Chemiluminescence, and Bioluminescence Detection

The major disadvantages of the colorimetric detection method are the additional handling time for the preparation of the substrate solutions and time consuming pipetting steps in the microtiter plate which prolongates the total assay time. To overcome this, other substrates have been selected to produce chemiluminescence, bioluminescence, or fluorescence. Nowadays, a large number of commercial substrates are available for the β-galactosidase. The following section will only present some examples that are used in yeast-based biosensors.

The work of Balsiger et al. (2010) introduces the use of the commercial chemiluminescent compound Tropix Gal-Screen, instead of CPRG. In chemiluminescence, chemical energy is transformed into light, and the enzymatic reaction can then be followed in a luminometer. Several new chemiluminescent substrates are now commercially available, each of them presenting advantages and disadvantages in terms of sensitivity, the duration of light emission, or compatibility with other reporter systems.

Besides chemiluminescence, it should be noted that substrates producing fluorescent products are also used in some yeast biosensors, for example, 4-methylumbelliferyl-β-D-galactopyranoside (MUG), a compound that reacts with β-galactosidase to form 4-methylumbelliferone. In contrast to chemiluminescence, excitation at 366 nm is required to detect the blue fluorescence of this product. This substrate is used in the recently developed planar YES assay (p-YES), a variant of the YES assay combining a yeast biosensor and thin-layer chromatography (Buchinger et al. 2013).

A third class of enzyme substrates is also described in the literature, namely, bioluminescent substrates. As an example, the firefly luciferase (Luc) enzyme is used as a sensitive reporter enzyme, and the respective gene has been integrated into the *S. cerevisiae* genome (Bovee et al. 2004). The produced luciferase converts the substrate D-luciferin into an excited oxyluciferin, and subsequently, a photon is emitted while falling back to the basic state. Light can be detected using light-sensitive detection devices with, e.g., charge-coupled device (CCD) sensor technology. A major drawback to using luciferase as a reporter enzyme is its susceptibility to inhibition caused by various substances, such as isoflavonoids, which may lead to a false interpretation of measurement data (Kenda et al. 2021).

Electrochemical Detection

Electrochemical yeast-based biosensors use an electrochemical transducer to produce a signal proportional to the analyte concentration. The electrochemical transducer converts a biochemical reaction between the reporter protein and a substrate into a measurable signal, such as current, potential, or conductance, which is then amplified and processed (Lei et al. 2006; Mulchandani and Rogers 1998; Riedel et al. 2002; Shantilatha et al. 2003; Turner et al. 1992). Depending on the type of transducer, electrochemical biosensors are divided into three types: amperometric, potentiometric, and conductometric.

Among electrochemical biosensors, the amperometric biosensor is the most commonly used. It operates at a fixed potential with respect to the reference electrode. Historically, it has been one of the preferred detection methods used for the construction of BOD biosensors and copper-sensing assays (Riedel et al. 2006). An interesting application of electrochemical detection is the development of the "EstraMonitor" device, which can determine hormonal activities in sewage treatment plants (Pham et al. 2012, 2013). The "EstraMonitor" was demonstrated to be a rapid detection method for estrogenic activity measurement in a semi-online, automated, and continuous format. In common with the A-YES assay, the "EstraMonitor" uses a modified *B. raffinosifermentans* strain constitutively producing the estrogen receptor and phytase K as a reporter protein. But instead of analyzing the phytase K enzymatic reaction by colorimetry, it is monitored by amperometry. The substrate is p-aminophenyl phosphate (pAPP), a compound that can be oxidized to p-aminophenol in the presence of phytase. This reaction produces electrons, which can be detected as a current in the EstraMonitor detection chamber. Measured current has been shown to be proportional to the level of phytase activity, thus allowing a correlation between estrogen concentration and the current.

13.3.2 Direct Measurement of Emitted Light

A new development in yeast biosensor reporter strategies is to use reporter proteins that produce fluorescence or bioluminescence without the need for a substrate. The advantages are obvious in terms of the convenient assay procedure. Neither a cell lysis nor an extra detection plate is required. The assay costs can be significantly reduced by dispensing with expensive chemiluminescent and bioluminescent substrates. Two main strategies have been followed in the past years to detect direct light emission: the utilization of bacterial luciferase (lux) and the utilization of self-fluorescing proteins such as GFP or DsRed.

The Lux System

The lux system is the basis of the *S. cerevisiae* BLYES and BLYAS biosensors, which aim to detect estrogens and androgens, respectively (Sanseverino et al. 2008). There are several different bacterial luciferases, which are heterodimers composed of two subunits, α and β, and which are encoded by two genes in an operon, *luxA* and *luxB*. This protein catalyzes the oxidation of long-chain aldehydes, a reaction that produces light at a wavelength of 490 nm. Situated in the same operon, there are three additional genes—*luxC*, *luxD*, and *luxE*—which encode for the synthesis of three enzymes involved in the reduction of fatty acids to aldehydes, regenerating the substrate for the light-emitting reaction (Meighen 1991). Having all five genes on the same operon allows bioluminescence without the need for an external substrate. In BLYES and BLYAS, these five genes have been inserted into the *S. cerevisiae* genome, and with a suitable promoter, the translocation of the nuclear receptor homodimer results in the expression of all five genes, thus producing bioluminescence.

Self-Fluorescing Proteins

Parallel to the development of the lux system, self-fluorescing proteins also have applications as promising reporter systems in the development of yeast biosensors. For instance, GFP is a reporter protein integrated into the estrogen-sensing biosensor developed by Beck et al. (2005). GFP, which was first isolated from the jellyfish *Aequorea victoria*, is nowadays one of the most commonly used proteins in biotechnology. It has the property of undergoing internal cyclization leading to fluorescence emission without the need for an external partner. The protein is relatively stable, and reliable fluorescence measurement can be performed after 6 h of incubation.

Another self-fluorescing protein, DsRed, was chosen as a reporter in *B. raffinosifermentans*-based biosensors dedicated to the detection of progesterone and glucocorticoids (Chamas et al. 2015; Pham et al. 2016). DsRed was isolated from *Discosoma* sp. and fluoresces in the orange-red section of the visible spectrum. The maturation time of DsRed is known to be longer than for GFP, thus leading to a delay in the signal measurement, but this maturation time can be influenced via targeted mutation in order to obtain more rapid maturation and more intense fluorescence.

A major advantage of self-fluorescing proteins, as a reporter, over a lux system is the extreme simplicity of the reporting strategy, which only necessitates the integration of one heterologous protein. The fact that GFP and DsRed have been studied in depth for biotechnological application also presents the advantage that several new variants of these proteins, emitting fluorescence at different wavelengths, have been created. Only a few amino acid substitutions led to the creation of a large number of new self-fluorescing proteins detectable in the blue, yellow, orange, and ultraviolet (UV) part of the light spectrum. Targeted mutagenesis also led to the creation of enhanced versions of GFP in terms of fluorescence yield, which have been incorporated into some yeast biosensors (Bovee et al. 2004).

This diversity of self-fluorescing proteins led to the idea of creating a yeast biosensor able to simultaneously sense different classes of hormones (Chamas et al. 2017). The *B. raffinosifermentans* biosensor is based on three different strains, each of them producing a different steroid hormone receptor coupled with a different self-fluorescing protein as a reporter. By incubating these three strains as a mix, the resulting biosensor can detect estrogenic, progesterone, and androgenic activities in a sample. The assay requires three fluorescence scans, one for each excitation and emission wavelength of the three reporter proteins.

This example highlights the potential of self-fluorescence proteins as reporter proteins. In comparison, the lux system needs the integration of a complete operon, and there are fewer options available to modify the emission wavelength produced by the luciferase in this system. One of the major limitations of self-fluorescing proteins as a reporter is the possibility that the sample itself fluoresces at the same wavelength, thus increasing the noise signal and lowering the sensitivity of the assay.

13.3.3 Direct Measurement of Pigments

Another recently described reporter strategy (Ostrov et al. 2017) is to use the enzymes responsible for the production of carotenoids, a class of natural pigments also produced by the yeast *Xanthophyllomyces dendrorhous*. The integration of only two enzymes—a phytoene synthase and a desaturase—into the genome of *S. cerevisiae* results in the transformation of endogenous geranylgeranyl pyrophosphate (GGPP) into a red pigment lycopene, and further gene integrations allow for the production of other carotenoids, like astaxanthin or β-carotene. This reporter system does not require the addition of any substrate, and detection can be performed by the naked eye, making it a low-cost alternative to most of the other reporters. The main drawback of this strategy, however, is its poor sensitivity as carotenoid production needs to be very high in order to be visible.

13.4 Applications of EDC Targeting Yeast Biosensors

13.4.1 Sensitivity of the Biosensors

In Table 13.1, two important characteristics for the selection of a biosensor are indicated, the EC_{50} value and the limit of detection toward a known ligand. The EC_{50} value corresponds to the concentration of a ligand at which half of the receptor molecules are bound to a ligand, whereas the limit of detection (LoD) is the lowest concentration of a ligand, which can be effectively distinguished from the noise signal and therefore the lowest concentration of a ligand a biosensor can detect. EC_{50} calculation is usually not contentious, but the method to determine LoD varies among researchers and can lead to confusion. For more information about this topic, we refer to the article by Shrivastava and Gupta (2011), which gives an overview of the existing guidelines and recommendations for establishing LoD. A comparison of LoD in different manuscripts is difficult because the method of calculation is often not clearly given and LoDs should be treated with caution. These values are presented in Table 13.1 as they are given in their manuscript.

13.4.2 Protocols

Historically, yeast biosensor assays were developed in a microtiter plate format for the detection of endocrine effects, especially in water samples. Thus, it is not surprising to observe that the majority of the published works are presenting protocols utilizing this format. Generally, the yeast cells undergo preliminary cultivation without a sample to reach a certain cell density, which is a critical parameter in order to obtain reproducible results. Then the cells are incubated with the sample in

Table 13.1 Description of several estrogen, androgen, progestogen, and glucocorticoid yeast biosensors

Target compound	Organism	Bioassay	Reporter gene	Detection method	EC_{50}	LoD	Source
Estrogens	*S. cerevisiae*	YES	*lacZ*	Colorimetry	17β-estradiol: 30 ng/l	17β-estradiol: 2.8 ng/l	Gehrmann et al. (2016)
		YES	*lacZ*	Luminescence	17β-estradiol: 40 ng/l	n.i.	Balsiger et al. (2010)
		YES	*luc*	Luminescence	17β-estradiol: 54.5 ng/l	n.i.	Bovee et al. (2004)
		BLYES	*lux*	Luminescence	17β-estradiol: 108.8 ng/l	17β-estradiol: 6.8 ng/l	Sanseverino et al. (2008)
		YES	*gfp*	Fluorescence	17β-estradiol: 163.2 ng/l	17β-estradiol: 27.2 ng/l	Beck et al. (2005)
		p-YES	*lacZ*	Fluorescence	17β-estradiol: 7.5 pg	17β-estradiol: 1.3 pg	Buchinger et al. (2013)
	B. raffinosifermentans	A-YES	*phyK*	Colorimetry	17β-estradiol: 17 ng/l	17β-estradiol: 1.8 ng/l	Gehrmann et al. (2016)
		EstraMonitor	*phyK*	Amperometric	17β-estradiol: 54.6 ng/l	17β-estradiol: 9.92 ng/l	Pham et al. (2013)
	Y. lipolytica		*lacZ*	Colorimetry	n.i.	17β-estradiol: 27.2 ng/l	Cho et al. (2010)
Androgens	*S. cerevisiae*	YAS	*lacZ*	Colorimetry	Testosterone: 1043 ng/l	Testosterone: 189 ng/l	Gehrmann et al. (2016)
		YAS	*luc*	Luminescence	5α-DHT: 1.6 μg/l	5α-DHT: 145 ng/l	Leskinen et al. (2005)
		BLYAS	*lux*	Luminescence	5α-DHT: 3.2 μg/l	5α-DHT: 290 ng/l	Sanseverino et al. (2008)
	B. raffinosifermentans	A-YAS	*phyK*	Biochemical	5α-DHT: 277 ng/l	5α-DHT: 56.5 ng/l	Gerlach et al. (2014)

Progestogens	*S. cerevisiae*	YPS	*gfp*	Fluorescence	Progesterone: 314 ng/l	Progesterone: 31.4 ng/l	Chatterjee et al. (2008)
	B. raffinosifermentans	A-YPS	*phyK*	Biochemical	Progesterone: 147 ng/l	Progesterone: 40.9 ng/l	Chamas et al. (2015)
		A-YPFS	*dsRED*	Fluorescence	Progesterone: 231 ng/l	Progesterone: 65.3 ng/l	Chamas et al. (2015)
Glucocorticoids	*S. cerevisiae*	YGS	*GFP*	Fluorescence	Dexamethasone: 120 µM	n.i.	Bovee et al. (2011)
	B. raffinosifermentans	A-YGS	*phyK*	Biochemical	Dexamethasone: 0.81 µM	Dexamethasone: 0.29 µM	Pham et al. (2016)

an incubation plate for a certain time, and detection is performed. Depending on the reporter protein used, the cell suspension can be directly utilized without further processing or an extra detection protocol is needed. An overview of these different protocols can be observed in Fig. 13.2.

Direct detection is made possible by the use of self-fluorescing proteins and lux reporter systems simply by scanning the incubation plate for fluorescence. If enzymes are used as reporters, these have to be separated from the cells—either by cell lysis or extracellular transport—and the enzyme solution is pipetted into a new microtiter plate for colorimetric or fluorescence measurement. Incubation parameters such as temperature, shaking, or the duration of each step depend on the organism chosen and the nature of the reporter protein. When working with an enzyme reporter in a microtiter plate format, besides the necessary devices for yeast incubation, the operator needs to have access to a spectrophotometer, a luminometer, or a fluorometer, depending on the substrate they use. With self-fluorescing proteins as reporter proteins, only a fluorometer is needed to quantify the response.

A notable modification of the basic microtiter plate protocol can be seen in the planar yeast estrogen assay (Buchinger et al. 2013; Klingelhöfer et al. 2014). In this assay, the detection of estrogens is coupled with a prior separation of the sample on a thin-layer chromatography plate. The cells are directly incubated on the thin-layer chromatography plate, and the reporter activity, in this case β-galactosidase reaction with MUG, is monitored via a fluorescence scan of the plate. For a complete identification of the potential endocrine disruptors, mass spectrometry can also be performed after biosensor detection. An overview of this protocol can be seen in Fig. 13.3.

For amperometric detection, another protocol was designed that allows semi-automatic estrogen detection. For this, a device containing all the necessary fluidic, heating, and measuring elements was constructed (Fig. 13.4). The immobilized *B. raffinosifermentans* cells used in EstraMonitor are easily handled, can function in nonsterile conditions, and can be reused up to 15 times without a significant loss of sensitivity (Pham et al. 2012). Furthermore, the use of immobilized cells and the amperometric detection method makes the EstraMonitor able to measure the total estrogenic activity in the sample in around 4 hours. Interestingly, the immobilized *B. raffinosifermentans* cells are fully functional in a wide range of samples, including samples that contained up to 5% sodium chloride (NaCl), and are able to determine estrogenic effects in the environmental sample without the need for sterilization, extraction, or concentration (Pham et al. 2013). These results were also confirmed using gas chromatography-mass spectrometry (GC-MS) analyses.

13.4.3 Applications with Real Samples

The ultimate goal of yeast biosensors is to be trialed with real samples in order to identify potentially dangerous compounds in the environment or to monitor physiological levels in humans and animals. A great majority of these real samples are

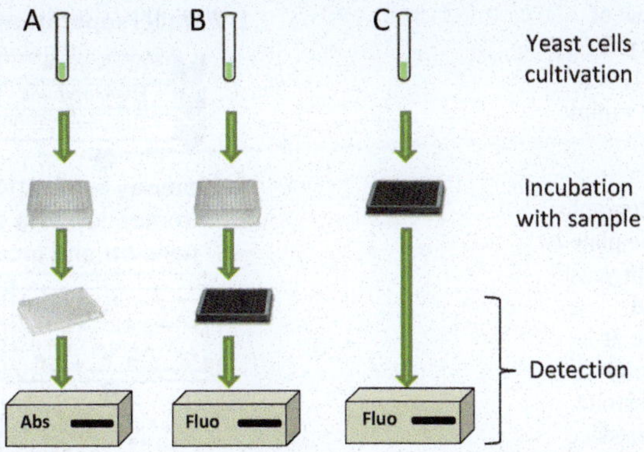

Fig. 13.2 Overview of three common microtiter protocols for the detection of hormones with yeast whole-cell biosensors. (**a**) A common protocol for enzyme-based yeast reporter assays with colorimetric detection. Incubation of the yeast cells with the sample takes place in multiwell incubation plates (also called deep-well plates). For the detection, cells are centrifuged, and the clear supernatant is transferred to a new microtiter plate for the enzyme assay. Absorbance measurement can be carried out with a spectrophotometer. (**b**) A common protocol for enzyme-based yeast reporter assays with fluorescence detection. Incubation of the yeast cells with the sample takes place in multiwell incubation plates. For the detection, cells are centrifuged, and the supernatant is transferred into a new black microtiter plate for the enzyme assay. Fluorescence measurement can be carried out with a fluorometer. (**c**) General fluorescence detection protocol, if a fluorescence protein is the reporter protein. Incubation of cells with the sample and the measurement of fluorescence are conducted directly in a black microtiter plate

wastewater samples because of their relative simplicity in comparison to bodily fluids or solid samples, for example. Yeast biosensors are particularly valuable in the analysis of wastewater samples because of the cells' robustness.

Table 13.2 presents some of the research applying yeast biosensors to real samples.

13.5 Emerging Biosensors Strategies

The field of whole-cell biosensors is developing rapidly, and many research teams around the world are proposing new strategies to establish innovative bioassays.

An example of an emerging strategy is glucose sensing using yeast as a biocatalyst. In fact, because of the frequency of fluctuations in blood glucose levels, patients need to monitor their glucose levels several times a day (Khadilkar et al. 2013; Hammond et al. 2016). Several studies (Wang et al. 2015) reported a novel electrochemical glucose biosensor based on glucose oxidase (GOx) displayed on the yeast surface. The entire *S. cerevisiae* cell catalyst (yeast-Gox) was immobilized with multiwalled carbon nanotubes on the electrodes, which allowed sensitive and selective glucose determination in a short time (Wang et al. 2015).

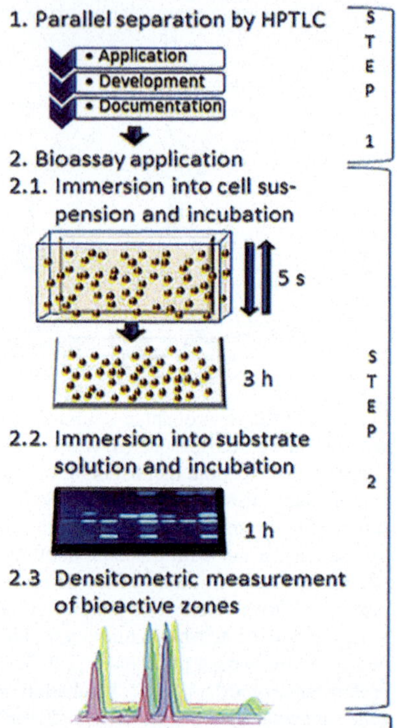

Fig. 13.3 Overview of planar YES assay (p-YES) first steps. In step 1, components of the sample are separated with high-performance thin-layer chromatography (HPTLC). Then the plate is shortly immersed in yeast cell suspension and allowed to incubate at 30 °C. After a second immersion in a substrate solution, the fluorescence of single zones can be detected in a densitometer. Modified after Klingelhöfer and Morlock (2015)

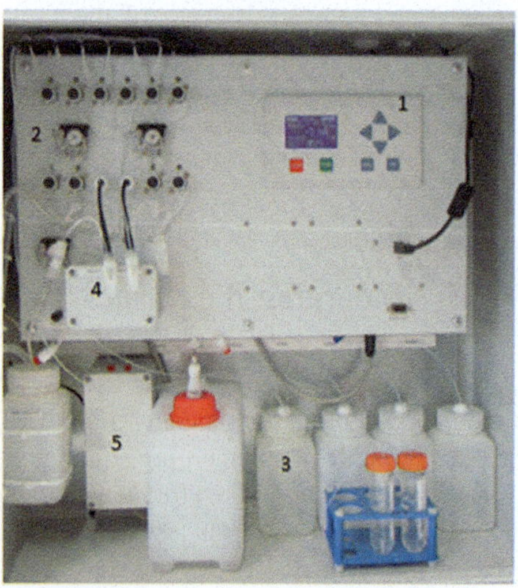

Fig. 13.4 Photograph of the EstraMonitor device. (1) Amperometric transducer, (2) pump module for waste and nutrient exchange, (3) reservoir, (4) measuring chamber with immobilized yeast cells. Modified after Pham et al. (2012)

Another very interesting strategy is the one followed by Fukuda et al. (2010). They developed a yeast whole-cell biosensor able to detect organophosphorus compounds, which are commonly used as pesticides. They constructed a *S. cerevisiae* strain that codisplays organophosphorus hydrolase (OPH) and enhanced green

Table 13.2 Some research applying yeast biosensors to real samples

Real sample	Real sample type	Hormone activity	Organism	Detection	Source
Waterbased	Wastewater effluent/ influent	Estrogens	*S. cerevisiae*	Luminescence	Balsiger et al. (2010)
	Wastewater effluent/ influent	Estrogens	*B. raffinosifermentans*	Colorimetry	Kaiser et al. (2010)
	Wastewater effluent/ influent	Progesterone	*B. raffinosifermentans*	Colorimetry	Chamas et al. (2015)
	River water, wastewater effluent/ influent	Glucocorticoids	*B. raffinosifermentans*	Colorimetry	Pham et al. (2016)
	Surface water, wastewater effluent/ influent	Estrogens	*S. cerevisiae*	Fluorescence (p-YES)	Klingelhöfer and Morlock (2015)
	Reservoir water	Thyroids	*S. cerevisiae*	Colorimetry	Li et al. (2014)
Bodily fluids	Swine urine	Androgens	*B. raffinosifermentans*	Colorimetry	Gerlach et al. (2014)
	Swine urine	Estrogens	*B. raffinosifermentans*	Colorimetry	Kaiser et al. (2010)
	Monkey serum	Progesterone	*B. raffinosifermentans*	Fluorescence	Chamas et al. (2017)
	Human serum	Androgens	*S. cerevisiae*	Luminescence	Michelini et al. (2005)
Solid-sample	Sediment	Organotins	*S. cerevisiae*	Luminescence	Kabierch et al. (2013)
	Moisturizing lotion	Estrogens	*S. cerevisiae*	Luminescence	Leskinen et al. (2005)
	Sunscreen, sediment	Estrogens	*S. cerevisiae*	Fluorescence (p-YES)	Buchinger et al. (2013)

fluorescent protein (EGFP) on the cell surface using a Flo1p anchor system. Organophosphorus compound degradation by the hydrolase releases protons and causes a pH change. This pH change induces conformational modification of EGFP, which triggers the quenching of its fluorescence, thereby making this cell useful for the visual detection of the presence of organophosphorus compounds.

Several studies have recently demonstrated the possibility of using *S. cerevisiae* as a biosensor for the in vivo monitoring of oxidative stress, which is a biomarker for colitis and cancer. This NADPH/NADP$^+$ redox biosensor utilizes the transcription factor Yap1p to activate GFP production (Zhang et al. 2016; Dacquay and McMillen 2021).

Besides the utilization of receptors to sense a particular cue, novel yeast-based biosensors relying on riboswitches have been developed. Riboswitches are ribonucleic acid (RNA) molecules containing a ligand-binding domain and an antisense domain responsible for interaction with a target messenger RNA (mRNA). The binding of the riboswitch with a specific ligand will induce a conformational change leading to the regulation of the target mRNA responsible for the production of a reporter protein. Following pioneering works from Bayer and Smolke in 2005, yeast riboswitch-based biosensors dedicated to the detection of theophylline (Michener and Smolke 2012) or thiamine (vitamin B1) (Donovan et al. 2018) were recently described.

13.6 Applications of *Blastobotrys raffinosifermentans* for the Effect-Based Determination of Hormonal Activity

The detection and quantification of hormonal activities caused, in particular, by endocrine disruptors continue to be of interest. In addition to the monitoring of hormonal effects in surface water bodies with regard to the achievement and sustainment of a good ecological and good chemical status (Water Framework Directive 2000/60/EC), the assessment of endocrine effects in food, feed, cosmetics, and articles of daily use is becoming increasingly important (Chantelouve and Ripoll 2022; Peivasteh-Roudsari et al. 2023). In recent years, various mammalian-cell-based bioassays and even in vivo assays utilizing amphibian embryos have been developed to provide sophisticated biological tools with advantageous properties. Nevertheless, yeast-based assays are used for the determination of hormonal activities because they are robust, quick, and easy to handle and have low technical requirements regarding the test procedure. These advantages make them applicable, e.g., for studies with complex samples or as simple screening tools prior to chemical analysis. Especially, the nonconventional yeast *B. raffinosifermentans*, formerly designated as *Arxula adeninivorans*, is used for more than two decades as a promising host for the development of hormone-sensitive biosensors (Böer et al. 2009).

Not least due to its high salt tolerance and high robustness against complex environmental samples, containing interfering matrices (Hahn et al. 2006), the well-established A-YES (*Arxula*-Yeast Estrogen Screen) with colorimetric detection was subjected to an ISO standardization process for the determination of the estrogenic potential of water and wastewater in 2011. As a result, the final standard ISO 19040-2:2018 was published in 2018 (International Organization for Standardization 2018) and provides a standardized effect-based method for routine applications of measuring estrogenicity in tap water, environmental water, and extracts thereof. The special osmotolerance of *B. raffinosifermentans* makes the use of the standardized A-YES particularly interesting for marine water samples (Deich et al. 2019).

It offers the possibility of evaluating estrogenicity in the original saline sample without the yeast cells being affected by the salt content of the sample.

A major challenge to using yeast-based assays to measure hormonal activity in various sample types is the very low concentration of the target compounds in combination with the low sensitivity of yeast biosensors in comparison to mammalian-cell-based assays (OECD 2018). Thus, sample preparations using solid-phase extraction (SPE) are almost exclusively utilized to enrich the target analytes. Along with this, inhibitory effects caused by the enriched sample matrix and the organic solvent may occur. The combination of effect-based yeast assays and thin-layer chromatography (TLC) was established more than 10 years ago and is able to overcome this undesirable effect, among others (Buchinger et al. 2013). Further developments in this field result in a technique specified as effect-directed analysis (EDA), which is done by combining high-performance thin-layer chromatography (HPTLC) with the biological detection step. Using a mix of *B. raffinosifermentans* biosensors and different self-fluorescing proteins enables the simultaneous detection of estrogenic, androgenic, and gestagenic effects in wastewater and surface water samples after separation by HPTLC (Baetz et al. 2021). This is a sophisticated approach to assessing multiple endocrine effects in one step and can be applied to various sample types.

A decisive aspect of the application of an effect-based bioassay for the determination of endocrine active zones on an HPTLC plate is its practical applicability and low complexity. A novel planar bioluminescence bioassay for the detection of estrogenic activity mediated by bisphenols (pA-YBS), based on the established microtiter assay A-YBS (*Arxula*-Yeast Bisphenol Screen), is capable of detecting suspicious estrogen activity induced by bisphenols in extracts of complex mixtures, such as tin can coatings, thermal papers, and botanicals, by using the sensitive reporter enzyme luciferase (Jaber et al. 2023). A distinct light signal could be detected already after 1 h of incubation of the yeast biosensor with a satisfactory sensitivity per zone, depending on the application technique of the yeast biosensor (Fig. 13.5). This feasibility approach demonstrates another application of the modified *B. raffinosifermentans* and reveals the biosensor potential of this yeast.

13.7 Conclusions

Yeast whole-cell biosensors have proven themselves to be very useful for the detection of bioactive substances. Constant improvements in bioengineering and biochemistry have allowed the construction of robust biosensors, some of which have already been commercialized. Most of the yeast biosensors that have been commercialized are dedicated to the detection of vertebrate hormones. This is mainly due to the fact that endocrine disruptors gained wide notoriety because of their negative impact on the environment and health. The range of possibilities is increasing as producing recombinant membrane G-protein-coupled receptors is no longer a technical obstacle. The different reporting strategies, which have been presented in this

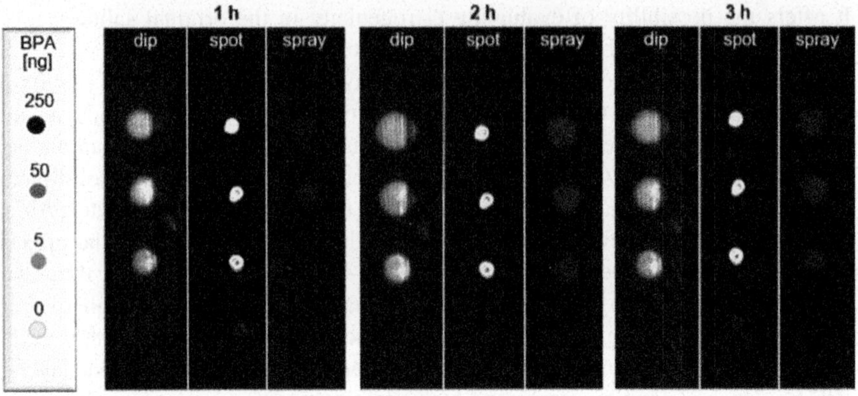

Fig. 13.5 Comparison of three yeast cell application techniques for the pA-YBS bioluminescence bioassay on TLC plates: manual dipping, spotting, and automated spraying of the yeast cell suspension of *Blastobotrys raffinosifermentans* on the TLC foil silica gel 60. The bioluminescence signal of BPA (0–250 ng/zone) was measured after 1, 2, and 3 h incubation time of the TLC plate at 37 °C in a humid chamber (bioluminescence depicted as a greyscale image)

chapter, also allow users to select the most appropriate biosensor for their application. To choose a suitable biosensor, the user should consider not only the sensitivity of the assay toward one ligand but also the time needed to perform the assay, which devices are necessary for detection, and for which real samples the assay has already been trialed with.

References

Adeniran A, Sherer M, Tyo KEJ (2015) Yeast-based biosensors: design and applications. FEMS Yeast Res 15:1–15

Adeniran A, Stainbrook S, Bostick JW, Tyo KEJ (2018) Detection of a peptide biomarker by engineered yeast receptors. ACS Synth Biol 7(2):696–705

Ault AD, Broach JR (2006) Creation of GPCR-based chemical sensors by directed evolution in yeast. Protein Eng Des Sel 19:1–8

Baetz N, Rothe L, Wirzberger V, Sures B, Schmidt TC, Tuerk J (2021) High-performance thin-layer chromatography in combination with a yeast-based multi-effect bioassay to determine endocrine effects in environmental samples. Anal Bioanal Chem 413:1321–1335

Balsiger HA, de la Torre R, Lee WY, Cox MB (2010) A four-hour yeast bioassay for the direct measure of estrogenic activity in wastewater without sample extraction, concentration, or sterilization. Sci Total Environ 408:1422–1429

Bardwell L (2005) A walk-through of the yeast mating pheromone response pathway. Peptides 26:339–350

Bayer TS, Smolke CD (2005) Programmable ligand-controlled riboregulators of eukaryotic gene expression. Nat Biotechnol 23:337–343

Beck V, Pfitscher A, Jungbauer A (2005) GFP-reporter for high throughput assay to monitor estrogenic compounds. J Biochem Biophys Methods 64:19–37

Bhattacharjee GB, Khurana SMP (2014) *In vitro* reporter assays for screening of chemicals that disrupt androgen signaling. J Toxicol. Article ID 701752. https://doi.org/10.1155/2014/701752

Böer E, Steinborn G, Florschütz K, Körner M, Gellissen G, Kunze G (2009) *Arxula adeninivorans* (*Blastobotry adeninivorans*)—a dimorphic yeast of great biotechnological potential, Chapter 27. In: Satyanarayana T, Kunze G (eds) Yeast biotechnology: diversity and application. Springer, Dordrecht, pp 615–638

Bovee TF, Helsdingen RJR, Koks P, Kuiper HA, Hoogenboom RLAP, Keijer J (2004) Development of a rapid yeast estrogen bioassay, based on the expression of green fluorescent protein. Gene 325:187–200

Bovee TF, Helsdingen RJ, Hamers AR, Brouwer BA, Nielen MW (2011) Recombinant cell assay for the detection of (gluco)corticosteroids and endocrine-disrupting potencies of several environmental PCB contaminants. Anal Bioanal Chem 401:873–882

Buchinger S, Spira D, Bröder K, Schlüsener M, Ternes T, Reifferscheid G (2013) Direct coupling of thin-layer chromatography with a bioassay for the detection of estrogenic compounds, application for effect-directed analysis. Anal Chem 85:7248–7256

Chamas A, Nieter A, Pham HTM, Giersberg M, Hettwer K, Uhlig S, Simon K, Baronian K, Kunze G (2015) Development of a recombinant *Arxula adeninivorans* cell bioassay for the detection of molecules with progesterone activity in wastewater. Anal Bioanal Chem 407:8109–8120

Chamas A, Pham HTM, Jähne M, Hettwer K, Uhlig S, Simon K, Einspanier A, Baronian K, Kunze G (2017) Simultaneous detection of three sex steroid hormone classes using a novel yeast-based biosensor. Biotech Bioeng. https://doi.org/10.1002/bit.26249

Chantelouve M, Ripoll L (2022) Endocrine disruptors in cosmetics: a review. Molecules. https://constellation.uqac.ca/id/eprint/8464/

Chatterjee S, Kumar V, Majumder CB, Roy P (2008) Screening of some anti-progestin endocrine disruptors using a recombinant yeast based in vitro bioassay. Toxicol In Vitro 22:788–798

Chen Y, Zheng H, Yang J, Cao Y, Zhou H (2023) Development of a synthetic transcription factor-based S-adenosylmethionine biosensor in *Saccharomyces cerevisiae*. Biotechnol Lett 45:255–262

Cho EM, Lee HS, Eom CY, Ohta A (2010) Construction of high sensitive system for endocrine disruptors with yeast n-alkane –assimilating *Yarrowia lipolytica*. J Microbiol Biotechnol 20:1563–1570

Chu WL, Shiizaki K, Kawanishi M, Kondo M, Yagi T (2009) Validation of a new yeast-based reporter assay consisting of human estrogen receptors alpha/beta and coactivators SRC-1: application for detection of estrogenic activity in environmental samples. Environ Toxicol 24:513–529

Dacquay LC, McMillen DR (2021) Improving the design of an oxidative stress sensing biosensor in yeast. FEMS Yeast Res 21:foab025

Deich C, Kanwischer M, Waniek J (2019) Patterns of estrogenic activity in the Baltic Sea. Chemosphere 240:124870

Donovan PD, Holland LM, Lombardi L, Coughlan AY, Higgins DG, Wolfe KH, Butler G (2018) TPP riboswitch-dependent regulation of an ancient thiamin transporter in Candida. PLoS Genet 14(5):e1007429

European Commission (2014) Directorate-General for Environment, The EU Water Framework Directive, Publications Office. https://data.europa.eu/doi/10.2779/75229

Fent K (2015) Progestins as endocrine disrupters in aquatic ecosystems: concentrations, effect and risk assessment. Environ Int 84:115–130

Fukuda T, Tsuchiya K, Makishima H, Tsuchiyama K, Mulchandani A, Kuroda K, Ueda M, Suye SI (2010) Organophosphorus compound detection on a cell chip with yeast coexpressing hydrolase and eGFP. Biotech J 5:515–519

Gehrmann L, Bielak H, Behr M, Itzel F, Lyko S, Simon A, Kunze G, Dopp E, Wagner M, Tuerk J (2016) (Anti-)estrogenic and (anti-)androgenic effects in wastewater during advanced treatment: Comparison of three *in vitro* bioassays. Environ Sci Pollut Res. https://doi.org/10.1007/s11356-016-7165-7174

Gerlach TK, Kaiser C, Körner M, Hettwer K, Uhlig S, Simon K, Baronian K, Kunze G (2014) Development and assessment of a novel *Arxula adeninivorans* androgen screen (A-YAS) assay and its application in analysis of cattle urine. Sci Total Environ 490:1073–1081

Hahn T, Tag K, Riedel K, Uhlig S, Baronian K, Gellissen G, Kunze G (2006) A novel estrogen sensor based on recombinant arxula adeninivorans cells. Biosensors and Bioelectronics 21(11):2078–2085

Hammond JL, Formisano N, Estrela P, Carrara S, Tkac J (2016) Electrochemical biosensors and nanobiosensors. Essays Biochem 60:69–80

International Organization for Standardization (2018) Water quality—determination of the estrogenic potential of water and waste water. Part 2: Yeast estrogen screen (A-YES, *Arxula adeninivorans*). ISO Standard No. 19040-2:2018

Jaber M, Jähne M, Oberle M, Morlock GE (2023) Screening bisphenols in complex samples via a planar *Arxula adeninivorans* bioluminescence bioassay. Anal Bioanal Chem 5193–5204

Kabiersch G, Rajasärkkä J, Tuomela M, Hatakka A, Virta M, Steffen K (2013) Bioluminescent yeast assay for detection of organotin compounds. Anal Chem 85:5740–5745

Kabir ER, Rahman MS, Rahman I (2015) A review on endocrine disruptors and their possible impacts on human health. Environ Toxicol Pharmacol 40:241–258

Kaiser C, Uhlig S, Gerlach T, Körner M, Simon K, Kunath K, Florschütz K, Baronian K, Kunze G (2010) Evaluation and validation of a novel *Arxula adeninivorans* estrogen screen (nAES) assay and its application in analysis of wastewater, seawater, brackish water and urine. Sci Total Environ 408:6017–6026

Kawanishi M, Ohnisi K, Takigami H, Yagi T (2013) Simple and rapid yeast reporter bioassay for dioxin screening: Evaluation of the dioxin-like compounds in industrial and municipal waste incineration plants. Environmental Science and Pollution Research 20:2993–3002

Kenda M, Vegelj J, Herlah B, Perdih A, Přemysl M (2021) Sollner MD (2021) Evaluation of firefly and renilla luciferase inhibition in reporter-gene assays: A case of isoflavonoids. Int J Mol Sci. 22(13):6927

Khadilkar KS, Bandgar T, Shivane V, Lila A, Shah N (2013) Current concepts in blood glucose monitoring. Indian J Endocrinol Metab 17(suppl 3):S643–S649

Klingelhöfer I, Morlock GE (2014) Sharp-bounded zones link to the effect in planar chromatography-bioassay-mass spectrometry. Journal of Chromatography A 1360:288–295

Klingelhöfer I, Morlock GE (2015) Bioprofiling of surface/wastewater and bioquantitation of discovered endocrine-active compounds by streamlined direct bioautography. Anal Chem 87:11098–11104

Kunze G, Kunze I (1994) Characterization of arxula adeninivorans strains from different habitats. Antonie van Leeuwenhoek 65:29–34

Kunze G, Gaillardin C, Czernicka M, Durrens P, Martin T, Böer E, Gabaldón T, Cruz JA, Talla E, Marck C, Goffeau A, Barbe V, Baret P, Baronian K, Beier S, Bleykasten C, Bode R, Casaregola S, Despons L, Fairhead C, Giersberg M, Gierski PP, Hähnel U, Hartmann A, Jankowska D, Jubin C, Jung P, Lafontaine I, Leh-Louis V, Lemaire M, Marcet-Houben M, Mascher M, Morel G, Richard GF, Riechen J, Sacerdot C, Sarkar A, Savel G, Schacherer J, Sherman DJ, Stein N, Straub ML, Thierry A, Trautwein-Schult A, Vacherie B, Westhof E, Worch S, Dujon B, Souciet JL, Wincker P, Scholz U, Neuvéglise C (2014) The complete genome of *Blastobotrys* (*Arxula*) *adeninivorans* LS3—a yeast of biotechnological interest. Biotechnol Biofuels 7:66

Lee HS, Cho EM, Jung JH, Ohta A (2007) Evaluation on antagonist activities of polycyclic aromatic hydrocarbons using the yeast two-hybrid detection system for endocrine disruptors. Environ Monit Assess 129:87–95

Lei Y, Chen W, Mulchandani A (2006) Microbial biosensors. Anal Chim Acta 566:200–210

Leskinen P, Michelini E, Picard D, Karp M, Virta M (2005) Bioluminescent yeast assay for detecting estrogenic and androgenic activity in different matrices. Chemosphere 61:259–266

Li J, Ma M, Wang Z (2008) A two-hybrid yeast assay to quantify the effects of xenobiotics on thyroid hormone-mediated gene expression. Environ Toxicol Chem 27:159–167

Li J, Ren S, Han S, Li N (2014) A yeast bioassay for direct measurement of thyroid hormone disrupting effects in water without sample extraction, concentration, or sterilization. Chemosphere 100:139–145

Meighen EA (1991) Molecular biology of bacterial bioluminescence. Microbiol Rev 55:123–142
Michelini E, Magliulo M, Leskinen P, Virta M, Karp M, Roda A (2005) Recombinant cell-based bioluminescence assay for androgen bioactivity determination in clinical samples. Clin Chem 51:1995–1998
Michener JK, Smolke CD (2012) High-throughput enzyme evolution in *Saccharomyces cerevisiae* using a synthetic RNA switch. Metab Eng 14(4):306–316
Miettinen K, Leelahakorn N, Almeida A, Zhao Y, Hansen LR, Nikolajsen IE, Andersen JB, Givskov M, Staerk D, Bak S, Kampranis SC (2022) A GPCR-based yeast biosensor for biomedical, biotechnological, and point-of-use cannabinoid determination. Nat Commun 13(1):3664
Miller CA (1999) A human aryl hydrocarbon receptor signaling pathway constructed in yeast displays additive responses to ligand mixtures. Toxicol Appl Pharmacol 160:297–303
Miller CA, Tan X, Wilson M, Bhattacharyya S, Ludwig S (2010) Single plasmids expressing human steroid receptors and a reporter gene for use in yeast signaling assays. Plasmid 63:73–78
Mulchandani A, Rogers KR (eds) (1998) Enzyme and microbial biosensors: techniques and protocols. Humana, Totowa
OECD (2018) OECD revised guidance document 150 on standardised test guidelines for evaluating chemicals for endocrine disruption. https://doi.org/10.1787/9789264304741-en
Ostrov N, Jimenez M, Billerbeck S, Brisbois J, Matragrano J, Ager A, Cornish VW (2017) A modular yeast biosensor for low-cost point-of-care pathogen detection. Sci Adv 3(6):e1603221
Peivasteh-Roudsari L, Barzegar-Bafrouei R, Sharifi KA, Azimisalim S, Karami M, Abedinzadeh S, Asadinezhad S, Tajdar-Oranj B, Mahdavi V, Alizadeh AM, Sadighara P, Ferrante M, Conti GO, Aliyeva A, Khaneghahl AM (2023) Origin, dietary exposure, and toxicity of endocrine-disrupting food chemical contaminants: a comprehensive review. Heliyon 9(7):e18140
Pham HTM, Giersberg M, Uhlig S, Hanke G, Simon K, Kunath K, Baronian K, Kunze G (2012) EstraMonitor—a monitor for amperometric detection of estrogenic activity with *Arxula adeninivorans* yeast cells as the biocomponent. Sens Actuator B Chem 161:137–145
Pham HTM, Kunath K, Gehrmann L, Giersberg M, Tuerk J, Uhlig S, Hanke G, Simon K, Baronian K, Kunze G (2013) Application of modified *Arxula adeninivorans* yeast cells in an online biosensor for the detection of estrogenic compounds in wastewater samples. Sens Actuator B Chem 185:628–637
Pham HTM, Giersberg M, Gehrmann L, Hettwer K, Tuerk J, Uhlig S, Hanke G, Weisswange P, Simon K, Baronian K, Kunze G (2015) The determination of pharmaceuticals using a recombinant *Arxula adeninivorans* whole-cell biosensor. Sens Actuator B Chem 211:439–448
Pham HTM, Chamas A, Nieter A, Giersberg M, Rutten T, Gehrmann L, Hettwer K, Tuerk J, Uhlig S, Simon K, Baronian K, Kunze G (2016) Determination of glucocorticoids using photometric (A-YGS) and spectrofluorometric (A-YGFS) bioassays based on modified *Arxula adeninivorans* cells: Applications in environmental analysis. Sens Actuator B Chem 223:540–549
Rajasärkkä J, Virta M (2013) Characterization of a bisphenol A specific yeast bioreporter utilizing the bisphenol A-targeted receptor. Anal Chem 85:10067–10074
Riedel K, Kunze G, König A (2002) Microbial sensors on a respiratory basis for wastewater monitoring. In: Scheper T (ed) Advances in biochemical engineering/biotechnology. Springer, Berlin, pp 81–118
Riedel K, Kunze G, Baronian KHR (eds) (2006) Biosensors and bioassays based on microorganisms. Research Signpost, Kerala
Routledge EJ, Sumpter JP (1996) Estrogenic activity of surfactants and some of their degradation products assessed using a recombinant yeast screen. Environ Toxicol Chem 15:241–248
Samsonova IA, Kunze G, Bode R, Böttcher F (1996) A set of genetic markers for the chromosomes of the imperfect yeast arxula adeninivorans. Yeast 12(12):1209–1217
Sanseverino J, Gupta RK, Layton AC, Patterson SS, Ripp SA, Saidak L, Simpson ML, Schulz TW, Sayler GS (2005) Use of *Saccharomyces cerevisiae* BLYES expressing bacterial bioluminescence for rapid, sensitive detection of estrogenic compounds. Appl Environ Microbiol 71:4455–4460

Sanseverino J, Elridge ML, Layton AC, Easter JP, Yarbrough J, Schultz TW, Sayler GS (2008) Screening of potentially hormonally active chemicals using bioluminescent yeast bioreporters. Toxicol Sci 107:122–134

Shantilatha P, Varma S, Mitra CK (2003) Designing a simple biosensor. In: Malhotra BD, Turner APF (eds) Advances in biosensors: perspectives in biosensors. JAI, Elsevier, Amsterdam, pp 1–17

Shaw WM, Yamauchi H, Mead J, Gowers GF, Bell DJ, Öling D, Larsson N, Wigglesworth M, Ladds G, Ellis T (2019) Engineering a model cell for rational tuning of GPCR signaling. Cell 177(3):782–796.e27

Shiizaki K, Asai S, Ebata S, Kawanishi M, Yagi T (2010) Establishment of yeast reporter assay systems to detect ligands of thyroid hormone receptors alpha and beta. Toxicol In Vitro 24:638–644

Shrivastava A, Gupta VB (2011) Methods for the determination of limit of detection and limit of quantification of the analytical methods. Chron Young Sci 2:21–25

Tag K, Hahn T, Riedel K, Kunze G (2006) Assays and microbial biosensors for detection of endocrine disruptors. In: Riedel K, Kunze G, Baronian KHR (eds) Biosensors and bioassays based on microorganisms. Research Signpost, Kerala, pp 211–228

Turner APF, Karube I, Wilson GS (eds) (1992) Biosensors: fundamentals and applications. Mir, Moscow

Viswanath G, Halder S, Divya G, Majumder CB, Roy P (2008) Detection of potential (anti)progestogenic endocrine disruptors using a recombinant human progesterone receptor binding and transactivation assay. Cell Mol Endocrinol 295:1–9

Wang H, Lang Q, Liang B, Liu A (2015) Electrochemical glucose biosensor based on glucose oxidase displayed on yeast surface. Methods Mol Biol 1319:233–243

Zhang J, Sonnenschein N, Pihl TP, Pedersen KR, Jensen MK, Keasling JD (2016) Engineering an NADPH/$NADP^+$ redox biosensor in yeast. ACS Synth Biol 5(12):1546–1556

Zhou T, Liang Z, Marchisio MA (2022) Engineering a two-gene system to operate as a highly sensitive biosensor or a sharp switch upon induction with β-estradiol. Sci Rep 12(1):21791

Chapter 14
Yeast-Based Biosensors for Clinical Diagnostics, Food Control, and Environmental Safety

G. Gayda, N. Stasyuk, O. Smutok, M. Gonchar, and Andriy A. Sibirny

Contents

14.1 Introduction 406
14.2 General Remarks Concerning Biosensors 407
  14.2.1 Biosensor as a Research Product and Analytical Instrument 407
  14.2.2 Biosensor as a Commercial Product 409
  14.2.3 Advantages and Peculiarities of Yeast-Based Biosensors 410
14.3 Yeast-Based Biosensors for Medicine 412
14.4 Yeast-Based Biosensors for Food Safety 415
14.5 Yeast-Based Biosensors for Environmental Monitoring and Bioremediation 420
14.6 Future Prospects and Trends in Yeast-Based Analytical Systems 426
References 427

Abstract Science, medicine, clinical diagnostics, biotechnologies, and environmental technologies all require methods that are highly selective, sensitive, rapid, and reliable for identifying key ingredients or metabolites. These methods determine product quality or serve as markers for diseases, physiological states, or environmental safety.

A wide range of dangerous pollutants, including heavy and transition metals, volatile toxicants, pesticides, plasticizers, pharmaceuticals, and natural and synthetic hormones, must be tested in various environmental media to predict potential hazards for humans and nature and to evaluate the effectiveness of wastewater treat-

G. Gayda (✉) · N. Stasyuk · M. Gonchar
Institute of Cell Biology, NAS of Ukraine, Lviv, Ukraine

O. Smutok
Department of Chemistry and Biomolecular Science, Clarkson University, Potsdam, NY, USA

A. A. Sibirny
Institute of Biotechnology, University of Rzeszow, Rzeszow, Poland

A. A. Sibirny (ed.), *Biotechnology of Yeasts and Filamentous Fungi*,
Grand Challenges in Biology and Biotechnology,
https://doi.org/10.1007/978-3-031-74726-7_14

ment. Many physicochemical analytical approaches have been developed; however, they are not suitable for the high-throughput monitoring of chemicals in air, soil, aquatic, and biological samples.

Biosensors are the most promising tools for these purposes. While enzymes have traditionally been used as biocatalytic elements in most biosensors, the development of microbial cell sensors has gained significant momentum only during the last decades. Microbial sensors, particularly yeast-based systems, show great promise in analytical practice. A microbial biosensor consists of a transducer combined with immobilized viable or nonviable microbial cells, offering an economical substitute for enzymes. The target analyte is typically either a substrate or an inhibitor of cell metabolism.

We highlight the main achievements in developing analytical systems based on yeast cells and discuss their potential applications in clinical diagnostics, food control, environmental monitoring, and detoxification.

14.1 Introduction

People in a peaceful world have the right to a healthy lifestyle: to drink clean water, eat quality food, breathe safe air, and receive accurate information about their health. Thus, the main goal of natural sciences is to provide adequate technologies to satisfy these demands.

Modern science, medicine, clinical diagnostics, biotechnologies (including those in the food and beverage industries), and environmental technologies demand highly selective, sensitive, rapid, and reliable methods. These methods are crucial for identifying key ingredients or metabolites that define product quality, act as markers for diseases, indicate the physiological state of the human organism, or ensure environmental safety (Turner et al. 1990).

Extensive research over the past decade has revealed the widespread presence of contaminants in wastewater, air, and soil worldwide. The increasing levels of pollutants in the environment pose a significant threat to ecosystems (Fiore et al. 2015). These hazardous substances, including heavy and transition metals, pesticides, plasticizers, pharmaceuticals, volatile toxicants, and natural and synthetic hormones, can enter surface waters through runoff from land application of biosolids, leaking sewer lines and septic systems, or incomplete removal by wastewater treatment systems (Eldridge et al. 2011; Kumar et al. 2013; Fiore et al. 2015). Being resistant to degradation, these pollutants pose a toxicological threat to wildlife and humans, adversely affecting various bodily systems, including the blood, cardiovascular, central nervous, respiratory, endocrine, reproductive, hormonal, and immune systems, potentially leading to premature mortality (Thurston et al. 2016; Watanabe et al. 2017).

The complex nature of these effects indicates the need for specialized methods to ensure the high quality and safety of human life. This includes advancements in biomedical diagnostics; environmental, agricultural, and food monitoring; as well as the detoxification and removal of hazardous chemicals. Developing and implementing these methods is crucial for effectively addressing the multifaceted challenges.

Among various analytical approaches, analytical biotechnology holds a special role by leveraging the principles of biomolecular recognition, which have been refined through evolution. Biomolecular recognition is responsible for storing information at the molecular level and reading it at the supramolecular level, making it a fundamental characteristic of life. This property of biomolecules and their complexes is harnessed in analytical biotechnology, which involves the use of tissues, cells, organelles, biomolecules, their complexes, and even the principles of their functioning for analytical purposes (Gonchar et al. 2017).

Biosensors are analytical devices converting energy from biospecific interactions into information about the nature and concentration of an analyte in a sample. They integrate a biological component, such as enzymes, antibodies, nuclcic acids, whole cell, or its fragments, with a physicochemical detector to provide a measurable signal in response to the presence of analytes.

This review is an expanded and updated version of our previous publications (Gonchar et al. 2017; Gayda et al. 2017). Here, the main achievements in the elaboration of microbial sensors, based on yeast cells, are described, and the perspectives of their usage in clinical diagnostics, food, and ecology control are discussed. It is worth mentioning that the distinction between the environmental, food, and medical applications of yeast-based biosensors is sometimes difficult to make and can appear arbitrary as environmental contaminants are harmful to humans, wildlife, and agriculture precisely because of their health effects (Battisti 2024, Mirmahdi et al. 2024; Martin-Yken 2020).

14.2 General Remarks Concerning Biosensors

14.2.1 Biosensor as a Research Product and Analytical Instrument

Biosensors represent a significant advancement in analytical biotechnology. While the International Union of Pure and Applied Chemistry (IUPAC) has not yet standardized an official definition, its electrochemical variant is described as "a self-contained integrated device capable of providing specific quantitative or semi-quantitative analytical information using a biological recognition element (biochemical receptor) that remains in direct spatial contact with an electrochemical transduction element" (Thévenot et al. 2001). Generally, a biosensor comprises two functional components: a bioelement (biorecognition unit)—which immobilizes biologically active material—and a physical transducer (signal converting unit). Metaphorically, biosensors are sometimes referred to as "chemical canaries," drawing on the historical use of canaries by miners to detect toxic methane in mines (Gonchar et al. 2017).

Essentially, biosensors function as information transducers where the energy from biospecific interactions is converted into information regarding the nature and concentration of an analyte in a sample. Various biorecognition elements can be employed in biosensors, including pieces of tissue, microbial cells, organelles, natural biomembranes or liposomes, receptors, enzymes, antibodies and antigens, nucleic acids, and biomimetics that mimic the structural and functional features of natural analogs. The bioelement serves as the recognition unit, selectively binding to or biochemically/metabolically converting the analyte, which in turn alters the physical or physicochemical characteristics of the transducer (Schmidt and Karube 1998). The output signal from the transducer is then typically amplified, processed, and displayed. Figure 14.1 illustrates the scheme of electrochemical biosensors.

The bioelement in biosensors is typically immobilized and often enclosed within an outer membrane (or positioned between two membranes in a sandwich configuration). This setup serves to prevent interfering substances from penetrating the sensitive bioselective layer and transducer surface, or it creates a diffusion barrier for the analyte. These membrane structures play a crucial role in enhancing the stability and selectivity of the biorecognition element, as well as providing diffusion limitations for biochemical reactions. Biosensor systems utilize various types of transducers, including electrochemical, optical, piezoelectric, thermoelectric, transistor-based, acoustic, and others. Each type of transducer converts the biochemical signal generated by the bioelement into a measurable output signal, facilitating the detection and quantification of analytes with high sensitivity and specificity.

The most essential advantages of biosensors are their high selectivity, high sensitivity, potential for miniaturization, and compatibility with computers. They also provide fast output due to the high activity and/or high local concentration of the catalytic or binding element in a biosensitive layer. However, their drawbacks include limited stability and the relatively complicated procedure required to prepare the biologically active material, especially the highly purified enzymes.

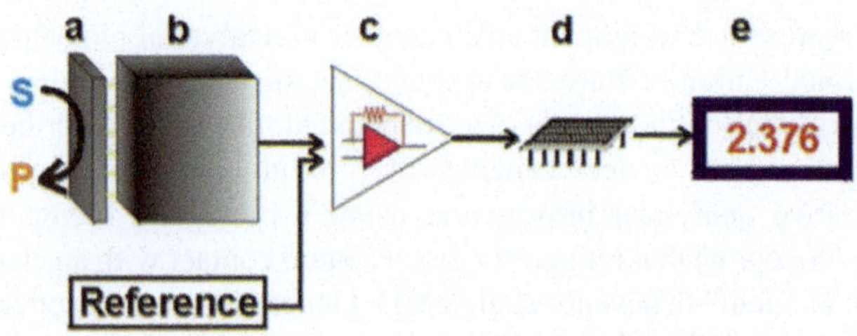

Fig. 14.1 Functional units of an electrochemical biosensor. The biocatalyst (**a**) converts the substrate (S) to the product (P). This reaction is determined by the transducer (**b**), which converts it to an electrical signal. The output from the transducer is amplified (**c**), processed (**d**), and displayed (**e**)

14.2.2 Biosensor as a Commercial Product

Biosensors, especially enzyme-based ones, are the most widespread devices; some are produced commercially. The global market for biosensors was estimated at approximately USD 28.44 billion in 2023, and it is expected to be worth around USD 58.44 billion by 2033 with a registered compound annual growth rate (CAGR) of 7.9%. The biosensor market may be described according to some principles: by region, by technology, by application, and by end user. The current state (in 2023) of the world biosensor market is demonstrated in Fig. 14.2 (https://www.precedenceresearch.com/biosensors-market).

According to Fig. 14.2c, d, biosensors for medicine, including point-of-care (PoC) testing and home care diagnostics, occupy the largest place in the world market. As per the World Health Organization, this fact may be explained by the rising prevalence of chronic disorders, especially diabetes.

In the healthcare sector, biosensors are extensively used to detect pathogens, antigens, and biomarkers for various diseases and ailments from biological samples such as blood serum, plasma, urine, saliva, and fecal matter. PoC biosensors are scaled down to compact devices that can detect diseases near the patient, reducing the therapeutic turnaround time. Although biosensors exist for various diseases, making them PoC remains challenging due to the specific requirements of bioreceptors, such as storage conditions and fabrication techniques. Cardiovascular diseases,

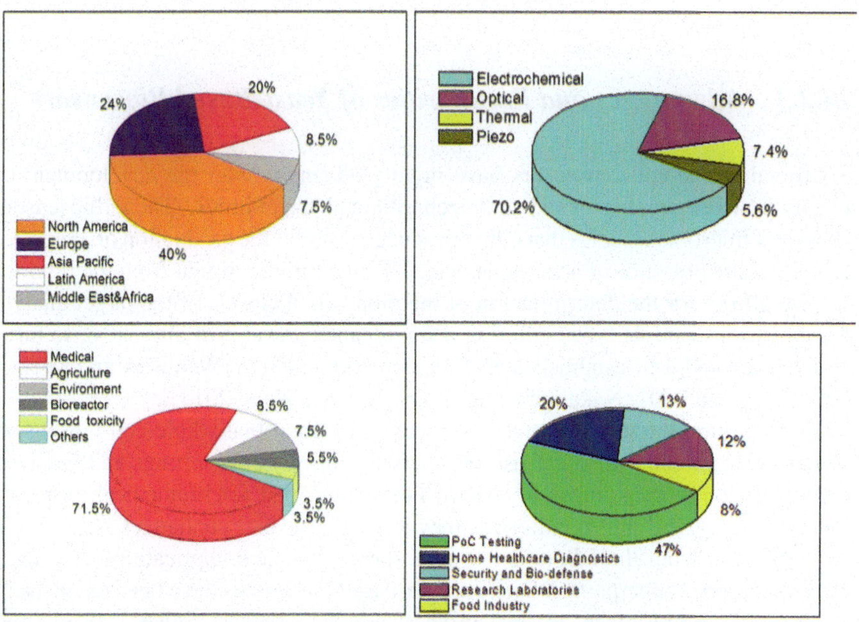

Fig. 14.2 Graphical illustration of the biosensor market in 2023: share by geography (**a**), technology (**b**), application (**c**), and end point (**d**)

neural diseases (stress), kidney diseases, urinary tract infections, and other viral infections are some of the conditions that can be detected using biosensors. PoC biosensors are available for detecting some of these diseases (Daurai and Gogoi 2022).

The most relevant analytes in the healthcare sector include glucose, cholesterol, and other metabolites; biomarkers of pregnancy and infectious diseases; drugs; and others. Biosensors are crucial in detecting blood glucose levels, which is essential for large-scale diabetes detection and management. This growing demand for diabetes monitoring is significantly supporting the growth and development of the global biosensor market. Analogously, chip-based biosensors were largely used for detecting the coronavirus among victims. As a result, due to the COVID-19 pandemic, the demand for biosensors increased gradually. Additionally, commercially produced biosensors are used for the online control of biotechnological processes, military or law enforcement applications, and fundamental research, such as in metabolomics. These devices provide real-time monitoring and analysis, which is crucial for optimizing production processes, ensuring security, and advancing scientific understanding in various fields (Nakamura et al. 2008; Shimomura-Shimizu and Karube 2010; Gupta et al. 2019; Martin-Yken 2020; Shepherd et al. 2022).

The main producers of biosensors are Bio-Rad Laboratories Inc., Abbott Laboratories, Biosensors International Group Ltd., DuPont Biosensor Materials, Pinnacie Technologies Inc., Johnson & Johnson, Molecular Devices Corporation, QTL Biodetection LLC, TDK Corporation, Zimmer & Peacock AS, etc.

14.2.3 Advantages and Peculiarities of Yeast-Based Biosensors

Microbial whole-cell biosensors have rapidly advanced with the development of twenty-first-century biological and technological capabilities. These biosensors consist of microbial species that either produce or inhibit the production of a reporter protein in the presence of a target analyte. The quantifiable signal from the reporter protein allows for the determination of the bioavailable levels of the target analyte across various sample types, offering a significantly lower cost alternative to most widely used and well-established analytical instrumentation (Walmsley and Keenan 2000; D'Souza 2001; Nakamura et al. 2008; Teo and Wong 2014).

Furthermore, the versatile and robust nature of microbial whole-cell biosensors shows great potential for their use in otherwise inaccessible settings and environments. Although these biosensors have been developed for biomedical, environmental, and agricultural monitoring, they still face several challenges before they can transition from the laboratory to widespread practical applications, like their enzyme-based counterparts. While bacteria-based biosensors have been extensively reported, there is a recent interest in yeast-based biosensors (YBBs), combining microbial with eukaryotic advantages, including the possession of specific receptors, stability, and high robustness (Nakamura et al. 2008; Moraskie et al. 2021; Shepherd et al. 2022; Wahid et al. 2023; Akor et al. 2024).

Yeast and mammalian cells are both eukaryotic organisms, which means they share similarities in cellular features and molecular mechanisms. This high level of conservation has long motivated scientists to use yeasts as model eukaryotes for studying a wide range of cellular biology processes. The best-studied among yeast species, *Saccharomyces cerevisiae*, was the first eukaryotic organism to have its genome entirely sequenced, making it easy to modify genetically. Yeasts are robust organisms with important practical advantages, including safety, stability, and ease of cultivation, transport, and conservation. These qualities make yeasts promising biological models for the development of biosensors that are suitable for the online monitoring of target analytes in samples of different natures, such as biological liquids, wastewater, and others. Additionally, from an ethical standpoint, using yeast cells allows researchers to avoid animal models when determining the potentially toxic effects of diverse compounds or screening for therapeutic molecules (Martin-Yken 2020; Wahid et al. 2023). Yeasts have been commonly used as biological elements of biosensors since the 1970s.

YBBs are prospective analytical instruments for several reasons: (1) they can be produced in an "active dry" form inexpensively and stored for long periods; (2) YBBs are well studied at both the technological and genetic levels; (3) they have the ability to accurately detect a wide range of compounds; (4) YBBs can directly measure organic compounds in samples without any pretreatment; (5) they can serve as a chassis for higher-eukaryotic sensing modalities (e.g., G-protein-coupled receptors (GPCRs)); (6) YBBs can tolerate rather harsh environments, especially compared to bacteria; (7) they are able to thrive in both aerobic and anaerobic conditions; (8) biosensing can be carried out in a micron-scale unit (the cell) without power requirements, functioning as a self-powered biosensor; (9) the analytical procedure does not require a highly trained operator, making it accessible and easy to use in various settings (Walmsley and Keenan 2000; Benton et al. 2007; Bovee et al. 2007, 2011; Shimomura-Shimizu and Karube 2010; Dragone et al. 2014; Gutiérrez et al. 2015; Adeniran et al. 2015; Jarque et al. 2016; Martin-Yken 2020; Wahid et al. 2023).

Yeasts have been used as biological elements of biosensors since the 1970s. The most common reporting systems for YBBs are based on optical and electrical methods, as well as cell growth (Nakamura et al. 2008; Vigneshvar et al. 2016). The advantages of optical methods include high sensitivity and simplicity, with the colorimetric variant being particularly promising due to its results being visible to the naked eye. Fluorescence and luminescence detection do not require substrate addition (Shetty et al. 2004; Leskinen et al. 2005; García-Alonso et al. 2009; Eldridge et al. 2011). The main advantages of electrical methods are high sensitivity and rapid procedure (Schwartz-Mittelman et al. 2005; Ino et al. 2009; Chelikani et al. 2012; Adeniran et al. 2015).

Although YBBs show promise as analytical tools, they still have some disadvantages, such as poor selectivity, low sensitivity, and a relatively slow response due to the moderate diffusion of substrates into the cell through natural barriers like the membrane or cell wall. Notably, most YBBs are still in the early developmental stage, with only a few prototypes tested for real applications. To overcome these drawbacks of the YBBs, numerous techniques are applied, such as genetic engineering to block undesired or induce desired metabolic pathways in the cells as well as micro- and nanotechnologies. Micro/nanofabrication has shown remarkable potential for YBBs due to enhanced optical and electrochemical measurements, improved immobilization and automated culture environments, as well as high portability, making them more practical compared to conventional analytical methods (Adeniran et al. 2015; Martin-Yken 2020; Romero and Picchio 2020; Bean et al. 2022; Wahid et al. 2023; Abu-Rmailah et al. 2024; Gregório et al. 2024; Lind et al. 2024).

14.3 Yeast-Based Biosensors for Medicine

YBBs have emerged as valuable tools in medical research, particularly for studying diseases and screening drugs. Their use is facilitated by the simplicity and versatility of yeast cells, which can be genetically engineered to express various proteins and pathways relevant to human diseases.

Yeast cells, especially *S. cerevisiae*, have been used as a model of mammalian cells to study cellular processes and pathways that are conserved in humans. This makes them ideal for investigating the molecular mechanisms underlying various diseases, including neurodegenerative disorders like Alzheimer's and Parkinson's disease, and for developing novel strategies against these contagious diseases. The list of tests that can be performed in the yeast *S. cerevisiae* as a model of Alzheimer's disease (AD) is the following: aging, drug screening, omics studies, assays of autophagy, protein aggregation, protein interaction, stress reporter, cell viability and growth, cell cycle, and life span (Chernoff et al. 2020; Mcdonald et al. 2020; Dhakal and Macreadie 2020; Dhakal and Macreadie 2022a).

A common approach for designing YBBs for studying diseases is the expression of fluorescent proteins (FPs) with disease-associated proteins (DPs). FPs are the best options for fluorescence reporters, whose genes have been cloned, sometimes further modified, and made available for exploitation. FPs offer a versatile and powerful tool for visualizing subcellular structures and studying cellular processes, providing critical insights into disease mechanisms and potential therapeutic targets. By tagging DPs with fluorescent markers, researchers can better understand protein interactions and dynamics, facilitating drug screening and the development of innovative therapeutic strategies (Dhakal and Macreadie 2020).

DPs such as huntingtin, alpha-synuclein, and amyloid-beta (Aβ), which are involved in Huntington's disease, Parkinson's disease, and AD, respectively, have been tagged with FPs to study their roles and screen drugs. Most commonly, green fluorescent proteins (GFPs) have been used as fluorescent markers, primarily due to the availability of several analytical platforms, including YBBs, to detect the goal DP, especially Aβ (Dhakal and Macreadie 2022b).

YBBs have been applied to search for drugs against the intracellular toxic protein Aβ, which could be effective in treating or preventing AD. AD is a highly prevalent, age-related, neurodegenerative disease associated with the accumulation of Aβ and tau proteins, which affect important cellular functions, such as aging, oxidative stress, cell cycling, and proteostasis. Strategies for the prevention of damage by these proteins can be explored due to YBB. In this way, YBBs may provide valuable information or aid in generating innovative ideas about the effects of Aβ on brain cells and help explore potential therapeutic targets for AD (Chernoff et al. 2020; Mcdonald et al. 2020; Dhakal and Macreadie 2020; Dhakal and Macreadie 2022a, b).

Besides tagging proteins directly involved in disease pathogenesis, YBBs are also designed to report on important cellular mechanisms involved in cellular defense systems, such as autophagy. For example, to monitor the autophagic flux or drug-induced activation of autophagy, Atg8, a protein sequestered in the autophagosomal membrane when autophagy is activated, has been tagged with GFP. The fluorescence detection of the fusion protein has been used as an indicator for the autophagy activation signal. Latrepirdine is one example of such a drug identified as an autophagy inducer following the usage of such YBB (Dhakal and Macreadie 2020).

McDonald et al. (2020) summarize the existing knowledge of AD and the work over the past two decades on yeast models for AD, highlighting how these YBBs contribute to the elucidation of the development of treatments and preventative strategies. The significant role of YBBs in translational research, particularly in the prescreening of compounds as potential drugs to treat and/or prevent AD, is emphasized. The authors discuss the future use of YBBs to provide further insights into the mechanisms of AD and to identify multifactorial drugs or therapeutics targeting this disorder.

YBBs are useful, too, for studying protein-protein interactions, which are critical in understanding disease pathways. For example, yeast two-hybrid systems can be employed to identify interactions between proteins that may contribute to disease progression. Protein interaction studies using yeast surface displays allow to detect interactions between two proteins of interest and are also useful for drug discovery. The yeast two-hybrid system is one of the most powerful techniques for identifying protein-protein interactions in vivo. In the reviews of Dhakal and Macreadie (2020, 2022a, b), the advances and prospects of this method are described in detail. Francino-Urdaniz and coauthors (2021) reported novel results that are useful for improving COVID-19 diagnostics and treatment. They used the SARS-CoV-2 spike receptor-binding domain surface display to identify escape mutants for the neutralizing antibodies. Uchański and coauthors (2019) reported on nanobodies against three distinct types of proteins of interest, including two membrane proteins. Nanobodies were selected from an immune library using the yeast surface display platform. Thus, yeast surface display has served multiple purposes, including its use in biosensing, and will prove to be instrumental in developing newer YBBs in the future.

YBBs can be used for genetic screening to identify genes involved in disease resistance or susceptibility. By introducing mutations or overexpressing certain genes, researchers can study the effects on cellular pathways and identify potential therapeutic targets (Cervelli and Galli 2021).

YBBs can be adapted for high-throughput drug screening. By incorporating a reporter system that indicates a cellular response to a drug, researchers can quickly assess the efficacy and toxicity of large libraries of compounds (Martin-Yken 2020; McDonald et al. 2020; Wahid et al. 2023).

Yeasts can be engineered to express human drug targets, allowing for the identification of compounds that interact with these targets. This is particularly useful for validating new drug targets and understanding the mechanisms of action of existing drugs, studying their cytotoxicity, and optimizing drug formulations (Bean et al. 2022).

Overall, the use of YBBs in medicine offers a cost-effective and efficient approach to studying diseases and screening drugs, providing insights that can lead to the development of new therapeutic strategies.

There is a growing need for suitable testing methods to screen compounds that may activate or inhibit human receptors and/or screen environmental samples for the presence of exogenous ligands or inhibitors, classified as hazardous endocrine-disrupting compounds (EDC). The relative ease, portability, and rapidity of the YBB assay, in addition to chemical methods, make it a cost-effective choice for preliminarily screening the EDC potential of chemicals on humans and wildlife. For this purpose, several *S. cerevisiae* yeast cell-based bioreporters utilizing different nuclear receptors have been developed (Bovee et al. 2007; Rajasärkkä 2013).

The most prolific class of YBBs is based on the heterologous expression of human receptors: steroid estrogen and its analogs, as well as thyroid, androgen, glucocorticoid, mineralocorticoid, and progesterone receptors, have all been recombinantly expressed, often multiple times with different reporting strategies (Kaiser et al. 2010). To assay estrogens, the sensitive analytical system BioLuminescent Yeast Estrogen Screen (BLYES) was proposed: the receptor is expressed continuously, and a plasmid contains reporter genes induced by the receptor-ligand dimer (Sanseverino et al. 2005).

Summarizing this information, it is worth concluding that yeast has been widely used as a model organism for studying various diseases due to its status as the simplest unicellular eukaryote, which is safe and easy to culture. It shares many fundamental processes with higher eukaryotes, including humans. Yeast's role as a biosensor extends from understanding disease pathogenesis to drug discovery studies. This is largely due to its conserved genetics, the ease with which its genome can be modified, and accessible analytical methods. Additionally, yeast's ability to survive with defective mitochondria makes it a highly adaptable microbe for designing whole-cell biosensing systems (Dhakal and Macreadie 2022b).

Other applications of yeast-based biosensors (YBBs) for analyzing biological liquids to detect various substances, such as glucose, ethanol, L-lactate, L-amino acids, hormones, heavy metals, other pollutants, and metabolites, were discussed in detail in our previous review (Gonchar et al. 2017) and in sections 4 and 5 of the current report.

14.4 Yeast-Based Biosensors for Food Safety

The food and beverage industry requires highly selective, sensitive, rapid, and reliable methods for monitoring key ingredients and metabolites, as well as contaminants, to ensure high product quality and safety for human health. The analytical parameters that are usually controlled in food products are sugar (glucose, fructose); alcohol; total acid, including organic acids (lactic, tartaric, malic, citric); amino; proteins (including enzymes); lipids; inorganic compounds, including metal ions; and microbiological components, including desirable yeasts and bacteria, as well as contaminating microorganisms (Mello and Kubota 2002; D'Souza 2001; Rotariu et al. 2002; Gayda et al. 2019).

Food safety is of great concern for many countries; e.g., the European Commission has identified it as a priority area of attention (Mello and Kubota 2002; Sankarankutty 2014). The development of biosensors for food safety applications has been an area of tremendous activity in recent years.

There are described a lot of biosensors for the assay of various components in food samples and during fermentation processes in winemaking and other biotechnological processes (Karube and Suzuki 1983; Karube and Tamiya 1987; Karube and Sode 1989; Tkac et al. 2003; Otles and Yalcin 2012; Lim et al. 2015; Chunhui and Choi 2013; Gayda et al. 2015b, 2019; Pundir et al. 2021; Smutok et al. 2021).

The determination of ethanol, sugars, organic acids, and amino acids is essential in many industries and biotechnologies; monitoring their content is crucial during their production in fermentation processes and when they are used as substrates in various biotechnologies, such as in the production of liquors, alcoholic beverages, milk products, and certain foodstuffs (Garjonyte et al. 2008; Garjonyte et al. 2009; Shkil et al. 2009; Smutok et al. 2019). Numerous enzyme biosensors and a few YBBs for the detection of these analytes have been developed and described in detail in our previous reviews (Gonchar et al. 2017; Gayda et al. 2019).

One of the indicators of food and quality is the absence of contaminants, which affect negatively to taste, odor or are dangerous for human organism, so they should be carefully detected. The appearance of the key chemicals and microbial contamination responsible for olfactory defects in food products is a result of the poor control of raw materials or nonstandard conditions in the production and storage of the products.

Toxic and carcinogenic ethyl carbamate (EC) is a product of the spontaneous reaction of urea with ethanol in beverages under nonstandard technology conditions (Uthurry et al. 2006). High-level countries control the EC level in some alcoholic drinks. To avoid the potential health hazards of EC in food (including wine), it is necessary to control the levels of EC or L-arginine (Arg), as EC's precursor, in all steps of the food (or beverage) production chain. Different types of biosensors, promising for the monitoring of the basic components and dangerous contaminants of wines and juices, were developed and summarized in our review (Gonchar et al. 2017; Gayda et al. 2015a, b, 2019, 2022).

Mycotoxins of the zearalenone family are common contaminants in milk that can cause mycotoxicoses (Bryden 2007). In order to ensure the quality of milk, genetically modified *S. cerevisiae* cells were used in the YBBs for the detection of zearalenone mycotoxins in milk (Valima et al. 2010).

YBBs with genetically modified *S. cerevisiae* cells were used for the analysis of cyanide in fruit brandies (Filipović-Kovačević et al. 2002) and methylamine in fish (Stasyuk et al. 2024). For the quality control of meat freshness, YBB using *Trichosporon cutaneum* was developed. This YBB detects polyamines and amino acids (Yano et al. 2001).

Due to the increasing demand for high and stable crop production, human populations are highly dependent on pesticide use for growing and storing food. Chen and Nielsen (2016) reported the identification of 27 different classes of pesticides in wines alone, highlighting the potential of YBBs for future investigations across a wide range of applications. The environmental monitoring of these agrochemicals is therefore of utmost importance because of their collateral effects on ecosystem and human health.

Mendes and coauthors (2024) described easy-to-use YBB for specific determination of tebuconazole (TEB), a common agrochemical fungicide. For this aim, *S. cerevisiae* cells were engineered with a reporter gene downstream of specific promoters that are expressed after exposure to TEB, and the sensitivity and specificity of this model system were characterized. After optimization, we found that this YBB consistently detects TEB at concentrations above 5 μg L^{-1} and does not respond to realistic environmental concentrations of other tested azoles, suggesting that it is specific. The use of this system was proposed as a complementary tool in environmental monitoring programs, namely, in high throughput scenarios requiring the screening of numerous samples.

Most of the developed YBBs exhibit good stability and fast response to the target analyte. The main analytical characteristics of several reported yeast-based biosensors (YBBs) for the analysis of key food components and contaminants are presented in Table 14.1.

A series of biosensor approaches, including YBBs, for the determination of organic and inorganic analytes, namely, ethanol, L-lactate, glycerol, arginine, and methylamine, was proposed by us and summarized in detail in our papers (Gonchar et al. 2017; Gayda et al. 2019, 2022; Smutok et al. 2021; Stasyuk et al. 2024). Whole-cell YBBs are simple and cost-effective to produce because they eliminate the need for enzyme purification. Additionally, enzymes tend to be more stable within their natural environment inside the cell. To construct these YBBs, a number of selected and genetic-engineering yeast cells and overproducers of analytically important enzymes were developed previously.

Nanotechnology has contributed significantly to biosensing technology by introducing materials with unique properties, such as increased surface area, improved conductivity, and enhanced catalytic activity. These advancements have led to the development of biosensors with greater sensitivity, specificity, and versatility. Naresh and Lee (2021) discuss how modern biosensors have evolved with the integration of various nanomaterials, enhancing their performance and application range.

Table 14.1 Food analysis using yeast-based biosensors

Sample of food	Target analyte	Yeast	Mode of signal registration	Detection limit, μM	Linearity range, mM	Reference
Alcoholic beverages	Ethanol	*S. ellipsoideus*	Amperometric	6	–	Rotariu and Bala (2007)
			Potentiometric	–	0.02–50	Rotariu et al. (2004a, b)
		S. cerevisiae	Conductometric	1000	5–100	Korpan et al. (1994)
	L-lysine		Amperometric	–	$1 \cdot 10^{-3}$–0.01	Akyilmaz et al. (2007)
Wines and juices	L-arginine	*H. polymorpha*		85	Up to 0.6	Stasyuk et al. (2014)
Model solution	Ethanol	*H. polymorpha*		–	0.3–2.7	Karkovska et al. (2017)
Milk, kefir, and yogurts	L-lactate	*S. cerevisiae*		–	Up to 1	Garjonyte et al. (2006)
Fruit brandies	Cyanide			8	–	Filipović-Kovačević et al. (2002)
Milk	Mycotoxin		Luminescence	–	$1 \cdot 10^{-6}$–$2.6 \cdot 10^{-4}$	Valima et al. (2010)
Fish	Methylamine		Amperometric	4	0.012–0.1	Stasyuk et al. (2024)
Model solution	Fungicide tebuconazole				5 μg L^{-1}	Mendes et al. (2024)

Combining nanobiotechnology with YBBs has become a crucially novel strategy for the development of simple and reliable monitoring systems for food quality and safety. Many electrochemical biosensors based on carbon nanotubes, graphene and its derivatives, various metal nanoparticles, and polymers were proposed and may be used in food analyses (Adeniran et al. 2015; Zhang et al. 2016; Stasyuk et al. 2014; Stasyuk et al. 2024). Nanotechnology approaches have been successfully used for the improvement of the functional properties of YBBs. Integrated approaches, including the best advantages of genetic engineering, nanotechnology, and synthetic biology, provided a better perspective for developing specific and sensitive biosensors with high regenerative potentials for diverse uses (Bereza-Malcolm et al. 2015; Adeniran et al. 2015; He et al. 2016). YBBs appear to be quite promising for online sample control. Their advantages lie in the fact that organic compounds in samples can be measured directly, without any pretreatment such as filtration and dialysis (Dragone et al. 2014).

The concept of using nanoparticles in biosensor technology to increase the output is rather fresh nowadays (Vigneshvar et al. 2016; Yudina et al. 2015; Jarque et al. 2016). A number of biosensors with immobilized nanoparticles on the electrode surface were constructed. On the other hand, we have recently reported improvement in the analytical parameters of YBBs by genetic manipulation with the yeast cell (Smutok et al. 2007; Karkovska et al. 2017).

The idea of the implementation of exogenous enzymes to increase the amount of the target enzyme in the cells to be used in YBBs was highlighted some years ago by Gonchar and colleagues (Karkovska et al. 2015, 2017). To enhance the sensitivity and shorten the response time of YBBs to their substrates, cells of the yeast *H. polymorpha*, both intact (icells) and permeabilized (pcell), were enriched with target yeast enzymes, which were immobilized on the surface of gold nanoparticles (nAu). In these experiments, both natural and recombinant cells (cells and rcells, respectively) were used, along with yeast enzyme alcohol oxidase (AO), recombinant arginase I, and flavocytochrome b_2 (Fcb_2) for the analysis of ethanol, L-arginine, and L-lactate, respectively. The permeability of icells and pcells to enzyme-nAu was analyzed via TEM and X-ray spectroscopy. It was demonstrated that using nAu coimmobilized with the target enzyme for enriching cells resulted in a twofold decrease in the time of YBB output and in significantly enhancing the sensitivity of the YBBs. The proposed approach was applied recently for the construction of YBBs on methylamine (Stasyuk et al. 2024) using recombinant *S. cerevisiae cells*, methylaminooxidase, and bimetallic nanoparticles. The analytical characteristics of the developed YBBs are presented in Table 14.2.

It was shown that the sensitivities of YBBs, which consist of enzyme-nanoparticles-pcells, are enhanced significantly in comparison with control intact cells (icell). For example, Fcb_2-nAu-prcells to L-lactate is 12-fold higher compared to icells and 2.6 times higher than for pcells. Similar results were demonstrated for other YBBs, presented in Table 14.2. The developed YBBs have been successfully used for the analysis of real samples of food and drinks.

Thus, the enrichment of the sensing cells by the target enzyme is achieved by a combination of two approaches: on the genetic level—by the overexpression of the

Table 14.2 Analytical characteristics of biosensors based on the cells of yeasts

Analyte	Biomembrane	Linearity range, mM	Response time, s (95%)	Stability, days (50%)	K_M^{app}, mM	Sensitivity, $A \cdot M^{-1} \cdot m^{-2}$	Reference
Ethanol	AO-nAu- pcells	Up to 2.0	20	30	1.93 ± 0.08		Karkovska et al. (2017)
L-lactate	Ircells	0.3–2.7	60	–	–	4.1	
	Prcells	0.3–2.7	30	–	–	18.7	
	Fc b_2-nAu-prcells	0.3–2.7	5	–	–	48.6	Karkovska et al. (2015)
	Prcells	Up to 1.6	20	Over 40	8.0 ± 0.66	–	Smutok et al. (2007)
L-arginine	[a]Prcells/urease	Up to 0.6	60	3	0.51 ± 0.05	14 ± 1.2	Stasyuk et al. (2014)
	nAu-prcells/urease	0.14–1.2	60	3	0.62 ± 0.01	37	Karkovska et al. (2017)
	[a]Arginase-nAu-prcells/urease	0.01–0.7	30	3	0.45 ± 0.09	357	
	Prcells/urease	Up to 0.6	–	3	0.62 ± 0.1	–	
Methylamine	Icells	0.024–0.09	60	5	670	56	Stasyuk et al. (2024)
	Pcells	0.012–0.10	60	5	105	272	
	Pcells-CdTe/GCE	0.024–0.10	35	5	100	320	
	Pcells-CeAu/GCE	0.024–0.09	35	5	150	189	
	[a]Pcells-MAO-nCdTe	0.001–0.02	10	7	20	5450	
	Pcells-MAO-nCdAu	0.005–0.02	20	6	30	2365	

[a]These YBBs have been successfully tested on real samples of food

corresponding gene in the recombinant cells—and with the nanotechnological approach—by the transfer of enzyme-bound nanoparticles into the cells. Immobilization of enzymes on the surfaces of nanoparticles was carried out to allow for an increase in the local concentrations of the enzymes and the enhancement of enzyme stability. These properties are critically important in biosensors.

YBBs offer a promising approach to managing fungal pathogens through innovative biotechnological strategies, contributing to sustainable agriculture and improved environmental health (Comitini et al. 2023; Dukare et al. 2019; Zhang et al. 2020). Antagonistic yeasts, also known as biocontrol yeasts, offer a promising alternative to chemical fungicides for controlling postharvest decay in fruit (Agirman et al. 2023; Wang et al. 2023). These yeasts are widely environmentally friendly and safe for humans. Their mechanisms include nutrition and space competition, mycoparasitism, and the induction of host resistance.

Yeast, as a model organism, provides a versatile platform for designing YBBs that can detect, interact with, and inhibit the growth of harmful fungi (Freimoser et al. 2019; Freimoser et al. 2024; He et al. 2024). YBBs can be used as antagonists of pathogenic fungi by leveraging the natural biological properties and genetic engineering capabilities of yeast (Liu et al. 2013; Oztekin et al. 2023). Some ways that YBBs can function as antagonists of pathogenic fungi are the following: yeast cells can be engineered to express specific receptors that recognize fungal pathogens, and upon detection, these yeast cells can produce antifungal compounds or other inhibitory substances to combat the pathogenic fungi; engineered yeast can outcompete pathogenic fungi for nutrients and space, effectively reducing their growth and proliferation in the environment; YBBs can be used as bioreactors, detecting and degrading mycotoxins or other harmful substances produced by pathogenic fungi; yeast can be engineered to interfere with the quorum sensing or signaling pathways of pathogenic fungi, disrupting their ability to coordinate and cause disease; yeast strains can be developed as biological control agents that naturally suppress the growth of pathogenic fungi in agricultural settings, reducing the need for chemical fungicides (Hilber-Bodmer et al. 2017; Zhang et al. 2020; Rueda-Mejia et al. 2023; Sipiczki et al. 2024).

14.5 Yeast-Based Biosensors for Environmental Monitoring and Bioremediation

To enhance environmental quality, various physiological and biological methods have been implemented worldwide. Although several in vivo and in vitro assays have been developed, these approaches are not suitable for high-throughput screening of chemicals in environmental samples. Chemical analysis methods, being sensitive, are labor-intensive and costly and require specialized equipment. As a result, robust microbiological methods have emerged as valuable tools for testing chemicals in environmental samples. The European Union's chemical legislation, REACH, has increased the demand for chemical testing methods while aiming to reduce the use of animals in these tests. It has been proposed that inexpensive high-throughput in vitro assays could be employed for the initial screening of chemicals before further testing with other methods. In this context, yeast-based systems are considered particularly promising for supporting environmental safety (Chatterjee et al. 2008; Shimomura-Shimizu and Karube 2010; Sibirny et al. 2011; Eldridge et al. 2011; Rajasärkkä 2013; Dragone et al. 2014; Sigawi et al. 2014; Bui et al. 2015; Sun et al. 2015; He et al. 2016; Guaragnella and Guaragnella 2023; Ukhurebor et al. 2024; Vigneshvar et al. 2016).

YBBs have the ability to accurately detect a lot of compounds (see Tables 14.1–14.3). YBBs incorporating various yeasts have been proposed as the effective approach of synthetic biology to detect an incredibly large range of environmental pollutants, including but not limited to metals, carcinogens, etc. (Hou et al. 2013; Bereza-Malcolm et al. 2015; Yudina et al. 2015).

The release of endocrine-disrupting compounds (EDCs) to the environment poses a health hazard to both humans and wildlife. EDCs can activate or inhibit endogenous endocrine functions by binding hormone receptors, leading to potentially adverse effects. Conventional analytical methods can detect EDCs at high sensitivity and precision but are blind to the biological activity of the detected compounds. To overcome this limitation, YBBs have previously been developed as a prescreening method, providing an effect-based overview of hormonal-disruptive activity within the sample prior to the application of analytical methods (Westlund and Yargeau 2017; Roszko et al. 2018; Cevenini et al. 2018; Battisti 2024; Chaves 2024). These YBBs express human endocrine-specific receptors, cotransfected with the relevant response element fused to the specific fluorescent protein reporter plate format, as well as in conjunction with sample separation using high-performance thin-layer chromatography (Moscovici et al. 2020; Baetz et al 2021). Abu-Rmailah and coauthors (2024) describe several molecular manipulations of the sensor/reporter circuit in a *S. cerevisiae* bioreporter strain that have yielded an enhanced detection of estrogenic-like compounds. Enhanced responses were observed both in liquid culture (96-well plate format) and in combination with sample separation via high-performance thin-layer chromatography. This latter method enables the evaluation of the biological effects of individual sample components without requiring their chemical identification during the screening stage.

The presence of EDCs in water poses a significant threat to human and animal health, as recognized by regulatory agencies throughout the world. The Yeast Estrogen Screen (YES) assay is an excellent method to evaluate the presence of these compounds in water due to its simplicity and capacity to assess the bioaccessible forms/fractions of these compounds (Kase et al. 2008). In the presence of a compound with estrogenic activity, *S. cerevisiae* cells, containing a lacZ reporter gene encoding the enzyme β-galactosidase, are induced; the enzyme is synthesized and released to the extracellular medium. Gregório et al. (2024) proposed a highly sensitive and rapid fluorescent method for 17β-estradiol determination, based on the described principle.

Two types of YBBs may be differentiated: dependent upon the transcription (YBB^{DT}) of a gene to report the detection of a desired molecule and independent (YBB^{IDT}) of this reporting mechanism (Adeniran et al. 2015).

$YBBs^{DT}$ frequently depend on the heterologous expression of sensing elements from nonyeast organisms, a strategy that has greatly expanded the range of

Table 14.3 Yeast-based biosensors for the detection of environmental pollutants

Group	Detected Compound(s)	Mode of detection	Limit of detection, range (μM)	Yeast	References
Metals	Copper	Amperometry Colorimetry Fluorescence Luminescence Puls stripping voltammetry	0.11–500 1–100 0.5 0.5 0.1–100	*Saccharomyces cerevisiae* *Rhodotorula mucilaginosa*	Tag et al. (2007); Jarque et al. (2016) Vopálenská et al. (2015) Shetty et al. (2004) Roda et al. (2011) Turdean (2011)
	Cadmium	Fluorescence	0.2	*S. cerevisiae*	Matsuura et al. (2013)
			1 to 900	*Hansenula polymorpha*	Park et al. (2007); Jarque et al. (2016)
	Methyl-mercury	Fluorescence	0.44 ng/L	*S. cerevisiae*	Tafurt-Cardona et al. (2015)
	Heavy metals	Fluorescence	12–45		Radhika et al. (2005)
Endocrine disruptors	Thyroids	Luminescence Colorimetry	3.7×10^{-3} 0.5×10^{-3} to 0.075		Li et al. (2014); Jarque et al. (2016) Li et al. (2014); Li et al. (2008a, b); Shiizaki et al. (2010)
	Retinoids	Colorimetry Luminescence	0.01 0.06		Shiizaki et al. (2014); Jarque et al. (2016) Kabiersch et al. (2013)
	Tributyltin	Luminescence	0.06		Kabiersch et al. (2013)
	Androgens	Colorimetry Luminescence Luminescence Fluorescence	0.2×10^{-3} 0.05×10^{-3} 0.5×10^{-3} 0.01×10^{-3}	*S. cerevisiae*	Gerlach et al. (2014) Michelini et al. (2008) Leskinen et al. (2005) Gregório et al. (2024)
	Estrogens	Fluorescence Colorimetry Luminescence Luminescence Fluorescence Amperometry	3×10^{-3} $(0.03–0.04) \times 10^{-3}$ 0.03×10^{-3} 0.01×10^{-3} 1×10^{-3}		Gregório et al. (2024) Bovee et al. (2011) Kaiser et al. (2010); Jarque et al. (2016) Sanseverino et al. (2005) Leskinen et al. (2005) Schwartz-Mittelman et al. (2005)
		Amperometry	5.3 ng/L	*Arxula adeninivorans*	Pham et al. (2012, 2013)
	Progestogens	Fluorescence Colorimetry	0.1×10^{-3} to 1 1		Chatterjee et al. (2008) Noguerol et al. (2006)
	Dioxin-like	Fluorescence	5×10^{-3}		Noguerol et al. (2006)
	Glucocorticoids	Fluorescence	0.5		Bovee et al. (2011)
	Metabolic disruptors	Amperometry	0.01		Chen et al. (2004)
	Bisphenol A	Bioreactor		*Pichia pastoris*	Guo et al. (2024)
	Bisphenol	Luminescence	0.11	*S. cerevisiae*	Rajasärkkä (2013)
	Serotonin Estrogens	Fluorescence	10		Nakamura et al. (2015)

(continued)

Table 14.3 (continued)

Group	Detected Compound(s)	Mode of detection	Limit of detection, range (μM)	Yeast	References
Genotoxins and cytotoxins	Chemicals	Fluorescence	50–90		Bui et al. (2015)
			3–50		
			50		
			10		Wei et al. (2013)
			3–45		Benton et al. (2007)
		Luminescence	NA		Leskinen et al. (2005)
	Tebuconazole				Mendes et al. (2024)
	Pharmaceuticals	Amperometry	83.65 μg/L	*A. adeninivorans*	Pham et al. (2015)
	Methyl methane sulfonate	Microfluidic system		*Saccharomyces, Yarrowia, Kluyveromyces, Rhodotorula*	Safarik et al. (2015)
		Fluorescence	0.28 μM ~ 450	*S. cerevisiae*	Lim et al. (2015); Knight et al. (1999)
			0.1%		García-Alonso et al. 2009)
	Formaldehyde	Amperometry	100	*H. polymorpha*	Sibirny et al. (2011)
	Methylamine	Amperometry	0.001–0.02	*S. cerevisiae*	Stasyuk et al. (2024)
	Inhibitors of the respiratory system	Amperometry	0.8–1.2	*S. cerevisiae*	Goldblum et al. (1990)
	ROS inducers	Fluorescence	0.1		Yang et al. (2005)
Biodegradable organics	Chemicals that support the growth of microorganisms	Biochemical oxygen demand (BOD)	1.24 mg/L	*A. adeninivorans*	Chan et al. (2000)
			2.4–80 mg/L	Artificial mix: *Pichia angusta, A. adeninivorans, Debaryomyces hansenii*	Yudina et al. (2015)
		Amperometry BOD	39–206	*S. cerevisiae*	Seo et al. (2009); Renneberg et al. (2004) Nakamura et al. (2007)

molecules available for detection by these biosensors. YBBs^{DT} recognize the ligand of interest, typically via a native or heterologous receptor protein. When activated, the protein interacts with the cell's transcription machinery, either directly or via a signaling cascade, to induce (or cease) the production of a reporter gene. A lot of different YBBs^{DT} include both receptors specific to organic compounds and heavy metals. In YBBs^{DT}, a fluorescence, luminescence, color, or current signal was usually detected (Eldridge et al. 2011; Adeniran et al. 2015).

YBBs^{IDT} are promising for sensing difficult-to-detect analytes by modifying yeast metabolism to generate easily detected molecules (redox potential, dissolved oxygen, etc.). YBBs^{IDT} convert signals that are difficult to measure into ones that can be easily detected, using either by-products of metabolism or reporting systems triggered directly by analyte recognition. Changes in fluorescence, luminescence, current, as well as pH or oxygen signal were detected (Adeniran et al. 2015).

G-protein-coupled receptors (GPCRs), being members of the largest family of human membrane proteins and having the same basic structure, are able to selectively detect an incredibly diverse range of molecules, including photons, ions, small molecules, and proteins (Miao and McCammon 2016). Many YBBs require the expression of heterologous GPCRs that interface with the *S. cerevisiae* pheromone mating pathway. The incorporation of GPCRs from higher organisms with functionalities not naturally available to yeast greatly expands the capability of YBBs (Jacobson 2015; Adeniran et al. 2015). To develop bionoses, two strategies were proposed: 1) replace or slightly modify the alpha subunit, which plays a key role in both activating and resetting the G protein after a ligand-binding event (Fukutani et al. 2012), and 2) replace the portions of the receptor itself, specifically the terminal domains and the first transmembrane regions (Hara et al. 2012).

The contamination of water by heavy metals represents a potential risk for both aquatic and terrestrial organisms, including humans (Teo and Wong 2014). Heavy metals in water resources can come from various industrial activities, and drinking water can be ex post contaminated by heavy metals such as Cu^{2+} from house fittings (e.g., water reservoirs) and pipes. A new copper optical YBB based on the cells of a specifically modified *S. cerevisiae* strain immobilized in alginate beads was developed and tested in real samples of water contaminated with copper ions. This optical biosensor was capable of detecting copper ions. This YBB does not require specific equipment and allows the quick detection of Cu^{2+} in the range of 1–100 μM (Vopálenská et al. 2015).

YBBs are well suited for monitoring heavy metals due to their availability to distinguish between bioavailability and concentration—an important distinction as only certain forms of ions have biological effects. Several yeast-based metal biosensors have been developed using either optical or electrical reporting systems (Bereza-Malcolm et al. 2015; Chaves 2024; Matsuura et al. 2013).

Two such systems measure copper, which is an essential cofactor for all life forms at low concentrations but is toxic and is an environmental concern at high concentrations. Both systems take advantage of the Cu^{2+} inducible promoter, which regulates the expression of the *CUP1* copper-binding protein. The electrical method uses the promoter to drive the expression of LacZ to enable lactose metabolism (Lehmann et al. 2000; Tag et al. 2007). The yeasts are incubated with lactose, and the increase in metabolism, and therefore oxygen consumption, associated with promoter activation is measured by an electrode, similar to the BOD systems. The optical approach uses the *CUP1* promoter to drive the firefly luciferase gene (Leskinen et al. 2003). The luciferase gene was modified from the wild type to remove peroxisome-targeting peptides, which improved cellular viability and signal strength. Both systems report a similar detection limit (1 μM). The electrical system requires cell immobilization and more specialized flow injection analysis equipment; however, the assay itself is faster as no incubation period is required. The electrical biosensor demonstrated agreement with chemical analyses on wastewater samples (Tag et al. 2007), while the optical biosensor was used to demonstrate copper bioavailability in a series of metal nanoparticle toxicity studies (Aruoja et al. 2009; Kasemets et al. 2009).

The advantages and disadvantages of using eukaryotic microorganisms to design YBBs for monitoring environmental heavy metal pollution in soil or aquatic habitats were described by Gutiérrez and coauthors (2015). A comparative analysis of the promoter genes used to design YBBs (*S. cerevisiae*), microalgae, and ciliated protozoa is carried out, and the sensitivity and reproducibility of the main reporter genes used are calculated.

A YBB optical sensing system for copper was developed based on recombinant *S. cerevisiae* cells harboring plasmid pYEX-GFPuv. The basis of this system was the ability of the transcriptional activator protein *Ace1p* present in *S. cerevisiae* to control the expression of the reporter protein GFPuv. When copper ions are present in the sample, the *Ace1* protein activates the *cup1* promoter located upstream from the gfpuv gene in the plasmid pYEX-GFPuv, thus inducing the production of GFPuv. The amount of GFPuv produced in the system was determined by monitoring the fluorescence emitted. This system can detect Cu^{2+} at concentrations of as low as 0.5 μM and is selective for Cu^{2+} over a variety of metal ions, with the exception of silver. The applicability of this YBB to different analytical platforms and in real samples is demonstrated (Shetty et al. 2004).

Tafurt-Cardona and coauthors (2015) have reported the development of *S. cerevisiae* cell-based YBB which was selective for methylmercury. Cells were immobilized in agarose gel as the binding phase and polyacrylamide as a diffusive layer in the diffusive gradient in the thin-film technique (DGT). When coupling the DGT technique with cold vapor atomic fluorescence spectrometry, the method has a LOD of 0.44 ng/L. The proposed approach was successfully employed for in situ measurements in the Negro River.

Chromium (Cr) can exist in several states of oxidation; Cr(III) and Cr(VI) are the most stable and widely used species of this element. The main representatives of chromium compounds in the hexavalent state—chromates/dichromates—are very toxic and display mutagenic and carcinogenic activity, and their pollution poses a serious environmental problem. Thus, the development of efficient methods for detoxification and the remediation of this hazardous chromium oxy-anion is of great importance. It is generally accepted that chromium(III) species are not highly toxic, and due to their limited solubility and biosorption, they can be easily removed by chemical and microbial approaches.

The reduction of chromium(VI) by some yeasts—baker's and nonconventional (*H. polymorpha*, *Yarrowia lipolytica*, *Rhodotorula pilimanae*, *Candida guilliermondii*)—followed by the production of chromium(III)-containing biocomplexes was, for the first time, described about 20 years ago (Puzon et al. 2005; Ksheminska et al. 2006). It was discovered that these yeasts and their metabolites are capable of reducing chromate extracellularly: Cr(VI) over Cr(V) to Cr(III). As a result, stable soluble Cr(III) biocomplexes or nanoparticles of Cr_2O_3 were obtained by green synthesis. The chromate-resistant mutants of the nonconventional yeast *Pichia guilliermondii* L2 were selected by different methods. The selected mutants demonstrate better growth and higher biomass yield at elevated chromate concentrations (1–2.4 mM) compared to the parent strain and thus bioremediate toxic chromate in cultural medium (Ksheminska et al. 2010; Ksheminska et al. 2011).

Microbial amperometric biosensors and microbial biofuel cells (MFCs) are bioelectronics-based devices with very similar principles in their design. Most microorganism-based amperometric biosensors show poor specificity, but this drawback can be exploited in the design of microbial biofuel cells because this enables them to consume a wider range of chemical fuels (Bruzaite et al. 2020; Rozene et al. 2021a, b). The efficiency of the charge transfer is among the most challenging and critical issues during the development of any kind of biofuel cell. In most cases, particular redox mediators and nanomaterials are applied for the facilitation of charge transfer from applied biomaterials toward biofuel cell electrodes. Some improvements in charge transfer efficiency can be achieved by the application of conducting polymers (CPs), which can be used for the immobilization of enzymes and, in some particular cases, even for the facilitation of charge transfers. In this review, charge transfer pathways and mechanisms, which are suitable for the design of biosensors and in biofuel cells, are discussed. The modification methods of cell walls/membranes by conducting polymers in order to enhance the charge transfer efficiency of microorganisms, which can be potentially applied in the design of microbial biofuel cells, are outlined. The biocompatibility-related aspects of conducting polymers with microorganisms are summarized (Žunar et al. 2022; Wang et al. 2024; Andriukonis et al. 2021).

According to Fig. 14.2 C, bioreactors occupy a significant segment (5.5%) in the world market of biosensors. Bioremediation stands out as the most promising strategy, utilizing microorganisms to degrade organic and inorganic pollutants. Biomolecular engineering approaches, such as rational design and directed evolution, have been developed to genetically modify microorganisms and their enzymes for the safe and efficient bioremediation of contaminated sites. These approaches enhance the natural capabilities of microorganisms, enabling them to break down pollutants more effectively and adapt to various environmental conditions, making bioremediation a powerful tool for environmental restoration (Wysocki and Tamas 2010; Sigawi et al. 2011; Wasilkowski et al. 2012; Kumar et al. 2013; Karkovska et al. 2016; Bian et al. 2020; Gayda et al. 2015a, b).

The advantages and disadvantages of using eukaryotic microorganisms to design YBBs for monitoring environmental pollutants have been described in detail in our previous review (Gayda et al. 2017). Selected examples of analytical results obtained using YBBs are presented in Table 14.3.

14.6 Future Prospects and Trends in Yeast-Based Analytical Systems

Yeast-based sensing technology is indeed a field in constant evolution, with increasingly sophisticated mechanisms currently being designed. Moreover, the rise of synthetic biology combined with computer-assisted structural biology is opening up exciting future prospects. Additionally, the adoption of innovative technologies, such as artificial intelligence and the Internet of Things, is creating growth opportunities for the global biosensor market. YBBs, specially engineered with specific receptors and highly sensitive reporter systems, are very promising for the rapid and sensitive monitoring of target analytes in samples of various natures. Due to a broad metabolic activity, YBBs can be used also for biodegradation/bioremediation processes of wastewater. It can be predicted that, after standardization, such YBBs will be candidates for commercialization and broad use in various fields related to practical and experimental medicine, as well as food and environmental safety. Compared with available diagnostic approaches, electrochemical and field-effect transistor biosensors have gained significant attention due to their advantages in miniaturization, point-of-care detection, on-site diagnosis, sensitivity, and low cost. Additionally, multiplex biosensors could be developed for the accurate detection of multiple biomarkers. Similarly, nanomaterial-based, smartphone-based, and wearable biosensors should be developed to control disease outbreaks to contain human loss as early as possible. Currently, a new horizon of nanobiosensors is established for clinical and nonclinical diagnoses.

These advancements highlight the crucial role of nanotechnology in pushing the boundaries of what biosensors can achieve, making them more effective tools for various applications.

Some commercially used antagonistic yeasts are effective against fruit pathogens.[1] However, it is important to note that yeasts may be less effective than chemical fungicides against certain pathogenic fungi.[2]

Acknowledgment This work was supported by the National Research Foundation of Ukraine (project No. 2021.01/0010), the "Presidential Discretionary-Ukraine Support Grants" from Simons Foundation, the IIE Scholar Rescue Fund, US National Science Foundation (NSF) grant CBET-2235349, including IMPRESS-U supplement (project STCU 7113).

References

Abu-Rmailah N, Moscovici L, Riegraf C et al (2024) Enhanced detection of estrogen-like compounds by genetically engineered yeast sensor strains. Biosensors (Basel) 14(4):193

Adeniran A, Sherer M, Tyo KEJ (2015) Yeast-based biosensors: design and applications. FEMS Yeast Res 15:1–15

Agirman B, Carsanba E, Settanni L, Erten H (2023) Exploring yeast-based microbial interactions: the next frontier in postharvest biocontrol. Yeast 40(10):457–475

Akor J, Nweze JE, Nweze JA et al (2024) Chapter 16 - Biological elements as important tools in the detection/monitoring of drug compounds in organic and environmental samples. In: Shah MP, Rodriguez-Couto S (eds) Development in wastewater treatment research and processes. Elsevier, pp 337–371. ISBN 9780323992787

Akyilmaz E, Erdogan A, Ozturk R et al (2007) Sensitive determination of L-lysine with a new amperometric microbial biosensor based on *Saccharomyces cerevisiae* yeast cells. Biosens Bioelectron 22(6):1055–1060

Andriukonis E, Celiesiute-Germaniene R, Ramanavicius S (2021) From microorganism-based amperometric biosensors towards microbial fuel cells. Sensors 21:2442

Aruoja V, Dubourguier H-C, Kasemets K, Kahru A (2009) Toxicity of nanoparticles of CuO, ZnO and TiO2 to microalgae Pseudokirchneriella subcapitata. Sci Total Environ 407:1461–1468

Baetz N, Rothe L, Wirzberger V et al (2021) High-performance thin-layer chromatography in combination with a yeast-based multi-effect bioassay to determine endocrine effects in environmental samples. Anal Bioanal Chem 413(5):1321–1335

Battisti A (2024) Editorial for special issue on biosensors for biomedical and environmental applications. Micromachines (Basel) 15(5):607

Bean BDM, Mulvihill CJ, Garge RK et al (2022) Functional expression of opioid receptors and other human GPCRs in yeast engineered to produce human sterols. Nat Commun 13:2882

Benton MG, Glasser NR, Palecek SP (2007) The utilization of a *Saccharomyces cerevisiae* HUG1P-GFP promoter–reporter construct for the selective detection of DNA damage. Mutat Res 633:21–23

Bereza-Malcolm LT, Mann G, Franks AE (2015) Environmental sensing of heavy metals through whole cell microbial biosensors: a synthetic biology approach review. ACS Synth Biol 4(5):535–546

Bian C, Wang H, Zhang X et al (2020) Sensitive detection of low-concentration sulfide based on the synergistic effect of rGO, np-Au, and recombinant microbial cell. Biosens Bioelectron 151:111985. e-coli

Bovee TFH, Helsdingen RJR, Hamers ARM et al (2007) A new highly specific and robust yeast androgen bioassay for the detection of agonists and antagonists. Anal Bioanal Chem 389:1549–1558

Bovee TFH, Helsdingen RJR, Hamers ARM et al (2011) Recombinant cell bioassays for the detection of (gluco)corticosteroids and endocrine-disrupting potencies of several environmental PCB contaminants. Anal Bioanal Chem 401:873–882

Bruzaite I, Rozene J, Morkvenaite-Vilkonciene I, Ramanavicius A (2020) Towards microorganism-based biofuel cells: the viability of *Saccharomyces cerevisiae* modified by multiwalled carbon nanotubes. Nanomaterials 10:954

Bryden WL (2007) Mycotoxins in the food chain: human health implications. Asia Pac J Clin Nutr 16(1):95–101

Bui VN, Nguyen TTH, Bettarel Y et al (2015) Genotoxicity of chemical compounds identification and assessment by yeast cells transformed with GFP reporter constructs regulated by the PLM2 or DIN7 promoter. Int J Toxicol 34:31–43

Cervelli T, Galli A (2021) Yeast as a tool to understand the significance of human disease-associated gene variants. Genes (Basel) 12(9):1303

Cevenini L, Lopreside A, Calabretta MM et al (2018) A novel bioluminescent NanoLuc yeast-estrogen screen biosensor (nanoYES) with a compact wireless camera for effect-based detection of endocrine-disrupting chemicals. Anal Bioanal Chem 410:1237–1246

Chan C, Lehmann M, Chan K et al (2000) Designing an amperometric thick-film microbial BOD sensor. Biosens Bioelectron 15(7–8):343–353

Chatterjee S, Kumar V, Majumder CB, Roy P (2008) Screening of some anti-progestin endocrine disruptors using a recombinant yeast based *in vitro* bioassay. Toxicol In Vitro 22:788–798

Chaves SR (2024) Novel yeast-based biosensor for environmental monitoring of tebuconazole. Appl Microbiol Biotechnol 108(1):10

Chelikani V, Downard AJ, Kunze G et al (2012) Investigating yeast cell responses to oestrogen by electrochemical detection. Electrochim Acta 73:136–140

Chen Q et al (2004) A yeast two-hybrid technology-based system for the discovery of PPAR?? Agonist and antagonist. Anal Biochem 335:253–259

Chen Y, Nielsen J (2016) Flux control through protein phosphorylation in yeast, FEMS Yeast Res 16(8):fow0969

Chernoff YO, Grizel AV, Rubel AA et al (2020) Application of yeast to studying amyloid and prion diseases. In: Kumar D (ed) Advances in genetics, vol 105. Academic Press, Cambridge, MA, pp 293–380

Chunhui D, Choi S (2013) Technology and applications of microbial biosensor. OJAB 2:83–93

Comitini F, Canonico L, Agarbati A, Ciani M (2023) Biocontrol and probiotic function of non-*Saccharomyces* yeasts: new insights in agri-food industry. Microorganisms 11(6):1450
D'Souza SF (2001) Microbial Biosensors. Biosens Bioelectron 16(6):337–353
Daurai B, Gogoi M (2022) Point-of-care biosensors for healthcare applications. In: Gogoi M, Patra S, Kundu D (eds) Nanobiosensors for point-of-care medical diagnostics. Springer, Singapore. https://doi.org/10.1007/978-981-19-5141-1_1
Dhakal S, Macreadie I (2020) Protein homeostasis networks and the use of yeast to guide interventions in Alzheimer's disease. Int J Mol Sci 21:8014
Dhakal S, Macreadie I (2022a) Chapter 1: Developing systems in yeast to address Alzheimer's disease. In: Gurtler V (ed) Methods in microbiology, vol 51. Academic Press, Cambridge, MA, pp 1–43
Dhakal S, Macreadie I (2022b) The use of yeast in biosensing. Microorganisms 10(9):1772
Dragone R, Frazzoli C, Grasso G, Rossi G (2014) Sensor with intact or modified yeast cells as rapid device for toxicological test of chemicals. J Agric Chem Environ 3:35–40
Dukare AS, Paul S, Nambi VE (2019) Exploitation of microbial antagonists for the control of postharvest diseases of fruits: a review. Crit Rev Food Sci Nutr 59(9):1498–1513
Eldridge M, Sanseverino J, Umbuzeiro GA, Sayler GS (2011) In: Ekundayo DE (ed) Analysis of environmental samples with yeast-based bioluminescent bioreporters, environmental monitoring. InTech. https://doi.org/10.5772/27139
Filipović-Kovačević Ž, Mikšaj M, Dean Š (2002) Cyanide determination in fruit brandies by an amperometric biosensor with immobilised *Saccharomyces cerevisiae*. Eur Food Res Technol 215(4):347–352
Fiore AM, Naik V, Leibensperger EM (2015) Air quality and climate connections. J Air Waste Manag Assoc 65(6):645–685
Francino-Urdaniz IM, Steiner PJ, Kirby MB et al (2021) One-shot identification of SARS-CoV-2 S RBD escape mutants using yeast screening. Cell Rep 36:109627
Freimoser FM, Rueda-Mejia MP, Tilocca B, Migheli Q (2019) Biocontrol yeasts: mechanisms and applications. World J Microbiol Biotechnol 35(10):154
Freimoser FM, Mahler M, McCullough M et al (2024) Heterologous pulcherrimin production in Saccharomyces cerevisiae confers inhibitory activity on botrytis conidiation. FEMS Yeast Res:foad053
Fukutani Y, Nakamura T, Yorozu M et al (2012) The N-terminal replacement of an olfactory receptor for the development of a yeast-based biomimetic odor sensor. Biotechnol Bioeng 109:205–212
García-Alonso J, Greenway GM, Hardege JD, Haswell SJ (2009) A prototype microfluidic chip using fluorescent yeast for detection of toxic compounds. Biosens Bioelectron 24:1508–1511
Garjonyte R, Melvydas V, Malinauskas A (2006) Mediated amperometric biosensors for lactic acid based on carbon paste electrodes modified with baker's yeast *Saccharomyces cerevisiae*. Bioelectrochemistry 68(2):191–196
Garjonyte R, Melvydas V, Malinauskas A (2008) Effect of yeast pretreatment on the characteristics of yeast-modified electrodes as mediated amperometric biosensors for lactic acid. Bioelectrochemistry 74:188–194
Garjonyte R, Melvydas V, Malinauskas A (2009) Amperometric biosensors for lactic acid based on baker's and wine yeast. Microchim Acta 164:177–183
Gayda G, Demkiv O, Klepach H et al (2015a) Formaldehyde: detection and biodegradation (chapter 6). In: Patton A (ed) Formaldehyde: synthesis, applications and potential health effects. Nova Science Publishers, Inc, New York, pp 117–142. ISBN: 978-1-63482-412-5
Gayda G, Stasyuk N, Klepach H et al (2015b) Arginase-based electrochemical sensors to L-arginine assay for prediction of ethyl carbamate formation in fermented beverages. In: Living organisms and bioanalytical approaches for detoxification and monitoring of toxic compounds, pp 73–88
Gayda G, Gonchar M, Sibirny A (2017) Yeast-based systems for environmental control. In: Sibirny AA (ed) Biotechnology of yeasts and filamentous fungi. Springer, pp 373–386

Gayda G, Stasyuk N, Klepach H et al (2019) Promising bioanalytical approaches to wine analysis. In: Grumezescu AM, Holban AM (eds) Quality control in the beverage industry. Woodhead Publishing, Elsevier, Duxford/Cambridge, pp 419–458. 17: The science of beverages, Ch. 12, ISBN 978-0-12-816681-91

Gayda G, Stasyuk N, Zakalskiy A et al (2022) Arginine-hydrolyzing enzymes for electrochemical biosensors. Curr Opin Electrochem 33:100941

Gerlach T et al (2014) Development and assessment of a novel *Arxula adeninivorans* androgen screen (A-YAS) assay and its application in analysis of cattle urine. Sci Total Environ 490:1073–1081

Goldblum DK, Holodnick SE, Mancy KH, Briggs DE (1990) Oxygen transport in biofilm electrodes for screening of toxic chemicals. AICHE J 36:19–28

Gonchar M, Smutok O, Karkovska M et al (2017) Yeast-based biosensors for clinical diagnostics and food control. In: Sibirny AA (ed) Biotechnology of yeasts and filamentous fungi. Springer, pp 392–400

Gregório BJR, Ramos II, Marques SS et al (2024) Microcarrier-based fluorescent yeast estrogen screen assay for fast determination of endocrine disrupting compounds. Talanta 271:125665

Guaragnella C, Guaragnella N (2023) Biological and technical challenges for implementation of yeast-based biosensors. Microb Biotechnol 16(1):54–66

Guo E, Zhao L, Li Z et al (2024) Biodegradation of bisphenol A by a *Pichia pastoris* whole-cell biocatalyst with overexpression of laccase from Bacillus pumilus and investigation of its potential degradation pathways. J Hazard Mater:134779

Gupta N, Renugopalakrishnan V, Liepmann D et al (2019) Cell-based biosensors: recent trends, challenges and future perspectives. Biosens Bioelectron 141:111435

Gutiérrez JC, Amaro F, Martín-González A (2015) Heavy metal whole-cell biosensors using eukaryotic microorganisms: an updated critical review. Front Microbiol 6:48. https://www.precedenceresearch.com/biosensors-market

Hara K, Inada Y, Ono T et al (2012) Chimeric yeast G-protein α subunit harboring a 37-residue C-terminal gustducin-specific sequence is functional in Saccharomyces cerevisiae. Biosci Biotechnol Biochem 76:512–516

He W, Yuan S, Zhong WH et al (2016) Application of genetically engineered microbial whole-cell biosensors for combined chemosensing. Appl Microbiol Biotechnol 100(3):1109–1119

He Y, Degraeve P, Oulahal N (2024) Bioprotective yeasts: potential to limit postharvest spoilage and to extend shelf life or improve microbial safety of processed foods. Heliyon 10(3):e24929

Hilber-Bodmer M, Schmid M, Ahrens CH, Freimoser FM (2017) Competition assays and physiological experiments of soil and phyllosphere yeasts identify Candida subhashii as a novel antagonist of filamentous fungi. BMC Microbiol 17(1):4

Hou QH, Ma AS, Zhuang XL, Zhuang GQ (2013) Advance in the bioavailability monitoring of heavy metal based on microbial whole-cell sensor. Huan Jing Ke Xue 34(1):347–356

Ino K, Kitagawa Y, Watanabe T et al (2009) Detection of hormone active chemicals using genetically engineered yeast cells and microfluidic devices with interdigitated array electrodes. Electrophoresis 30:3406–3412

Jacobson KA (2015) New paradigms in GPCR drug discovery. Biochem Pharmacol 98(4):541–555

Jarque S, Bittner M, Blaha L, Hilscherova K (2016) Yeast biosensors for detection of environmental pollutants: current state and limitations. Trends Biotechnol 34(5):408–419

Kabiersch G, Rajasärkkä J, Tuomela M et al (2013) Bioluminescent yeast assay for detection of organotin compounds. Anal Chem 85:5740–5745

Kaiser C, Uhlig S, Gerlach T et al (2010) Evaluation and validation of a novel *Arxula adeninivorans* estrogen screen (nAES) assay and its application in analysis of wastewater, seawater, brackish water and urine. Sci Total Environ 408:6017–6126

Karkovska M, Smutok O, Stasyuk N et al (2015) L-lactate-selective microbial sensor based on flavocytochrome *b2*-enriched yeast cells using recombinant and nanotechnology approaches. Talanta 144:1195–1200

Karkovska M, Smutok O, Gonchar M (2016) Laboratory prototype of bioreactor for oxidation of toxic D-lactate using yeast cells overproducing D-lactate-cytochrome *c* oxidoreductase. BioMed Res Intern 2016:1–5
Karkovska M, Stasyuk N, Gayda G et al (2017) Nanomaterials in the design of biosensors for biomedical purposes. In: Stoika RS (ed) Multifunctional nanomaterials for biology and medicine: molecular design, synthesis and application, pp 165–177
Karube I, Sode K (1989) Biosensors for fermentation process control. In: Ghose TK (ed) Bioprocess engineering: the first generation. Chapman Hall, London, p 207
Karube I, Suzuki S (1983) Application of biosensor in fermentation processes. In: Tsao GT (ed) Annual reports on fermentation processes, vol 6. Academic, New York, p 203
Karube I, Tamiya E (1987) Biosensors for the food industry. Food Biotechnol 1:147–165
Kase R, Hansen PD, Fischer B et al (2008) Integral assessment of estrogenic potentials of sediment-associated samples. Part 1: The influence of salinity on the in vitro tests ELRA, E-Screen and YES. Environ Sci Pollut Res 15:75–83
Kasemets K, Ivask A, Dubourguier HC, Kahru A (2009) Toxicity of nanoparticles of ZnO, CuO and TiO2 to yeast Saccharomyces cerevisiae. Toxicol In Vitro 23:1116–1122
Knight AW, Goddard NJ, Fielden PR et al (1999) Development of a flow-through detector for monitoring genotoxic compounds by quantifying the expression of green fluorescent protein in genetically modified yeast cells Meas Sci Technol 10:211
Korpan YI, Dzyadevich SV, Zharova VP et al (1994) Conductometric biosensor for ethanol detection based on whole yeast cells. Ukr Biokhim Zh 66:78–82
Ksheminska HP, Honchar TM, Gayda GZ, Gonchar MV (2006) Extra cellular chromate reducing activity of the yeast cultures. Cent Eur J Biol 1(1):137–149
Ksheminska H, Honchar T, Usatenko Y et al (2010) The chromate resistance phenotype of some yeast mutants correlates with a lower level of Cr(V)-species generated in the extra-cellular medium. Biometals 23:633–642
Ksheminska HP, Gayda GZ, Ivash MF, Gonchar MV (2011) Chromate resistant mutants of the yeast *Pichia guilliermondii*: selection and properties. Microbiology 80(3):314–325
Kumar S, Kumar Dagar V, Khasa YP, Kuhad RC (2013) Genetically modified microorganisms (GMOs) for bioremediation. In: Kuhad RC, Singh A (eds) Biotechnology for environmental management and resource recovery. Springer, New Delhi, pp 191–218
Lehmann M, Riedel K, Adler K, Kunze G (2000) Amperometric measurement of copper ions with a deputy substrate using a novel Saccharomyces cerevisiae sensor. Biosens Bioelectron 15:211–219
Leskinen P, Virta M, Karp M (2003) One-step measurement of firefly luciferase activity in yeast. Yeast 20:1109–1113.
Leskinen P et al (2005) Bioluminescent yeast assays for detecting estrogenic and androgenic activity in different matrices. Chemosphere 61:259–266
Li J, Ma M, Wang Z (2008a) A two-hybrid yeast assay to quantify the effects of xenobiotics on retinoid X receptor-mediated gene expression. Toxicol Lett 176:198–206
Li J, Ma M, Wang Z (2008b) A two-hybrid yeast assay to quantify the effects of xenobiotics on thyroid hormone-mediated gene expression. Environ Toxicol Chem 27:159–167
Li J, Ren S, Han S, Li N (2014) A yeast bioassay for direct measurement of thyroid hormone disrupting effects in water without sample extraction, concentration, or sterilization. Chemosphere 100:139–145
Lim JW, Dogyeong H, Lee J et al (2015) Review of micro/nanotechnologies for microbial biosensors. Front Bioeng Biotechnol 3(61):1–13
Lind AC, David F, Siewers V (2024) Evaluation and comparison of colorimetric outputs for yeast-based biosensors in laboratory and point-of-use settings. FEMS Microbiol Lett:fnae034
Liu J, Sui Y, Wisniewski M et al (2013) Review: utilization of antagonistic yeasts to manage post-harvest fungal diseases of fruit. Int J Food Microbiol 167(2):153–160
Martin-Yken H (2020) Yeast-based biosensors: current applications and new developments. Biosensors (Basel) 10(5):51

Matsuura H, Yamamoto Y, Muraoka M et al (2013) Development of surface-engineered yeast cells displaying phytochelatin synthase and their application to cadmium biosensors by the combined use of pyrene-excimer fluorescence. Biotechnol Prog 29:1197–1202

Mcdonald JB, Dhakal S, Macreadie IG (2020) Yeast contributions to Alzheimer's disease. J Human Clin Gen 2(2):1–19

Mello LD, Kubota LT (2002) Review of the use of bio- sensors as analytical tools in the food and drink industries. Food Chem 77(2):237–256

Mendes F, Miranda E, Amaral L et al (2024) Novel yeast-based biosensor for environmental monitoring of tebuconazole. Appl Microbiol Biotechnol 108(10):10

Miao Y, McCammon JA (2016) G-protein coupled receptors: advances in simulation and drug discovery. Curr Opin Struct Biol 41:83–89

Michelini E, Cevenini L, Mezzanotte L et al (2008) A sensitive recombinant cell-based bioluminescent assay for detection of androgen-like compounds. Nat Protoc 3:1895–1902

Mirmahdi RS, Mahoozi T, Zoghi A et al (2024) The roles of Saccharomyces cerevisiae on the bioaccessibility of phenolic compounds. World J Microbiol Biotechnol 40(7):221

Moraskie M, Or Roshid HM, O'Connor G et al (2021) Microbial whole-cell biosensors: current applications, challenges, and future perspectives. Biosens Bioelectron 191:113359

Moscovici L, Riegraf C, Abu-Rmailah N et al (2020) Yeast-based fluorescent sensors for the simultaneous detection of estrogenic and androgenic compounds, coupled with high-performance thin layer chromatography. Biosensors (Basel) 10(11):169

Nakamura H, Suzuki K, Ishikuro H et al (2007) A new BOD estimation method employing a double-mediator system by ferricyanide and menadione using the eukaryote *Saccharomyces cerevisiae*. Talanta 72:210–216

Nakamura H, Shimomura-Shimizu M, Karube I (2008) Development of microbial sensors and their application. Adv Biochem Eng Biotechnol 109:351–394

Nakamura Y, Ishii J, Kondo A (2015) Applications of yeast-based signaling sensor for characterization of antagonist and analysis of site-directed mutants of the human serotonin 1A receptor. Biotechnol Bioeng 112(9):1906–1915

Naresh V, Lee N (2021) A review on biosensors and recent development of nanostructured materials-enabled biosensors. Sensors 21(4):1109

Noguerol T-N, Boronat S, Jarque S et al (2006) Detection of hormone receptor ligands in yeast by fluorogenic methods. Talanta 69:351–358

Otles S, Yalcin B (2012) Review on the application of nanobiosensors in food analysis. Acta Sci Pol Technol Aliment 11(1):7–18

Oztekin S, Dikmetas DN, Devecioglu D et al (2023) Recent insights into the use of antagonistic yeasts for sustainable biomanagement of postharvest pathogenic and Mycotoxigenic fungi in fruits with their prevention strategies against mycotoxins. J Agric Food Chem 71(26):9923–9950

Park JN, Sohn MJ, Oh DB et al (2007) Identification of the cadmium-inducible *Hansenula polymorpha SEO1* gene promoter by transcriptome analysis and its application to whole-cell heavy-metal detection systems. Appl Environ Microbiol 73(19):5990–6000

Pham HTM, Giersberg M, Uhlig S et al (2012) EstraMonitor – a monitor for amperometric detection of estrogenic activity with *Arxula adeninivorans* yeast cells as the biocomponent. Sensors Actuators B Chem 161(1):137–145

Pham HTM, Kunath K, Gehrmann L et al (2013) Application of modified *Arxula adeninivorans* yeast cells in an online biosensor for the detection of estrogenic compounds in wastewater samples. Sensors Actuators B Chem 185:628–637

Pham HTM, Giersberg M, Gehrmann L et al (2015) The determination of pharmaceuticals in wastewater using a recombinant *Arxula adeninivorans* whole cell biosensor. Sensors Actuators B Chem 211:439–448

Pundir CS, Nohwal B, Chaudhary R (2021) A comprehensive review of methods for determination of L-lysine with detailed description of biosensors. Int J Biol Macromol 186:445–461

Puzon GJ, Roberts AG, Kramer DM, Xun L (2005) Formation of soluble organo-chromium(III) complexes after chromate reduction in the presence of cellular organics. Environ Sci Technol 39:2811–2817

Radhika V, Milkevitch M, Audige V et al (2005) Engineered *Saccharomyces cerevisiae* strain BioS-1, for the detection of water-borne toxic metal contaminants. Biotechnol Bioeng 90:29–35

Rajasärkkä J (2013) Development of high-throughput yeast-cell-based bioreporter assays for specific monitoring of bisphenol A and chemical testing of endocrine disrupting compounds. Academic Dissertation, Helsinki, 50 pp

Renneberg T, Kwan RCH, Chan C et al (2004) A salt-tolerant yeast-based microbial sensor for 24 hour community wastewater monitoring in coastal regions. Microchim Acta 148:235–240

Roda A, Roda B, Cevenini L et al (2011) Analytical strategies for improving the robustness and reproducibility of bioluminescent microbial bioreporters. Anal Bioanal Chem 401:201–211

Romero MR, Picchio ML (2020) Biosensors based on nanomaterials: transducers and modified surfaces for diagnostics. In: Chandra P, Prakash R (eds) Nanobiomaterial engineering. Springer, Singapore. https://doi.org/10.1007/978-981-32-9840-8_2

Roszko MŁ, Kamińska M, Szymczyk K et al (2018) Endocrine disrupting potency of organic pollutant mixtures isolated from commercial fish oil evaluated in yeast-based bioassays. PLoS One 13(5):e0197907

Rotariu L, Bala C (2007) New type of ethanol microbial biosensor based on a highly sensitive Amperometric oxygen electrode and yeast cells. Anal Lett 36(11):2459–2471

Rotariu L, Bala C, Magearu V (2002) Yeast cells sucrose biosensor based on a potentiometric oxygen electrode. Anal Chim Acta 458(1):215–222

Rotariu L, Bala C, Magearu V (2004a) Microbial biosensors for ethanol determination in alcoholic beverages. Annals ii Universities in Bucharest I-II:69–76

Rotariu L, Bala C, Magearu V (2004b) New potentiometric microbial biosensor for ethanol determination in alcoholic beverages. Anal Chim Acta 513:119–123

Rozene J, Morkvenaite-Vilkonciene I, Bruzaite I et al (2021a) Baker's yeast-based microbial fuel cell mediated by 2-Methyl-1, 4-naphthoquinone. Membranes 11:182

Rozene J, Morkvenaite-Vilkonciene I, Bruzaite I et al (2021b) Yeast-based microbial biofuel cell mediated by 9, 10-phenantrenequinone. Electrochim Acta 73:137918

Rueda-Mejia MP, Bühlmann A, Ortiz-Merino RA et al (2023) Pantothenate Auxotrophy in a naturally occurring biocontrol yeast. Appl Environ Microbiol 89(7):e0088423

Safarik I, Maderova Z, Pospiskova K et al (2015) Magnetically responsive yeast cells: methods of preparation and applications. Yeast 32:227–237

Sankarankutty KM (2014) Biosensors and their applications for ensuring food safety. Glob J Pathol Microbiol 2:15–21

Sanseverino JA, Gupta RKA, Layton ACA et al (2005) Use of *Saccharomyces cerevisiae* BLYES expressing bacterial bioluminescence for rapid, sensitive detection of estrogenic compounds. Appl Environ Microbiol 71:4455–4460

Schmidt RD, Karube I (1998) Biotechnology. In: Rehm HJ, Reed G (eds) Biosensors and bioelectronics, vol 6b. VCH Verlagsgesellschaft, Weinheim, pp 317–365

Schwartz-Mittelman A et al (2005) Electrochemical detection of xenoestrogenic and antiestrogenic compounds using a yeast twohybrid-17-b-estradiol system. Bioelectrochemistry 65:149–156

Seo KS, Choo KH, Chang HN et al (2009) A flow injection analysis system with encapsulated high-density Saccharomyces cerevisiae cells form rapid determination of biochemical oxygen demand. Appl Microbiol Biotechnol 83:217–223

Shepherd HAM, Bondarenko EMA, Jennings KM et al (2022) Whole cell yeast-based biosensors. In: Darvishi Harzevili F (ed) Synthetic biology of yeasts. Springer, Cham. https://doi.org/10.1007/978-3-030-89680-5_4

Shetty RS, Deo SK, Liu Y, Daunert S (2004) Fluorescence-based sensing system for copper using genetically engineered living yeast cells. Biotechnol Bioeng 88(5):664–670

Shiizaki K, Asai S, Ebata S et al (2010) Establishment of yeast reporter assay systems to detect ligands of thyroid hormone receptors alpha and beta. Toxicol In Vitro 24:638–644

Shiizaki K, Yoshikawa T, Takada E et al (2014) Development of yeast reporter assay form screening specific ligands of retinoic acid and retinoid X receptor subtypes. J Pharmacol Toxicol Methods 69:245–252

Shimomura-Shimizu M, Karube I (2010) Yeast based sensors. Adv Biochem Engin/Biotechnol 117:1–19. https://doi.org/10.1007/10_2009_18

Shkil H, Stoica L, Dmytruk K et al (2009) Bioelectrochemical detection of L-lactate respiration using genetically modified *Hansenula polymorpha* yeast cells overexpressing flavocytochrome b_2. Bioelectrochemistry 76:175–179

Sibirny V, Demkiv O, Sigawi S et al (2011) Formaldehyde oxidizing enzymes and genetically modified yeast Hansenula polymorpha cells in monitoring and removal of formaldehyde. In: Einschlag FSG (ed) Waste water - evaluation and management. InTech Open Access Publisher, Rijeka, Ch. 6, ISBN 978-953-307-233-3, pp 115–154

Sigawi S, Smutok O, Demkiv O et al (2011) Immobilized formaldehyde-metabolizing enzymes from *Hansenula polymorpha* for removal and control of airborne formaldehyde. J Biotechnol 153:138–144

Sigawi S, Smutok O, Demkiv O et al (2014) Detection of waterborne and airborne formaldehyde: from amperometric chemosensing to a visual biosensor based on alcohol oxidase. Materials 7(2):1055–1068

Sipiczki M, Czentye K, Kállai Z (2024) High intragenomic, intergenomic, and phenotypic diversity in pulcherrimin-producing *Metschnikowia* yeasts indicates a special mode of genome evolution. Sci Rep 14:10521

Smutok O, Dmytruk K, Gonchar M et al (2007) Permeabilized cells of flavocytochrome b_2 over-producing recombinant yeast *Hansenula polymorpha* as biological recognition element in amperometric lactate biosensors. Biosens Bioelectron 23:599–605

Smutok O, Karkovska M, Prokopiv T et al (2019) D-lactate-selective amperometric biosensor based on the mitochondrial fraction of *Ogataea polymorpha* recombinant cells. Yeast 36(5):341–348

Smutok O, Dmytruk K, Kavetskyy T et al (2021) Flavocytochrome b_2 of the methylotrophic yeast *Ogataea polymorpha*: construction of overproducers, purification, and bioanalytical application. Methods Mol Biol 2280:249–260. ISSN 1064-3745: and 1940-6029; In Book: Flavins and Flavoproteins

Stasyuk NYE, Gayda GZ, Gonchar MV (2014) L-arginine-selective microbial amperometric sensor based on recombinant yeast cells over-producing human liver arginase I. Sens Actuat B Chem 204:515–521

Stasyuk N, Zakalskiy A, Nogala W et al (2024) Genetically engineered yeast cells enriched with nanocomposites containing a natural enzyme and nanozyme for the construction of microbial sensors. Electrochim Acta 497:144605

Sun JZ, Peter Kingori G, Si RW et al (2015) Microbial fuel cell-based biosensors for environmental monitoring: a review. Water Sci Technol 71(6):801–809

Tafurt-Cardona M, Eismann CE, Suárez CA et al (2015) In situ selective determination of methylmercury in river water by diffusive gradient in thin films technique (DGT) using baker's yeast (Saccharomyces cerevisiae) immobilized in agarose gel as binding phase. Anal Chim Acta 887:38–44

Tag K, Riedel K, Bauer H-J et al (2007) Amperometric detection of Cu^{2+} by yeast biosensors using flow injection analysis (FIA). Sensors Actuators B Chem 122:403–409

Teo SC, Wong LS (2014) Whole cell-based biosensors for environmental heavy metals detection. Ann Res Rev Biol 4(17):2663–2674

Thévenot DR, Toth K, Durst RA, Wilson GS (2001) Electrochemical biosensors: recommended definitions and classification. Biosens Bioelectron 16(1–2):121–131

Thurston GD, Burnett RT, Turner MC et al (2016) Ischemic heart disease mortality and long-term exposure to source-related components of U.S. fine particle air pollution. Environ Health Perspect 124(6):785–794

Tkac J, Vostiar I, Gorton L et al (2003) Improved selectivity of microbial biosensor using membrane coating. Application to the analysis of ethanol during fermentation. Biosens Bioelectron 18:1125–1134

Turner BL, Kasperson RE, Meyer WB et al (1990) Two types of global environmental change Glob Environ Chang 1(1):14-22

Turdean GL (2011) Design and development of biosensors for the detection of heavy metal toxicity. Res Int J Electrochem. Article ID 343125:15
Uchański T, Zögg T, Yin J et al (2019) An improved yeast surface display platform for the screening of nanobody immune libraries. Sci Rep 9:382
Ukhurebor KE, Hossain I, Pal K et al (2024) Applications and contemporary issues with adsorption for water monitoring and remediation: a facile review. Top Catal 67:140–155
Uthurry CA, Lepe JA, Lombardero J et al (2006) Ethyl carbamate production by selected yeasts and lactic acid bacteria in red wine. Food Chem 94:262–270
Valima A, Kivisto AT, Leskinen PI et al (2010) A novel biosensor for the detection of Zearalenone family mycotoxins in milk. J Microb Met 80(1):44–48
Vigneshvar S, Sudhakumari CC, Senthilkumaran B, Prakash H (2016) Recent advances in biosensor technology for potential applications - an overview. Front Bioeng Biotechnol 4:11
Vopálenská I, Váchová L, Palková Z (2015) New biosensor for detection of copper ions in water based on immobilized genetically modified yeast cells. Biosens Bioelectron 72:160–167
Wahid E, Ocheja OB, Marsili E et al (2023) Biological and technical challenges for implementation of yeast-based biosensors. Microb Biotechnol 16(1):54–66
Walmsley RM, Keenan P (2000) The eukaryote alternative: advantages of using yeasts in place of bacteria in microbial biosensor development. Biotechnol Bioprocess Eng 5:387–394
Wang S, Tan Z, Wang C et al (2023) Iron competition as an important mechanism of Pulcherrimin-producing *Metschnikowia* sp. strains for controlling postharvest fungal decays on citrus fruit. Foods 12(23):4249
Wang Z, Dong Z, Shen X, Wu B (2024) Molecularly imprinted polymers using yeast as a supporting substrate. Molecules 28(20):7103
Wasilkowski D, Swedziol Z, Mrozik A (2012) The applicability of genetically modified microorganisms in bioremediation of contaminated environments. Chemik 66(8):817–826
Watanabe M, Noma H, Kurai J et al (2017) A panel study of airborne particulate matter composition versus concentration: potential for inflammatory response and impaired pulmonary function in children. Allergol Int 66(1):52–58
Wei T, Zhang C, Xu X et al (2013) Construction and evaluation of two biosensors based on yeast transcriptional response to genotoxic chemicals. Biosens Bioelectron 44:138–145
Westlund P, Yargeau V (2017) Investigation of the presence and endocrine activities of pesticides found in wastewater effluent using yeast-based bioassays. Sci Total Environ 607-608:744–751
Wysocki R, Tamas MJ (2010) How Saccharomyces cerevisiae copes with toxic metals and metalloids. FEMS Microbiol Rev 34:925–951
Yang H-C, Lim B, Lee Y (2005) In vitro assessment of biocompatibility of biomaterials by using fluorescent yeast biosensor. Curr Appl Phys 5:444–448
Yano Y, Numata M, Hachiya H et al (2001) Application of a microbial sensor to the quality control of meat freshness. Talanta 54:255–262
Yudina NY, Arlyapov VA, Chepurnova MA et al (2015) A yeast co-culture-based biosensor for determination of waste water contamination levels. Enzym Microb Technol 78:46–53
Zhang J, Sonnenschein N, Pihl TPB et al (2016) Engineering an NADPH/NADP + Redox Biosensor in Yeast. ACS Synth Biol 5(12):1546–1556
Zhang X, Li B, Zhang Z et al (2020) Antagonistic yeasts: a promising alternative to chemical fungicides for controlling postharvest decay of fruit. J Fungi (Basel) 6(3):158
Žunar B, Ito T, Mosrin C et al (2022) Confocal imaging of biomarkers at a single-cell resolution: quantifying 'living' in 3D-printable engineered living material based on Pluronic F-127 and yeast *Saccharomyces cerevisiae*. Biomater Res 26(1):85

The manufacturer's authorised representative in the EU is Springer Nature Customer Service Centre GmbH, Europaplatz 3, 69115 Heidelberg, Germany. If you have any concerns regarding our products, please contact ProductSafety@springernature.com

ed and bound by CPI Group (UK) Ltd, Croydon, CR0 4YY
30/09/2026
02225454-0011